Parental Care
in Mammals

Parental Care
in Mammals

Edited by
David J. Gubernick

and
Peter H. Klopfer

Duke University
Durham, North Carolina

PLENUM PRESS • NEW YORK AND LONDON

Library of Congress Cataloging in Publication Data

Main entry under title:

Parental care in mammals.

Includes index.
1. Mammals—Behavior. 2. Parental behavior in animals. I. Gubernick, David J. II.
Klopfer, Peter H. [DNLM: 1. Behavior, Animal. 2. Maternal behavior. 3. Paternal
behavior. QL763 R228]
QL739.3.P37   599.05'6   80-36692
ISBN 0-306-40533-4

© 1981 Plenum Press, New York
A Division of Plenum Publishing Corporation
227 West 17th Street, New York, N.Y. 10011

Printed in the United States of America

Contributors

Marc Bekoff, Department of Environmental, Population, and Organismic Biology, Behavioral Biology Group, University of Colorado, Boulder, Colorado 80309

Bennett G. Galef, Jr., Department of Psychology, McMaster University, Hamilton, Ontario, Canada L8S 4K1

David J. Gubernick, Department of Zoology, Duke University, Durham, North Carolina 27706. Present address: Department of Psychology, Indiana University, Bloomington, Indiana 47405

Lawrence V. Harper, Department of Applied Behavioral Sciences, University of California, Davis, Davis, California 95616

Myron A. Hofer, Albert Einstein College of Medicine at Montefiore Hospital, Bronx, New York 10467

Devra G. Kleiman, Department of Zoological Research, National Zoological Park, Smithsonian Institution, Washington, D.C. 20008

Peter H. Klopfer, Zoology Department, Duke University, Durham, North Carolina 27706

James R. Malcolm, Department of Zoological Research, National Zoological Park, Smithsonian Institution, Washington, D.C. 20008

James J. McKenna, Department of Anthropology and Sociology, Pomona College, Claremont, California 91711

Jay S. Rosenblatt, Institute of Animal Behavior, Rutgers University, Newark, New Jersey 07102

Leonard A. Rosenblum, Primate Behavior Laboratory, Department of Psychiatry, Downstate Medical Center, Brooklyn, New York 11203

Harold I. Siegel, Institute of Animal Behavior, Rutgers University, Newark, New Jersey 07102

Bruce B. Svare, Department of Psychology, State University of New York at Albany, Albany, New York 12222

Karyl B. Swartz, Department of Psychological Sciences, Purdue University, West Lafayette, Indiana 47907

Foreword

The editors of this volume have honored me by their invitation to write its Foreword, an invitation extended because of my editing a book on the maternal behavior of mammals in 1963. Much as I would like to think that I had opened a new area of study—and so played some part in the appearance of this fine new collection of chapters—the facts are quite otherwise. That in 1963 I could assemble the efforts of many distinguished investigators shows that the topic had already engaged their attention, and had for some years past. But even then, the topic had origins extending much farther into the past, to mention only Wiesner and Sheard's book *Maternal Behavior in the Rat* of 1933. Nevertheless, in 1963 it seemed to me that the study of maternal care in mammals had lagged behind the study of other kinds of social behavior.

The present volume does much to establish parental care of the young as a topic central to an understanding of the relation between ontogeny and phylogeny, to the development of the young, to the social organization of the species, and to its preservation. It may now be seen not only as interesting but as a most significant pattern of behavior among mammals.

This volume clearly shows that in the intervening years the topic has indeed been a lively one, and the progress made must be characterized as truly remarkable. Some of the contributions to knowledge were foreshadowed in the earlier volume, such as the process of weaning, evidence of maternal aggression, the part the young play in obtaining the care they require, the role of fathers and unrelated members of the group in ministering to the needs of the young, and the role of parental care in the social organization of the group. But what a wealth of new information has been amassed in these few years when each of these topics is made the primary focus of study!

Several of the contributions wrestle with the concept of attachment, a term used only by Harlow in the 1963 volume but which since then has received much attention. Now, however, several contributors to the current volume are questioning its usefulness as an explanatory concept. Proximity of the young to its caretakers provides much more than protection from predators; it provides food, warmth, and so on, but more important, stimulation, opportunities to learn, and models for becoming members of the social group. The disrupting effects of separation of the young from mother and other members of the group are also shown to vary with many other factors. Under the hands of these skilled investigators, the seemingly magical property of the social bond is yielding to sharply analytic attempts at explanation.

The advances in neurology, biochemistry, and endocrinology from 1963 on have also provided new techniques for the study of processes. Further, the ultrasonic sounds of rodent pups represent a new finding, as does knowledge about pheromones and their effects on both rodent pups and their mothers. Last but by no means least, a new vocabulary of evolutionary concepts now distinguishes the present collection of chapters. That is not to say that the contributors of the 1963 volume were unaware of the evolution of parental care and its adaptiveness, but they did not have at hand the explicit organizing concepts introduced in the intervening years. I refer to such concepts as parental investment, inclusive fitness, and kin selection, as well as cost-benefit analysis, the proximate and ultimate functions of behavior, and optimal strategies.

The chapters of this book, then, provide not only new information on an important class of behaviors, but also a fresh and stimulating framework in which they can be examined. Rich indeed are the delineations of questions still to be answered. The chapters, furthermore, provide models not only for the investigation of parental care in other mammals (dare I now include the human mammal?), but also for the investigation of other related areas of social behavior. I am confident that all readers will profit as have I, for in these chapters they will find challenges to inform and sustain their own efforts for years to come.

Harriet L. Rheingold
University of North Carolina at Chapel Hill

Preface

Since the publication in 1963 of Harriet L. Rheingold's classic volume *Maternal Behavior in Mammals,* the study of parental care has witnessed remarkable progress in theoretical orientation, conceptual advances, and an empirical data base.

Past research had focused predominantly on the immediate causation of maternal behavior and the development of maternal–filial relationships. While these topics continue to stimulate fruitful investigations, a new generation of complimentary research questions has emerged. First, we have witnessed a new awareness of the infant as an active, potent force in the shaping, maintenance, and expression of parental behavior. Gone are the simpler notions that mothers autonomously deliver care to their passive, ineffectual infants. In fact, we now recognize that the allocation of parental care is not restricted to the mother alone and that other individuals can also exert a profound influence on the developing infant. These newly recognized sources include male parental care, allomothering, sibling interactions, and the social context of parental care.

A significant factor in the emergence of these "new" questions and issues has been the application of principles and concepts of evolutionary biology toward understanding the nature of parental care and the offspring's relationships with other individuals. Indeed, the profound influence of the evolutionary perspective can be seen in the original contributions presented in this volume.

In the first chapter, Klopfer discusses why parental care evolved and what factors might influence the evolution of various parental care strategies. He also raises the question of whether, and in what sense, patterns of parental care are even adaptive (a question also raised by McKenna, Chapter 10, in his consideration of allomothering).

Chapters 2–7 are primarily concerned with the mother–infant relationship, while Chapters 8–11 deal with the broader social network of the infant. The hormonal and nonhormonal factors involved in the onset and maintenance of maternal behavior are examined in Chapter 2 by Rosenblatt and Siegel. They discuss and review an extensive literature about these functions and present a theoretical framework for regarding the principal phases in the regulation of maternal behavior.

Next, Hofer (Chapter 3) examines the contributions parents make to the development of their offspring. He provides compelling evidence that the sensory stimulation provided by the mother regulates the physiology and behavior of the developing infant.

Although parents obviously contribute to the development of their young, the singular view of the infant as a passive recipient of parental care is no longer acceptable. As an active participant in parent–infant relationships, the young can affect the caregiver's behavior. Harper (Chapter 4) draws together a vast and diverse literature of such offspring effects and discusses these within an evolutionary, ecological, and developmental framework.

One dramatic example of such offspring effects is that the suckling stimulation of the young can regulate the mother's aggression toward conspecifics. Despite the fact that defense of the young is a common and widespread feature of parental care, it has received little systematic study. The psychobiology of maternal aggression in mammals is an emerging field which is discussed and reviewed by Svare (Chapter 5).

The process of weaning has also received comparatively little analysis or theoretical treatment despite its importance in the transition of an infant from a dependent to an independent existence. Galef (Chapter 6) presents a provocative analogy of the infant as a parasite of its mother and discusses weaning within this context.

The specificity of parent–infant relationships (i.e., attachment) is discussed next by Gubernick in Chapter 7. He addresses the questions of "What is attachment?" and "Why form attachments?" The presence of parent and infant attachment in only some mammals raises the question of what conditions favor attachment, and these conditions are also examined.

In Chapter 8, Bekoff considers alloparenting among siblings and discusses the conditions favoring sibling interactions and sibling recognition. In many mammalian groups, males provide some form of care for the young, yet such male parental investment has not been well studied. Kleiman and Malcolm (Chapter 9) discuss the evolution of male parental care, the various types of such male parental behavior, and its influence on social organization. They also present a summary of those mammals displaying male parental investment.

Like male parental behavior, allomothering is also widespread among mammals. McKenna (Chapter 10) discusses the evolution and presumed functions of allomothering and uses the common Indian langur monkey as a case study. He also raises the issue that allomaternal behavior, as well as parental behavior in general, is not necessarily adaptive, and in some cases allomothering may actually be harmful to the infant.

In the final chapter, Swartz and Rosenblum discuss the social context of parental behavior with special emphasis on infant socialization in primates. They examine the influences various social agents, such as those included in other chapters, may have on the social development of the infant.

We would like to express our appreciation to all the authors for their effort and contributions and to the publishers for their encouragement and support. We hope that this volume will stimulate further research and the evolution of new questions.

David J. Gubernick and Peter H. Klopfer

Durham

Contents

Chapter 3

Parental Contributions to the Development of Their Offspring

Myron A. Hofer

Chapter 4

Offspring Effects upon Parents

Lawrence V. Harper

Chapter 5

Maternal Aggression in Mammals

Bruce B. Svare

Chapter 6

The Ecology of Weaning: Parasitism and the Achievement of Independence by Altricial Mammals

Bennett G. Galef, Jr.

Chapter 7

Parent and Infant Attachment in Mammals

David J. Gubernick

Chapter 8

Mammalian Sibling Interactions: Genes, Facilitative Environments, and the Coefficient of Familiarity
Marc Bekoff

Chapter 9

The Evolution of Male Parental Investment in Mammals

Devra G. Kleiman and James R. Malcolm

Chapter 10

Primate Infant Caregiving Behavior: Origins, Consequences, and Variability with Emphasis on the Common Indian Langur Monkey

James J. McKenna

Chapter 11

The Social Context of Parental Behavior: A Perspective on Primate Socialization

Karyl B. Swartz and Leonard A. Rosenblum

Origins of Parental Care

Peter H. Klopfer

Vater werden ist nicht schwehr.
Vater sein dagegen sehr.
—attributed to W. Busch

1. Parental Care Is Widespread

Caring for one's offspring is so widespread among birds and mammals that it is the exceptions that attract our attention. Among birds, megapodes are a special curiosity precisely because their precocial young are unattended by their elders, even though the latter may have labored long maintaining an optimal temperature in their solar-heated nests (Frith, 1962). Mammals, by virtue of the nursing relationship between mothers and young, are never quite so distant from their offspring, but in some species the contact between the generations ends abruptly with weaning while in others it persists long after. Why is parental care not universal and uniform?

Global questions of this sort have come increasingly to occupy biologists, who were heretofore content with the construction of phylogenetic sequences. It is a happy development, for it has both helped enlighten us as to the factors that have dictated evolutionary changes as well as helped us to reject the simplistic, deterministic schemes that have in the past been so hobbling. However, an important caveat must be interjected at the outset of comparisons: likenesses may be deceptive.

Similarities between organisms can be due to convergence as well as to kinship, i.e., they may be analogues or homologues. The green color of many forest-dwelling insects and amphibians can arise from very different sources: hairs,

PETER H. KLOPFER • Zoology Department, Duke University, Durham, North Carolina 27706.

refracting scales, or pigments. Even pigments with similar spectral reflectances may be composed of different molecules. In order to make the distinction between analogous and homologous resemblances, however, we need prior knowledge of the phyletic relationship and physiology of the animals in question. If one already knows bats to be mammals and flies to be insects, then one can classify their wings as analogous structures. But, if one has no basis for knowing that relationship, one cannot conclude that the wings are homologies (Klopfer, 1973b). Behavioral features follow the same rule: two similar acts cannot in and of themselves justi-fiably be classed as either homologues or analogues. Thus, the comparative approach of classical ethology must be deemed suspect. Behavior at the organismic level is especially malleable and subject to convergence. But, if the techniques of the comparative ethologists are inappropriate, what one can do is to generate "the rules of the game." That is, one can specify both the ecological factors and evo-lutionary accidents that have provided the significant constraints on particular pat-terns of behavior. Having done this for an array of species, we can expect gener-alizations applicable to others to emerge. For instance, Crook (1970) and Eisenberg *et al.* (1972) have examined the relation between habitat and social structure in various primates. Arboreal, diurnal, leaf-eating species tend toward small home ranges and troops with but a single adult male, irrespective of taxo-nomic relationships. The common problems raised by predation by cats and of foraging in treetops have encouraged the evolution of similar adaptations (Eisen-berg *et al.,* 1972).

An appropriate example of the application of a functional approach akin to those alluded to above is seen in studies on the origins of sexuality. While Thurber and White (1957) had asked "Is Sex Necessary?" (it isn't!), proper Darwinians have posed the question somewhat differently. Granted that asexual reproduction can occur, what factor(s) favors the sexual mode? Attempts to answer this query have been based on economic, cost/benefit considerations. "Costs" include those generated by the additional genetic manipulations entailed in sexual reproduction, recombination, and meiosis, plus the energetic and other costs in mating (including increased risks of predation). "Benefits" may include a higher genetic variability (compared with that of asexually reproduced offspring). In variable environments or colonizing species, variability may offer a useful hedge against the extinction of a particular lineage.

Many species include both sexually and asexually reproducing populations. The geographical distribution of sexual and asexual forms have been examined by Glesener and Tilman (1978). Their findings:

> When parathenogenetic populations of terrestrial animals have different geographic distributions than closely related bisexual populations, the sexual forms tend to occupy regions of higher biotically imposed stress, generally being found at lower latitudes and altitudes, in mesic rather than xeric areas, on the mainland as opposed to islands, and in undisturbed as opposed to disturbed habitats. If sexual reproduction is favored by unpredictably chang-

ing conditions, then it would appear that regions of greater biotic stress are more unpredictable than those where abiotic factors are a major source of stress. We suspect that this greater unpredictability stems from interspecific interactions (competition, predation). We further suggest that the changing genotypes of the organisms with which an individual interacts are the major source of this biotic uncertainty; that, once evolved, sex in one population may lead to the contagious spread and persistence of sex in a community of highly interacting individuals. (p. 669)

The question as to why parental care should exist is considered amenable to a similar approach. Care is costly. There must be a benefit. Ultimately, of course, this is the preservation of the individual's genotype, and a relative increase in its frequency. But why does it appear that in some species it is of benefit to minimize parental investment in any one offspring for the sake of producing many more offspring, while in others the investment is maximized to the exclusion of a repetitive reproductive effort? Is it a better strategy for a bird to have but a single brood each year, to care for it tenderly, or to have two or three broods, even though each will then necessarily receive shorter shrift? There is, apparently, no best choice; different strategies have been adopted in different situations and by different species. One reason is clearly the relativity of "fitness." Fitness is a measure of the proportion of one's genes that are represented in future generations. This proportion can be increased either by increasing one's number of surviving offspring or by decreasing the number of one's conspecifics' offspring. Different environmental circumstances will favor one tactic over the other (see also Harper, this volume).

2. Factors Influencing Parental Care Strategy

2.1. r and K Selection

Perhaps the most basic strategic consideration, which influences tactics, is whether selection is of the "K" or "r" variety. This is ecological jargon for the distinction between species that produce large numbers of offspring at frequent intervals (resource-limited or r types) and those that are less fecund because of density-dependent controls (K types). Fruit flies are a typical r species, and man, despite the evidence of exploding world populations, is of the K type.

The r pattern is clearly called for where resources are sporadic in their availability, either regularly pulsed or unpredictable, and, when available, present to excess. High reproductive capacity then assures maximum ability to exploit what would otherwise be lost. "r" species, characteristically, have short generation times that allow them to fluctuate in synchrony with favorable seasons. "K" species, on the other hand, play the game of survival by investing more time and energy in the production of fewer, less exposed young. Some species may even alternate modes. Some copepods will produce a large number of small-yolked eggs after the summer season, then a smaller number of larger, better-endowed (nutritionally) eggs for the fall (Hutchinson, 1951). Table 1 (from Daly and Wilson, 1978, p. 125) illustrates other relationships.

Ectothermic vertebrates are, as a class, more often r than K strategists. Their basic metabolism is already closely tied to environmental conditions, so it is reasonable to expect that their reproductive effort will also track environmental changes—either by being always maximized, so no opportunities are lost, or pulsing to take advantage of propitious moments. In the more equitable conditions of the tropics, this tendency ought to be at its nadir. Comparably, endothermic vertebrates, more often K than r, ought to expand their brood size and number (i.e., shift toward r) as they range to the more extreme latitudes and altitudes.

The differences between K and r strategies are relative, of course, and the related ectotherm–endotherm or precocial–altricial distinctions are far from absolute. This notion has been specifically tested by Ar and Yom-Tov (1978), who examined the geographic distributions of precocial and altricial birds. Precocial birds are essentially r strategists, producing large broods, while the altricial species are K strategists:

> Under steady state conditions each pair of adult animals is replaced on the average by two offspring, and the rest of their progeny is exterminated. Hence, the relatively larger clutch of precocial birds is, on the average, an indication of heavier losses to their offspring between hatching and maturation and particularly during fledging, in comparison to altricial birds. In the altricial mode of reproduction the squabs are less exposed to unpredictable environmental conditions. This might be one of the reasons why the altricial strategy of "individual care" vs. the "mass production" strategy of the precocial birds evolved and became commoner than the precocial one. (p. 661)

Over 80% of bird species are altricial, but their prevalence diminishes in the tropics, where climatic and feeding conditions are more predictable than in temperate regions (Klopfer, 1973a). And, as the hypothesis predicts, Ar and Yom-Tov find that the percentages of precocial species in the tropics (East Africa) is about one-half that in the temperate latitudes of Central Europe:

> It is concluded that the two main evolutionary strategies of parental care in birds differ in their energetic budget, altricial birds being more economic, partly by being K-selectionists in comparison to the r-selection precocial mode of reproduction. Semiprecocial birds represent a transitional position. (p. 663)

Parental care may or may not be functional irrespective of the overall reproductive strategy, as we shall see, so that brood or clutch size is not necessarily predictive of parental behavior. It is, however, likely that among the more pronouncedly K strategists, the reproductive effort per individual offspring is so large as to dictate post-hatching care or protection. When an investment is large, it

Table 1. Some of the Reproductive and Life-Historical Differences between r and K Strategies[a]

r strategist	K strategist
Many offspring	Fewer offspring
Low parental investment in each offspring	High parental investment in each offspring
High infant mortality (mitigated during population explosions)	Lower infant mortality
Short life	Long life
Rapid development	Slow development
Early reproduction	Delayed reproduction
Small body size	Large body size
Variability in numbers, so that population seldom approaches K	Relatively stable population size, at or near K
Recolonization of vacated area and hence periodic local superabundance of resources	Consistent occupation of suitable habitat, so that resources more consistently exploited
Intraspecific competition often lax	Intraspecific competition generally keen
Mortality often catastrophic relatively nonselective, and independent of population density	Mortality steadier, more selective, and dependent upon population density
Higher productivity (maximization of r)	High efficienty (maximization of K)

[a]Adapted from Daly and Wilson (1978).

ought to be protected (Ghiselin, 1979). The extreme r types can, so to speak, take their chances.

2.2. Phyletic Continuity and Morphological Constraints

Related species may, despite their relatedness, differ greatly in appearance and behavior. This can beguile one to minimize the importance of phyletic, morphological, or historical constraints, by which we refer to developmental mechanisms that promote evolutionary conservatism. The problem is further compounded because an evolutionarily stable or conservative trait in one species, unchanging through many generations because it was unrelated to functions shaped by selection, may prove highly adaptive and labile to another. Bill shape and size in passerine birds may stand as one example. Among some species of woodpeckers, bill size parameters vary according to whether the birds occupy mainland or island habitats (Wallace, 1978). This is, apparently, a morphological response to changing competitive pressures. In other birds, however, such as the woodpecker finch of the Galápagoes (Lack, 1947), it is the behavior of the bird that yields to environmental pressures, not its morphology. *Camarhynchus pallidus* retained its finchlike bill, discovering instead that a stick could be manipulated so as to substitute for a probing bill. It is reasonable to interpret the relative fixity of the bill size and shape as due to phyletic constraints.

Basically, this notion of "phyletic" constancy assumes that different features of an animal's phenotype are integrated to differing degrees into the overall developmental schemes. Eye color may be attributed to a few alleles which assort independently, but axial symmetry clearly cannot reside in any fixed number of loci. It must be an expression of an intricately and diffusely operating system of control. Thus, axial symmetry is more tightly constrained by events in the species' distant past.

The relation between the phenotypes of ancestral and descendant species is a complex topic, recently addressed by Gould (1977) in his reexamination of Haeckel's and Baer's theories of "recapitulation." Gould's conclusions include the recognition of heterochony as a major evolutionary factor. Heterochony is a change in the onset of timing of development, so as to retard or accelerate the time of appearance of a given character, relative to its time of appearance in organisms of previous generations. That prescient geneticist, R. Goldschmidt (1955), anticipated current notions of gene action, which entail regulatory genes controlling the timing and activity of structural genes. Goldschmidt's explanations of evolutionary changes, through changes in the hierarchical organization of genes, provides a spatial analogue to the notion of heterochrony. The two views are entirely complementary, of course. What makes them relevant here is that both view "new" features in evolution as due largely to temporal changes in the timing of development, spatial alterations in the positions of genes relative to one another, or both, and only infrequently to "new" genes or mutations. In both Goldschmidt's and Gould's schemes, phenotypic characters are bound to cluster. Furthermore, the earlier in development a character develops, the more thoroughly buffered it will inevitably be from changes—hence the greater conservatism of axial symmetry as compared with eye color.

In selected species, heterochrony is likely to entail precocious maturation or neotony, since this maximizes reproductive ability. K-selected species are most likely to display a retardation of maturation since their morphology must more closely match ecological demands. This is seen in most extreme form in the recent primates. Neotony is probably, says Gould, the major factor in human evolution.

In summary, the nature of the environment, changing, stable, predictable, or variable, determines a species' overall reproductive strategy. Those species opting for a high turnover shorten developmental periods. There is then, of necessity, reduced opportunity for parental care. In contrast, the young of K-selected species can more readily rely on sustained care by their parents. But, while K strategists are often finely tuned to respond to environmental vagaries, their development is highly canalized. Once characteristics are established during development, they tend to resist change. Hence, structures (and ultimately their behavior) could persist even when they have ceased to provide a specific selective advantage: they are phyletically constrained.

Not all the phenotypic features of even the most slowly evolving lineage are necessarily the products of more ancient developmental events. There is a great deal of evidence that animals, particularly mammals, can adjust within a single generation to new demands, varying their behavior in striking fashion (Klopfer, 1973*a*, for example). However, Gubernick (this volume) records what may be reasonably regarded as the model situation: stable, but not inflexible, adjustments to overall ecological relationships. Thus, ungulates that must live with mobile herds and cannot readily isolate themselves for parturition must rely on imprinting-like mechanisms if a firm bond between mother and young is to be forged. Where the herd can be less mobile, as among Canadian moose, for example, the mother can give birth in solitude and keep her offspring secreted for its earlier days; other mechanisms can then operate to assure that mother and young will recognize and respond to one another (Altman, 1958).

In short, parental care is not universal nor is it infallibly and exclusively associated either with endothermy or a K-type reproductive stratdgy. It shows no clear phyletic continuity, either, for even within one class, all extremes can be seen—again, compare megapodes with geese, or hares with rabbits, and salmon with mouth-breeding cichlids. Similarities in the patterns of parental care are thus more probably due to similarities in the constraining ecological forces than to similar physiologies or phyletic origins (note Gould, 1977).

3. Patterns of Parenting

Probably the most conspicuous features that distinguish different kinds of parent–offspring relationships are the extent of postweaning care, of participation by both parents in infant care, and the degree to which alien infants are tolerated. This last includes "aunting" or the care of infants by individuals that are not parents (see McKenna, this volume). E. Klopfer (unpublished notes) has surveyed the mammalian literature with regard to these attributes, to discover whether any of these features cluster. Among primates, she found, most species display post-weaning care and tolerate alien young, at least in some circumstances. The prosimians appear least frequently to tolerate alien young. The participation of the male in the rearing of the primate infant, however, varies greatly from one species to the next, even within a genus (see also Kleiman and Malcolm, this volume). *Macaca nemestrina* males do not usually handle their young, though *Macaca fascata* do. A similar variability extends to the Insectivora, among which the hedgehogs (*Erinaceous* sp.) neither extend care beyond weaning, attend to aliens, nor involve males in care of the young. Tenrecs (*Hemicentites* sp.), however, while otherwise similar in their parental behavior, do involve the male. The Rodentia and Carnivora also vary from species to species, with all three features varying independently (see Kleiman and Malcolm, this volume). Within an order, the most consistent feature is the extent of postweaning care. In part, this probably reflects the maturation rates of locomotory and allied systems, which are evolutionarily conservative. However, the lack of intraordinal consistency argues against any of these features of parental care being taxonomic indicators, i.e., phyletically constrained. At least, they are less likely to be evolutionarily stable and diagnostic characters than, for instance, the bony processes of vertebrate. They are, rather, responses to the fugitive demands of particular habitats and life styles. A detailed survey of paternal behavior by Ridley (1978) results in comparable conclusions.

> Paternal care usually correlates with external fertilization. There exist two hypotheses to explain this: certainty of paternity and order of gamete release. It is likely that each is the final (evolutionary) cause in different phylogenetic lineages. Another suggestive correlation is between paternal care and male territoriality. The data do not tell us whether male territoriality is the authentic final cause of paternal care or *vice versa;* but either way territoriality must have been an important influence on the evolution of paternal care. . . . Female fecundity is also likely to have been an influence during the evolution of paternal care but there is a general absence of data for establishing suggestive correlations between the reproductive effort of the female in eggs and paternal care. (pp. 927–928)

Even an effort to examine the broader issue of the relation between parental care and the altricial condition of the young has not produced a clear pattern. Altricial young are generally small and ectothermic, and thus face problems in maintaining body temperatures. This ought to dictate a high level of parental care. But Case (1978) notes there is not a consistent correlation between body size and the degree of maturity (and thus need for care) at birth. And, where we do find endothermy, there is a similar lack of correlation with degree of parental care.

To repeat, all behavior is ultimately subject to a cost/benefit analysis. If it "pays," it persists; otherwise, it does not. The currency of the payment is genetic: a large coin (enhanced "fitness") means a proportionately larger contribution to the gene pools of future generations. In some instances, the alternative modes may have equal ratios. Kummer (1971) has pointed out that some cud-chewing ungulates alternate the direction of the lower jaw with each chomp. Others will first chew n times to the right, then n times to the left. It is clearly important that tooth wear (and thus chewing directionality) be equalized, but both patterns of chewing are similarly effective in achieving this. Thus, "choice" of how to chew may then be left to the "inclinations" of each individual, or be consistent for all members of a species, i.e., constrained by developmental factors ("hard programs"). The species' "choice," in turn, may have been based on purely fortuitous circumstances or on advantages of one particular pattern that obtained at some point in its past. Evolutionary history thus generates an inertial effect which may persist in the absence of any continuing impetus.

4. Adaptiveness and Patterns of Parental Care

Despite an evolutionary bias, biologists are still inclined to assume that structures, limbs, and neurons undergo changes more slowly than do their outputs, particular movements, or behavior. The latter may change rapidly in response to specific and even seemingly trivial ecological demands, as noted above. Consider further three species of the genus *Lemur, L. catta, fulvus,* and *variegatus.* They are quite similar in their morphology. Ecological relations differ only in degree, i.e., *L. catta* may spend up to 35% of its feeding time on the ground, while *L. fulvus,* in the same forest, feeds almost entirely arboreally (Sussman, 1979). Differences in parental care, however, appear more dramatic.

L. catta mothers readily pass their infants on to the other females in their troop utilizing baby sitters even when the infants are no more than seventy-two hours old. In contrast, *L. fulvus* mothers are much more zealous in their guardianship. Their infants, as a general rule, are almost a month old before being regularly cared for by a sitter. *L. variegatus,* a third species, leaves its infants in a nest, visiting them only at intervals. The infant is rarely carried by the mother, and then only in the mouth (Klopfer and Boskoff, 1979).

Are these differences immediately adaptive, or are they the results of selective forces that operated only at some distant time past?

A persuasive case for the adaptiveness of the specific differences can be made. The terrestrial behavior of *L. catta* is likely to expose them more often to potential predators. In the rush to escape, infants might be dropped; how advantageous, then, were others to assist in their retrieval? For that matter, sitters might even allow mothers to forage unencumbered by infants. *L. variegatus* solves the dilemma by depositing its young in a nest. The subsequent problem of maintaining the infants' body heat is resolved by multiple births. Only *L. variegatus,* of this genus, regularly gives birth to twins or triplets (Klopfer and Boskoff, 1979).

There is a subtle difference, too, in the manner in which *L. catta* and *L. fulvus* carry their infants: body axis parallel to their own or at right angles. The orientation of the infant may affect locomotion on ground and in treetops differently. These suppositions could be subjected to experimental verification, though this has not been done. But, if we are compelled to test the view that the patterns that prevail are the best possible from the standpoint of maximizing fitness, how do we proceed? Is it the thesis of Voltaire's Pangloss: a fit topic for comforting speculation and of no heuristic value?

Our dilemma stems in part from a too simplistic view of causality. *A* may be the necessary and sufficient condition for the occurrence of *B.* More usually, it appears that *A* or *A'* or *A''* may be equally quasi-sufficient (and only under specified circumstances, necessary) conditions for *B.* Behavior not only has multiple causes, but a given "cause" may produce different results. This may reflect a high degree of idiosyncrasy: the uniqueness of each individual, or at least species. More likely (and this is an expression of faith), it reflects our ignorance of the boundary conditions which determine whether and when a given cause may be operative. An analogy with the process of early development as postulated by Davidson and Britten (1971) is instructive. In connection with a theory of the control of gene expression during development, they write:

> ... interaction [of sensor structures associated with the genome and extra-genic agents] ... leads to the transcription of integrator genes and their products in turn effect control of the transcription of many genes and estab-lish patterns of gene activation. ... We assume that there exists an initial divergence in genetic activity due to an unequal distribution of egg cyto-plasmic regulatory elements among the different cells in early cleavage. As a result, particular sensor structures could be synthesized in some cells and not in others. Therefore, individual cells would differ in their response to external inductive agents arising, for example, in nearby cell layers. Each specific cell or cell type would then be characterized by the integrated acti-vation of a proper set of genes and be capable of carrying out its role in subsequent developmental events. (p. 123)

As an analogy, suppose that the neurons of particular regions of the brain and nervous system correspond to "genes," and behavior patterns to "cell types." The "regulatory elements," of course, are hormones!

The system implied by Davidson and Britten has a high degree of internal stability—it is buffered at many points—yet it also is susceptible to alterations. Similar circuits can produce widely disparate outputs and vice versa.

Since the details of the control of gene expression are far from clearly under-stood, it is not embarrassing that the inevitably more complex behavioral systems still elude comprehension. Nonetheless, it is our suspicion that to begin to under-stand maternal behavior in a general way and to be able to predict the relations between particular species, their behavior, ecology, and neuroendocrinology, we shall have to fundamentally alter our approach and the form of our questions. Analysis of systems as functional wholes may be less appealing than mere descrip-tions of motherhood, or analysis of isolated elements thereof, but they may be more revealing (Klopfer *et al.*, 1973).

Nor ought we, I believe, be overly disdainful of the possibility that what matters most in the parent–young relationship is the predictability of interactional patterns. By analogy, "permissive" and "authoritarian" (human) parents may be equally successful in rearing adjusted children. It is the inconsistently demanding parents who rear the problem child. In another context, but also in repudiation of the adage "as the twig is bent so grows the limb," Kagan (1972, unpublished) writes:

> It does not seem to be the case that there is a set of fixed characteristics that lead inexorably to a particular pattern of psychological growth. Nor is the behaviorist view that the child is soft putty in the hands of experience a useful alternative. As indicated earlier, Darwin's prejudice for a totally environmental interpretation of speciation prevented him from considering

the organism's inherent contribution to its own evolution. A large number of physiological environmentalists in the U.S. are in a similar position. They have made the elaboration of mental capacity too dependent on continued environmental input from the outside, and have failed to see the brain's own contribution to its autochothonous growth. (pp. 59–60)

In all events, we are left with a difficult and unresolved problem for whose solution no adequate methodology exists. Major features of parental care vary. The variations, however, appear not to be simply related to taxonomic status or ecological demands whose influence can be tested. They may or may not be adaptive, *per se*. Their adaptive value of particular patterns may be simply in their predictability. While the adaptive significance of specific features of parental care ought still be considered, we might search with equal diligence for intraspecific differences. Where they persist within lineages they may offer a tool to test the thesis that orderliness in parent–young relations is the most important adaption.

ACKNOWLEDGMENTS. My studies of parental care have been supported by grants from the NIMH (04453) and the Duke University Research Council. I acknowledge with gratitude the criticism and support of the Duke Behavior Group, and in particular my colleagues, David Gubernick and Robert Wallace.

References

Altman, M., 1958, Social integration of the moose calf, *Anim. Behav.* **6**:155–159.

Ar, A., and Yom-Tov, Y., 1978, The evolution of parent care in birds, *Evolution* **32**:655–669.

Case, T. J., 1978, Endothermy and parental care in the terrestrial vertebrates, *Am. Nat.* **112**:861–874.

Crook, J. H., 1970, Social organization and the environment: Aspects of contemporary social ethology, *Anim. Behav.* **18**:197–209.

Daly, M., and Wilson, M., 1978, *Sex, Evolution, and Behavior,* Duxbury Press, N. Scitvate, Mass.

Davidson, E. H., and Britten, R. J., 1971, Note on the control of gene expression during development, *J. Theor. Biol.* **32**:123–130.

Eisenberg, J. F., Muckenhirn, N. A., and Rudran, R., 1972, The relation between ecology and social structure in primates, *Science* **176**:863–874.

Frith, H. J., 1962, *The Mallee-fowl: The Bird That Builds an Incubator,* Angus & Robertson, Sydney.

Ghiselin, M., 1979, *The Economy of Nature and the Evolution of Sex,* University of California Press, Berkeley and Los Angeles.

Glesener, R. R., and Tilman, D., 1978, Sexuality and the components of environmental uncertainty: Clues from geographic parthenogenesis in terrestrial animals, *Am. Nat.* **112**:659–673.

Goldschmidt, R., 1955, *Theoretical Genetics,* University of California Press, Berkeley and Los Angeles.

Gould, S. J., 1977, *Ontogeny and Phylogeny*, Harvard University Press, Cambridge, Mass.

Hutchinson, G. E., 1951, Copepodology for the ornithologist, *Ecology* 32:571–577.

Klopfer, P. H., 1973a, *Behavioral Aspects of Ecology*, Prentice-Hall, Englewood Cliffs, N.J.

Klopfer, P. H., 1973b, Does behavior evolve? *Ann. N.Y. Acad. Sci.* **222**:113–125.

Klopfer, P. H., McGeorge, L., and Barnett, R., 1973, *Maternal Care in Mammals*, Addison-Wesley (Module), Reading, Mass.

Klopfer, P. H., and Boskoff, K., 1979, Maternal behavior in prosimians, in: *The Study of Prosimian Behavior* (G. A. Doyle and R. D. Martin, eds.), pp. 123–156, Academic Press, New York.

Kummer, H., 1971, *Primate Societies*, Aldine-Atherton, Chicago.

Lack, D., 1947, *Darwin's Finches*, Cambridge University Press, Cambridge.

Ridley, M., 1978, Paternal care, *Anim. Behav.* **26**:904–932.

Sussman, R., 1979, *Primate Ecology*, Wiley, New York.

Thurber, J., and White, E. B., 1957, *Is Sex Necessary? Or, Why You Feel the Way You Do*, Harper and Row, New York.

Wallace, R. A., 1978, Social behavior on islands, in: *Perspectives in Ethology*, Vol. 3 (P. P. G. Bateson and P. H. Klopfer, eds.), pp. 167–204, Plenum Press, New York.

2

Factors Governing the Onset and Maintenance of Maternal Behavior among Nonprimate Mammals

The Role of Hormonal and Nonhormonal Factors

Jay S. Rosenblatt and Harold I. Siegel

This chapter will review what is known about the factors which regulate the onset and maintenance of maternal behavior among nonprimate mammals. Maternal behavior is an outgrowth of the endocrinological processes which regulate pregnancy and parturition; therefore, in order to understand the factors which govern its onset, it is necessary to study the endocrine control of pregnancy in each species. In addition, many of the procedures used to investigate the onset of maternal behavior (e.g., hormone administration, hysterectomy, caesarean-section delivery, prostaglandin administration, ovariectomy, etc.) alter the normal course of pregnancy in specified ways that can only be understood with reference to the normal endocrine control of pregnancy.

This is not intended as a comprehensive review of all aspects of maternal behavior. We shall not review the many descriptive studies of maternal behavior nor studies on the sensory stimuli which elicit maternal care and aggression. The main purpose of this review is to present the current status of research in this field within a theoretical framework that arose from our research on the maternal behavior of the laboratory rat (Rosenblatt, 1975; Rosenblatt *et al.*, 1979). In this review it is proposed that there are two principal phases in the regulation of

JAY S. ROSENBLATT AND HAROLD I. SIEGEL ● Institute of Animal Behavior, Rutgers University, Newark, New Jersey 07102.

maternal behavior: the first phase is the *onset of maternal behavior* which is governed by endocrine processes. The second phase is the *maintenance of maternal behavior;* during this phase maternal behavior is governed mainly by sensory stimulation and psychological processes (i.e., perceptual and motivational processes). There is a *transition* period joining these two phases, centered around parturition, during which the regulation of maternal behavior undergoes a transition with respect to the factors which govern it.

The first section of this review will discuss the *onset of maternal behavior* in a variety of species. The hormonal regulation of pregnancy in these and other species will be presented together with studies aimed specifically at the question of hormonal stimulation of maternal behavior. The second main section will deal with the *regulation of postpartum maternal behavior* and will discuss hormonal and sensory factors during the maintenance of maternal behavior. The third section will present studies on *maternal behavior in nonpregnant females,* that is, sensitization or the stimulation of maternal behavior in females as a result of continuous exposure to offspring. The fourth section will discuss the *transition in the regulation of maternal behavior between the onset and maintenance phases.* The fifth section will review studies on maternal agression within the framework of the above theory.

1. Onset of Maternal Behavior

1.1. Hormonal Regulation of Pregnancy

The hormones regulating pregnancy and parturition are quite similar among a large variety of mammals. Figure 1 shows circulating levels of the principal hormones: estrogen, progesterone, and prolactin for many species and, in addition, the gonadotropins, follicle-stimulating hormone (FSH), and luteinizing hormone (LH) for a few species. For some species only the terminal phase of pregnancy is shown. Throughout most of pregnancy plasma progesterone is maintained at a high level; then it declines slowly after midpregnancy and precipitously just before parturition. Plasma estrogen (estradiol-17β) is maintained at low levels throughout most of pregnancy; then it rises just before parturition. Circulating prolactin is also maintained at low levels throughout most of pregnancy and rises sharply just before parturition shortly after the rise in estrogen. In one species, the hamster, progesterone and estrogen both decline shortly before parturition while prolactin rises, and in several other species the rise in estrogen is not as sharp as in others (Fig. 1d).

The gonadal hormones and prolactin have traditionally been viewed as the hormones most likely to be implicated in the onset of maternal behavior, largely because of the changes in circulating levels of these hormones which occur during pregnancy, and because of their involvement in parturition and the initiation of lactation. However, other hormones may also prove to be important in the onset of maternal behavior (e.g., oxytocin; see below). While our discussion will deal mainly with the above hormones, we shall discuss other hormones where this is appropriate.

The similarity of patterns of hormone secretion during pregnancy in the different species does *not* imply that similar mechanisms regulate the secretion of these hormones. The four principal sources of these hormones are the maternal ovaries, the anterior pituitary gland, the fetal–maternal placental unit, and the fetus. The relative contribution of these three endocrine organs (to the regulation of pregnancy) varies considerably in different species (Bedford *et al.,* 1972; Hilliard, 1973), and the specific factors which regulate the ovary also vary. A discussion of these factors can best be presented, however, in the context of studies on the hormonal basis of maternal behavior in the various species.

It is worth noting, in preparation for our later discussion of postpartum maternal behavior, that the hormonal picture changes completely after parturition in each of the species depicted in Fig. 1. In a number of these species the female undergoes an estrous period immediately after parturition ("postpartum estrus") during which she can become pregnant again if mated. In all of the species the high prepartum level of circulating estrogen drops sharply postpartum (except during the postpartum estrus), and remains low throughout lactation (not shown in Fig. 1), while progesterone, which is also low immediately postpartum, either remains low throughout lactation or rises after the first few days. Prolactin remains high or increases during the lactation period as a consequence of suckling. In several species it has been shown that suckling both stimulates the release of prolactin and other pituitary hormones of the "lactogenic complex" (i.e., ACTH, TSH, GH) and inhibits the release of the pituitary hormones, FSH, and LH, thereby preventing ovarian follicular development and estrous cycling throughout most of lactation [rat (Rothchild, 1960)]. The hormone picture after parturition is best described as one which supports lactation and suppresses estrous cycling (except for the postpartum estrus), and in species in which the female becomes pregnant during postpartum estrus, there is a delay in the onset of the pregnancy by not permitting implantation of the blastocyst ("delayed implantation").

1.2. Hormonal Stimulation of Maternal Behavior

The earliest efforts to stimulate maternal behavior hormonally were in the rat and utilized chiefly prolactin and ovarian hormones (Fig. 1a). These have been summarized in Table 1 with the test conditions also included (Rosenblatt and Siegel, 1980). Although there were claims for the involvement of prolactin (Riddle *et al.,* 1934), these claims could not be substantiated in later, more carefully done experiments (Beach and Wilson, 1963; Lott, 1962; Lott and Fuchs, 1962), and ovarian hormones were also tested without success. The first successful demonstrations that humoral (as contrasted with hormonal) factors could stimulate

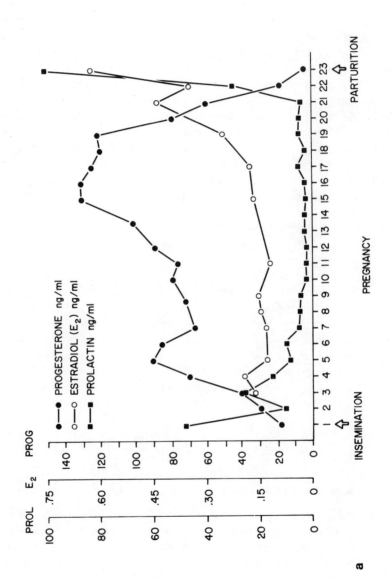

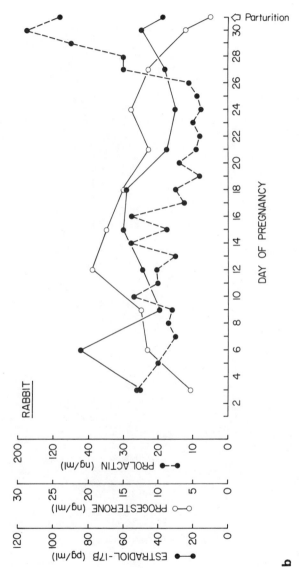

Figure 1. Plasma hormone concentrations during pregnancy and postpartum lactation for estradiol-17β, progesterone, prolactin, luteinizing hormone (LH), and follicle-stimulating hormone (FSH). Not all of these hormones are shown for each species, and hormone levels during postpartum lactation are not shown for all species. (a) Rat: progesterone (Pepe and Rothchild, 1974), estradiol (Shaikh, 1971), prolactin (Morishige et al., 1973). (b) Rabbit: estradiol-17β, progesterone (Challis et al., 1973), prolactin (McNeilly and Friesen, 1978). (Figures c–n on pages 18–24.) (c) Mouse: progesterone, estradiol-17β (McCormack and Greenwald, 1974), prolactin, LH, FSH (Murr et al., 1974). (d) Hamster: estradiol, progesterone (Baranczuk and Greenwald, 1974), prolactin, FSH, LH (Bast and Greenwald, 1974). (e) Guinea pig: estrogen, progesterone (Challis et al., 1971). (f) Sheep: estradiol, progesterone, prolactin (Chamley et al., 1973). (g) Goat: estrogen (Challis and Linzell, 1971), progesterone (Irving et al., 1972), prolactin (Buttle et al., 1972). (h) Cow: estrogen, progesterone, prolactin (Hoffman et al., 1973). (i) Pig: estrogens, progesterone (Ash and Heap, 1975). (j) Dog: estradiol-17β, progesterone, prolactin (Graf, 1978). (k) Cat: estradiol, progesterone (Verhage et al., 1976). (l) Rhesus monkey: estradiol, progesterone, prolactin (Weiss et al., 1976). (m) Chimpanzee: estradiol, progesterone, prolactin (Reyes et al., 1973). (n) Human: estradiol, progesterone (Turnbull et al., 1974).

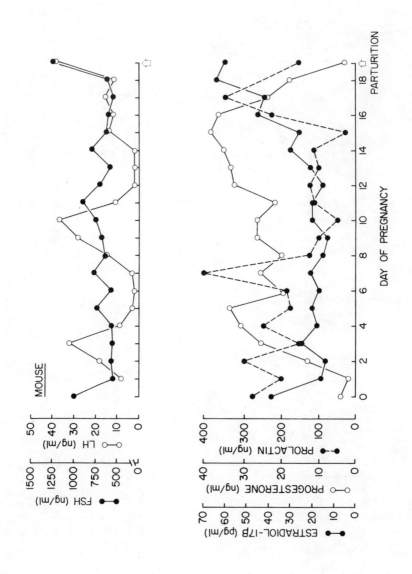

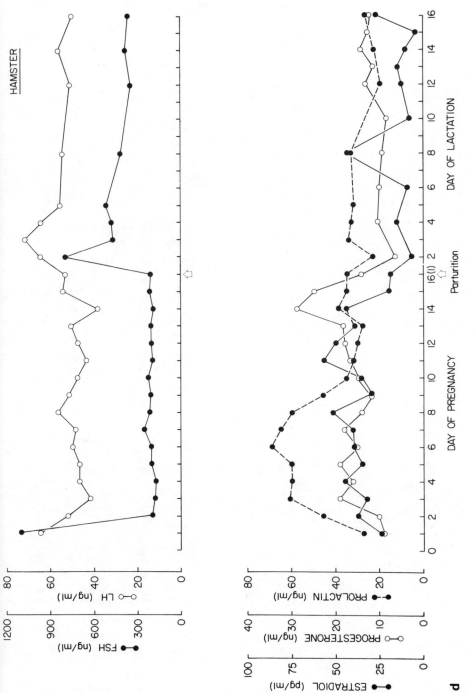

Figure 1 (*continued*)

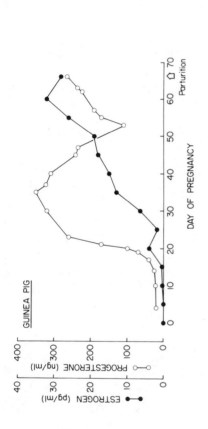

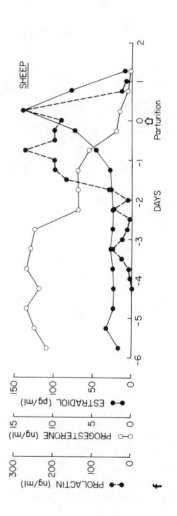

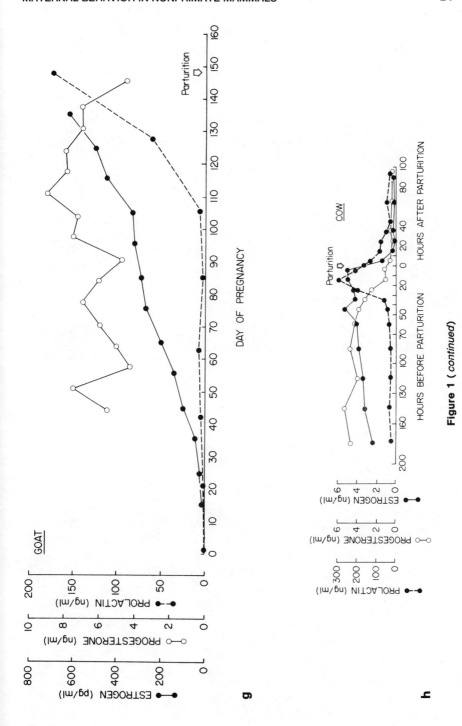

Figure 1 (continued)

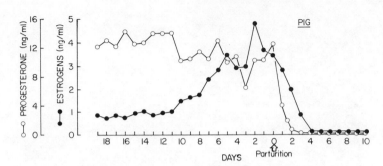

i

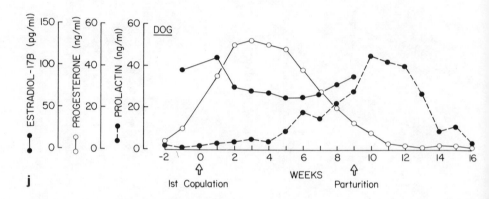

j

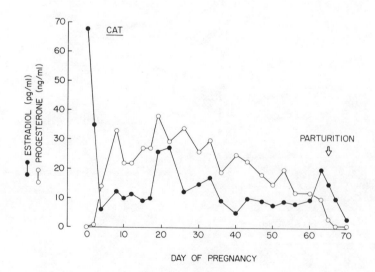

k

Figure 1 (*continued*)

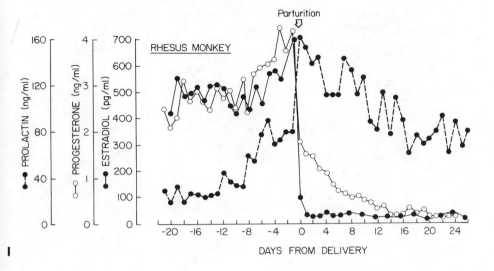

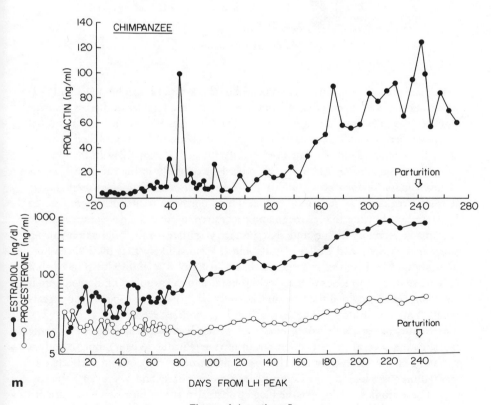

Figure 1 (*continued*)

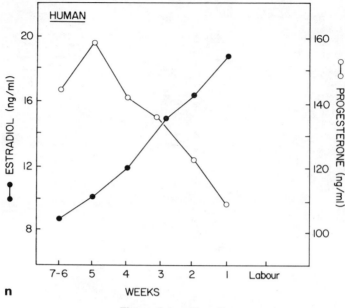

Figure 1 (*continued*)

maternal behavior in the rat were reported by Terkel and Rosenblatt (1968, 1972) using injections of blood plasma from newly maternal females and cross transfusion between new mothers and virgins. The latencies for the onset of maternal behavior were about 2 days using injection and about 14 hr using blood transfusion. These studies did not, however, specify the active component in the blood.

Hormone administration has been used successfully in the rat by two teams of investigators, Moltz *et al.* (1970) and Zarrow *et al.* (1971), who used similar hormone regimens as shown in Fig. 2. These hormone treatments were derived from studies showing that estrogen and progesterone cause the development of the mammary secretory tissue and that prolactin initiates milk production during pregnancy (Cowie and Folley, 1961), and it seemed likely that these three hormones might be involved in maternal behavior. Moltz *et al.* (1970) showed that only these three hormones in the specific temporal order would stimulate maternal behavior, but there are difficulties in this study with respect to control group treatments that should have been effective but were not (see Rosenblatt *et al.*, 1979). The treatment used by Zarrow *et al.* (1971) has been used successfully by these investigators several times in their strain of rats (Purdue Wistar), but recently a variation of this procedure and that of Moltz *et al.* (1970) proved ineffective in shortening onset latencies in Long–Evans rats (Krehbiel and LeRoy, 1979).

These studies, again, like the blood transfusion study, but with much greater specification of the relevant hormones, indicate that the onset of maternal behavior

Table 1. Previous Hormone Treatments in the Study of Hormonal Stimulation of Maternal Behavior in the Rat

Hormone and other treatments	Procedure	% Exhibit maternal behavior		
		Intact	Ovariectomized	Reference
Prolactin		71	79	
Intermedin		60	77	
Luteinizing hormone	Pretest days	59	60	
Desoxycorticosterone	1–10 (10 min)	56	42	
Progesterone	Inject days 11–20	68	66	Riddle *et al.* (1942)
Testosterone	Test days 21–23	63	85	
Phenol		46	55	
Thyroxine		31	53	
Estrone withdrawal		78	50	
Progesterone	Inject days 1–10	21		Lott (1962)
Oil	Test days 11–13	14		
Prolactin + sensitization[a]		0		
Prolactin + 10 min exposure to pups	Inject and test days	0		Lott and Fuchs (1962)
Control + sensitization	1–10	33		
Control + 10 min. exposure to pups		17		
Prolactin (10 mg)		0		
Prolactin (20 mg)	Inject days 1–5	0		Beach and Wilson (1963)
Control	Test days 6–8	67[b]		
Estrogen (days 1–20) + prolactin (days 21–30)		25		
Estrogen (days 1–20) + water (days 21–30)	Test days 31–33	60		Beach and Wilson (1963)

[a]Continuous exposure to pups.
[b]Partial pattern of maternal behavior.

in the rat is hormonally based. How the hormonal stimulus arises during pregnancy and which components of the above treatments are really effective require further study, as we shall see.

Methods for studying the onset of maternal behavior in the rabbit were introduced by Zarrow and his colleagues (Zarrow *et al.*, 1961, 1962*b*) which involved terminating pregnancies prematurely either by hormone treatment, ovariectomy, or caesarean-section delivery, the latter having been used earlier in the rat in a

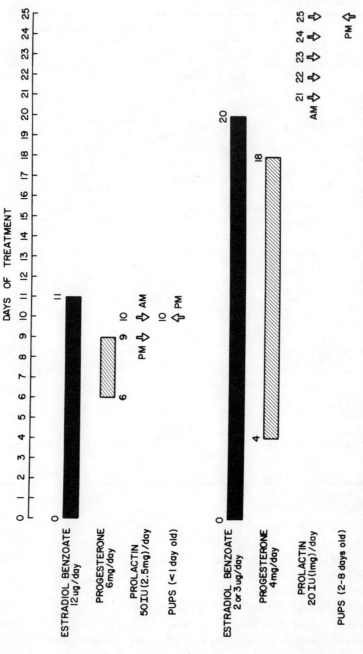

Figure 2. Hormone treatments and testing conditions used by (a) Moltz *et al.* (1970) and (b) Zarrow *et al.* (1971) to stimulate maternal behavior in ovariectomized rats.

preliminary study by Wiesner and Sheard (1933). The procedures were successful (see below) and were adopted by Rosenblatt and Siegel (1975) in their studies of maternal behavior in the rat. Using hysterectomy performed at various times during pregnancy, Rosenblatt and Siegel (1975) showed that there is a rapid onset of maternal behavior with latencies becoming shorter the later in pregnancy that the hysterectomies are done. What was most important was that females remaining pregnant had latencies that were considerably longer than those of the hysterectomized animals. Therefore, the hysterectomy had stimulated a change in the pattern of hormonal secretions during pregnancy and one or more of the components of this change was responsible for the rapid appearance of maternal behavior.

It became necessary, therefore, to determine what hormonal changes occurred after hysterectomy and which of these changes was instrumental in stimulating maternal behavior (Rosenblatt and Siegel, 1975). One hypothesis was that by removing the fetal placentas, ovarian progesterone secretion would decline and this would, in turn, allow ovarian estrogen secretion to rise, stimulated by anterior pituitary gonadotropins. This would be consistent with the findings of Pepe and Rothchild (1972) and Rothchild et al. (1973) that ovarian progesterone secretion during the second half of pregnancy is regulated by the fetal placentas. This hypothesis was largely confirmed by the following: plasma progesterone declined rapidly after hysterectomy (Bridges et al., 1978a; Rosenblatt et al., 1979), and the appearance of lordosis behavior and ovulation around 48 hr after hysterectomy were indirect evidence that there was a rise in ovarian estrogen secretion. Plasma estrogen and pituitary gonadotropins have not yet been measured directly after hysterectomy.

The above hypothesis and its virtual confirmation relates the effects of hysterectomy, an artificial procedure, admittedly, to the hormonal changes which normally occur at the end of pregnancy (Fig. 1a). Ovarian secretion of progesterone declines, and the secretion of estrogen rises. Prostaglandin $F_{2\alpha}$ is involved in producing these hormonal changes (Behrman et al., 1971; Buckle and Nathanielsz, 1975; Carmanati et al., 1975; Strauss et al., 1975). We have therefore used this uterine product instead of hysterectomy to stimulate maternal behavior (Rodriguez-Sierra and Rosenblatt, unpublished) and found that it is equally as effective as hysterectomy and also is accompanied by lordosis behavior and ovulation.

On the basis of the above analysis of the possible role of estrogen, it was predicted, based upon the early rise in plasma estrogen and the decline in progesterone (Fig. 1a) that maternal behavior might, in fact, arise before parturition. This prediction was confirmed in our laboratory (Rosenblatt and Siegel, 1975) and in another laboratory as well (Slotnick et al., 1973).

These hysterectomy studies led to two lines of research: one was to study the role of estrogen given to hysterectomized–ovariectomized late-pregnant females and later to nonpregnant similarly operated females, and the other was to study the role of progesterone decline ("progesterone withdrawal") as a stimulus for maternal behavior. The first of these lines of research (Siegel and Rosenblatt,

1975*a,b*) established that estradiol benzoate stimulated short-latency maternal behavior in both kinds of females, and the second line of research (Bridges *et al.*, 1978*b*) established that progesterone withdrawal has a short-term facilitory effect on maternal behavior.

We have shown that progesterone also blocks the action of estrogen on maternal behavior if given 24 hr after estrogen is administered to hysterectomized–ovariectomized nonpregnant females (Siegel and Rosenblatt, 1975*c*) and this has now also been shown in similarly treated pregnant females (Siegel and Rosenblatt, 1978*a*).

Several studies have ruled out the possibility that estrogen stimulates maternal behavior via the release of prolactin (Numan *et al.*, 1977; Rodriguez-Sierra and Rosenblatt, 1977) or that prolactin is involved in the normal onset of maternal behavior (Baum, 1978; Stern, 1977; Zarrow *et al.*, 1971). A recent study has raised the possibility that oxytocin, a hormone released by the posterior pituitary gland, may either supplement estrogen or be the intermediary between estrogen and the neural substrate of maternal behavior (Pedersen and Prange, 1979). In this study oxytocin was infused directly into the lateral and third ventricles of the brain, a normal path of release of this hormone (Dogterom *et al.*, 1977). These ovariectomized, nonpregnant females had been primed with estrogen given systemically 48 hr earlier. The onset latencies for estrogen–oxytocin treated females were very rapid (i.e., 10–45 min after oxytocin infusion), and the behavior pattern was complete in nearly all animals. Fewer estrogen-vehicle-treated females and none of the nonprimed oxytocin-treated animals exhibited the complete pattern within the 2-hr test period. Preliminary studies have shown, however, that oxytocin is not effective if given systemically rather than directly into the ventricles.

Oxytocin is normally present at the onset of maternal behavior and is quite active at that time. The number of oxytocin receptors in the uterus rises dramatically during the last 6–8 hr before parturition, in preparation for oxytocin's action upon uterine contractions during parturition; their increase in mammary tissue is more gradual from the 5th day of pregnancy onward (Soloff *et al.*, 1979). In the uterus the rise in the number of oxytocin receptors appears to be dependent upon circulating estrogen, and there is evidence that reflex release of oxytocin is also dependent upon prior action of estrogen (Catalá and Deis, 1973; Roberts, 1973). These facts would lend credence to the possibility of oxytocin stimulation of maternal behavior and to the dependence of this effect on prior treatment with estrogen (Pedersen and Prange, 1979). However, the experiment reported does not distinguish between oxytocin facilitation of the effects of estrogen and, the contrary, estrogen facilitation of the effect of oxytocin. There is already evidence that estrogen implanted at a specific brain site (medial preoptic area) stimulates maternal behavior, whereas when it is implanted at other sites it has no effect on maternal behavior (Numan *et al.*, 1977). It is difficult to see how oxytocin could mediate this effect.

In the *rabbit* (Fig. 1b), like the rat, terminating pregnancy after midterm (pregnancy lasts 31 days) results in the onset of *maternal nest building*—hair loosening and pulling to incorporate hair into the nest, the principal items of maternal behavior that have been studied (Zarrow *et al.*, 1961, 1962*a,b*). The effects of hysterectomy performed on various days of pregnancy and the associated latencies for onset of maternal nest building are shown in Fig. 3. The figure also shows the effects of ovariectomy, which is equally successful in causing maternal nest building. Further, high doses of estrogen, which also terminate pregnancy in most cases, result in maternal nest building, and this also occurs in those females that maintain their pregnancies (Fig. 3). Finally, the termination of pseudopregnancy is associated with the onset of maternal nest building.

The specific effects of each of these treatments on hormonal changes during pregnancy have not been studied, but it is evident that pregnancy depends upon the presence of functioning ovaries (Hilliard, 1973) and that the fetal placentas in some way regulate ovarian function during the latter half of pregnancy. In addition to the role of pregnancy termination in stimulating maternal behavior, these studies point to the need for prolonged stimulation by estrogen and progesterone. As an alternative, the response to pregnancy termination that is responsible for the onset of maternal nest building may not be present until midpregnancy.

Based upon these pregnancy-termination studies Zarrow and his colleagues

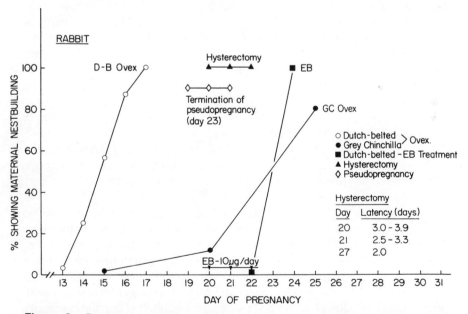

Figure 3. Pregnancy-termination effects of ovariectomy (Ovex), estradiol benzoate (EB), and hysterectomy on maternal nest building in two strains of pregnant rabbits. Termination of pseudopregnancy also shown. See text for explanation. (From Zarrow *et al.*, 1961, 1962*a,b*.)

(Zarrow *et al.*, 1961, 1962*a,b*; Ross *et al.*, 1963) initiated hormone treatments using, at various times, different combinations of estrogen and progesterone, and occasionally prolactin. Two features of estrogen/progesterone treatment appeared crucial for stimulating maternal nest building: *prolonged treatment with estrogen* lasting 18–31 days, treatment with *progesterone starting either at the same time or shortly afterward, and, most important, terminating 3 days before estrogen treatment* (Zarrow *et al.*, 1962*a,b*). Maternal nest building was stimulated in more than 80% of intact and ovariectomized females using this treatment schedule.

The hormone treatment proved ineffective, however, when females were hypophysectomized (Anderson *et al.*, 1971; Zarrow *et al.*, 1971). Moreover, rabbits that were hypophysectomized during the last third of pregnancy also failed to build nests following the induced abortions. These results implied that the pituitary hormone prolactin might be involved in the hormonal stimulation of maternal nest building. In a subsequent study, therefore, ergocornine hydrogen maleate was injected into 26-day-pregnant females to block the release of prolactin (Zarrow *et al.*, 1971). At higher doses of ergocornine none of six females built nests at parturition (which was advanced by 1 day). If prolactin was given together with ergocornine, nest building was restored at parturition.

In the rabbit, in contrast to the rat, prolactin does not play a direct role in the regulation of progesterone secretion. It is responsible for the storage of cholesterol in the interstitial tissue where it is available to be converted into progestins. Treatment with ergocornine would be expected to eventually lead indirectly to a decline in progesterone secretion, terminating pregnancy (Hilliard *et al.*, 1968*a,b*). Progesterone decline (progesterone withdrawal) is not a sufficient stimulus for maternal nest building as the hormone administration studies and the study with hypophysectomized females have shown. It is more likely that the withdrawal of progesterone acts in two ways: it removes an inhibitory influence which permits prolactin to act (Zarrow *et al.*, 1961), and it may play either a direct or indirect role in the release of prolactin (Vermouth and Deis, 1974).

There remain several questions about the hormonal basis of maternal behavior in the rabbit: (1) What are the hormonal sequelae of ovariectomy, hysterectomy, and estrogen treatment during late pregnancy which result in maternal nest building and how are these related to the normal events of pregnancy termination at parturition? (2) For prolactin to act is it required that there be a prior period of estrogen/progesterone stimulation and progesterone withdrawal and how long a period is necessary? (3) Are other components of maternal behavior (e.g., maternal nursing and aggression toward intruders) also stimulated by these or other hormones?

Until recently, studies of the hormonal basis of maternal behavior in *mice* (except for nest building) have not been an important aspect of the study of maternal behavior in this group (Fig. 1c). This is because nearly all maternal behavior (except aggression; see below) can readily be elicited in only a few minutes by exposing females and males to appropriately aged young (Noirot, 1972). All components of maternal care appear (except lactation, of course), and they resemble in their performance and their patterning the maternal behavior that appears immediately postpartum in parturient females (Beniest-Noirot, 1958; Noirot, 1969). Recently, however, Jakubowski and Terkel (personal communication) reported that the wild cousins (F_1) of domesticated mice used in laboratories cannibalize newborn or older young unless the female has just given birth and the male has either lived with the female during pregnancy or has had previous experience during an earlier lactation period. The implication is that hormones play a much more important role in the maternal behavior of female mice that have not been "domesticated."

Koller (1952, 1956) was the first to report a sharp increase in nest building around the 4th–5th day of pregnancy at a time when there is a corresponding rise in circulating levels of progesterone and a decline in earlier high levels of estrogen (Fig. 1c). Using intact and ovariectomized females he was able to show that progesterone could increase nest building to a level similar to that of the pregnant female. Lisk et al. (1969) have confirmed these early findings and have gone on to show that nest building during pregnancy is dependent upon a synergism between low levels of estrogen and high levels of progesterone (Lisk, 1971) but that at high levels of estrogen, nest building is inhibited, an effect pointed to by Lehrman (1961) in his early review of this area.

Although Koller (1952) had not been able to detect any effect of prolactin on maternal nest building in mice, Voci and Carlson (1973) have found that prolactin implanted directly into the hypothalamus increases the amount of nest building and also affects retrieving behavior. Latencies were shorter to initiate retrieving and the speed of retrieving was greater in females with prolactin implants or systemic injections of this hormone. This study is not easy to interpret because *intact* females were used and the administered prolactin might have acted via the gonads and gonadal secretions. In favor of an interpretation that prolactin was the effective hormone were the findings that this hormone implanted in the cortex or in the female's neck under the skin was not effective as indicated above. Thus, at the low levels of prolactin typical of brain implants this hormone could act if placed directly near the neural site of its action, or when given systemically at a high dose level but not when given at a low dose level via the systemic route.

Among mice there is also indirect evidence that hormones of pregnancy play a role in maternal behavior in addition to their role in nest building (Noirot, 1972). Noirot and Goyens (1971) found an initial reduction in maternal behavior from nonpregnant levels soon after conception which they attributed to the high levels of progesterone [which could equally well be attributed to decreasing estrogen levels (Gandelman, 1973a); Fig. 1c]. However, later in pregnancy they found an increase in licking pups but no increases in crouching, carrying young to the nest, and nest building behavior [as compared with amounts of nest material used to build a nest, measures used by Koller (1952, 1956), Lisk et al. (1969), and Lisk (1971)]. Fraser and Barnett (1975) compared the retrieving, gathering of nest material, and open-field activity and defecation of virgin and late-pregnant mice (4 days before parturition) as indirect measures of hormonal influences during pregnancy. The late-pregnant females retrieved more rapidly and in a larger percentage of cases than the virgins, and they gathered more nest materials; they were also less active in the open field and defecated more. Fraser and Barnett (1975) suggest that this latter change in responsiveness to stimulation may be the basis of an earlier finding (Barnett and McEwan, 1973) that late-pregnant mice spend more time in the nest than virgins.

The short-duration (i.e., 16 days) pregnancy of the *hamster* is unique among mammals as is the decline in both plasma estrogen and progesterone at the termination of pregnancy (Fig. 1d). Although the principal items of maternal care *normally* begin at parturition, it has been shown by Siegel and Greenwald (1975) that retrieving and crouching over young are exhibited toward foster young on the morning of the 16th day, several hours before parturition has begun. Earlier Richards (1969) and more recently Daly (1972) and Swanson and Campbell (1979a) have shown that maternal nest building begins on the 4th day of pregnancy and increases from then on.

The nature of the hormonal basis of maternal behavior in the hamster has been difficult to establish (Siegel and Rosenblatt, 1980). Richards (1969) was successful in stimulating the maternal nest building that normally appears in early pregnancy in ovariectomized females with large-sized subcutaneous implants of estrogen and progesterone over a period of 18 days. However, those items of maternal behavior which appear later during pregnancy (retrieving, crouching over young, etc.) cannot be stimulated in ovariectomized females by daily injections of estradiol benzoate and progesterone for 14 days (Siegel and Rosenblatt, unpublished data). What is difficult to understand is the fact that the maintenance of high circulating levels of estradiol and progesterone at the end of pregnancy fail to *prevent* the onset of maternal behavior in hamsters. Siegel and Greenwald (1975) administered either estradiol, progesterone, or a combination of both hormones on the 15th day of pregnancy without reducing the percentage of animals becoming maternal or the timing of the prepartum onset of maternal behavior.

The Mongolian gerbil has a gestation period of about 25 days (Elwood, 1977), but no assays of circulating ovarian hormones and prolactin during pregnancy and lactation have been reported. Wallace et al. (1973) have verified the occurrence of postpartum estrus by finding ova in the oviducts and corpora lutea in the ovaries. Nevertheless, studies on the hormonal basis of maternal behavior and descriptions of the onset of maternal behavior have been reported.

In the gerbil an interesting and somewhat unique behavior, namely, scent marking, which normally occurs at low frequency in females, increases 30% by the second week of pregnancy and continues to increase until parturition (Wallace et at., 1973). The increase in marking behavior is accompanied by an increase in the size of the ventral scent gland, which is the source of the material that the female deposits on the marking posts placed in the test arena. Scent marking qualifies as a component of maternal behavior in this species because of its functions, which are not only to mark the area surrounding the nest but to mark the pups as well and, in turn, to preferentially retrieve pups marked with the female's own sebum (Wallace *et al.*, 1973). In addition, the increase in scent marking during gestation is associated with an increase in nest building, and after parturition scent marking is associated with other components of maternal behavior in the gerbil (retrieving, nest building, nursing, licking). The decline of scent marking occurs after the 4th week, somewhat later than other components of maternal behavior. Nest building, for example, declines after the 8th postpartum day.

The fact that scent marking in nonpregnant females is regulated by ovarian hormones (Owen *et al.*, 1974; Wallace *et al.*, 1973) suggests that it is also regulated by these hormones during pregnancy and lactation, and this is supported by the finding that females ovariectomized two days after parturition fail to show the postpartum increase in scent marking (Wallace *et al.*, 1973). In nonpregnant, high-marking females ovariectomy reduces this behavior to low levels and estradiol benzoate restores the original high level, but in low-marking females, both progesterone and estrogen are required to stimulate high levels of scent marking (Owen and Thiessen, 1973, 1974; Yahr, unpublished, cited in Owen *et al.*, 1974).

In addition to the rise in nest building and scent marking during pregnancy, reported by Wallace *et al.* (1973), both of which can occur in the absence of pups, Elwood (1977) has shown that cannibalism by females, the prevailing pattern of nonpregnant females, begins to decrease and by the last 6 days of pregnancy only 50% of the females continue to eat pups. Those that do not eat pups in late pregnancy exhibit mixed responses, suggesting some avoidance and fear of the single test pup, but more often they explore the pup, sniff and lick it, build a nest, and, on rare occasions, assume a suckling position. The change from cannibalism to maternal care, begun during pregnancy in half of a group of females, is completed in the remaining half at parturition.

Wallace (1973) hysterectomized or hysterectomized and ovariectomized female gerbils within 2 days of parturition and found that, compared with the hysterectomized females, which were considered as representing normal maternal behavior, hysterectomized–ovariectomized females were deficient in licking, sniffing, and anogenital licking and that they ignored the pups for longer periods. On the other hand, the hysterectomized females ate and attacked pups more, were not significantly better as nest builders, retrievers, or nursers, and in general spent very little time engaging in these three activities.

These findings indicate that ovarian secretions are important stimuli for the onset of maternal behavior at the termination of pregnancy, either directly or indirectly through their effects on the secretion of other hormones (e.g., prolactin). However, without a normal, untreated control group it is difficult to evaluate the effects of hysterectomy alone; the short durations spent in all but licking, sniffing, and anogenital licking suggest there might have been deficiencies in the maternal behavior of these females: the large amount of eating and attacking of pups that occurred in these females appears to confirm this. The records of maternal behavior published by Elwood (1975) and Kaplan and Hyland (1972) do indicate higher levels of maternal behavior (crouching or contact with pups, nest building) and an absence of pup eating compared with hysterectomized females of the Wallace report, but conditions of rearing and observation (i.e., size of litters, presence or absence of male) may not make these observations comparable. Scent marking of pups was reduced by ovariectomy + hysterectomy, but scent marking in the open field did not differ between the two groups. Absolute values of these activities were not given.

In a second study (Wallace, 1973) high levels of plasma progesterone were maintained to the end of pregnancy to observe its effect on maternal behavior. Beginning on the 19th day of gestation, females were given three daily injections of progesterone, then were hysterectomized to avoid problems of delivery. Progesterone treatment was maintained for 7 days after hysterectomy. One day after surgery, 2 days before parturition would have taken place normally, the females were tested with two foster pups. The author noted "little if any deficiencies in maternal behavior" (Wallace, 1973), but only one of the progesterone-treated females successfully reared her foster pups. The reason for the high mortality was not given; very likely lactation was blocked. Measures of maternal behavior and scent-marking frequencies did not differ between control and hormone-treated groups.

This experiment is difficult to interpret with respect to the hormonal stimulation of maternal behavior. High levels of plasma progesterone could have affected the secretion or the action of several other hormones (estrogen and prolactin).

A third experiment employed, with slight modification, the hormonal regimen used by Moltz *et al.* (1970) to stimulate maternal behavior in ovariectomized rats. Estradiol benzoate was given to ovariectomized gerbils at a low dose from days 1 to 5 and at a stepped-up dose from days 6 to 9; the dose was again increased on days 9 and 10. Progesterone was given on days 6 through 9 and prolactin on days 9 and 10. On day 10, females were tested with two pups; one was then left overnight, and the other returned the next day for the next test. This procedure was continued for 10 days.

The hormone-treated females again showed mixed responses: licking and sniffing pups and retrieving were greater than among oil-treated females, but anogenital licking and nest building did not differ; both groups ignored the pups equally and for long periods. Eating and attacking pups were *greater* among the hormone-treated females, and this increased during testing. Neither group exhibited crouching over the young as in nursing. Scent marking was greater among hormone-treated females.

These studies represent a worthwhile beginning in the study of the hormonal basis of maternal behavior in the gerbil, but before we can plan further investigation, studies are needed on the pattern of hormone secretions that maintain pregnancy in the gerbil.

In summary, there is no uniformity in the hormonal basis of maternal behavior in the rodent and lagomorph species which we have reviewed, thus far. Despite similar hormone profiles during pregnancy and at parturition among rats, rabbits, and mice, the hormonal stimulus for maternal behavior differs in these species: among rats it is very likely estrogen; among rabbits, prolactin; and among mice, nest building is stimulated by a combination of estrogen and progesterone, early in pregnancy. The hamster differs in the hormonal picture during pregnancy, and only nest building has been stimulated hormonally by prolonged estrogen/progesterone treatment.

Among the *ungulates* the hormonal basis of maternal behavior has been studied most systematically and intensively in sheep by the French investigators Poindron and Le Neindre (1979, 1980). The pattern of circulating hormones, chiefly estrogen, progesterone, and prolactin, closely resembles that found in the rat and other small mammals (Fig. 1f). In an initial attempt to induce maternal behavior hormonally, intact, multiparous ewes were treated hormonally using two methods to induce lactation and at intervals during the course of this treatment the ewes were tested with newborn lambs for two hours (Le Neindre *et al.,* 1979). The treatments consisted of a long-term administration of progesterone and estradiol benzoate every third day for 30 days and a single large dose of estradiol benzoate on the 30th day. In this group ewes were tested for maternal behavior on the 31st day. A short-term treatment consisted of 7 days of injections of the same hormones with tests conducted on the 14th and 21st days. In both groups the latest dates of testing coincided with measures of induced lactation.

At the end of 14 days the short-term-treated ewes exhibited maternal behavior in 56% of the cases (7/12), and on the 21st day, 61% (14/23) were maternal. The ewes receiving long-term treatment were maternal in 47% of the cases (9/19). The groups did not differ substantially, indicating that 7 days of estrogen/progesterone treatment was sufficient. While this study established that maternal behavior could be induced by some combination of these hormones, it did not determine the respective roles of each of the hormones; and even more important, the treatment caused the release of large amounts of prolactin, which could have played a role in maternal behavior (Head *et al.,* 1975).

Following this successful induction of maternal behavior with exogenous gonadal hormones an attempt was made (Poindron and Le Neindre, 1979, 1980) to observe maternal behavior during various reproductive states: during estrus when estrogen levels are high and progesterone levels low, during pregnancy on the 45th, 75th, and 105th days (pregnancy lasts 145 days in ewes) when progesterone levels are high and estrogen levels are low, and when this is reversed around the 135th day of pregnancy (Fig. 1f). Newborn lambs were used for testing. Ewes responded maternally during estrus (25% of the ewes) and again at 135 days' gestation (35% of the ewes). High estrogen levels and low progesterone levels at parturition resulted in maternal behavior in all of the ewes. On the other hand, when the reverse hormonal condition existed, on the 45th, 75th, and 105th days of pregnancy, none of the females responded maternally to the lambs.

The specific roles of estrogen and progesterone in the onset of maternal behavior were investigated by administering these hormones separately to multiparous, intact, and ovariectomized ewes (Poindron and Le Neindre, 1979, 1980). In one study, intact ewes were injected daily for 7 days with either estradiol-17β or progesterone and tested 14 days after the start of treatment. Maternal behavior was elicited in 50% (estradiol) and 45% (progesterone) of the ewes. In a second study, ovariectomized ewes were given a single injection of either estradiol-17β or progesterone and tested between 6 and 10 hr after injection and again at 24 hr. As a control for repeated testing, a group of ovariectomized ewes was not treated but was presented with test lambs three times in 24 hr. At the first test, 72% (8/11) of the estrogen-treated ewes were maternal and only 45% (5/11) of the progesterone-treated ewes accepted the newborn lambs. In subsequent tests after 10 hr and up to 24 hr from the start of the hormone injections an additional ewe that was treated with estradiol became maternal and by the end of 24 hr, therefore, 80% had exhibited maternal behavior, while no additional progesterone-treated ewes became maternal. In the untreated control group of ewes, 30% (6/20) were maternal at the end of 24 hr after three periods of exposure to lambs. Estrogen treatment was therefore much more effective than progesterone treatment in eliciting maternal behavior, and progesterone treatment was only slightly better than no treatment and repeated exposure to newborn lambs. Because of the small number of ewes in both of the hormone-treated groups, these groups did not differ significantly; on the other hand, the estrogen-treated group differed significantly from the untreated controls, while the progesterone-treated group did not.

In these studies, as in parallel studies in the rat in which estrogen has been shown to elicit maternal behavior, it is necessary to rule out prolactin as playing a role, since estrogen causes the release of large amounts of prolactin in ewes (Fulkerson *et al.*, 1976*a,b*; Head *et al.*, 1975). The experiments by Poindron and Le Neindre (1979, 1980) in which the role of prolactin was studied require some explanation because they impinge on the topics of postpartum regulation of maternal behavior in the ewe and the nature of the transition period to which we referred in the introduction to this chapter.

The postpartum period during which newly parturient ewes will accept newly born lambs that are introduced to them at intervals has been studied. By 4 hr postpartum only 50% of the ewes are still willing to accept the lambs, and by 12 hr this has declined to 25%, where it remains at the end of 24 hr. The immediate postpartum period during which ewes are still ready to accept newborn from whom they were separated at birth has been called the "sensitive period," and while its duration can be defined only with respect to each individual ewe, in a group of ewes it appears to last for approximately the first 12 hr.

Poindron and Le Neindre (1979, 1980) have shown that when high levels of circulating estradiol-17β are maintained during the first 23 hr after parturition, during a period when estrogen normally falls to a low level (Fig. 1f), then the sensitive period is extended from 12 hr to at least 24 hr: at 12 hr ewes with high circulating levels of estrogen accept newborn in 60% of the cases (9/15), and at 24 hr in 56% of the cases (9/16). Females with low levels of circulating estrogen accept lambs in 30% and 17% of the cases at 12 and 24 hr, respectively (Fig. 4).

In this situation, when circulating estrogen remains elevated, prolactin is also elevated, in the postparturient ewes. In a second study Poindron and Le Neindre (1979, 1980) therefore administered dibromoergocryptine to block the release of prolactin while maintaining high levels of estrogen (the resulting low levels of prolactin were confirmed by RIA measurements over the first 24 hr postpartum). A second group of postparturient ewes had high levels of both of these hormones. An essentially normal group had high levels of prolactin and low levels of estrogen (Poindron and Le Neindre, 1980).

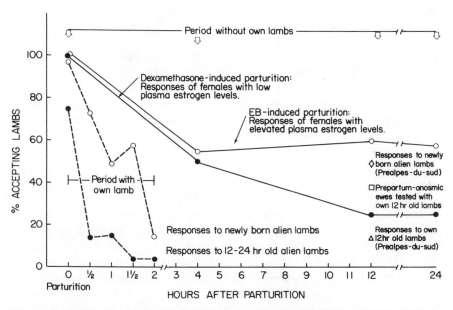

Figure 4. Maternal responsiveness during the transition period in the ewe. Effects of various periods without own young on acceptance of own young at parturition, 4, 12, and 24 hr postpartum. Dexamethasone- and estradiol-benzoate-induced parturition groups shown for comparison. Different groups of ewes tested at each interval. During the first 2 hr the development of an exclusive bond with own lamb is shown by tests with alien lambs, either newborn or 12- to 24-hour-olds, while the ewe remained with her own lamb except for the test. Acceptance of newborn and own lambs (Prealpes-du-sud strain) at 12 hr shown for comparison; all other ewes were Merino strain. Prepartum anosmic ewes tested with own lambs at 12 hr shown for comparison; ewes were without own lambs for 12 hr. (Data from Poindron and Le Neindre, 1980.)

The ewes were separated from their lambs at birth and reunited with them at 24 hr to test their acceptance of their young. Blocking the release of prolactin did *not* reduce the percentage of ewes that accepted their lambs (77%, 17/22), compared with ewes that did not receive the blocking agent (61%, 15/23). Ewes with essentially normal hormonal conditions postpartum (low levels of estrogen and high levels of prolactin) accepted their young in only 15% of the cases (3/22) at the end of 24 hr separation from them. Thus it was clear that prolactin did not play a role in the estrogen-stimulated extension of the postpartum sensitive period.

Apart from our studies on the rat and the studies of Zarrow and his colleagues on the rabbit, these studies on ewes are the most extensive series that has been done on any other mammal. It is significant, we believe, that although all three species (rat, rabbit, sheep) have similar circulating ovarian hormone and prolactin profiles during pregnancy and especially at the termination, they differ somewhat in the hormonal basis of maternal behavior. The principal difference may be the importance of prolactin for the onset of maternal nest building in the rabbit. In the ewe there is a suggestion that progesterone may be somewhat effective in stimulating maternal behavior, and it would be interesting to see whether, as in the rat, it is the decline in progesterone after reaching a high level that provides a short-term facilitating effect.

The research on maternal behavior among goats and cows has focused mainly on the sensitive period for initiating maternal behavior or accepting young during the immediate postpartum period (Collias, 1956; Klopfer *et al.,* 1964; Le Neindre and Garel, 1976). The circulating ovarian hormone and prolactin profiles during pregnancy are quite similar in both of these ungulates, and they resemble the profile of the ewe and are quite similar to that of the rat (Fig. 1g,h). The sow may also be included in this group of ungulates as having a similar hormonal profile during pregnancy and at parturition (Fig. 1i), but there is no evidence whether a sensitive period exists in this species.

In the goat (Hemmes, 1969; Rosenblatt *et al.,* unpublished) tests of the prepartum onset of maternal behavior have proven negative over the period from 6½ days to 1 hr before parturition. In both studies, however, the test young may not have been optimal stimuli for eliciting maternal behavior, since Poindron and Le Neindre (1979, 1980) have shown among ewes that newly born lambs are more effective in this respect than even 12-hr-old lambs (see below). In the two studies cited above the young were 1–5 days of age (Hemmes, 1969) and 12–48 hr old (Rosenblatt *et al.,* unpublished) except for one 3-hr-old test kid.

P. H. Klopfer and M. S. Klopfer (1968) have suggested that the onset of maternal behavior in the goat may be triggered by oxytocin and that the rapid waning of maternal behavior in does separated from their kids immediately postpartum may be due to the rapid decline in the secretion of this hormone once parturition is completed. High levels of oxytocin are released during parturition in the goat in response to cervical stimulation by the fetus, and it aids in parturition by causing vigorous uterine contractions. This is, therefore, a reasonable hypothesis, but it has not yet been adequately tested. Hemmes (1969) perhaps had in mind to test this theory when he studied the effects of vaginal distention on the maintenance of maternal behavior 2½ hr after parturition in does whose kids had been removed at parturition. The results were not conclusive because control females retained their responsiveness, and therefore the experimental females, which also were responsive to kids at 2½ hr postpartum, could not be distinguished from them. Hemmes has suggested a number of other possible candidates responsible for the onset of maternal behavior in addition to simple vaginal distention, but none has yet been tested (i.e., prolactin and neurophysin).

Rosenblatt *et al.* (unpublished) recently tested two hormonal regimens for their effects on the induction of maternal behavior. These were modeled after similar studies among ewes (see above). In the first study intact does were treated to induce lactation by the administration of estradiol-17β and progesterone for 7 days, no hormone treatment for the following 10 days, and 5 days of treatment with hydrocortisone. During the course of this treatment they were tested either with newborn kids or kids less than 2 days of age, after 3 or 6 days of estrogen/progesterone treatment, and again on the 21st or 22nd day during the administration of hydrocortisone. Although lactation was successfully induced by this regime, due largely to the endogenous release of prolactin between days 7 and 18 of treatment, none of the 9 does tested at 3 or 6 days exhibited maternal behavior and only 2 of 11 of the females accepted kids at 21–22 days. Most of the females rejected the kids or ignored them.

In a second study three ovariectomized does were tested before hormone treatment for their responses to newly born kids. They were then treated with estradiol-17β twice daily and at the end of either 3½ or 4 days they were tested again with wet newborn. One of the does changed from rejecting a newly born kid to accepting it, but the other two continued to either reject or ignore the kids. Untreated controls were not employed in this study.

These two studies hint at the possibility that the gonadal hormones, perhaps estrogen, may be involved in the onset of maternal behavior in the goat, but the evidence is too scanty to draw any conclusions at this time.

In Fig. 1 are shown a number of additional circulating ovarian hormone and prolactin profiles during pregnancy or near term of guinea pigs (Fig. 1e), carnivores (dog and cat, Fig. 1j,k), ungulates (cow and pig, Fig. 1h,i), and primates (rhesus monkey, chimpanzee, human female, Fig. 1l–n) for which no studies have been done on the hormonal basis of the onset of maternal behavior. These profiles are quite similar among the different mammalian orders with the exception of the chimpanzee, in which progesterone does not decline at term as it does in other primates and in nearly all nonprimates. The concept that this characteristic terminal pattern of hormone change is associated with the onset of maternal behavior is supported by the finding in the beagle bitch that when this circulating hormone pattern occurs at the end of pseudopregnancy the female exhibits maternal behavior even though no pups have been delivered (Smith and McDonald, 1974).

Whether cats exhibit maternal behavior when pseudopregnancy terminates 2 weeks before normal pregnancy would be an interesting problem to pursue (Fig. 1k).

2. The Regulation of Postpartum Maternal Behavior

In many of the species we have discussed, parturition is followed almost immediately by a single estrous cycle during which the female exhibits estrous behavior, mates, and may become pregnant again. In these species (i.e., rat, mouse, rabbit, gerbil) there is clearly a sharp change in the pattern of hormones secreted postpartum by the pituitary gland and ovaries. Even in those species in which there is no postpartum estrus, plasma estrogen and progesterone levels decline and prolactin is maintained at a high level or increases. The hormonal stimulus that is responsible for the onset of maternal behavior, therefore, is not maintained postpartum, and this raises the question of whether or not maternal behavior postpartum is maintained by any hormones. In the rabbit, in which prolactin stimulates maternal nest building, hormonal regulation of maternal behavior postpartum may not pose a problem if this hormone also stimulates nursing and other aspects of maternal behavior. In the rat and ewe, where estrogen is believed to stimulate maternal behavior, the decline to low levels or its absence in circulation (during lactation) poses a problem for the hormonal regulation of postpartum maternal behavior.

In the *rat* it has been shown that removal of the ovaries, adrenal glands or pituitary gland (Bintarningsih *et al.*, 1958; Erskine, 1978) postpartum does not interfere with the maintenance of maternal behavior (see review by Rosenblatt *et al.*, 1979). Moreover, blocking the release of prolactin with ergocornine hydrogen maleate or ergocryptine (Numan *et al.*, 1972; Stern, 1977; Zarrow *et al.*, 1971) during the first few days after parturition, or administering progesterone at dose levels that are effective in blocking the prepartum onset of maternal behavior (Herrenkohl and Lisk, 1973; Moltz *et al.*, 1969), also does not interfere with postpartum maternal behavior.

Of course, lactation is dependent upon prolactin, and milk letdown is stimulated by oxytocin, but the performance of nursing *behavior* (i.e., crouching over the young) does not depend upon these hormones and may even occur in females that have been thelectomized (i.e., nipples have been removed) before they were mated or whose mammary glands have been removed (mammectomy). In the latter case one can be sure that prolactin release is not stimulated by pup contact with the female's ventrum, but in the former case, prolactin release does occur despite the absence of suckling (Moltz *et al.*, 1967; Zarrow *et al.*, 1973). Thelectomized and mammectomized mothers perform all items of maternal behavior in ways that are indistinguishable from intact females, at least quantitatively and when evaluated qualitatively in a gross fashion (Moltz *et al.*, 1967; Wiesner and Sheard, 1933).

Leon *et al.* (1978) and Woodside (1978) have recently shown that one component of maternal behavior is modulated mainly by adrenocortical hormone (cortisone) and partly by prolactin. In earlier studies Leon (see review, 1978) has shown that production of excess maternal pheromone (contained in the volatile caecotrophic portion of feces) by 2-week postpartum females is a consequence of the high levels of prolactin that are secreted during late pregnancy and lactation, which, in turn, increase food intake of the lactating female (Fleming, 1976, 1977). Increased food intake is the basis for the production of excess caecotrophe.

The role of adrenocortical hormone, cited above, is mediated through its effects on heat production, which plays an important role in regulating bout durations of contact between mother and young during the first 2 weeks postpartum. Normally, in intact females, bout durations decline after the 1st week because the growing ability of pups to thermoregulate accelerates the attainment of high ventral and core temperatures, making it necessary for the female to leave the young (see also Galef, this volume). In adrenalectomized lactating mothers, heat production is reduced, enabling them to remain on the pups for longer bout durations, thus preventing the normal decline in the second postpartum week.

Ovarian, adrenocortical, and pituitary hormones have been shown to play no role in postpartum maternal behavior. Pup stimulation appears to play a crucial role in its regulation. The removal of pups at parturition results in a rapid decline in maternal responsiveness at the end of 4 days (Rosenblatt, 1965; Rosenblatt and Lehrman, 1963). Pup stimulation appears not to act through stimulating the release of pituitary hormones (except with respect to lactation and milk letdown) but in a more direct fashion (see below), since removing the pituitary gland has no apparent effect on the basic maintenance of maternal behavior (Rosenblatt, 1967).

We shall not review the specific sensory stimuli by means of which the pups regulate the female's behavior during lactation (cf. Harper, Svare, this volume). It has been shown, however, that the size of the litter and its age affect the duration of maternal care and the time of weaning (Bruce, 1961; Grota, 1973; Wiesner and Sheard, 1933). Vocalizations (sonic and ultrasonic calls), olfactory stimuli, visual and contact–thermal stimuli, and pup activity regulate the moment-to-moment interactions between the mother and her litter (Rosenblatt and Siegel, 1980; Smotherman *et al.*, 1974). Of particular importance are the effects on the young caused by stress which, in turn, influence the mother's response to them and her maternal behavior in general (Lee and Williams, 1974). Pathological conditions in the young, such as those caused by nutritional deficiency, hypothyroidism, and various genetic defects, are also known to affect the mother's behavior toward offspring (see also Hofer, this volume).

It has not been shown in each of the instances in which pup stimulation affects maternal behavior whether this effect is mediated by the release of hormones. However, there is sufficient evidence to allow us to say that should such hormonal influences be found, they are likely to act as modulators of one or another aspect of maternal care but not as basic conditions for the performance of maternal behavior.

In other mammalian species we cannot readily eliminate the possibility that stimulation by the offspring regulates maternal behavior by causing the release of hormones. Lehrman (1961), in his extensive review of the hormonal basis of maternal behavior among infrahuman mammals, distinguished between the *endogenous release* of pituitary hormones that regulate reproductive behavior, gonadal activity, and other hormone-dependent processes (e.g., follicular growth and ovulation) and the *exogenous* release of these hormones via external stimulation from various sources including the young. He was able to find only a few examples in which it had been shown that external stimuli might act on reproductive behavior *without the intervention of the pituitary gland and the subsequent release of gonadal hormones*. We shall discuss these examples, but not many examples have been added except the findings in the rat discussed above. Yet the possibility that maternal behavior postpartum may not be regulated by hormones is a persistent theme in the literature on maternal behavior with no strong evidence to contradict it. In fact, a new body of evidence, to which we shall refer later in this article, not available to Lehrman when he wrote his review, has given direct support to this idea. This is the finding in the rat and increasingly in other species that maternal behavior can be elicited from nonpregnant females by continuous exposure to young, a process which is referred to as "sensitization" or "concaveation" (Wiesner and Sheard, 1933). The potency of pup stimulation to induce maternal behavior certainly argues for its ability to maintain maternal behavior once it has been stimulated by hormones. By itself this evidence is not conclusive, unless accompanied by further evidence of the absence of hormonal mediating processes.

The concept of lack of hormonal regulation of maternal behavior postpartum arose from Koller's (1952, 1956) studies on nest building in *mice* in which ovariectomized females built maternal nests in response to stimulation by pups. Leblond and his colleagues (Leblond, 1940; Leblond and Nelson, 1937) have shown that postpartum hypophysectomy does not eliminate any components of maternal behavior, and they have expressed their conclusion in the following terms: "the action of various hormones during pregnancy and parturition might have stimulated the nervous system in such a way that the maternal instinct developed under these hormonic stimuli would not disappear after their removal by hypophysectomy" (Leblond and Nelson, 1937, p. 170). There is support for this view in the findings of Noirot (1964*a,b,* 1969) that the patterning of maternal behavior in nonpregnant females closely resembles that in postpartum mothers and that pups of advancing age given to nonpregnant females elicit age-dependent changes in maternal behavior, including a decline when they reach 2 weeks of age, which resembles that seen in lactating mothers rearing their own litters.

In the *hamster*, maintenance of maternal behavior is dependent upon stimulation from the young: there is a decline in maternal responsiveness soon after removal of the pups if this occurs either 1 hr or 24 hr after birth (Siegel and Greenwald, 1978). Conversely, maternal behavior may be maintained without decline long beyond its normal time, but within limits, by replacing the mother's own pups with younger pups (Swanson and Campbell, 1978, 1980). Whether these effects of pup stimulation are based upon hormonal stimulation is not known.

Poindron and Le Neindre (1979, 1980) have proposed that postpartum regulation of maternal behavior among sheep is not based upon hormonal stimulation once the critical period for attachment to the lamb has passed. As already shown (see Section 1.2), this period can be extended by hormones (estrogen or progesterone) even in the absence of stimulation by the kid, but in nontreated females at the end of 24 hr postpartum only 25% of the ewes respond maternally to lambs. Even more suggestive is the contrary finding that if ewes are separated from their lambs starting 2–4 days after parturition, all of them accept their offspring immediately upon its return 24 hr later. None of these studies is crucial, however, for answering the question whether postpartum maternal behavior is hormonally regulated, since it cannot be excluded that hormones are involved in lamb-maintained maternal behavior in ewes.

The same situation pertains in the goat, in which there is also a critical period immediately postpartum during which the doe must receive kid stimulation if maternal behavior is to be maintained (Klopfer *et al.*, 1964). The fact that it is much more difficult to have a doe accept a kid after the critical period has passed and that it requires forced contact lasting many days suggests that a different mechanism is involved from that which is hormonally based during the critical period (Hersher *et al.*, 1963*a,b*). In this case it is not the prior formation of an exclusive bond between the mother and her own kid that prevents the new kid from being adopted but the low level to which maternal behavior has declined in the absence of kid stimulation immediately postpartum (see below).

3. Maternal Behavior in Nonpregnant Females

The theoretical importance of maternal behavior induced in nonpregnant females of many species by exposure to suitably aged offspring lies in its relation to the regulation of postpartum maternal behavior. Apart from the use of "sensitized" female rats to serve as nonlactating mothers to pups in various nutritional and other kinds of studies, these sensitized females are not useful for mothering foster young since they do not lactate. Additionally, the long latencies required to induce maternal behavior in these females would present problems for the survival of the young. They are useful "experimental preparations" for studying some of the properties of the substrate of maternal behavior upon which hormones and sensory stimuli act during the reproductive cycle. These preparations are particularly useful because they avoid some of the added complications of the rapidly changing hormonal conditions of pregnancy and parturition. Recently we and others have used these females to study the neural basis of maternal behavior (Numan et al., 1977) and factors which influence the ontogeny of maternal behavior in both females and males (Bridges et al., 1974; Mayer et al., 1979; Mayer and Rosenblatt, 1979a,b).

In rats, exposure to pups induces all components of maternal behavior in females, the latencies varying according to the amount of contact with the young which the females have (Cosnier and Couturier, 1966; Rosenblatt, 1967; Terkel and Rosenblatt, 1971). The onset is fairly abrupt, anticipated several days earlier by an increase in licking and sniffing pups (Fleming and Rosenblatt, 1974). Pups below 3 days of age more rapidly induce maternal behavior than older pups, and pups above 12 days of age are least effective (Stern and MacKinnon, 1978).

Several lines of evidence suggest that sensitization does not involve hormones as the basis for the onset of maternal behavior, that is, it is *nonhormonal*. The evidence for this has been summarized and discussed extensively in Rosenblatt *et al.* (1979) and will not be given in its entirety here, but the main points will be enumerated.

1. Hypophysectomized–ovariectomized females can be sensitized as easily as, or perhaps more easily than, intact females (Rosenblatt, 1967).

2. The onset of maternal behavior during sensitization is not affected by the phase of the estrous cycle at the start of testing nor does sensitization alter estrous cycling except after maternal behavior has been in progress for some time (Erskine, 1978; Korányi et al., 1976; Rosenblatt, 1967; Stern and Siegel, 1978).

3. Although there have been claims that estrogen and progesterone can modulate onset latencies (Leon et al., 1973, 1975) we have not been able to validate these findings (Mayer and Rosenblatt, 1979a; Siegel and Rosenblatt, 1975b); moreover, these claims do not challenge the major concept that the induction process is, in the main, nonhormonal in nature.

4. Spontaneous retrieving and pup-induced maternal behavior very likely share a nonhormonal basis (Terkel and Rosenblatt, 1971, 1972); neither is affected by changes in hormone conditions associated with pregnancy, and both are elicited with ease at this time (Bridges, 1978). Moreover, cross-transfused blood from spontaneous retrievers and from sensitized females has no effect upon recipient females (Terkel and Rosenblatt, 1971, 1972).

In *mice* the onset of maternal behavior in naive females is so rapid following initial exposure to pups, occurring in less than 5 min in laboratory strains, that it is difficult to conceive of hormonal stimulation acting so rapidly (see review, Noirot, 1972). With respect to maternal nest building, Gandelman (1973b) reported that ovariectomized females built nests as large as those of intact virgins when both were presented with 1-day-old pups. Nest building was largely determined, in these virgin mice, by the nature of the pup stimulation and was greater with live 1-day-old pups than with freshly killed 1-day-olds, or with 19-day-old pups. Combined with the previously mentioned fact that hypophysectomy immediately postpartum does not disrupt maternal behavior, we can conclude that nonhormonally based maternal behavior is readily elicited in mice (Leblond and Nelson, 1937).

Until recently, in *hamsters,* it has been difficult to elicit maternal behavior in virgin females (Richards, 1966; Rowell, 1961; Siegel and Greenwald, 1975). There were indications that nonpregnant females retrieved and cared for pups older than 6 days of age without prior exposure but typically killed younger pups (Noirot and Richards, 1966; Rowell, 1961). Recently, however, Siegel and Rosenblatt (1978b), Swanson and Campbell (1979b), and Buntin *et al.* (1979) have shown that younger pups can also induce maternal behavior in females if exposures are repeated several times a day for 2-3 days. Average latencies are, in fact, shorter than those found in rats, ranging between 18 hr and 43 hr in these three studies. Marques and Valenstein (1976) have reported an unusually high percentage of females spontaneously retrieving pups on their first exposure to them if tests are conducted under their experimental conditions. Siegel and Rosenblatt (1978b) used ovariectomized females and found latencies as short (43 hr) as those of intact females, suggesting, therefore, that pup-induced maternal behavior is independent of ovarian hormones.

Among *sheep,* Poindron and Le Neindre (1979, 1980) have shown that repeated exposure of nonparturient ovariectomized ewes to lambs can stimulate maternal behavior in less than 24 hr. Thirty percent of a group of 20 nonpregnant ewes became maternal, and the fact that they were ovariectomized, in a species in which estrogen and progesterone have been shown to stimulate the onset of maternal behavior, strongly suggests that the basis of this induction was nonhormonal.

In summary, the existence of nonhormonally based maternal behavior has been established in the rat and mouse, and it is likely that the young-induced maternal behavior in the hamster and in ewes is also nonhormonally mediated. The theoretical significance of this phenomenon, which has little practical value, lies in the strong possibility that postpartum maternal behavior in these species may be nonhormonally based.

4. Transition in the Regulation of Maternal Behavior between the Onset and Maintenance Phases

The proposal that there are two phases in the regulation of maternal behavior during the reproductive cycle, the phase of onset of maternal behavior and the phase of postpartum maintenance, implies that there is a transition period during which regulation of maternal behavior undergoes a change from the first type of regulation to the second. These phases very likely correspond to the hormonal onset of maternal behavior and the nonhormonal maintenance by stimulation received from the offspring. During the transition phase hormonal influences on maternal responsiveness reach a maximum then decline and nonhormonal factors are initiated and gain in strength. It remains an open question in many species whether postpartum maintenance of maternal behavior is nonhormonal in nature. What can be shown, nevertheless, in several species is that a transition in regulation occurs. By removing the offspring for a period immediately postpartum it can be shown that maternal behavior wanes; this can be prevented by leaving the offspring with the mother or replacing them before maternal responsiveness has disappeared entirely.

In the *rat,* maternal behavior wanes if pups are removed during parturition and kept away. By the end of 2 days only 30% of a group of females exhibit components of maternal behavior, and by the end of the 4th day, none are immediately responsive to pups (Rosenblatt and Lehrman, 1963; Rosenblatt, 1965). On the other hand, a 4-day separation from the young started at 3 days, 9 days, or 14 days postpartum results in a decline of only 30–40% in the percentage of females exhibiting maternal behavior. These females recover their responsiveness shortly after pups are returned to them, while the females whose pups were removed immediately at parturition remain largely unresponsive.

Bridges (1975, 1977) has studied the amount of contact with her young which the postpartum rat requires to retain the influence of the young after the young have been removed. Retention is tested by measuring the latencies to sensitization in mothers 25 days later when they are undergoing estrous cycling. We have interpreted the retention as evidence that the mothers had made the transition from hormonal to nonhormonal regulation of maternal behavior at the time of contact with their pups. This interpretation is supported by Bridges' recent study (Bridges, 1978), which showed that retention is not affected by the hormones of pregnancy, and by earlier studies (Fleming and Rosenblatt, 1974; Mayer and Rosenblatt, 1975) suggesting retention only occurred if the transition had been made earlier.

Bridges' findings are summarized in Fig. 5: if females have contact with their pups for as little as 4–6 hr postpartum (upper panel) they exhibit short latencies upon sensitization 25 days later (compare with lowest panel). During parturition itself, therefore, they make significant progress in the transition if allowed as little as the amount of contact involved in cleaning and licking half the number of pups in their litters (upper middle panel). However, without this amount of contact they fail to make the transition and 25 days later their latencies are unaffected and are as long as those of inexperienced females. Caesarean-section-delivered mothers (lower middle panel) are like postparturient mothers, depending upon whether they are given pups or remain without pups after the delivery.

In the postparturient *hamster,* Siegel and Greenwald (1978) have shown that

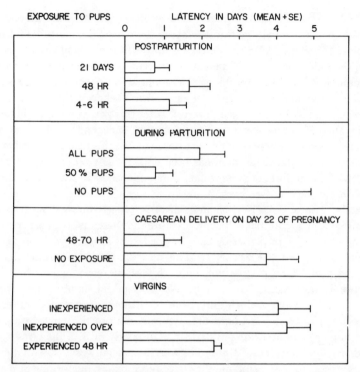

Figure 5. Studies on the *transition period* in the maternal behavior of the rat. The effects of parturition and postparturition contact with pups on latencies for induction of maternal behavior 25 days later. Period of exposure to pups shown on ordinate and latencies on the upper abscissa. Caesarean-section-delivered females with virgins shown for comparison. (From Bridges, 1975, 1977.)

the mother requires 2 days of contact with her pups before she can withstand separation and still maintain her responsiveness to them. Females were permitted either 1, 24 or 48 hr of pup contact postpartum prior to removal of their litters. Test pups (1–2 days old) were then offered daily to independent groups of animals. The group of females allowed 1 hr of litter experience following parturition was maternal for 3 days (days 2, 3, and 4 of lactation) but cannibalized pups on the 5th and 6th days. Litter contact during the first 24 hr increased the interval during which animals remained maternal by 1 day. However, groups with 48 hr of pup exposure were able to sustain their maternal behavior from day 4 to day 15 and only began to kill the test young on days 16 and 17.

It is characteristic of the transition period when hormonal effects are waning and nonhormonal effects are only weakly established that interruption of the mother–young relationship at this time has more deleterious effects on maternal behavior than a similar interruption occurring at a later time, when nonhormonal regulation is better established.

The "critical period" for maternal acceptance of offspring among goats, sheep, and cows resembles the transition period we have described. Poindron and Le Neindre (1980, p. 113) have described this critical or sensitive period among sheep in the following terms: "The few hours post-partum during which the ewe can remain maternal in the absence of young—in other words the sensitive period—probably make up the time when maternal behavior changes from a hormonal regulation to a neurosensory control. After this post-partum period, maternal acceptance of young seems to become largely independent of the hormonal state of the mother." These investigators have studied this transition period in the ewe perhaps in greater detail than has been done for any other species. It would be worthwhile, therefore, to present their data and discuss in some detail the transition period in the ewe (see Figs., 4 and 6).

The behavioral measure that most directly reflects the waning of the hormonal influence on maternal behavior, according to these investigators, is the percentage of acceptance responses by ewes whose lambs are removed at parturition, and returned at either 4, 12, or 24 hr. This measure (Fig. 4) indicates that at 12 hr and again at 24 hr postpartum, only 25% of the ewes accept their lambs. Acceptance is influenced, however, by the ages of the lambs when they are returned to their mother, which differed at the different test periods. Presenting newly born lambs at each test interval is typically not practical, but this was done in one group of ewes at 12 hr postpartum, and nearly 50% of the ewes accepted the newborn at a time when 12-hr lambs were accepted by only 25% of the mothers. Thus, as the authors point out, the curve representing the waning of maternal responsiveness in Fig. 4 probably overestimates the rate at which maternal responsiveness wanes after parturition.

Very likely this curve depicts the waning influence of the hormonal stimulus, which was shown to be estradiol (see above). If this is so, then waning should be slowed by maintaining high circulating levels of estradiol (i.e., terminating pregnancies with estradiol benzoate; see above). Estradiol levels were maintained in three groups of ewes whose lambs were removed at parturition and returned at either 4, 12, or 24 hr. As shown in Fig. 4, at 12 and 24 hr 55–60% of these mothers accepted their young, compared with only 25% in the groups of ewes in which estradiol was not maintained at a high-level postpartum.

Bouissou (1968) has distinguished between maternal responsiveness, which we have been discussing above, and individual recognition as two components of maternal behavior in sheep (see also Smith et al., 1966). The development of individual recognition is measured by testing ewes that are taking care of their own lambs for their responses to alien lambs of the same or different ages. In Fig. 4 the development of individual recognition or, at it is often called, the exclusive bond with the mother's own lamb, is shown. When 12- to 24-hr-old lambs are presented to mothers at parturition, they are accepted by 75% of the ewes, but 30 min later when the mother has been alone with her own lamb, she accepts older lambs in only 15% of the cases and, of course, rejects them in 85% of the trials. Undoubtedly it is the discrepancy between her own 30-min-old lamb and the older one with respect to a number of sensory features which is the basis of the rejection, but olfactory stimuli probably play a major role (see below). Newly born lambs resemble her own offspring more closely and are accepted by ewes in at least 58% of the cases until 90 min postpartum, but then newborn too no longer are accepted at 2 hr postpartum.

The rejection of alien lambs on the basis of the formation of an exclusive bond (i.e., individual recognition) does not provide a good measure of the course of postpartum maternal responsiveness in ewes. It measures the combined influences of the ewe's level of maternal responsiveness and the degree to which she has formed an exclusive bond based upon the specific stimulus characteristics of her own lamb. It is therefore already a *product* of the transition and, while important, is only of limited usefulness in analyzing the transition period.

Bouissou (1968) and Poindron (1974, 1976) have prevented the development of individual recognition on the basis of olfaction by making females anosmic before parturition using either olfactory bulbectomy or intranasal infusion of zinc sulfate, respectively. Poindron (see Poindron and Le Neindre, 1980) then separated the anosmic females from their lambs at birth and tested them with newborn at 12 hr. The anosmic ewes accepted the newborn in more than 35% of the cases. This is nearly the same rate of acceptance of newborn by nonanosmic ewes at that time (Fig. 4). This indicates that the rejection of newborn by more than 50% of a group of 12-hr-postpartum ewes, separated from their lambs at birth, is not equivalent to rejection of an "alien" lamb but is the direct expression of the current low level of maternal responsiveness of these ewes.

Under normal conditions, of course, the postparturient ewe remains in contact with her lamb and maternal responsiveness does not wane at 12 hr postpartum. Poindron and Le Neindre (1980) have investigated the stimuli from the lambs which play a role in the maintenance of maternal responsiveness in ewes. The various degrees and kinds of lamb stimuli which ewes were exposed to are shown in Fig. 6: they range from full contact with the lamb, including suckling, to complete separation from it, with intermediate forms of contact excluding suckling and licking. In tests at 12 hr, ewes were permitted full contact with their own lambs and all the mothers that had been able to smell their lambs (groups 1–3) accepted their lambs; but only 20–50% of the ewes that either could not smell their lambs from nearby (less than 1 m), or whose lambs were enclosed in a sealed box or who were without lambs (groups 4–6), accepted their lambs. Of course, those ewes that responded positively to their own young (groups 1–3) rejected alien lambs of the same age (Fig. 6). More interesting is the finding that in the group of ewes that was kept at 1 m from their lambs (group 4), 50% accepted their own lambs and 50% also accepted alien lambs. This suggests that the absence of close olfactory stimulation from their lambs during the first 12 hr postpartum led to some degree of waning of maternal responsiveness and also prevented the formation of an exclusive bond. This interpretation is supported by the finding that a second group of similarly treated ewes in which estrogen levels were maintained at a high level during the first 12 hr postpartum (Fig. 6, 4EB) showed a

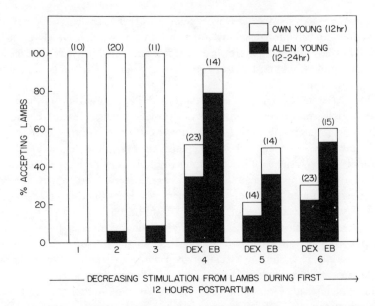

Figure 6. Early postpartum maintenance of maternal behavior in the ewe. During the first 12 hr postpartum ewes had the following kinds of contact or no contact with their own lambs: (1) full contact, including suckling; (2) full contact without suckling; (3) lambs nearby, no licking or suckling; (4) lambs at 1 m distance; (5) lamb enclosed in transparent box allowing sight and hearing; (6) completely separated from lamb. Dex, dexamethasone-induced parturition, normal levels of plasma estradiol; EB, estradiol-benzoate-induced parturition, elevated levels of plasma estradiol. Tests with own and alien lambs were conducted at 12 hr postpartum. *Only ewes that spontaneously accepted their own lambs were tested with alien lambs afterward.* Numbers in parentheses are sample sizes. (Data from Poindron and Le Neindre, 1980.)

higher percentage of acceptance of own young (92%; Fig. 6); the same was true for groups 5EB and 6EB of Fig. 6. Poindron (1976) and Poindron and Le Neindre (1980) have shown that waning of maternal responsiveness does not occur in ewes that are anosmic but are allowed to make contact with their lambs, lick them, and suckle them immediately postpartum. Therefore maternal responsiveness can be maintained by nonolfactory stimuli, but a totally exclusive bond does not develop between the ewe and her lamb.

In summary, these studies on ewes during the transition period between the hormonal onset of maternal behavior and the maintenance of maternal responsiveness clearly show the component processes and their interaction. As the hormonal influence wanes, stimulation from the lamb, chiefly olfactory stimulation, maintains maternal responsiveness, and at the same time it forms the basis for an exclusive bond between the mother and her lamb. Nonolfactory stimuli may also function in both capacities. As Bouissou (1968) has shown, ewes may shift from olfactory to nonolfactory stimuli if they are made anosmic before parturition, and under these circumstances, although the bond is no longer exclusive, maternal behavior does not decline. When anosmia is performed *after* ewes have established an exclusive bond, the shift may still occur but with a lower rate of success (Poindron and Le Neindre, 1980).

Studies on the transition period in the goat are much less definitive than those in the sheep, and moreover, there has been some confusion about the term "critical period" as to whether it refers to maternal responsiveness (Klopfer *et al.*, 1964) or to the formation of an exclusive relationship with the mother's own kid (Gubernick *et al.*, 1979; Gubernick, 1980). In the original studies (Collias, 1956; Klopfer *et al.*, 1964), mothers were either immediately separated from their kids at parturition or allowed short periods of contact (i.e., 5 min) with them before they were separated. When the kids that were separated at birth were returned to their does after 15 min, all were accepted (Collias, 1956), but at the end of 1 hr, only 40% of the does accepted their kids, and at 2 hr and later, none of the does accepted their kids (Klopfer *et al.*, 1964). By contrast, with only 5 min of prior contact, does accepted their kids in nearly every case following separations lasting as long as 3 hr. Thus, brief contact immediately after parturition has lasting effects on maternal responsiveness, and the absence of such contact results in its rapid waning.

Data on the development of individual recognition, measured by presenting alien kids to does that have been with their own kids long enough to respond positively to them, are difficult to find among the published studies by Klopfer and his colleagues (Klopfer *et al.*, 1964). Various statements made by these investigators suggest, however, that it is present at 1–3 hr in all (5/5) females that have had contact with their own kids for only 5 min immediately postpartum (Klopfer, 1971; P. H. Klopfer and M. S. Klopfer, 1968; Klopfer and Gamble, 1966; Klopfer *et al.*, 1964). Rosenblatt *et al.* (unpublished) found that alien kids were rejected in nearly all instances in does that had had from 15 min to 16 hr contact with their own kids.

Klopfer and Gamble (1966) have shown that does made anosmic before parturition fail to develop individual recognition during 5 min of postpartum contact with their kids. At the end of 3 hr without their kids these does readily accepted their own kids in 89% of the cases (8/9) and they also accepted alien kids in all instances in which tests were made (6/6). What is particularly interesting about the results of this study is that the level of maternal responsiveness of the *anosmic* does was as high as that of the *normal* does of a previous study (Klopfer *et al.*, 1964) undergoing the same procedures, but which exhibited individual recognition of their kids. This study indicates, therefore, that although olfactory stimulation may be necessary (within the 5-min exposure period) for the development of individual recognition, it is not necessary for maintenance of maternal responsiveness. It suggests also that these are two separate processes, as in the ewe, one process which maintains maternal responsiveness after parturition and another which is concerned with individual recognition. Klopfer and Gamble (1966) proposed a similar distinction, but they believe that both processes are dependent upon chemical stimulation. The above findings suggest that this need not be the case.

Recently Gubernick *et al.* (1979) reported findings that suggest individual recognition may not develop as rapidly as indicated by previous studies. Does tested 1 hr after separation from their kids, with which they had had 5 min post-partum contact, did not reject alien kids in 62% of the cases. The alien kids in this study had spent little time with their own mothers, and this might have been the basis for the new findings. In a subsequent series of studies Gubernick (1980) has shown that whether or not alien kids have been with their own mother for any length of time before the test determines whether they are rejected or accepted by a mother rearing her own kid. Older kids are rejected, in part, because they have spent more time with their own mothers. Gubernick has suggested that each mother marks her young in several possible ways and that mothers reject kids with the markings of other mothers. If kids are unmarked then mothers will accept them and subsequently mark them only if they remain in contact long enough (approximately 24 hr) with that mother to be labeled (see Gubernick, this volume).

One additional fact was noted: in several previous studies it has been found that mothers that had not had contact with their own kids except during the process of parturition (but presumably had not licked them even then) accepted them after a 1-hr separation, even though during the interim they had adopted an unmarked alien newborn. Gubernick *et al.* (1979) have suggested that mothers mark their kids and then respond to this mark at a later time. The marking or labeling occurs after birth and within the first 24 hr of contact between the mother and the kid. The means by which a label is provided is likely through licking the kid and through the kid's milk intake. Unlabeled aliens will be accepted by a mother after 5 min contact with her own kid. This idea of labeling helps to explain the earlier findings of Klopfer that a mother will accept her own kid even without prior contact with it, not because it shares a similar odor with the mother but because it is unlabeled (i.e., has not been in contact with another goat); it is not that she recognizes the kid as her own but rather that she accepts unlabeled kids.

These new studies deal specifically with the problem of individual recognition and do not bear on the question of maternal responsiveness. In our view they do not change the significance of earlier findings that individual recognition can develop very rapidly during the first 5 min postpartum and that it will be exhibited 1 hr later *provided the alien kid possesses stimulus characteristics that enable the mother to distinguish it from her own kid.* The above studies indicate that these normally arise during contact between the kid and its own mother. Perhaps what these studies have revealed is that the development of individual recognition in the doe is initially based upon rather gross differences between kids but that it is gradually refined with further contact between mothers and their kids.

The most direct measure of the waning of maternal responsiveness during the postpartum period in the goat is based upon the separation studies cited above, which indicate that there is a rapid waning during the first hour or two postpartum (Rosenblatt *et al.*, unpublished). Hemmes (1969) has reported retention of responsiveness at near 100% at 2½ hr postpartum in the same strain of goats studied by Klopfer and his associates. Several procedural differences may account for the late acceptance of the kids in this study compared with the earlier study of Klopfer *et al.* (1964). Mothers were isolated during parturition, and tests were conducted later at the parturition site rather than outdoors at some other site. Observations lasted 20 min instead of 10 min, and there were significant changes in the does' behavior during the last 10 min. Whether the different results are based upon the last 10 min of observation can be evaluated by scoring only the first 10 min of the 20-min observations and comparing them with the earlier scores.

There is one final finding that deserves mention: Klopfer and Gamble (1966) reported that does made anosmic 2 hr 40 min after parturition, following an earlier contact with their kids for 5 min immediately postpartum, rejected their own kids at 3 hr in 55% of the cases. It is not clear if this represents a waning of maternal responsiveness that had been supported by olfactory stimulation (and visual stimulation) or removal of all cues for recognizing its kid, since the does were blindfolded as well as anosmic during testing.

It appears that maternal responsiveness wanes more rapidly after parturition in goats than in sheep and that individual recognition also is established, at least grossly, more rapidly in the goat. The onset of maternal behavior at parturition also may be more abrupt, but adequate tests have not been conducted. Further research needs to be done, but it is already clear that the waning of maternal responsiveness does not coincide with the establishment of individual recognition (see above) and that a clear distinction should be made with respect to these two components of maternal behavior in the goat.

5. Maternal Agression: Pre- and Postpartum Regulation

In this section we shall discuss another aspect of maternal behavior, namely, maternal aggression particularly toward conspecifics, rather than toward interspecific predators. Our interest will be in whether this aspect of maternal behavior is regulated, during the pre- and postpartum periods (to the extent that it occurs prepartum), by the same factors as the items of maternal care that we have discussed thus far. The display of maternal aggression is widespread among mammals, varying in intensity in different species. Maternal goats, for example, are highly aggressive toward other females and alien kids; they threaten or drive away by butting any female or alien kid that comes near their kids or attempt to nurse (i.e., alien kids), especially during the first few days postpartum. Maternal sheep, on the other hand, are only mildly aggressive, and most of this aggression is directed at alien young that attempt to suckle. However, the topic of maternal aggression is covered more fully in Chapter 5. Our inclusion of this topic in the present chapter has as its main aim to see whether the theoretical approach we have been presenting with respect to items of maternal care of young is applicable to a behavior pattern which occurs at the same time but is directed at adult conspecifiecs.

In *rats,* Erskine *et al.* (1978*a*) were unable to find a prepartum onset of maternal aggression as late as the 22nd day of pregnancy* (less than 1 day before parturition), and maternal aggression was at a low level even after parturition, rising slowly during the 1st week postpartum, then abruptly reaching a maximum on the 9th day postpartum. From then on it declined rapidly and was at prepartum levels by the 13th day postpartum. Aggression was measured directly by observing the frequency of attacking, the latency of attacking, and the number of bites inflicted on a male intruder placed inside the female's cage that also contained her nest and litter. These measures correlated highly with indirect measures of the female's aggression reflected in the male's behavior when threatened and attacked. These consisted of prolonged state of immobility in an upright posture following an attack by the female and the adoption of an upright subordinate posture: latencies and durations of these male reactions were closely correlated with measures of maternal aggression in the resident mothers.

Erskine *et al.* (1978*a,b*) and Erskine (1978) have reported that attainment of the high level of maternal aggression on day 9 postpartum is peculiar to females that have undergone parturition and have reared their litters from parturition. Sensitized virgins that were induced through exposure to pups to exhibit maternal behavior (i.e., nest building, retrieving, crouching over young, and licking young) and were maintained for an additional 9 days with pups, were no more aggressive towards intruders than nonsensitized virgins that were exposed to pups only during testing on the 9th day. The lactating females had the typical high levels of aggression of 9-day lactating mothers, and male intruder behavior corroborated the data obtained directly from the female's behavior. Females of all of the groups exhibited aggression (91%) regardless of their reproductive state, but only the lactating females exhibited high levels of aggression.

Price and Belanger (1977) have confirmed, in both wild and domesticated rats (the former, laboratory-reared F_1 of trapped Norway rats), high levels of

---

*Recent observations by Mayer and Rosenblatt (unpublished observations) suggest that maternal aggression may be initiated prepartum among rats of the Charles River CD strain. Females were tested (2-min tests) during the morning of day 22 of pregnancy by placing a young male into their cages (Plexiglas, 23.4 × 35.6 × 24.1 cm), in which they had been housed 3–5 days before their due dates. Parturition was monitored closely at 15- to 30-min intervals. There was a sharp onset of maternal aggression around 3½ hr before delivery. Among 20 mixed-parity females only 2 exhibited maternal aggression earlier than 3½ hr prepartum, while 8 did not; 9 were aggressive after 3½ hr prepartum, and only 1 was not (median test, chi square-9.9, p <0.01).

aggression towards adult female intruders on the 6th–8th days postpartum, but they did not compare these postpartum mothers with either sensitized or nonsensitized virgins. LeRoy and Krehbiel (1978) compared the maternal aggression of lactating females on day 2 or 3 postpartum with that of both sensitized intact females on the 4th day following the onset of retrieving and sensitized ovariectomized females 3 weeks after ovariectomy. Adult male intruders were attacked and bitten by lactating mothers about 15 times per 30-min test and only about once or twice by the sensitized females of both groups. Male immobility was highly correlated with female attacking.

It is quite clear from these studies that maternal aggression requires hormonal stimulation and that despite the appearance of maternal care in sensitized females the fact that this maternal behavior is nonhormonally based precludes the appearance of enhanced levels of maternal aggression. Several attempts have been made to establish experimentally the hormonal conditions for the onset of maternal care either by hysterectomy–ovariectomy on day 16 of pregnancy and treatment with estradiol benzoate (LeRoy, 1977) or by prolonged treatment of ovariectomized females with estradiol benzoate (days 1–20, at 3 μg daily), progesterone (days 3–17, at 1 mg daily), and prolactin (day 21, at 1 mg) (Krehbiel and LeRoy, 1979), a treatment combining features of the Moltz et al. (1970) and Zarrow et al. (1971) treatments that successfully stimulated maternal behavior in ovariectomized females (see Section 1 and Fig. 2).

Estrogen treatment of hysterectomized–ovariectomized, late-pregnant females stimulated high levels of aggression comparable to those of lactating mothers, and the prolonged hormone treatment stimulated maternal aggression in a similar proportion of females as natural parturition (3/8 vs. 4/9), although this proportion was low in both groups despite the fact that testing of lactating mothers was done when they normally exhibit their highest levels of aggression (Krehbiel, 1978, cited in Krehbiel and LeRoy, 1979). Both groups, however, had larger proportions of females exhibiting aggression than vehicle-treated ovariectomized females of the control group.

A number of difficulties arise in attempting to interpret these findings, chief among which is the problem of attributing maternal aggression to the estrogen treatment alone. In both experiments progesterone withdrawal may have played a role and the release of prolactin by these treatments also might have been instrumental. Observations by Siegel and Rosenblatt (unpublished observations) during the prepartum testing of maternal behavior (Rosenblatt and Siegel, 1975) indicated that females might be ready to exhibit maternal aggression prepartum if they are stimulated to exhibit maternal behavior (i.e., retrieving and nest building). They found that when prepartum females were exposed to 5- to 10-day-old pups at 2-hr intervals, for 15 min each time, they began to exhibit maternal behavior and simultaneously began to attack the experimenter's hand when he attempted to remove the test pups or insert them into the home cages. These

observations suggest that the hormonal basis for maternal aggression may be similar to that for maternal behavior but that the female needs to be stimulated by
older pups than newborn in order to exhibit the aggression.

Postpartum regulation of maternal aggression appears to be dependent upon
pup stimulation: removal of the pups for 4 hr before tests of maternal aggression
with an intruder, on the 9th and 10th days of lactation when aggression is normally at its highest level, completely prevented the rise in aggression normally
seen when females remain with their litters (Erskine *et al.* 1978*a*). On the other
hand, hypophysectomy performed on the 5th day postpartum does not prevent the
rise in aggression that normally occurs on the 9th day (Erskine, 1978). Moreover,
ovariectomy done on day 1 postpartum does not affect levels of aggression in mothers on the 5th day. These experiments suggest that the pup maintenance of maternal aggression may not be mediated by exteroceptive release of hormones but
directly through the sensory stimulation they provide (i.e., nonhormonally), as we
have found is the case with respect to other aspects of maternal behavior.

Once it has been established in females that have undergone parturition and
reared their litters for several days (i.e., 2 or 9 days), maternal aggression can
readily be stimulated following sensitization begun 6 days after removal of their
original litters (Erskine, 1978). Tests for maternal aggression in these (primiparous) sensitized females were conducted 9 days after they had begun to exhibit
maternal behavior and were being continuously exposed to pups. At the time of
testing, estrous cycling was present in 6 of 10 females that had had 2 days' postpartum rearing of their litters, and 4 of 10 females that had had 9 days. An
additional group of postparturient females was allowed to lick each pup, individually, during parturition, but the pups were then removed and none were present
after parturition was completed. Of these females 7 of 10 were exhibiting estrous
cycles.

Maternal aggression at normal lactating levels was stimulated by the pups
on the 9th day of sensitized maternal behavior in all three primiparous groups of
females, although females without postpartum contact with their litters exhibted
lower levels of aggression with respect to some components (e.g., hip throws during fighting, frequency and duration of lateral posture, latency to attack).

A surprising finding was that females that failed to exhibit increased levels
of aggression on the 9th day after sensitization, when resensitized and given an
additional 9 days of exposure to pups, exhibited levels of aggression that were as
high as those of currently lactating females in the groups described above. In some
measures of maternal aggression they showed higher levels: hip throws during
fighting, and the frequency and duration of lateral posture. In this group, at the
beginning of testing for aggression, only one of eight females was exhibiting
estrous cycling; the remainder of the group was exhibiting prolonged diestrous
vaginal smears.

One might assume, as we have with regard to retention tests for maternal behavior by Bridges (1975, 1977; see Section 4) using a similar experimental schedule, that this study indicates a nonhormonal basis for the maintenance of maternal aggression once it has been stimulated by hormones at parturition and maintained for a period postpartum. The shorter interval between the initial hormonal stimulation of maternal behavior and the later reestablishment of this behavior by pup stimulation which was used in Erskine's (1978) study, however, introduced additional factors which do not allow such an interpretation at this time. The fact that among the 2- and 9-day postparturient experienced females 60–70% of the females were lactating during reinduction of maternal behavior indicates that prolactin relase was stimulated by the pups before tests of maternal aggression were made. Among the resensitized virgins 87% had ceased cycling, which may indicate that they too were releasing large amounts of prolactin. Only among the females that had minimal postpartum experience with their litters was estrous cycling present in 70% of the females, and lactation occurred in only 20%; however, large amounts of prolactin can be secreted during nocturnal surges for several days before estrous cycling is disrupted by pseudopregnancy, and lactation may not yet be evident (Korányi et al., 1977; Stern and Siegel, 1978; Terkel et al., 1978, 1979). Thus prolactin may have played a role in the maternal aggression displayed by all of these groups.

In summary, under normal circumstances the onset of maternal aggression is stimulated by hormones, but by the 5th day at the latest, and most likely earlier (see ovariectomy study above), it is no longer dependent upon hormones but is maintained on both a long- and short-term basis by pup stimulation. The short-term basis may involve prolactin; this has not been ruled out by the studies by Erskine (1978) and Erskine et al. (1978a,b). A remaining question is whether there may also be a nonhormonal onset in sensitized females given sufficiently long exposure to pups (i.e., 18 days). Closely linked to this question is whether the long-term retention of prior hormonal stimulation on maternal aggression is mediated by hormones or is based mainly on the retention of pup-stimulation effects.

Among mice Noirot and her students (Beniest-Noirot, 1958; Noirot, 1969, 1974; Noirot et al., 1975) have described changes in maternal aggression during pregnancy and lactation: during pregnancy maternal aggression towards males and females rises during the first few days and is maintained at a high level. It rises again at parturition, then falls gradually during lactation unless the female is mated and becomes pregnant at postpartum estrus. Svare and Gandelman (1976b) have reported only irritable behavior during pregnancy in their strain of mice (Rockland–Swiss) but have confirmed Noirot's description of postpartum decline in maternal aggression after the high level attained following parturition. This strain difference may prove to be an important one with respect to understanding the factors related to the onset of maternal aggression.

Noirot el al. (1975) found a similar increase and maintenance of high levels of maternal aggression during pregnancy and pseudopregnancy, both of which share common hormonal events until the 10th day after mating. The initial rise in maternal aggression, however, also occurred in the control groups of virgins, but these groups did not establish the high level of aggression of the pregnant and pseudopregnant females. After the 10th day the pseudopregnant females showed a decline in aggression to virgin control levels while the pregnant females maintained their high levels of aggression until parturition (day 19) and afterward, as noted above. Pseudopregnancy is terminated by the onset of estrous cycling, which implies that there is a decline in the high levels of progesterone characteristic of pseudopregnancy and the rise in circulating levels of estrogen and accompanying changes in gonadotropic and prolactin secretion. The rise in estrogen is important, since Svare and Gandelman (1975) have shown that estrogen inhibits the aggression shown by postpartum mothers.

As Noirot *et al.* (1975) note, the maintained high levels of aggression during pregnancy and pseudopregnancy may be based upon the elevated levels of circulating progesterone; the initial rise may be due either to a decline in estrogen at the start of pregnancy or to the effects of repeated testing, since it also appeared in the nonpregnant virgins. The decline in aggression at the end of pseudopregnancy is an important feature of this study since it indicates that once the hormonal support for aggression disappears—in this case the progesterone decline at the end of pseudopregnancy, as well as the rise in estrogen—then aggression declines. Of course, pseudopregnancy does not result in the delivery of newborn, but similar hormonal changes at the end of normal pregnancy might have similar effects, the difference being that newborn appear and they are able to maintain maternal aggression.

Gandelman and his student (Svare and Gandelman, 1976a; Svare, 1977) have tried various procedures for stimulating the onset of maternal aggression in mice (see Svare, this volume). In one study (Gandelman, 1972; Svare, 1977) different groups of pregnant females were hysterectomized on days 10–15, and given 1-day-old pups immediately afterwards. The females were tested for maternal aggression 48 hr later. Pups remained with them a total of 7 days, during which maternal behavior was displayed by all females but maternal aggression was exhibited by only a proportion of the females. Only hysterectomies performed on day 12 or later stimulated two-thirds to all of the females to lactate, and after day 12 more than half the females exhibited aggression. There was a close association between aggression and lactation: from day 12 onward nearly all females who exhibited aggression also lactated. However, there was *not* a close association between lactation and aggression: only a little more than half of the females that lactated exhibited maternal aggression. After normal pregnancy, following a similar period of contact with pups, the same proportion of lactating females exhibit aggression (i.e., 54–65%).

This study obscures the direct effect of hormonal changes caused by hysterectomy on the onset of maternal aggression, since aggression was not tested until the females had already been in contact with pups for 3 days. In a second study (Gandelman, 1972), therefore, pregnant females were hysterectomized only on day 14 and pups were withheld from them until 96 hr after surgery. Following 24-hr contact with the pups (4 days after surgery) the females were tested and none exhibited aggression nor did they lactate. As in the previous study, females given young immediately after hysterectomy lactated in nearly all cases and exhibited aggression in nearly two-thirds of the cases. Again it is difficult to interpret this study, since it was 96 hr after hysterectomy that the females first received pups in the group described above and they were not tested at that time but only after they had been with the pups for 24 hr. One thing is certain, after such a long interval whatever aggression might have been stimulated by the hysterectomy had waned and could not be revived by pup stimulation; the same, of course, can be said for lactation. Females of both groups, however, displayed all other aspects of maternal behavior (except nest building) in all cases.

There persists good reason to believe maternal aggression is stimulated by hormones: it occurs normally immediately after parturition and during pregnancy as we have seen. Moreover, virgin females that are exposed to pups continuously and exhibit maternal behavior do not exhibit aggression (Svare, 1977) (see below, however). Several studies have tested preparturient females (day 19) for aggression (Svare and Gandelman, 1976a), and none of the females have exhibited maternal aggression, although all exhibited maternal behavior.

Unfortunately, in none of the studies have females been allowed to complete parturition, had their pups removed before any significant contact (e.g., suckling) occurred, and been tested for maternal aggression immediately. This is particularly significant because an important increase in the level of aggression occurs over parturition. The closest to this procedure was reported in a study by Svare and Gandelman (1976a, experiment 1) in which a group of females had their litters removed at parturition. However, they were not tested for maternal aggression for 24 hr; none displayed maternal aggression, and this continued for an additional 7 days of testing during which they were now exposed to pups continuously. In each of the other studies, which we shall describe, females tested before parturition were not tested again after parturition, but they already had been exposed to pups for nearly 24 hr in several experiments and for 48 hr in others (Svare and Gandelman, 1976a). Among these postparturient females aggression was displayed by 54–65% of the animals. In other studies (Svare and Gandelman, 1976a) females that gave birth were not tested until 48 hr later; in the interim they were exposed to pups with and without suckling throughout this period.

There is, however, a reason for the procedures described above and for those to be described: these investigators (Svare, 1977; Svare and Gandelman, 1976a,b) believe that aggression is evoked in postparturient females only after they have been suckled. To test this hypothesis postparturient females were tested for aggression 2 days after parturition following sham thelectomy at 24 hr postpartum in one group, and thelectomy at 24 hr in a second group, thus preventing suckling during the 24th–48th hr postpartum. Whereas 63% of the sham-thelectomized females exhibited aggression at 48 hr, only 25% of the thelectomized females did so; in both cases these percentages held over 8 days of testing. Suckling during the first 24 hr was not sufficient to establish aggression, and pup contact during 48 hr was also not sufficient, despite the fact that maternal behavior was displayed by the thelectomized females. Only when thelectomy was done at 48 hr, allowing 2 days of suckling, was aggression displayed by 63% of the females at 72 hr and in tests during the following 8 days.

Further studies using thelectomized females deprived of all suckling stimulation until the time of testing for aggression strongly support the idea that suckling is required for at least 48 hr postpartum in order for aggression to be displayed (Svare and Gandelman, 1976a). It should be noted that in all of these studies there was no evidence of lactation in the thelectomized females upon examination of their dissected mammary glands at the termination of 8 days of testing.

To enable pups to suckle, nipple development is required, and Svare (1977) and Svare and Gandelman (1976a) have suggested that the hormones of pregnancy have as their major role, with respect to the onset of aggression, the enlarging of the mother's nipples. A hormone regimen given over 19 days to ovariectomized females consisting of estradiol benzoate (0.2 μg/day) and progesterone (500 μg/day) was successful in producing near-normal nipple enlargement, but by itself this hormone treatment did not stimulate the display of aggression 24 hr after the termination of treatment and before pups had suckled the females. However, within about a day and a half after pups had begun to suckle them these females began to exhibit aggression toward male intruders in 50% of the cases, which compares favorably with the 54% value for the postparturient control group. Finally, if a similar group of hormone-treated ovariectomized virgins was thelectomized and given pups, only 10% of the group displayed aggression, indicating that the hormone treatment alone was not sufficient to stimulate aggression without suckling.

While this is an impressive series of studies that certainly establishes the importance of suckling during the first 48 hr for the appearance of maternal aggression, it leaves open the question of whether hormones play a role in maternal aggression in the mouse other than through their effects on nipple development. Certainly combined treatment with estrogen and progesterone, as employed, is not effective, but this treatment does not fully duplicate the hormonal pattern of late pregnancy (see Fig. 1c) in which progesterone declines before parturition and estrogen rises. Moreover, the role of prolactin has not been investigated; although the thelectomized females that exhibited aggression did not lactate, no direct measurement of circulating levels of prolactin has yet been made in these females. Moreover, the prepartum and early postpartum increase in prolactin release may not depend upon suckling stimulation and their effect on milk production might not still be evident 9 days after parturition or later, when the mammary glands were actually examined for milk production.

Relevant to this point is the recent study by McDermott and Gandelman (1979) in which 67% (four of six females) of virgin females that were exposed continuously to 1-day-old pups began to exhibit aggression toward male intruders starting on the 9th day of exposure. These females had exhibited maternal behavior almost immediately when given pups, and therefore the pups had an opportunity to suckle from them. It was found that only those females that exhibited nipple development and in which there was evidence of suckling, displayed aggression; those without such nipple development remained maternal but did not exhibit aggression. Moreover, when pups were removed from the aggressive females, maternal aggression waned after 1 hr but was revived 30 min after the pups were returned. In most respects, these females, therefore, resembled postpartum lactating females, with which they compared favorably in this study. The authors interpret their results as support for the importance of suckling for maternal aggression, and while this may be so, there is a good likelihood that prolactin was released during suckling, as has been found in virgin rats after a similar interval of the display of maternal behavior due to sensitization (Korányi et al., 1977; Stern and Siegel, 1978). There may also have been changes in ovarian hormone secretion, since suckling often causes the suspension of estrous cycling.

Whether the lack of maternal aggression during pregnancy in the Rockland–Swiss mice and its presence in the strain used by Noirot represents a strain difference or some difference in procedure which is not evident in the reports of these investigations, it is an important difference, theoretically. If maternal aggression appears prepartum and also in pseudopregnant females, it cannot be based upon suckling stimulation but must be based upon hormonal stimulation. As it stands, with reservations cited above, Gandelman and Svare are essentially proposing that maternal aggression is established nonhormonally by the sensory effects of suckling, and that this may turn out to be a unique situation among mammals.

After its onset, maternal aggression continues to be strongly dependent upon pup stimulation for its maintenance. On a long-term basis, the increasing age and associated changes in behavior and physical appearance of the pups are responsible for the decline in aggression as it normally occurs between parturition and day 20 postpartum. If females are maintained only with 1-day-old pups, maternal aggression is maintained longer and at higher levels, although it eventually declines. If they are maintained only with older, 12-day-old pups, starting on day 4 postpartum, the decline occurs slightly earlier (Svare, 1977; Svare and Gandelman, 1973). On a short-term basis, maintenance depends upon almost continuous contact with pups: removing pups for as little as 5 hr on day 4 postpartum causes a reduction in aggression from 100% to 16% of the females. Upon replacing the pups after a 5-hr separation, aggression reappears in all of the females in 10 min.

Several studies indicate that once suckling has played a role in the establishment of maternal aggression, it is no longer crucial for its maintenance. Females thelectomized after 48 hr postpartum and tested from 72 hr onward for 8 days continue to exhibit high levels of aggression. Postpartum females thelectomized on day 5 of lactation and tested either on day 6 or 12 continue to exhibit aggression in 54–71% of the cases (Svare and Gandelman, 1976a). Moreover, females separated from their young for 5 hr on day 4 postpartum but permitted exposure to them behind a wire partition during this period, maintain their aggression in 80% of the cases; allowing sucking and other contact increases this to 100% (Svare, 1977; Svare and Gandelman, 1976a).

Thus, although hormonal factors have not yet been implicated directly in the stimulation of maternal aggression, there is probably a clear period of onset shortly after parturition during which suckling stimulates maternal aggression. The period during which suckling can stimulate aggression is a limited one and does not extend as long as 96 hr, since pup exposure starting at that time was not effective in stimulating maternal aggression in 14-day hysterectomized females, whereas earlier exposure to pups was effective. In postpartum females, at least 2 days of suckling is required to establish aggression, but following that period, aggression is maintained by nonsuckling stimuli from the pups, as well as suckling when it occurs.

Among *hamsters* maternal aggression *already* is present at parturition and continues at a high level throughout lactation, rising perhaps during the 3rd week (Wise, 1974). It begins early in gestation and increases throughout pregnancy to reach postpartum levels at midpregnancy (day 10 of the 16-day period of gestation). A similar rise in aggression does not occur during pseudopregnancy: levels of aggression are only slightly above those of estrous cycling females.

The onset of maternal aggression either during pregnancy or at parturition has not been investigated in the hamster, but its maintenance postpartum appears to depend upon high levels of prolactin that are secreted in response to pup suckling. Wise and Pryor (1977) studied the effect of blocking prolactin release with ergocornine hydrogen maleate on maternal aggression starting on the 4th day postpartum. Three groups of lactating females were used: all were ovariectomized and hysterectomized on day 4 of lactation to preclude the possibility that ovarian hormones were involved in the response to ergocornine. One group was then administered ergocornine alone, whereas a second group received, in addition, replacement therapy with prolactin. A final group was given the vehicle only. Ergocornine and other treatments were administered on days 4–7 of lactation, and tests for aggression, conducted in an arena without the litter present, were carried out on the same days.

Maternal aggression was almost completely eliminated by blocking prolactin release, and prolactin replacement restored aggression to levels comparable with, but slightly lower, in some items, then those of vehicle-treated control females (Wise and Pryor, 1977). The decline in aggression was rapid, and only a single female performed one item over the test days.

It is significant that ovariectomy on day 4 of lactation did not affect maternal aggression in the lactating control group.

One question that can be asked is whether blocking prolactin release affected only maternal aggression or whether its effect on this behavior was a consequence of broader effects on maternal behavior as a whole. Among females given ergo-cornine 55% (7/13 females) killed an entire litter at least once compared with only 7% of the control females that killed their litters (1/14 females). Of those treated with ergocornine alone, with almost no maternal aggression, 70% (4/6) killed their litters at least once and prolactin replacement did not entirely prevent this, since 42% (3/7 females) of these females also killed their litters at least once; the difference between the two ergocornine-treated groups was not significant. In addition to this effect on pup killing the authors noted that "ergocornine-treated females tended to maintain poorer nests and to spend less time in nests than did normally lactating females. The attacks on pups may have been associated with a disruption of maternal behavior following interference with PRL scretion" (Wise and Pryor, 1977, pp. 38–39).

While pup killing is a normal feature of maternal behavior in hamsters (Day and Galef, 1977), it is generally confined to the first 2 or 3 days after parturition, tapering off by the 5th day, and only a small number of pups are cannibalized, not entire litters. This was noted by Wise and Pryor (1977) in their females before surgery, and treatments were begun on day 4 postpartum. The basis for pup kill-ing is not known, and whether it is hormonally stimulated or, in fact, inhibited by hormones after the first few days, is also not known. Since untreated females did not continue to cannibalize pups after the 4th-day removal of the ovaries and uteri, ovarian hormones are not crucial for this behavior. Blocking prolactin release appears to increase cannibalism, but again, it is not clear whether this is a specific effect or part of a broader change in maternal responsiveness.

With the reservations noted, it does appear that maternal aggression in the hamster requires hormonal stimulation for its maintenance. It has not yet been determined whether the onset of maternal aggression during gestation is based upon prolactin as well. Prolactin levels are high during gestation, but progesterone levels are also high (Fig. 1d) and Ciaccio et al. (1979) have suggested that pro-gesterone may be the basis for the high level of aggression seen. Wise and Pryor (1977) tested the effects of ergocornine on aggression in a nonreproductive situa-tion using the same females described above. Males, introduced into their home cages, were attacked regularly, and ergocornine given as above over 4 days had no effect on this behavior. Not all aggression toward males is stimulated by hormones: Ciaccio et al. (1979) have also reported that ovariectomized females exhibit high levels of aggression, and the present study rules out prolactin. Estrogen reduces aggression in ovariectomized females, but progesterone overcomes this inhibition (after stimulating lordosis in estrogen-primed females) either by blocking the effects of estrogen (Ciaccio and Lisk, 1967, 1971; DeBold et al., 1978) or by directly stimulating aggression. In either case the hormonal basis for the onset of aggression during gestation is not yet clear, and it is also not clear whether postpartum maternal aggression has a similar hormonal basis.

6. Concluding Remarks

There has been a rapid growth in the study of maternal behavior of nonprimate mammals since the senior author of this review began his research on the rat in 1958. An increasing number of laboratory, domestic, and wild animals have come under investigation. Research has advanced from descriptive reports to the study of hormonal and nonhormonal causative factors in maternal behavior, and this has been aided by an equally rapid growth of research on the endocrine regulation of pregnancy and postpartum lactation and ovarian function. More recently the ontogeny of maternal behavior has begun to be studied and analyzed with respect to genetic, hormonal, and psychological influences (see reviews in Mayer *et al.*, 1979; Mayer and Rosenblatt, 1979*a,b*; Rosenblatt *et al.*, 1979).

One aim of this review has been to examine the extent to which the theoretical integration of findings in the rat with respect to phases in the regulation of maternal behavior, hormonal, and nonhormonal mechanisms and the existence of a transition period are applicable to the organization of maternal behavior in other species. Some have disputed its applicability to the rat itself, pointing to the inability of the nonhormonally induced maternal behavior of sensitized females to adequately account for the qualitative and sometimes quantitative characteristics of the maternal behavior of the postpartum lactating female (Krehbiel and LeRoy, 1979). However, nonhormonal maternal behavior, under natural conditions, never occurs without initially being stimulated hormonally; this is the significance of the transition period, an aspect of the theory which is often ignored in discussions of this problem. What needs to be studied is whether pup stimulation is able to maintain the qualitative characteristics of hormonally induced maternal behavior, and what studies exist on this problem suggest that it is capable of doing so (LeRoy, 1977; Mayer and Rosenblatt, 1979*a*; Stern and MacKinnon, 1976).

The concept of the transition period throws a new light on the problem of the "critical period" in the mother's attachment to her young among the ungulates (e.g., sheep and goats). It distinguishes two problems where previously only one was perceived. The formation of the selective attachment of the mother to her own young is an aspect of the larger problem of the transition from the hormonal to a nonhormonal regulation of maternal behavior, as the analysis of maternal behavior in the ewe by Poindron and Le Neindre (1980) shows.

While clearly there are two phases in the regulation of maternal behavior in all species in which there is evidence related to this problem, it is not yet clear that in all species the postpartum phase is nonhormonal in nature. In all species studied, however, the postpartum phase is strongly dependent upon stimulation from the young; often it is mainly suckling stimulation, but other stimuli play an important role as well.

Maternal aggression is a more or less prominent feature of maternal behavior in all species, but its relationship to maternal care and to other forms of aggression is not yet clear. Together with nest building it is a pattern of behavior that is not directed toward the young; nevertheless, it is dependent upon stimulation received from them, even more, perhaps, than other items of maternal care which do not wane as rapidly as aggression when the young are removed. Nest building also wanes rapidly when pups are removed (Rosenblatt and Lehrman, 1963) and revives quickly upon their return, as occurs with aggression.

Our aim in writing this review will have been achieved if the proposed theory of the organization of maternal behavior points investigators towards significant problems in the study of this interesting and most significant pattern of behavior among animals.

The omission of primates from this review is not based solely upon the need to limit the scope. The authors are not aware of any studies on the hormonal basis of maternal behavior among primates, including humans. Among humans, there is a growing literature on the transition period, i.e., the effects of a mother's contact with her infant during delivery and the first hours afterward, on her subsequent responsiveness to it (Carlsson et al., 1978; de Chateau and Wiberg, 1977; Klaus et al., 1972; Schaffer, 1977; Whiten, 1977). Also, the close relationship between the mother and infant among rhesus monkeys in the postpartum period has been admirably analyzed by Hinde and his co-workers, among others (Hinde, 1975; Hinde and Spencer-Booth, 1968; Hinde and White, 1974; White and Hinde, 1975). The emphasis, however, has been on the *interaction* between mother and infant; the hormonal onset and postpartum regulation of maternal behavior have not yet been subjected to special study. The opportunities for studying the physiological (hormonal) basis of maternal behavior among women are becoming increasingly available, under both normal conditions of delivery and following various forms of premature termination of pregnancy using different abortifacient agents (e.g., prostaglandins and saline) (Klopper, 1971; Lahteenmaki and Luukkainen, 1978). Also there are opportunities to study the growth of maternal responsiveness in women who adopt newly born and older infants. These studies require methods appropriate to humans—a combination of objective behavioral and clinical psychological techniques—and would benefit from team research with endocrinologists, obstetricians, and allied specialists in addition to those capable of studying the behavioral and psychological aspects of maternal behavior.

ACKNOWLEDGMENTS. The research reported in this article and the writing of it were supported by USPHS Grant MH-08604 (JSR). We wish to acknowledge with gratitude the secretarial work of W. Cunningham and graphic art of C. Banas. This is publication number 351 of the Institute of Animal Behavior. We wish to thank P. Poindron for his critical comments on the section on sheep and goats and P. Poindron and P. Le Neindre for making available to us their unpublished research on sheep.

References

Anderson, C. O., Zarrow, M. X., Fuller, G. B., and Denenberg, V. H., 1971, Pituitary involvement in maternal nest-building in the rabbit, *Horm. Behav.* **2**:183–189.

Ash, R. W., and Heap, R. B., 1975, Oestrogen, progesterone and corticosteroid concentrations in peripheral plasma af sows during pregnancy, parturition, lactation and early weaning, *J. Endocrinol.* **64**:141–154.

Baranczuk, R., and Greenwald, G. S., 1974, Plasma levels of oestrogen and progesterone in pregnant and lactating hamsters, *J. Endocrinol.* **63**:125–135.

Barnett, S. A., and McEwan, I. M., 1973, Movements of virgin, pregnant and lactating mice in a residential maze, *Physiol. Behav.* **10**:741–746.

Bast, J. D., and Greenwald, G. S., 1974, Daily concentrations of gonadotrophins and prolactin in the serum of pregnant or lactating hamsters, *J. Endocrinol.* **63**:527–532.

Baum, M. J., 1978, Failure of pituitary transplants to facilitate the onset of maternal behavior in ovariectomized virgin rats, *Physiol. Behav.* **20**:87–89.

Beach, F. A., and Wilson, J., 1963, Effects of prolactin, progesterone, and estrogen on reactions of nonpregnant rats to foster young, *Psychol. Rep.* **13**:231–239.

Bedford, C. A., Challis, J. R. G., Harrison, F. A., and Heap, R. B., 1972, The rôle of oestrogens and progesterone in the onset of parturition in various species, *J. Reprod. Fertil.* (Suppl.) **16**:1–23.

Behrman, H. R., Yoshinaga, K., Wyman, H., and Greep, R. O., 1971, Effects of prostaglandin on ovarian steroid secretion and biosynthesis during pregnancy, *Am. J. Physiol.* **221**:189–193.

Beniest-Noirot, E., 1958, Analyse du comportement di 'maternel' chez la souris, *Monogr. Fr. Psychol.* No. 1.

Bintarningsih, Lyons, W. R., Johnson, R. E., and Li, C. H., 1958, Hormonally-induced lactation in hypophysectomized rats, *Endocrinology* **63**:540–547.

Bouissou, M. F., 1968, Effet de l'ablation des bulbes olfactifs sur la reconnaissance du jeune par sa mère chez les Ovins, *Rev. Comp. Anim.* **3**:77–83.

Bridges, R., 1975, Long-term effects of pregnancy and parturition upon maternal responsiveness in the rat, *Physiol. Behav.* **14**:245–249.

Bridges, R. S., 1977, Parturition: Its role in the long term retention of maternal behavior in the rat, *Physiol. Behav.* **18**:487–490.

Bridges, R. S., 1978, Retention of rapid onset of maternal behavior during pregnancy in primiparous rats, *Behav. Biol.* **24**:113–117.

Bridges, R. S., Zarrow, M. X., Goldman, B. D., and Denenberg, V. H., 1974, A developmental study of maternal responsiveness in the rat, *Physiol. Behav.* **12**:149–151.

Bridges, R. S., Rosenblatt, J. S., and Feder, H. H., 1978a, Serum progesterone concentrations and maternal behavior in rats after pregnancy termination: Behavioral stimulation after progesterone withdrawal and inhibition by progesterone maintenance, *Endocrinology* **102**:258–267.

Bridges, R. S., Rosenblatt, J. S., and Feder, H. H., 1978b, Stimulation of maternal responsiveness after pregnancy termination in rats. Effect of time of onset of behavioral testing, *Horm. Behav.* **10**:235–245.

Bruce, H. M., 1961, Observations on the suckling stimulus and lactation in the rat, *J. Reprod. Fertil.* **2**17–34.

Buckle, J. W., and Nathanielsz, P. W., 1975, A comparison of the characteristics of parturition induced by prostaglandin $F_{2\alpha}$ infused intra-aortically, with those following ovariectomy in the rat, *J. Endocrinol.* **64**:257–266.

Buntin, J. D., Jaffee, S., and Lisk, R. D., 1979, Physiological and experiential influences on pup-induced maternal behavior in female hamsters, Presented at Eastern Conference on Reproductive Behavior, New Orleans, La.

Buttle, H. L., Forsyth, I. A., and Knaggs, G. S., 1972, Plasma prolactin measured by radioimmunoassay and bioassay in pregnant and lactating goats and the occurrence of a placental lactogen, *J. Endocrinol.* **53**:483–491.

Carlsson, S. G., Fagerman, H., Horneman, G., Hwang, P., Larrson, K., Rödholm, M., Schaller, J., Danielsson, B., and Gundewall, C., 1978, Effects of amount of contact between mother and child on the mother's nursing behavior, *Dev. Psychobiol.* **11**:143–150.

Carminati, P., Luzzani, F., Soffientini, A., and Lerner, L. J., 1975, Influence of day of pregnancy on rat placental, uterine, and ovarian prostaglandin synthesis and metabolism, *Endocrinology* **97**:1071–1079.

Catalá, S., and Deis, R. P., 1973, Effect of oestrogen upon parturition, maternal behaviour and lactation in ovariectomized pregnant rats, *J. Endocrinol.* **56**:219–225.

Challis, J. R. G., and Linzell, J. L., 1971, The concentration of total unconjugated oestrogens in the plasma of pregnant goats, *J. Reprod. Fertil.* **26**:401–404.

Challis, J. R. G., Heap, R. B., and Illingworth, D. V., 1971, Concentrations of oestrogen and progesterone in the plasma of nonpregnant, pregnant and lactating guinea-pigs, *J. Endocrinol.* **51**:333–345.

Challis, J. R. G., Davies, I. J., and Ryan, K. J., 1973, The concentrations of progesterone, estrone and estradiol-17 beta in the plasma of pregnant rabbits, *Endocrinology* **93**:971–976.

Chamley, W. A., Buckmaster, J. M., Cerini, M. E., Cumming, I. A., Goding, J. R., Obst, J. M., Williams, A., and Winfield, C., 1973, Changes in the levels of progesterone, corticosteroids, estrone, estradiol-17 beta and prolactin in the peripheral plasma of the ewe during late pregnancy and at parturition, *Biol. Reprod.* **9**:30–35.

Ciaccio, L. A., and Lisk, R. D., 1967, Facilitation and inhibition of estrous behavior in the spayed female golden hamster (*Mesocricetus auratus*), *Am. Zool.* **7**:712.

Ciaccio, L. A., and Lisk, R. D., 1971, Hormonal control of cyclic estrus in the female hamster, *Am. J. Physiol.* **221**:936–942.

Ciaccio, L., Lisk, R., and Reuter, L., 1979, Prelordotic behavior in the hamster: A harmonally modulated transition from aggression to sexual receptivity, *J. Comp. Physiol.* **93**:771–780.

Collias, N. E., 1956, The analysis of socialization in sheep and goats, *Ecology* **37**:228–239.

Cosnier, J., and Couturier, C., 1966, Comportement maternel provoqué chez les rattes adultes castreés, *C. R. Seances Soc. Biol. Paris* **160**:789–791.

Cowie, A. T., and Folley, S. J., 1961, The mammary gland and lactation, in: *Sex and Internal Secretions*, 3rd ed. (W. C. Young, ed.), pp. 590–642, Williams and Wilkins, Baltimore.

Daley, M., 1972, The maternal behaviour cycle in golden hamsters, *Z. Tierpsychol.* **31**:298–299.

Day, C. S. D., and Galef, B. G., 1977, Pup cannibalism: One aspect of maternal behavior in golden hamsters, *J. Comp. Physiol. Psychol.* **91**:1179–1189.

DeBold, J. F., Morris, J. L., and Clemens, J. G., 1978, The inhibitory actions of progesterone: Effects on male and female sexual behavior in the hamster, *Horm. Behav.* **11**:28–41.

de Chateau, P., and Wiberg, B., 1977, Long-term effect on mother-infant behaviour of extra contact during the first hour post partum. I. First observations at 36 hours, *Acta Paediatr. Scand.* **66**:137–144.

Dogterom, J., van Wimersma Greidanus, Tj. B., and Swaab, D. F., 1977, Evidence for the release of vasopressin and oxytocin in cerebrospinal fluid: Measurements in plasma and CSF of intact and hypophysectomized rats, *Neuroendocrinology* **24**:108–118.

Elwood, R. W., 1975, Paternal and maternal behaviour in the Mongolian gerbil, *Anim. Behav.* **23**:766–772.

Elwood, R. W., 1977, Changes in the responses of male and female gerbils *(Meriones unguiculatus)* towards test pups during pregnancy in the female, *Anim. Behav.* **25**:46–51.

Erskine, M. S., 1978, Hormonal and experiential factors associated with the expression of aggression during lactation in the rat, Ph.D. dissertation, University of Connecticut.

Erskine, M. S., Barfield, R. J., and Goldman, B. D., 1978a, Intraspecific fighting during late pregnancy and lactation in rats and effects of litter removal, *Behav. Biol.* **23**:206–218.

Erskine, M. S., Denenberg, V. H., and Goldman, B. D., 1978b, Aggression in the lactating rat: Effects of intruder age and test arena, *Behav. Biol.* **23**:52–66.

Fleming, A. S., 1976, Control of food intake in the lactating rat: Role of suckling and hormones, *Physiol. Behav.* **17**:841–848.

Fleming, A. S., 1977, Effects of estrogen and prolactin on ovariectomy-induced hyperphagia and weight gain in female rats, *Behav. Biol.* **3**:417–423.

Fleming, A., and Rosenblatt, J. S., 1974, Maternal behavior in the virgin and lactating rat, *J. Comp. Physiol. Psychol.* **86**:957–972.

Fraser, D. G., and Barnett, S. A., 1975, Effects of pregnancy on parental and other activities of laboratory mice, *Horm. Behav.* **6**:181–188.

Fulkerson, W. J., Hooley, R. D., McDowell, G. H., and Fell, L. R., 1976a, Progesterone and induction of lactation in ewes, *J. Reprod. Fertil.* **46**:512–513.

Fulkerson, W. J., Hooley, R. D., McDowell, G. H., and Fell, L. R., 1976b, Artificial induction of lactation in ewes: The involvement of progesterone and prolactin in lactogenesis, *Aust. J. Biol. Sci.* **29**:357–363.

Gandelman, R., 1972, Mice: Maternal aggression elicited by the prescence of an intruder, *Horm. Behav.* **3**:23–28.

Gandelman, R., 1973a, Maternal behavior in the mouse: Effect of estrogen and progesterone, *Physiol. Behav.* **10**:153–155.

Gandelman, R., 1973b, Induction of maternal nest building in virgin female mice by presentation of young, *Horm. Behav.* **4**:191–197.

Gandelman, R., and Svare, B., 1974, Pregnancy termination, lactation and aggression, *Horm. Behav.* **5**:379–405.

Graf, K. J., 1978, Serum oestrogen, progesterone and prolactin concentrations in cyclic, pregnant and lactating beagle dogs, *J. Reprod. Fertil.* **52**:9–14.

Grota, L. J., 1973, Effects of litter size, age of young, and parity on foster mother behaviour in *Rattus norvegicus, Anim. Behav.* **21**:78–82.

Gubernick, D. J., 1980, Maternal "imprinting" or maternal "labelling" in goats? *Anim. Behav.* **28**:124–129.

Gubernick, D. J., Jones, K. C., and Klopfer, P. H., 1979, Maternal "imprinting" in goats? *Anim. Behav.* **27**:314–315.

Head, H. H., Delouis, C., Terqui, M., Kann, G., and Djiane, J., 1975, Hormonal induction of lactation in sheep, *J. Dairy Sci.* **58**:140.

Hemmes, R. B., 1969, The ontogeny of the maternal–filial bond in the domestic goat, Doctoral dissertation, Duke University.

Herrenkohl, L. R., and Lisk, R. D., 1973, Effects on lactation of progesterone injections administered before and after parturition in the rat, *Proc. Soc. Exp. Biol. Med.* **142**:506–510.

Hersher, L., Richmond, J. B., and Moore, A. U., 1963a, Modifiability of the critical period for the development of maternal behavior in sheep and goats, *Behaviour* **20**:311–320.

Hersher, L., Richmond, J. B., and Moore, A. U., 1963b, Maternal behavior in sheep and goats, in: *Maternal Behavior in Mammals* (H. L. Rheingold, ed.), pp. 203–232, Wiley, New York.

Hilliard, J., 1973, Corpus luteum function in guinea pigs, hamsters, rats, mice and rabbits, *Biol. Reprod.* **8**:203–221.

Hilliard, J., Spies, H. G., Lucas, L., and Sawyer, C. H., 1968a, Effect of prolactin on progestin release and cholesterol storage by rabbit ovarian interstitium, *Endocrinology* **82**:122–131.

Hilliard, J., Spies, H. G., and Sawyer, C. H., 1968b, Cholesterol storage and progestin secretion during pregnancy and pseudopregnancy in the rabbit, *Endocrinology* **82**:157–165.

Hilliard, J., Scaramuzzi, R. J., Penardi, R., and Sawyer, C. H., 1973, Progesterone, estradiol and testosterone levels in ovarian venous blood of pregnant rabbits, *Endocrinology* **93**:1235–1238.

Hinde, R. A., 1975, Mothers' and infants' roles: Distinguishing the questions to be asked, in: *Parent–Infant Interaction,* pp. 5–16, *Ciba Found. Symp.* No. 33.

Hinde, R. A., and Spencer-Booth, Y., 1968, The study of mother-infant interactions in captive group-living rhesus monkeys, *Proc. Roy. Soc. London Ser. B* **169**:177–201.

Hinde, R. A., and White, L. E., 1974, Dynamics of a relationship: Rhesus mother-infant ventro-ventral contact, *J. Comp. Physiol. Psychol.* **86**:8–23.

Hoffman, B., Schans, D., Gimenez, T., Ender, M. L., Herrmann, Ch., and Karg, H., 1973, Changes of progesterone, total oestrogens, corticosteroids, prolactin and LH in bovine peripheral plasma around parturition with special reference to the effect of exogenous corticoids and a prolactin inhibitor respectively, *Acta Endocrinol. (Copenhagen)* **73**:385–395.

Irving, G., Jones, D. E., and Knifton, A., 1972, Progesterone concentration in the peripheral plasma of pregnant goats, *J. Endocrinol.* **53**:447–452.

Kaplan, H., and Hyland, S. O., 1972, Behavioural development in the Mongolian gerbil *(Meriones unguiculatus), Anim. Behav.* **20**:147–154.

Klaus, M. H., Jersauld, R., Kreger, N. C., McAlpine, W., Steffa, M., and Kennell, J. H., 1972, Maternal attachment: Importance of the first postpartum days, *New Engl. J. Med.* **286**:460–463.

Klopfer, P. H., 1971, Mother love: What turns it on? *Am. Sci.* **59**:404–407.

Klopfer, P. H., and Gamble, J., 1966, Maternal "imprinting" in goats: The role of the chemical senses, *Z. Tierpsychol.* **23**:588–593.

Klopfer, P. H., and Klopfer, M. S., 1968, Maternal "imprinting" in goats: Fostering of alien young, *Z. Tierpsychol.* **25**:862–866.

Klopfer, P. H., Adams, D. K., and Klopfer, M. S., 1964, Maternal imprinting in goats, *Proc. Natl. Acad. Sci. U.S.A.* **52**:911–914.

Klopper, A., 1971, Endocrine factors in abortion and premature labor, in: *Endocrinology of Pregnancy* (F. Fuchs and A. Klopper, ed.), pp. 328–348, Harper and Row, New York.

Koller, G., 1952, Der Nestbau der weissen Mäus und seine hormonale Auslösung, *Verh. Dtsch. Zool. Ges. Freiburg* 160–168.

Koller, G., 1956, Hormonale und psychische Steuerung beim Nestbau weiser Mäuse, *Zool. Anz.* **19**(Supple.) (*Verh. Dtsch. Zool. Ges. Freiburg* 1955):123–132.

Korányi, L., Lissák, K., Tamásy, V., and Kamarás, L., 1976, Behavioral and electrophysiological attempts to elucidate central nervous system mechanisms responsible for maternal behavior, *Arch. Sex. Behav.* **5**:503–510.

Korányi, L., Phelps, C. P., and Sawyer, C. H., 1977, Changes in serum prolactin and corticosterone in induced maternal behavior in rats, *Physiol. Behav.* **18**:287–292.

Krehbiel, D. A., 1978, The influence of behavioral measures on models of the role of hormones in induction of maternal behaviour in the rat, Ph.D. dissertation, University of Wisconsin.

Krehbiel, D. A., and LeRoy, M. L., 1979, The quality of hormonally stimulated maternal behavior in ovariectomized rats, *Horm. Behav.* **12**:243–252.

Lahteenmaki, P., and Luukkainen, T., 1978, Return of ovarian function after abortion, *Clin. Endocrinol.* **8**:123–132.

Leblond, C. P., 1940, Nervous and hormonal factors in the maternal behavior of the mouse, *J. Genet. Psychol.* **57**:327–344.

LeBlond, C. P., and Nelson, W. O., 1937, Maternal behavior in hypophysectomized male and female mice, *Am. J. Physiol.* **120**:167–172.

Lee, M. H. S., and Williams, D. I., 1974, Changes in licking behaviour of rat mother following handling of the young, *Anim. Behav.* **22**:679–681.

Lehrman, D. S., 1961, Hormonal regulation of parental behavior in birds and infrahuman mammals, in: *Sex and Internal Secretions,* 3rd ed. (W. C. Young, ed.), pp. 1268–1382, Williams and Wilkins, Baltimore.

Le Neindre, P., and Garel, J.-P., 1976, Existence d'une période sensible pour l'établissement du comportement maternel de la vache après la mise-bas, *Biol. Behav.* **1**:217–222.

Le Neindre, P., Poindron, P., and Delouis, C., 1979, Hormonal induction of maternal behavior in non-pregnant ewes, *Physiol. Behav.* **22**:731–734.

Leon, M., 1978, Filial responsiveness to olfactory cues in the laboratory rat, in: *Advances in the Study of Behavior,* Vol. 8 (J. S. Rosenblatt, R. A. Hinde, C. Beer, and M.-C. Busnel, eds.), pp. 117–153, Academic Press, New York.

Leon, M., Numan, M., and Moltz, H., 1973, Maternal behavior in the rat: Facilitation through gonadectomy, *Science* **179**:1018–1019.

Leon, M., Numan, M., and Chan, A., 1975, Adrenal inhibition of maternal behavior in virgin female rats, *Horm. Behav.* **6**:165–171.

Leon, M. L., Croskerry, P. G., and Smith, G. K., 1978, Thermal control of nurtural behavior, *J. Comp. Physiol. Psychol.* **21**:793–811.

Le Roy, L. M., 1977, Induction of maternal behavior in the rat, Presented at Eastern Conference on Reproductive Behavior, University of Connecticut, Storrs, Conn. June 5–8.

Le Roy, L. M., and Krehbiel, D. A., 1978, Variations in maternal behavior in the rat as a function of sex and gonadal state, *Horm. Behav.* **11**:232–247.

Lisk, R. D., 1971, Oestrogen and progesterone synergism and elicitation of maternal nest-building in the mouse *(Mus musculus)*, *Anim. Behav.* **19**:606–610.

Lisk, R. D., Prelow, R. A., and Friedman, S. A., 1969, Hormonal stimulation necessary for elicitation of maternal nest building in the mouse, *Anim. Behav.* **17**:730–738.

Lott, D., 1962, The role of progesterone in the maternal behavior of rodents, *J. Comp. Physiol. Psychol.* **55**:610–613.

Lott, D., and Fuchs, S., 1962, Failure to induce retrieving by sensitization or the injection of prolactin, *J. Comp. Physiol. Psychol.* **55**:1111–1113.

Marques, D. M., and Valenstein, E. S., 1976, Another hamster paradox: More males carry pups and fewer kill and cannibalize young than do females, *J. Comp. Physiol. Psychol.* **90**:653–657.

Mayer, A. D., and Rosenblatt, J. S., 1975, Olfactory basis for the delayed onset of maternal behavior in virgin female rats: Experiential effects, *J. Comp. Physiol. Psychol.* **89**:701–710.

Mayer, A. D., and Rosenblatt, J. S., 1979a, Hormonal influences during the ontogeny of maternal behavior in famale rats, *J. Comp. Physiol. Psychol.* **98**:879–898.

Mayer, A. D., and Rosenblatt, J. S., 1979b, Ontogeny of maternal behavior in the laboratory rat: Early origins in 18 to 27 day old young, *Dev. Psychobiol.* **12**:407–424.

Mayer, A. D., Freeman, N. G., and Rosenblatt, J. S., 1979, Ontogeny of maternal behavior in the laboratory rat: Factors underlying changes in responsiveness from 30 to 90 days, *Dev. Psychobiol.* **12**:425–439.

McCormack, J. T., and Greenwald, G. S., 1974, Progesterone and oestradiol-17β concentrations in the peripheral plasma during pregnancy in the mouse, *J. Endocrinol.* **62**:101–107.

McDermott, N. J., and Gandelman, R., 1979, Exposure to young induced postpartum-like fighting in virgin female mice, *Physiol. Behav.* **23**:445–448.

McNeilly, A. S., and Friesen, H. G., 1978, Prolactin during pregnancy and lactation in the rabbit, *Endocrinology* **102**:1548–1554.

Moltz, H., Geller, D., and Levin, R., 1967, Maternal behavior in the totally mammectomized rat, *J. Comp. Physiol. Psychol.* **64**:225–229.

Moltz, H., Levin, R., and Leon, M., 1969, Differential effects of progesterone on the maternal behavior of primiparous and multiparous rats, *J. Comp. Physiol. Psychol.* **67**:36–40.

Moltz, H., Lubin, M., Leon, M., and Numan, M., 1970, Hormonal induction of maternal behavior in the ovariectomized nulliparous rats, *Physiol. Behav.* **5**:1373–1377.

Morishige, W. K., Pepe, G. J., and Rothchild, I., 1973, Serum luteinizing hormone (LH), prolactin and progesterone levels during pregnancy in the rat, *Endocrinology* **92**:1527–1530.

Murr, S. M., Bradford, G. E., and Geswind, I. I., 1974, Plasma luteinizing hormone, follicle-stimulating hormone and prolactin during pregnancy in the mouse, *Endocrinology* **94**:112–116.

Noirot, E., 1964*a*, Changes in responsiveness to young in the adult mouse. I. The problematical effect of hormones, *Anim. Behav.* **12**:52–58.

Noirot, E., 1964*b*, Changes in responsiveness to young in the adult mouse. II. The effect of external stimuli, *J. Comp. Physiol. Psychol.* **57**:97–99.

Noirot, E., 1969, Serial order of maternal responses in mice, *Anim. Behav.* **17**:547–550.

Noirot, E., 1972, The onset and development of maternal behavior in rats, hamsters and mice, in: *Advances in the Study of Behavior,* Vol. 4 (D. S. Lehrman, R. A. Hinde, and E. Shaw, eds.), pp. 107–145, Academic Press, New York.

Noirot, E., 1974, Le comportement territorial des souris, Actes du Colloque: "La communication sociale et la guerre," 20–22 Mai 1974, pp. 185–211, Bruylant, Bruxelles.

Noirot, E., and Goyens, J., 1971, Changes in maternal behavior during gestation in the mouse, *Horm. Behav.* **2**:207–215.

Noirot, E., and Richards, M. P. M., 1966, Maternal behavior in virgin female golden hamsters: Changes consequent upon initial contact with pups, *Anim. Behav.* **14**:7–10.

Noirot, E., Goyens, J., and Buhot, M., 1975, Aggressive behavior of pregnant mice toward males, *Horm. Behav.* **6**:9–17.

Numan, M. M., Leon, M., and Moltz, H., 1972, Interference with prolactin release and the maternal behavior of female rats, *Horm. Behav.* **3**:29–38.

Numan, M., Rosenblatt, J. S., and Komisaruk, B. R., 1977, The medial preoptic area and the onset of maternal behavior in the rat, *J. Comp. Physiol. Psychol.* **91**:146–164.

Owen, K., and Thiessen, D. D., 1973, Regulation of scent marking in the female Mongolian gerbil *(Meriones unguiculatus)*, *Physiol. Behav.* **11**:441–445.

Owen, K., and Thiessen, D. D., 1974, Estrogen and progesterone interaction in the regulation of scent marking in the female Mongolian gerbil *(Meriones unguiculatus)*, *Physiol. Behav.* **12**:351–356.

Owen, K., Wallace, P., and Thiessen, D. D., 1974, Effects of intracerebral implants of steroid hormones on scent marking in the ovariectomized female gerbil *(Meriones unguiculatus)*, *Physiol. Behav.* **12**:755–760.

Pedersen, C. A., and Prange, A. J., 1979, Induction of maternal behavior in virgin rats after intracerebroventricular administration of oxytocin, *Proc. Natl. Acad. Sci. U.S.A.* **76**:6661–6665.

Pepe, G. J., and Rothchild, I., 1972, The effect of hypophysectomy on day 12 of pregnancy on the serum progesterone level and time of parturition in the rat, *Endocrinology* **91**:1380–1385.

Pepe, G., and Rothchild, I., 1974, A comparative study of serum progesterone levels in pregnancy and various types of pseudopregnancy in the rat, *Endocrinology* **95**:275–279.

Poindron, P., 1974, Methods de suppression réversible de l'odorat chez la brebis et vérification de l'anosmie au moyen d'une épreuve comportementale, *Ann. Biol. Anim. Biochim. Biophys.* **14**:411–415.

Poindron, P., 1976, Effets de la suppression de l'odorat, sans lésion des bulbes olfactifs sur la sélectivité du comportement maternel chez la brebis, *C. R. Acad. Sci. Paris Ser. D* **282**:489–491.

Poindron, P., and Le Neindre, P., 1979, Hormonal and behavioural basis for establishing maternal behaviour in sheep, in: *Psychoneuroenocrinology in Reproduction* (L. Zichella and P. Pancheri, eds.), pp. 121–128, Elsevier/North-Holland Biomedical Press, Amsterdam.

Poindron, P., and Le Neindre, P., 1980, Endocrine and sensory regulation of maternal behaviour in the ewe, in: *Advances in the Study of Behavior,* Vol. 11 (J. S. Rosenblatt, R. A. Hinde, C. Beer, and M.-C. Busnel, eds.), pp. 75–119, Academic Press, New York.

Price, E. O., and Belanger, P. L., 1977, Maternal behavior of wild and domestic stocks of Norway rats, *Behav. Biol.* **20**:60–69.

Reyes, F. I., Winter, J. S. D., Faiman, C., and Hobson, W. C., 1973, Serial serum levels of gonadotropins, prolactin and sex steroids in the nonpregnant and pregnant chimpanzee, *Endocrinology* **96**:1447–1455.

Richards, M. P. M., 1966, Maternal behaviour in virgin female golden hamsters (*Mesocricetus auratus* Waterhouse): The role of the age of the test pup, *Anim. Behav.* **14**:303–309.

Richards, M. P. M., 1969, Effects of oestrogen and progesterone on nest-building in the golden hamster, *Anim. Behav.* **17**:356–361.

Riddle, O., Lahr, E. L., and Bates, R. W., 1942, Maternal behavior induced in virgins by prolactin, *Am. J. Physiol.* **137**:299–317.

Roberts, J. S., 1973, Functional integrity of the oxytocin-releasing reflex in goats: Dependence on estrogen, *Endocrinology* **93**:1309–1314.

Rodriguez-Sierra, J., and Rosenblatt, J. S., 1977, Does prolactin play a role in estrogen-induced maternal behavior in rats: Apomorphine reduction of prolactin release, *Horm. Behav.* **9**:1–7.

Rosenblatt, J. S., 1965, The basis of synchrony in the behavioural interaction between the mother and her offspring in the laboratory rat, in: *Determinants of Infant Behaviour,* Vol. III (B. M. Foss, ed.), pp. 3–45, Methuen, London.

Rosenblatt, J. S., 1975, Prepartum and postpartum regulation of maternal behaviour in the rat, Parent-Infant Interaction, *Ciba Found. Symp.* No. 33.

Rosenblatt, J. S., 1967, Nonhormonal basis of maternal behavior in the rat, *Science* **156**:1512–1514.

Rosenblatt, J. S., and Lehrman, D. S., 1963, Maternal behavior of the laboratory rat, in: *Maternal Behavior in Mammals* (H. L. Rheingold, ed.), pp. 8–57, Wiley, New York.

Rosenblatt, J. S., and Siegel, H. I., 1975, Hysterectomy-induced maternal behavior during pregnancy in the rat, *J. Comp. Physiol. Psychol.* **89**:685–700.

Rosenblatt, J. S., and Siegel, H. I., 1980, Maternal behavior in the laboratory rat, in: *Maternal Influences on Early Behavior* (W. P. Smotherman and R. W. Bell, eds.), Spectrum, New York (in press).

Rosenblatt, J. S., Siegel, H. I., and Mayer, A. D., 1979, Progress in the study of maternal behavior in the rat: Hormonal, nonhormonal, sensory, and developmental aspects, in: *Advances in the Study of Behavior,* Vol. 10 (J. S. Rosenblatt, R. A. Hinde, C. Beer, and M.-C. Busnel, eds.), pp. 225–311, Academic Press, New York.

Ross, S., Sawin, P. B., Zarrow, M. X., and Denenberg, V. H., 1963, Maternal behavior in the rabbit, in: *Maternal Behavior in Mammals* (H. L. Rheingold, ed), pp. 94–121, Wiley, New York.

Rothchild, I., 1960, The corpus luteum-pituitary relationship: The association between the cause of luteotrophin secretion and the cause of follicular quiescence during lactation; the basis for a tentative theory of the corpus luteum-pituitary relationship in the rat, *Endocrinology* 67:9–41.

Rothchild, I., Billiar, R. B., Kline, I. T., and Pepe, G., 1973, The persistence of progesterone secretion in pregnant rats after hypophysectomy and hysterectomy. A comparison with pseudopregnant, and lactating rats, *J. Endocrinol.* 57:63–74.

Rowell, T. E., 1961, Maternal behaviour in non-maternal golden hamsters *(Mesocricetus auratus)*, *Anim. Behav.* 9:11–15.

Schaffer, H. R., 1977, *Studies in Mother-Infant Interaction,* Academic Press, New York.

Shaikh, A. A., 1971, Estrone and estradiol levels in the ovarian venous blood from rats during the estrous cycle and pregnancy, *Biol. Reprod.* 5:297–307.

Siegel, H. I., and Greenwald, G. S., 1975, Prepartum onset of maternal behavior in hamsters and the effects of estrogen and progesterone, *Horm. Behav.* 6:237–245.

Siegel, H. I., and Greenwald, G. S., 1978, Effects of mother-litter separation on later maternal responsiveness in the hamster, *Physiol. Behav.* 21:147–149.

Siegel, H. I., and Rosenblatt, J. S., 1975a, Hormonal basis of hysterectomy-induced maternal behavior in the rat, *Horm. Behav.* 6:211–222.

Siegel, H. I., and Rosenblatt, J. S., 1975b, Estrogen-induced maternal behavior in hysterectomized-ovariectomized virgin rats, *Physiol. Behav.* 14:465–471.

Siegel, H. I., and Rosenblatt, J. S., 1975c, Progesterone inhibition of estrogen-induced maternal behavior in hysterectomized-ovariectomized virgin rats, *Horm. Behav.* 6:223–230.

Siegel, H. I., and Rosenblatt, J. S., 1978a, Duration of estrogen stimulation and progesterone inhibition of maternal behavior in pregnancy-terminated rats, *Horm. Behav.* 11:12–19.

Siegel, H. I., and Rosenblatt, J. S., 1978b, Short-latency induction of maternal behavior in nulliparous hamsters, Presented at Eastern Conference on Reproductive Behavior, Madison, Wis.

Siegel, H. I., and Rosenblatt, J. S., 1980, Hormonal and behavioral aspects of maternal care in the hamster: A review, *Neurosci. Biobehav. Rev.* 4:17–26.

Slotnick, B. M., Carpenter, M. L., and Fusco, R., 1973, Initiation of maternal behavior in pregnant nulliparous rats, *Horm. Behav.* 4:53–59.

Smith, F. V., Van Toller, C., and Boyes, T., 1966, The "critical period" in the attachment of lambs and ewes, *Anim. Behav.* 14:120–125.

Smith, M. S., and McDonald, L. E., 1974, Serum levels of luteinizing hormone and progesterone during the estrous cycle, pseudopregnancy and pregnancy in the dog, *Endocrinology* 94:404–412.

Smotherman, W. P., Bell, R. W., Starzec, J., Elias, J., and Zachman, T. A., 1974, Maternal responses to infant vocalizations and olfactory cues in rats and mice, *Behav. Biol.* 12:55–56.

Soloff, M. S., Alexandrova, M., and Fernstrom, M. J., 1979, Oxytocin receptors: Triggers for parturition and lactation, *Science* 204:1313–1314.

Stern, J. M., 1977, Effects of ergocryptine on postpartum maternal behavior, ovarian cyclicity and food intake in rats, *Behav. Biol.* 21:134–140.

Stern, J. M., and MacKinnon, D. A., 1976, Postpartum, hormonal, and nonhormonal induction of maternal behavior in rats: Effects on t-maze retrieval of pups, *Horm. Behav.* **7**:305–316.

Stern, J. M., and MacKinnon, D. A., 1978, Sensory regulation of maternal behavior in rats: Effects of pup age, *Dev. Psychobiol.* **11**:579–586.

Stern, J., and Siegel, H. I., 1978, Prolactin release in lactating, primiparous and miltiparous thelectomized, and maternal virgin rats exposed to pup stimuli, *Biol. Reprod.* **19**:177–182.

Strauss, J. F., Sokoloski, J., Caploe, P., Duffy, P., Mintz, G., and Stambaugh, R. L., 1975, On the role of prostaglandins in parturition in the rat, *Endocrinology* **96**:1040–1043.

Svare, B., 1977, Maternal aggression in mice: Influence of the young, *Biobehav. Rev.* **1**:151–164.

Svare, B., and Gandelman, R., 1973, Postpartum aggression in mice: Experiential and environmental factors, *Horm Behav.* **4**:323–334.

Svare, B., and Gandelman, R., 1975, Postpartum aggression in mice: Inhibitory effect of estrogen, *Physiol. Behav.* **14**:31–36.

Svare, B., and Gandelman, R., 1976a, Postpartum aggression in mice: The influence of suckling stimulation, *Horm. Behav.* **7**:407–416.

Svare, B., and Gandelman, R., 1976b, A longitudinal analysis of maternal aggression in mice, *Dev. Psychobiol.* **9**:437–446.

Swanson, L. J., and Campbell, C. S., 1978, Weaning behavior in the female hamster, Presented at Animal Behavior Society meetings, Seattle, Wash.

Swanson, L. J., and Campbell, C. S., 1979a, Maternal behavior in the primiparous and multiparous golden hamster, *Z. Tierpsychol.* **50**:96–104.

Swanson, L. J., and Campbell, C. S., 1979b, Induction of maternal behavior in nulliparous golden hamsters *(Mesocricetus auratus)*, *Behav. Neural. Biol.* **26**:364–371.

Swanson, L. J., and Campbell, C. S., 1980, Weaning in the female hamster: Effect of pup age and days postpartum, *Behav. Neural. Biol.* **28**:172–182.

Terkel, J., and Rosenblatt, J. S., 1968, Maternal behavior induced by maternal blood plasma injected into virgin rats, *J. Comp. Physiol. Psychol.* **65**:479–482.

Terkel, J., and Rosenblatt, J. S., 1971, Aspects of non-hormonal maternal behavior in the rat, *Horm. Behav.* **2**:161–171.

Terkel, J., and Rosenblatt, J. S., 1972, Humoral factors underlying maternal behavior at parturition: Cross transfusion between freely moving rats, *J. Comp. Physiol. Psychol.* **80**:365–371.

Terkel, J. Yogev, L., and Jakuboloski, M., 1978, Neuroendocrine aspects of maternal behaviour in rats with emphasis on endocrine response of maternal virgins to pups, Presented at Bat-Sheva Seminar, Rehovot, Israel.

Terkel, J., Damassa, D. A., and Sawyer, C. H., 1979, Ultrasonic cries from infant rats stimulate prolactin release in lactating mothers, *Horm. Behav.* **12**:95–102.

Turnbull, A. C., Flint, A. P. F., Jeremy, J. Y., Patten, P. T., Keirse, M. J. N. C., and Anderson, A. B. M., 1974, Significant fall in progesterone and rise in oestradiol levels in human peripheral plasma before onset of labour, *Lancet* **101**:101–104.

Verhage, H. G., Beamer, N. B., and Brenner, R. M., 1976, Plasma levels of estradiol and progesterone in the cat during polyestrus, pregnancy and pseudopregnancy, *Biol. Reprod.* **14**:579–585.

Vermouth, N. T., and Deis, R. P., 1974, Prolactin release and lactogenesis after ovariectomy in pregnant rats: Effect of ovarian hormones, *J. Endocrinol.* **63**:13–20.

Voci, V. E., and Carlson, N. R., 1973, Enhancement of maternal behavior and nest building following systemic and diencephalic administration of prolactin and progesterone in the mouse, *J. Comp. Physiol. Psychol.* **83**:388–393.

Wallace, P., 1973, Hormonal influences on maternal behavior in the female Mongolian gerbil *(Meriones unguiculatus)*, Ph.D. dissertation, University of Texas at Austin.

Wallace, P., Owen, P., and Thiessen, D. D., 1973, The control and function of maternal scent marking in the Mongolian gerbil, *Physiol. Behav.* **10**:463–466.

Weiss, G., Butler, W. R., Hotchkiss, J., Dierschke, D. J., and Knobil, E., 1976, Periparturitional serum concentrations of prolactin, the gonadotropins, and the gonadal hormones in the rhesus monkey, *Proc. Soc. Exp. Biol. Med.* **151**:113–116.

Wiesner, B. P., and Sheard, N. M., 1933, *Maternal Behavior in the Rat,* Oliver and Boyd, London.

White, L. E., and Hinde, R. A., 1975, Some factors affecting mother-infant relations in rhesus monkeys, *Anim. Behav.* **23**:527–542.

Whiten, A., 1977, Assessing the effects of perinatal events on the success of the mother-infant relationship, in: *Studies in Mother–Infant Interaction* (H. R. Schaffer, ed), pp. 403–425, Academic Press, New York.

Wise, D. A., 1974, Aggression in the female golden hamster: Effects of reproductive state and social isolation, *Horm. Behav.* **5**:235–250.

Wise, D. A., and Pryor, T. L., 1977, Effects of ergocornine and prolactin on aggression in the postpartum golden hamster, *Horm. Behav.* **8**:30–39.

Woodside, B., 1978, Thermoendocrine influences on the duration of mother-litter contact in the Norway rat, Ph.D. dissertation, McMaster University.

Zarrow, M. X., Sawin, P. B., Ross, S., Denenberg, V. H., Crary, D., Wilson, E. D., and Farooq, A., 1961, Maternal behavior in the rabbit: Evidence for endocrine basis of maternal-nest building and additional data on maternal-nest building in the Dutch-belted race, *J. Reprod. Fertil.* **2**:152–162.

Zarrow, M. X., Farooq, A., and Denenberg, V. H., 1962a, Maternal behavior in the rabbit: Critical period for nest building following castration during pregnancy, *Proc. Soc. Exp. Biol. Med.* **111**:537–538.

Zarrow, M. X., Sawin, P. B., Ross, S., and Denenberg, V. H., 1962b, Maternal behavior and its endocrine basis in the rabbit, in: *Roots of Behavior* (E. L. Bliss, ed), pp. 187–197, Harper and Row, New York.

Zarrow, M. X., Gandelman, R., and Denenberg, V. H., 1971, Prolactin: Is it an essential hormone for maternal behavior in the mammal? *Horm. Behav.* **2**:343–354.

Zarrow, M. X., Johnson, N. P., Denenberg, V. H., and Bryant, L. P., 1973, Maintenance of lactational diestrum in the postpartum rat through tactile stimulation in the absence of suckling, *Neuroendocrinology* **11**:150–155.

3

Parental Contributions to the Development of Their Offspring

Myron A. Hofer

1. Introduction

This chapter will consider the question of what mammalian parents contribute to the development of their offspring and how they do it. Of course, the very designation of an animal as a mammal depends on how the mother nourishes the young, and this aspect of the relationship still dominates our thinking on the nature of early parental care. The mother is there to feed the infant so it can grow. Before birth she does this through the placenta and after birth through her mammary system. The contribution of the mother, then, is to supply the metabolic substrate for the biochemical synthetic factories in the muscles, bones, skin, and internal organs of the fetus. But the infant has another obvious need—to be protected. The evolutionary origins of these two basic characteristics of parental care in mammals have been discussed by Klopfer (this volume).

Parents use their bodies as shields between the infant and the potentially harmful environment. In this aspect also there is a continuum between fetal and neonatal life. Even after birth, by their continuous close proximity and by shutting out the world, the parents literally become the environment of the infant. And it is in this way that the parents' impact on the infant's development takes on a whole new dimension. For environment (from the cellular to the social levels) is thought to be an integral and necessary component of development. In the embryonic stage, the physical and biochemical environment of growing neural tissues has been found to organize brain structure and to determine its function (Lund, 1978). Subsequently, in fetal development, as sensory receptors become function-

MYRON A. HOFER ● Albert Einstein College of Medicine at Montefiore Hospital, Bronx, New York 10467.

ally adequate (Bradley and Mistretta, 1975), the effective enviroment is greatly enlarged to include sources of stimulation in a number of modalities. After birth, the parent, as the major component of the infant's environment, interacts through all major sensory pathways. The roles of this stimulation in the regulation of the infant's behavior and physiology will be the main concern of this chapter (see also Galef, this volume, for a related view regarding weaning).

The general viewpoint which I would like to propose goes something like this. In the fetal period, while the mother's tissues filter out most environmental stimuli, she strongly affects the biochemical environment of her offspring, and may provide some stimulation of newly developing fetal sensory receptors. After birth, the mother's direct biochemical influence is reduced to the milk she supplies, but every sensory characteristic of the parents and everything that they do as parents impinges on the infant's sensory systems. Evidence is accumulating that these patterns of stimulation exert long-term as well as immediate regulatory effects on offspring development. It should be emphasized that the parents, and the mother in particular, are regulated as well by the infant. I use the term "regulate" in the sense of "direct or control according to a rule or principle" *(Webster's New World Dictionary)*, rather than in the physiological systems analysis sense of preserving a relatively constant value through detector feedback and control mechanisms. The synchrony of this relationship should be borne in mind, although the focus of this chapter will be on one aspect of the interaction. It is not a one-way street, but a dance, as is made clear in other chapters of this volume that deal with the powerful effects the young have on parental behavior (see Harper, this volume).

The parental interaction with the developing young can, thus, be divided into two phases. First, during pregnancy, the mother can affect the biochemical environment of the embryo and fetus by foods and drugs that she takes, by hormones and antibodies which she produces, and by her ability to regulate placental nutrient, oxygen, and waste exchange with the fetus. The second phase of the interaction begins even before birth, as the parent begins to impinge on the newborn through its developing sensory receptors. These interactions can be organized around seven major sensory systems: tactile, olfactory, thermal, vestibular, nutrient-interoceptive, auditory, and visual. The first four dominate the earliest postnatal interactions, while visual and auditory systems develop later in most mammals. Activation of the infant's sensory receptor systems is clearly not a simple matter of "on" or "off" and not limited to one system at a time. Even within a single modality, a single stimulus can vary in intensity, quality, dynamics (rising or falling intensity), and duration. Furthermore, most stimuli occurring within the parent–infant interaction are going to be repeated many times. Combinations of these elements produce the patterns and rhythms of parental care. But it is only by recognizing the constituent elements that we are able to analyze the processes which are at work. Because of the inapparent (to us) nature of some of these elements (e.g., olfactory, thermal), many of these processes are hidden from ordinary observation and their regulatory effect on the infant can only be made apparent by experimental intervention. What the parent does may be concomitant with infant behavior rather than causal or regulatory. The question of causality is most directly settled by experiment.

How can we test such a hypothesis as is outlined above, and in particular, how can we take the relationship apart in order to see if it works according to such a notion of regulatory interaction? First we need careful naturalistic observations. This greatly narrows the range of possibilities and gives us something to attack experimentally. Once the general plan of the development of the relationship is known, one can go in one of two directions. The first approach to be tried has been to make major alterations in the parents' role and assess the long-term effects of these on developmental outcome in the infant (see Section 2 below). For example, can the young be brought up artificially, in the absence of the mother, and if so, will they act any differently as a result? If the answer to this question is yes (and it generally has been), one can ask whether gross differences in parental behavior will have long-term effects. Such differences can be produced by stressing the mothers and thereby altering their maternal caretaking patterns, or by cross-fostering the infants from birth to mothers of a different species. Taking this approach has the advantage of getting an answer right away to one's doubts about whether different patterns of parental care have any differential influences on development of the young. The disadvantage is that we can not learn anything from such studies about the developmental interactions under *normal* conditions.

The successful demonstration of the importance of the parent–infant behavioral interaction for development of the young provided the impetus for the second, more molecular and analytic approach, based on the seven major sensory systems as described above. Stimulation delivered to these systems of the infant may elicit or modify hormonal or neural states in both parties which act as intervening behavioral mechanisms.

Since all these systems are involved continuously in the early mother–infant interaction, how are we to carry out an experimental analysis? Two approaches are available, one altering the sensory systems and the other, the signals. In the first, one studies the relationship after one or another system has been selectively lesioned or ablated in one member of the dyad (e.g., blind infants or anosmic mothers). If a particular aspect of the social relationship is selectively altered at a certain age, one is led to further studies of that process as a regulator of that aspect of the relationship. If no changes are observed, the system can be only partially responsible. The danger here is that techniques of sensory deafferentation may render the individual abnormal (e.g., deaf, blind) and different social behavior may result because of nonspecific effects of injury or the mobilization of compensatory central neural adjustments, rather than because this was a crucial pathway of regulation of social behavior. Thus, other approaches should be used to confirm inferences drawn from lesion studies.

The other approach, the altering of signals, is based on separation of the two members of an established relationship. According to the line of reasoning developed above, this deprives each member of all the regulatory influences previously exerted by the other. Individual aspects or factors (e.g., warmth, tactile stimulation) can then be selectively presented in the absence of other aspects of the relationship. Or surrogates lacking one or another aspect of the intact animal can be provided to test for their capacity to reverse or prevent a given change which had been observed to follow sudden complete separation.

The experimental control over individual variables which can be achieved by these analytic methods appears promising. But this approach depends on the concept that an infant's responses to separation are the result of withdrawal of regulating influences previously exerted by the mother before the separation occurred. Such a view would appear to be reasonable in the case of the responses of a fetus to removal from the uterus by caesarean section, but does not appear so reasonable when used to explain the responses of a 1-year-old child to the parents' leaving for a vacation. In the second instance, the behavioral and physiological responses of the 1-year-old are considered to be part of an emotional response to the stress imposed by disruption of the social bond, rather than to be responses to withdrawal of preexisting parental regulation of behavior and physiology. (This question is considered below in Section 5; a brief sketch will suffice here.) The capacity for perceptual and cognitive functioning develops enormously from fetus to juvenile in all species, as do systems of emotional expression. Younger infants just do not yet have the capacity to respond with an integrated behavioral and physiological pattern to a signal or cue in advance of the experience itself. Rather, they seem to depend on one channel of sensory input from the mother for maintenance of one function, in a piecemeal fashion. In such instances, if one sensory element is provided, despite the absence of the mother, that function can be maintained at "normal" levels while other functions remain disrupted by the separation. This piecemeal maintenance of the infant's sensory functioning by the mother lends itself to experimental analysis, and the changing nature of these interactions at different periods of postnatal life can be worked out. Gradually the infant comes to assume more and more of the job of regulating its own physiological functioning and to shift to a wider circle of conspecifics and other environmental stimuli which come to elicit and modulate its behavior (see Galef, this volume). Tendencies and susceptibilities have been built in by the early regulation which carry over and affect the life of the offspring as adults. These long-term effects are the most complex and difficult to understand since they involve series of interactional processes in an organism that is changing over time and developing through a series of different environments.

Having described how important the parents are to an infant by virtue of their providing so much of its environment in early infancy, it should be emphasized that infants appear to be capable of responding to a wide variety of other environmental stimuli and that stimuli from other sources (e.g., peers, inanimate environment) can often substitute for those of parental origin if necessary (see McKenna's chapter on allomothering, and also Rosenblum and Schwartz, this volume). In addition, infants have self-correcting mechanisms built in which tend to restore a developmental trajectory after a displacement. This allows infants to recover from premature withdrawal of parental regulation, so long as the environment provides the necessary support. Examples of these homeostatic tendencies wil be given in the last section.

2. Strategies for Demonstrating Long-Term Parental Effects on Development of Young

2.1. Maternal Deprivation

The major contribution of the artificial rearing of infants has been to prove that mothers provide essential ingredients for normal development in addition to milk (Harlow, 1958). Purely artificial rearing has actually been difficult to accomplish because of the special needs of neonatal mammals. The experimenters have had to substitute for necessary species-specific maternal behaviors [e.g., stimulating urinary and fecal elimination reflexes in infant rodents by substituting a moist camel's-hair brush for the mother's tongue (Thoman and Arnold, 1968)]. Discovering how to make neonatal animals survive has taught us a good deal about crucially important aspects of parental care, including the provision of warmth and moisture and the conditions necessary for feeding. But the infants have received a great deal of undefined stimulation from the experimenters' efforts to keep them alive, thus blurring interpretation of the original question.

Other studies have relied on the mother to raise the infant through the most demanding neonatal stages and then separate and place the infant into an artificial environment (e.g., Seitz, 1958). This procedure raises the question of whether it was the sudden change, the novelty of the new environment, or the absence of something ordinarily provided by the mother which accounted for any long-range effects found (see below, Sections 4 and 5).

Despite these drawbacks, such studies have established that animals raised without parents show a variety of severe deficits in emotional regulation, social interaction, and complex goal-directed behaviors. Sexual and maternal behavior in adulthood were particularly devastated. Disagreement exists as to the extent to which these effects can be explained on the basis of generalized sensory and/or nutritional deprivation and how much depends on the necessity for highly specific forms of interaction which only a single species-specific caretaker can provide. The necessity of some critical amount of postnatal stimulation for normal behavior development has, however, been clearly established by these studies.

2.2. Selective Breeding

It is often assumed that behavioral characteristics produced and maintained by selective breeding will be produced in each generation by an unfolding of "instructions" derived from genes within the germ cells of the offspring generation itself. However, selection does not operate on genes but on the phenotype. The implication of this principle is that the behavioral trait may be transmitted and produced by the influence of the mother's uterine environment or by parental behavior acting postnatally. The potential of this approach for studying parental influences on development has not been sufficiently exploited, possibly because of initial negative results (Ginsberg and Allee, 1942; Broadhurst, 1961). Specific behavior traits can be selected by this method and then traced to specific prenatal or postnatal parental influences. By cross-fostering infants between (and within) strains which show heritable behavioral differences, we can determine if the behavior is transmitted prenatally or postnatally. If found to be prenatal, ova transplantation (in inbred strains) can determine whether the difference is due primarily to the uterine environment or to the parental germ cells.

That heritable behavior differences can indeed be transmitted through the postnatal maternal interaction has been demonstrated by a number of studies (Reading, 1966; Ressler, 1963; Southwich, 1968). The most recent of these (Flandera and Novakova, 1974) is particularly impressive. They divided their Wistar-derived strain into two subpopulations, one which regularly attacked and killed mice and the other which did not. These traits were heritable in each subpopulation. When infants born to each type of mother were cross-fostered at birth, however, the infants developed the characteristics of the foster mother rather than of their biological parent, even though mothers and infants were not exposed to mice until the young were first tested at 30 days of age. Although watching their mothers kill mice further enhanced mouse killing in the young, this was not a possible explanation for the original effect. Clearly, the mothers shaped the development of their offspring powerfully enough to overcome whatever prenatal determinants may have existed. Some prenatal effect was evident in the pups born to mothers who were mouse killers. These animals began to show increased mouse killing at 90 days, although they had not shown it at 30 days, after having been raised by a non-mouse-killing mother. Offspring of the non-mouse-killing strain continued to show mouse killing at both 30 and 90 days after rearing by a "killer" mother. It seemed unlikely that the mother transmitted only one specific kind of behavior, and we know of no processes which could explain this sort of effect, in the absence of the opportunity for specific learning. And, in fact, Flandera and Novakova report that the mouse-killing rats tended to be generally more irritable, with high levels of nonspecific excitability. Killers have been found to consume more water and salt solution than nonkillers, to fight with each other more intensively, and to engage in more precopulatory activity. Thus, a whole dimension of their behavior has been affected, of which mouse killing is but one part.

The study by Southwick (1968) showed postnatal maternal effects on conspecific aggression in A/J mice fostered to CFW mothers but not in the reverse case. Southwick observed that CFW mothers spent one-half to one-third as much time on their litters as A/J mothers after the 4th postnatal day, suggesting a behavioral interactional or thermoregulatory mechanism. Differences in milk should also be considered, either hormonal or nutrient mechanisms being possibilities. Further research will be needed to reveal which aspects of the parent–infant relationship and which sensory systems of the young are responsible for these developmental effects.

2.3. Stressed Parents

Another approach is to attempt to alter the parental influence experimentally by stressing the pregnant mother. This can produce behavioral effects in the offspring. Subsequent cross-fostering of half the litters to normal mothers at birth reveals whether the effect is transmitted primarily by prenatal or postnatal interactions.

An example of prenatal development interaction is the work of Thompson et al. (1962). These investigators stressed pregnant rats by a signal which had previously been associated with electric shock. Whether raised by their own or normal mothers, the offspring of stressed mothers differed in their latency to venture into an unfamiliar test area, and in their subsequent activity and defecation rate in the novel environment. One of the mediating mechanisms for this effect may be a stress-induced increase in maternal epinephrine, since injection of the mother with this hormone altered the open-field activity of her offspring. Other hormones likely to be increased in stressed pregnant mothers, hydrocortisone and norepinephrine, when given to the pregnant female mouse, tend to mimic some of the behavioral effects on the offspring of crowding the mothers prior to parturition (Lieberman, 1963). There is little consistency in the direction of effects observed in these studies, the offspring showing evidences of increased and/or decreased emotionality from experiment to experiment, depending on the type of stress, its timing, the type of hormone given, and of course the genetic makeup of the animals used.

Genetic analysis by deFries (1964) has shown that the effects of prenatal stress are a function of both the maternal and the fetal genotype. But we have no clear evidence whether stress or its associated hormonal changes are acting directly on the developing fetus or by altering placental exchange systems (see Section 3). Although cross-fostering controls have generally confirmed that the effects are exerted during the prenatal period, persistent changes in the behavior of prenatally stressed mothers, carried over into the postnatal period, can have important modifying effects on offspring behavior (Becker and Kowall, 1977).

Maternal stress during the *postnatal* period has been shown to adversely affect conditioned avoidance behavior of offspring and to raise plasma corticosterone levels during adulthood (Levine and Thoman, 1969). In addition, the well-known long-term effects of handling of the developing rat may well be mediated by changes induced in the mother–infant relationship (Lee and Williams, 1974) in addition to the effects of the stimulation on the offspring themselves. Mothers of young which are repeatedly handled by experimenters are more active imme-

diately after the return of their pups, possibly stimulated by the increased ultrasound emitted by the cooled pups. Furthermore, they build less well-formed nests
and their pups tend to be dispersed more widely, even 24 hr. after the daily handling session. Mothers are slower to retrieve pups with slightly reduced skin temperature. These or other changes in the mother–infant relationship may be the
crucial variables (see review by Russell, 1971).

     This approach demonstrates that stressed parents raise different offspring
than do controls and can point to prenatal and/or postnatal periods as responsible
for the effect. The pity is that none of these studies have been followed by systematic attempts to trace down and isolate the processes by which the stressed mother
affects the development of her offspring.

2.4. Premature Separation

     Infants normally separate from their parents; this is a gradual process and
is accomplished by events taking place within both parents and offspring. (see
Galef's chapter on weaning, this volume). If parent and infant are separated prior
to this gradual process, a reasonably predictable series of behavioral responses
takes place. Both mother and infant become agitated, vocalize, and attempt to
reunite. (see Gubernick's chapter 7, on attachment, this volume). Physiological
changes have been recorded in these separated infants which are similar to those
elicited by moderately severe physical pain. This immediate response gradually
subsides and in some primates is followed by a longer but transient phase of
reduced social interaction and decreased spontaneous behavior (Kaufman and
Rosenblum, 1969). The less predictable series of longer-acting changes varies
more widely across species and between individuals than does the acute response.
The long-term effects of such an occurrence can tell us something about the
importance of a continued parental relationship to the development of the young.
As in the case when no relationship had been established (maternal deprivation),
premature separation cannot give us direct information on precisely how parental
care normally interacts with and influences the offspring. The presence and nature
of long-term effects following premature separation *can* tell us something about
the importance of a continuous parental relationship during the early development
of young.

     There are a very large number of studies on premature separation in young
mammals, most of which are directed at an understanding of pathological behavior
and depression in particular (for recent reviews, see Bowlby, 1973; Akiskal and
McKinney, 1973; and Lewis *et al.*, 1976). Relatively few studies have looked systematically for the kind of long-term effects which we are concerned with in this
section of the chapter. The best of these is Hinde and Spencer-Booth's study
(1971) on rhesus monkeys. These investigators studied the effects of one or two
brief (6-day) separations carried out when the infants were capable of feeding
themselves and had begun to take the initiative in maintaining proximity with

their mothers (16–32 weeks postnatal). When the infants were studied 1 and 2½ years later, they showed little or no differences from controls when in familiar surroundings and social groups, but when confronted with strange objects in a strange cage, the infants which had not been allowed a constant parental relationship showed less exploratory activity, less time playing with peers, and more time sitting quietly or in nonsocial play. They were slow to initiate play with novel toys and, instead, tended to remain with their mothers. The depression of locomotor and play activity elicited by novel surroundings in these older offspring was reminiscent of the behavior they had shown during the earlier separations in the home cage after the first agitated response to the mother's removal had waned. The severity of the acute response to separation could be related to measures of tension in the relationship prior to separation, measures involving frequency of maternal rejection and the infant's persistence in attempting to maintain proximity with the mother.

This study demonstrated that even short and temporary breaks in the early mother–infant relationship of the rhesus monkey had measurable effects on behavior as long as two years later. The role of parental substitutes during the separation could not be studied well because there was little "adopting" of infants by members of the group in this study. Other studies suggest that adopting or surrogate care can have dramatic ameliorating effects, at least on the immediate response to separation (Kaufman and Rosenblum, 1969; Kaufman and Stynes, 1978; see also Rosenblum and Swartz, this volume) and raise interesting questions about exactly what aspects of parental care are needed to maintain normal development. This question will be addressed at a later point in Section 5 below.

In addition to behavioral developmental effects, we also have evidence that biological function and disease susceptibliity can be modified by alterations of the early social relationship. Ackerman et al. (1975; 1978a,b) have found that if infant rats are separated from their mothers 1–2 weeks early, at 15 days of age, they develop an extraordinary susceptibility to gastric erosions induced by restraint stress at 30 days of age. At this age 100% of early separated rats develop ulcers and more than 70% bleed massively from them, a lethal complication almost never seen in adults. Only 5–10% of normally reared 30-day-old rats show ulceration at this age. By 100 days of age, normally reared rats are susceptible, but early-separated ones are still more susceptible. At 200 days, no differences persist. A crucial variable appears to be the nutritional deficit suffered in the first few days after separation. However, provision of a nonlactating mother from days 15 to 30 markedly reduces the incidence and severity of ulcers in 30-day-old rats, showing that the mother–infant behavioral interaction itself, irrespective of nutritional effects, can modify the course of development of susceptibility in this stress disease model.

From these and other experiments it becomes clear that an uninterrupted early parental relationship is essential to the normal development of physiological systems concerned with promoting resistance to stress-induced pathology. Whether the long-term effects in these studies are mediated by the disruption of the separation, the altered relationship after reunion, or the absence of some essential regulators of development during the separation are questions which are discussed below (Sections 4 and 5).

2.5. Interspecies Parenting

Since there are widespread differences in patterns of mothering between different species of mammals, one way to confirm the role of species-specific patterns in the development of the offspring is to foster neonatal infants to mothers of another species. Often there are problems with this approach, for example, primates of closely related species will not accept young of another species even if the substitution takes place at the time of parturition (Kaufman and Rosenblum, 1969). Romulus and Remus, the legendary founders of Rome, are the most famous of a number of accounts of successful cross-species parenting of human infants. Usually the results are less successful from the human point of view (Malson, 1972).

Even nonlactating adult female rats will adopt and mother newborn mice. Whether as sole mothers or as "aunts" in the same cage as the mouse mother, female rats' maternal attentions alter the behavior and physiological response of the developing mice so that in adulthood they are less active in an open field, show less ability to inhibit behavior in a passive avoidance task, have lower adrenal corticosterone responses to novelty, and show a marked decrease in intraspecific aggression compared with mice fostered to different mothers of their own species (Denenberg, et al., 1964, 1969a,b). These effects are not likely to be mediated by factors in the mother's milk because nonlactating aunts are also effective, nor by pheromonal communication (see below, Section 4.4) because when behavioral interaction was prevented by a wire mesh partition, the rat aunt had no effect on the mice. Therefore, some aspect of the behavioral interaction or close proximity of the rat altered the developing mouse so that it was different in several ways as an adult.

The specific processes mediating these powerful effects are not known, but this interesting question could be approached along the lines described in Section 4 below.

2.6. Effects Visited upon Subsequent Generations

It is not only by cultural evolution or genetic selection that behavior can be passed down from generation to generation. Wehmer et al. (1970) showed not only that stress (i.e., shock avoidance conditioning) given prior to mating affected the offspring of the stressed animal, but that the effect was transmitted to the next generation as well. The grandpups of the grandmothers stressed prior to mating showed higher levels of activity in a novel test area (open field) than grandpups of control grandmothers. Judging from cross-fostering evidence, this effect appears likely to be transmitted prenatally. More recent experiments have implicated the maternal adrenal cortical hormones as the intervening variable, since direct manipulation of these in the pregnant mother *does* affect offspring open-field activity (Smith *et al.*, 1975).

Denenberg and Whimbey (1963) showed a number of years earlier that experiences which mothers had as infants (handling) affected the open-field behavior of their offspring and that pre- and postweaning housing conditions interacted with the infantile handling effect. Skolnick *et al.* (1980), working with Ackerman and Weiner in our lab, has found that the susceptibility to stress ulcer is "handed down" to the offspring of mothers which had been prematurely separated from their own mothers as infants. This increased susceptibility in the second generation occurs even if these pups are themselves weaned normally and raised by normal foster mothers, thereby indicating prenatal transmission of susceptibility. Nothing is yet known about the processes by which this might be brought about.

It is not too hard to accept the inference that an early experience which alters the mother's physiology may alter her uterine environment, even weeks later, so that the offspring will be affected. What is hard to explain is the apparent *specificity* of the effect. Why, for example, should premature separation, which increases stress ulcer susceptibility of the mother, have effects on the same system in the offspring and in the same direction? We know about several prenatal and postnatal processes (see Sections 3 and 4 below) which could possibly account for transmission of acquired behaviors across generations. But no one example has been analyzed, and no specific mechanism is known.

These experimental results provide modern-day examples of the sort which supported the ideas of Lamarck, only we think we can explain them on the basis of alterations in phenotype, not of the genetic material. These observations challenge our preconceptions, and indicate how much we still have to learn.

3. The Prenatal Relationship

Parental behavior does not begin with birth, but comes into being during pregnancy (see Chapters 2 and 5). From the foregoing section, it is clear that maternal interactions with the fetus during pregnancy can have long-term effects on offspring development, but we still know relatively little about the specific processes and mechanisms involved.

We do know that fetal sensory receptors become functionally adequate as early as the first third of pregnancy (Bradley and Mistretta, 1975), so that there is only the insulating nature of the intrauterine position to buffer the infant from whatever stimulation an active pregnant mother provides. The most obvious form of stimulation is to vestibular and tactile receptors of the fetus by the physical activity of the mother and her rhythms and patterns of motor behavior. There is some accumulating evidence (Neal, 1968; Solkoff *et al.*, 1969) that this form of stimulation can in fact affect prenatal behavior development. Prematurely born infants of 28–32 weeks' gestational age were given 5–15 min. of extra stimulation each hour by to-and-fro horizontal displacement of the incubator surface or by tactile stimulation of the skin on limbs and trunk. After 4–6 weeks, the stimulated infants showed significantly more advanced scores on sensorimotor tests and on tests of visual responsiveness and had gained more weight. Observation of their sleep–wake cycles showed that the stimulated infants showed greater increases in quiet sleep time and greater declines in active wakefulness. A small number of these infants were followed up at 1 year and showed a tendency toward more advanced functioning, particularly in expressive language development.

These results suggest that the vestibular or tactile stimulation provided to the infant by the mother plays a role in facilitating the infant's behavior development. A specific role of the mother's heart sounds for human infants (Salk, 1962) is controversial, since most forms of repetitive stimulation have a quieting effect on young infants (Brackbill, 1971). The sound level in the human uterus is dominated by a low-frequency constant "masking noise" at about 95 db, rising to rhythmic peaks about 10 db higher, 0.3 sec after contraction of the mother's left ventricle (Walker *et al.*, 1971). The sound is loudest at 20 Hz frequency (or lower) and appears to be produced by turbulence of pelvic blood flow and uterine muscle movement. The contribution of outside environmental sound higher than 1000 Hz is markedly attenuated by the tissues of the uterine and abdominal walls, but even below this level, 20–30 db attenuation occurs. So the infant is effectively masked against all environmental sounds below 115 db, that is, anything quieter than nearby trains, pneumatic hammers, and jet planes. The only really effective contribution to the fetal auditory system is generated by the mother.

Taste receptors can readily be stimulated because the infant swallows amniotic fluid from midpregnancy on. Taste preferences or aversions for foreign substances such as drugs can theoretically become established this way, although I know of no data on this. It is known that infants markedly reduce swallowing when certain substances such as iodinated dye are injected into the amniotic fluid (Liley, 1972).

Of course, the main pathway for maternal interaction with the fetus is via the placenta. The metabolic and nutrient exchange vital for maintenance of normal physical development of the fetus is not, however, our concern here, for we are interested in behavioral effects. There are four major known ways by which a mother's behavioral state can influence fetal behavior development across the placental barrier: first, be neuroendocrine control of hormonal patterning (for example, under stress); second, by altered autonomic neural control over the effective circulation in the intervillous space of the placenta; third, by altered amount or composition of maternal diet; and fourth, by maternal ingestion of drugs or other chemicals which affect the developing brain.

The major hormones known to affect the early development of the brain and behavior are thyroxine, the adrenal corticoids, and testosterone. Epinephrine may affect the fetus directly or act by altering placental blood supply (see above). A good example of maternal stress effects mediated by prenatal hormones is given by Ward (1972). Pregnant rats were stressed by restraint under bright light every day for the last week of gestation. Their male offspring, when grown to sexual maturity, showed marked reduction in attempted copulation and ejaculatory responses. When castrated, and given fixed doses of estrogen, males whose mothers had been stressed gave significantly more female (lordotic) responses to a vigorous male "stud" than did offspring of nonstressed mothers. Thus, the behavioral state elicited in their mothers during pregnancy demasculinized and feminized her male offspring. The increased feminine behavior was observed while the concurrent hormone level was kept identical to that of controls, suggesting that the modification induced in the offspring was in the central nervous system. Testosterone is known to be necessary during fetal and early neonatal periods for the later development of masculine sexual behavior (Money and Ehrhardt, 1972), and the mechanism for these effects is likely to involve an interference with fetal testosterone levels by the action of maternal hormones such as corticosterone secreted in response to stress (Dahlhof et al., 1978).

The fetus is unable to compensate for changes in maternal blood supply to the placenta, since its umbilical vessels and their villous branches have no autonomic neural innervation. In addition, on the maternal side, there is no direct feedback (as in other organs) from the smaller tissue vessels to the arteries supplying them. This places the fetus at the mercy of the mother's uterine arteries. Maternal autonomic vasoconstriction and uterine muscle contractions lead to decreased fetal oxygenation and increased levels of metabolic by-products in the fetal blood (Gruenwald, 1975). Transient metabolic changes, such as occur during maternal psychological stress, are known to produce short-term changes in fetal behavior (Barcroft, 1938), and maternal emotional distress can produce extreme increases in objectively recorded fetal movements (Sontag, 1966).

The mother can alter her diet as a part of her behavioral adaptation to being pregnant, or in response to intercurrent events in her life. This need not involve serious degrees of malnutrition in order to affect fetal development. For example, Erway et al., (1966), studying a strain of mice with congenital ataxia, found that if the mothers were given a manganese supplement during pregnancy, their offspring developed normally even though the offspring were fed the stock diet. If these offspring continued to eat a stock diet during their pregnancy, however, their infants developed ataxia at the expected rate for this strain. This study illustrates how alterations in consumption of a specific dietary factor by the mother during pregnancy can control the expression of the genetic constitution of her offspring.

Finally, mammals can radically affect the development of their offspring by excessive alcoholic intake, smoking, or drugs or by exposure to chemical pollutants during pregnancy. Whereas development of most organs is affected by ingested chemicals only during a short period early in pregnancy, and most of these structural abnormalities result in death of the fetus, the brain of the fetus continues to be affected throughout the whole period of pregnancy by a wide variety of substances ingested by the mother. Functional or ultrastructural abnormalities may be produced by substances which are not teratogens (i.e., not capable of producing structural abnormalities). Usually these effects are not lethal, but alter the development of the child's behavior. Hutchings (1978) has pointed out the alarming extent of this problem and reviewed the recent experimental work on animals in which the effects of one substance at a time can be assessed clearly and the possible confounding by altered postnatal maternal behavior can be removed by cross-fostering to nontreated mothers. Particularly implicated, in addition to alcohol and methadone (or heroin), are the tranquilizers, barbiturates, and amphetamines, all of which have been shown to produce long-term effects on behavior of the offspring weeks and months after the intrauterine period of exposure to these substances.

These several kinds of interactions between maternal behavior and development of the fetus undoubtedly do not complete the list. A great deal still waits to be learned about this relationship. For example, Sterman (1972) gives an example of an interaction which defies explanation by present knowledge of physiological regulatory pathways between mother and fetus. By time series analysis of continuous 8-hr fetal activity recordings in humans (using pressure-sensitive electrodes on the maternal abdomen), he was able to extract two clear periodicity peaks from the records of fetuses 27–38 weeks of gestational age. One peak represented an average cycle length of 40 min and the other of 96 min. These two peaks were present in even the oldest fetuses, but shortly following birth the longer cycle length was no longer present while the shorter cycle length persisted unchanged. The normal adult REM cycle is very close to the longer fetal period of 96 min, and concurrent recordings of fetal activity and maternal REM sleep gave some suggestive support to the interpretation that the longer of the two fetal cycles has its origin in this ultradian physiologic rhythm of the mother, possibly in some aspect of her REM sleep physiology. We have no clue how the mother may be inducing this fetal activity rhythm, and the results remind us how little we know about this formative period of parental interaction.

4. The Postnatal Relationship

Throughout much of this volume runs the theme of how the parents and the infant initiate and maintain their relationship and how they eventually separate from each other. The elicitation of responses, their molding, and their change with time are the result of a subtle communication involving patterns of repeated reciprocal stimulation over known sensory systems. How these processes are involved in the development of attachment, in the postpartum regulation of maternal endocrine activity, in the initiation and maintenance of maternal aggression, in the formation of peer relationships, and in the transitions of weaning, are the subjects of separate chapters. It is clear that the relationship is guided by the interplay of the behavior that parent and infant direct at each other, the sensory stimulation received as a result of their interaction, and the alterations of hormonal and neural states induced in both by this interplay. One could say that the relationship is "regulated" by these processes, although the mechanisms involved may be different from biological regulatory mechanisms operating at the cellular level.

In this section I will describe some recent evidence that the development of the young is regulated by the effects of this parent–infant interaction on the behavioral and physiological systems of the infant. The forces which are at work between parent and infant in the period between birth and weaning are not confined to the reciprocal actions of observable behavior. Transactions such as the exchange of heat, the provision of nutrient, even activation of the infant's vestibular system by the parent must be considered as well. However pervasive these regulatory effects of the early social relationship may be, they must act within the genetic potential of each member and be influenced by the genotypic characteristics of both parents and infant (e.g., Scott and Fuller, 1965). Indeed, some behavioral and many physiological processes are relatively stable and unaffected by the interactional processes of the relationship. In this paper I will be focusing upon those which are consistently influenced by definable aspects of the parent–infant interaction.

I will begin with the long-term behavioral effects attributable to visual perception by the human infant during its complex behavioral interactions with the mother, move on to the regulation of the rat infants' behavior by less complex aspects of maternal stimulation, then to the timing of puberty in mice by olfactory stimulation by father or littermates, followed by regulation of the infant rats' physiological and biochemical systems by maternal interoceptive and exteroceptive stimulation.

4.1. Facial Communication in Blind Infants

As a first example of the developmental effects of the parent–infant interaction, I have chosen the observations on blind human infants made by Selma Fraiberg and co-workers (Fraiberg, 1974). Here, a discrete sensory deficit allows us to study the contribution of a single component of the parent–infant interaction.

It is well known that humans who have been blind from birth have marked deficits in their ability to communicate by facial expression their mood, intent, and attention, as well as the finer gradations of their emotions. The deficiency becomes immediately apparent even on the basis of casual observation. And yet, other adults who have been blind for just as long but whose blindness occurred after the first year or two of postnatal life are virtually unimpaired in this important communicative skill. Adults blind from birth develop compensatory means of communicating by voice inflection and hand and body movement as well as with language, but the ordinary sighted person at first feels severely impaired in relating to them.

The impaired development of facial-expressive skills in blind infants can be understood as a result of the absence of those elements of the early parent–infant interaction which depend on a single sensory system: reciprocal visual interaction. This unfortunate experiment of nature allows us to get a much closer look at a single element of the numerous processes at work during complete maternal deprivation, for example, Harlow's experiments (1958) or the institutional child syndrome described by Spitz (1945).

The blind neonate develops normal smiling in response to its parents' voices at the 4th postnatal week. At this point, blind and sighted infants are no different in their smiling or in other facial signs. A number of stimuli can elicit smiling irregularly at this age. But by 2 months of age, differences begin to appear. The sighted infant regularly smiles to the configuration of the human face, but there is no one stimulus which has this predictable effect on blind infants. Mothers of blind infants find that gross tactile and kinesthetic stimulation increases the predictability of smiling responses to her voice at this age, but the response often must be "coaxed" to appear. The overall incidence of smiling is less frequent in blind infants at this age, and by 4–5 months of age, clear qualitative differences become apparent. The smile is muted and lacks the extreme or exaggerated form which we might label "ecstatic." In sighted infants of this age other differentiable expressions have developed, such as "attentive," "doubtful," "quizzical," "coy," and "bored," which are totally absent from the blind infants' repertoire. All that can be discerned in the face of a 6-month-old blind infant are variations in intensity of either smiling or crying, conveying only relative positions on an undifferentiated continuum from happy to unhappy. This is not to say that the blind infant necessarily has an impoverished range of internal affective states. And, in fact, by attending to hand "language" and other nonfacial cues, Fraiberg was able to learn how to interpret the moods and needs of blind infants. But it is not easy, and sighted mothers of blind infants experience much difficulty until language establishes a pathway for clear communication.

The failure of development of a differentiated facial-sign repertoire in blind infants is likely to be due to the absence of modeling stimuli for imitation and the absence of a means by which the mother can reinforce facial expressions quickly enough over other sensory pathways, in order to allow highly selective shaping of specific responses. The mother's smile is a strong reinforcer and can act in milliseconds while the infant's gaze is fixated on the mother's face, as it so frequently is. Other infant facial signs may be reinforced by the mother's imitating them or through signaling by her expression that she will respond to a need expressed by the infant's facial sign. All these interactional processes are unavailable to a blind infant.

The remarkable thing is that a wide variety of facial expressions are emitted "spontaneously" by all infants during the late prenatal and early neonatal periods.

Gesell and Amatruda (1945) give photographic evidence of the facial virtuosity of the neonate. So it is not a question of the infant having to build up its facial neuromuscular systems during the first 6 months of postnatal life. Rather, it seems that infants *lose* their capacity to emit differentiated facial expressions if they do not receive necessary visual stimulation and necessary patterns of visual and behavioral interaction based on their facial expressions. The relative importance of unpatterned and patterned visual stimulation, the availability of specific facial configurations for imitation, and the shaping of facial responses by moment-to-moment visual reinforcement have not yet been worked out and would be a fascinating area for further study.

4.2. Levels of Stimulation and Infant Behavioral Arousal

The kinds of effects I will be discussing in this section are those which are exerted over long periods of time by the cumulative actions of the repeated episodes of stimulation delivered by the mother to the infant during their ongoing social relationship. In this sense, they represent the tonic effects of phasic stimulation. In the cases described, the relationship between the particular form of stimulation and the behavioral effect cannot be deduced from observation of the ongoing relationship. Only by removal of the mother for a period of hours does the behavioral effect become evident. It is only upon prevention of that behavioral consequence of separation by provision of a specific form of repeated stimulation that the regulatory process can be identified. These processes appear to be slow-acting ones, requiring a period of hours for behavioral change to occur after offset or onset of the altered stimulation. And the cumulative or tonic effects of a given repeated stimulation can be quite different from, even opposite to, the phasic effects of the stimuli.

4.2.1. Thermal Influences

In most altricial mammals, the parents maintain the body temperature of the young by conveying their own body heat through conduction and radiation within the sheltered nest environment they have selected or constructed. The amount of thermal input is directly related to the amount of time the parents spend in contact with the young. This is a familiar example of a physiological regulatory process originating in the parent–infant social relationship; since the young lack fully developed thermoregulatory capacity, their body temperature is regulated by how much time the parents spend with them. But can the parents regulate other behavioral and developmental processes by regulating the body temperature of their young? A partial answer to this question has been available for a number of years in data on markedly decreased heart rate, oxygen consumption, etc., in infant rats separated from their mothers and placed in refrigerators (Fairfield, 1948). But these data only told us why a mother's abandoning infants in a cold climate could be lethal; it did not deal with the effects of relatively minor changes in body temperature which might occur within variations of an ongoing social relationship.

In our studies (Hofer, 1973a) we found that separating 2-week-old infant rats from their mother and housing them at room temperature for 18 hr significantly reduced most behaviors which are customarily elicited by placing the infants alone in an unfamiliar observation area (e.g., self-grooming, rearing, total activity count). Only a 3°C drop in body temperature occurred on the average, and the correlation between body temperature and measures of locomotion was quite high. The decreases in body temperature which so affected behavioral activity levels were not sufficient to affect heart rate in these experiments, a fall in heart rate being generally considered to be a sensitive indicator of the point at which hypothermia has reached a biologically significant level (cooling slows sinoatrial pacemaker activity).

The effects of several days' exposure to room temperature without the mother are to cause a profound motor deficit (Hofer, 1975b), to markedly slow biological maturation of brain, internal organs, and skeleton, and to hasten death, rather than to have an adaptive value in prolonging survival (Stone et al., 1976). The motor deficit prevents adequate foraging for food, but the developmental retardation is not simply a consequence of decreased food intake.

For these two experiments, the comparison groups were also separated from their mothers, but provided with enough thermal input to maintain warm nest temperatures (35°C). The absence of these changes in the comparison groups demonstrates that it is not the separation from the mother *per se*, but the release from her thermal input which is responsible for the changes observed in the experimental group.

We can infer from these studies that the thermal input regularly provided by the repeated periodic visits of the mother serves to maintain the level of behavioral activity of the infants and to determine the rate of biological maturation of body and brain. The physiological mechanisms are not yet established by which small changes in thermal input regulate activity levels of the 2-week-old rat. But we have evidence that this is not simply the result of sluggish and unresponsive muscles or nerves, since *d*-amphetamine in small doses completely overcomes the motor deficit, without reversing the hypothermia.

4.2.2. Tactile Stimulation

Using the provision of body heat (above) as a model, can we identify other aspects of the mother–infant behavioral interaction which also regulate behavioral activity? To answer this question, we studied the behavior of 2-week-old infants after being separated from the mother for 18 hr. During this time they were kept at nest temperature, and provided with normal nutrient intake by constant gastric infusion (Hofer, 1973b). At the end of the period of separation, these infants were markedly *more* active than normally mothered pups, when placed alone in an unfamiliar observation area, as judged by such behaviors as locomotion, rearing, self-grooming, and defecation/urination and by the total activity count. The onset of sleep was also delayed in separated infants (Hofer, 1975a). Furthermore, if infants were deprived of milk, but allowed to interact with their mother for 24 hr after she had undergone mammary ligation, their behavior closely resembled that of normally mothered pups (Hofer, 1973c).

Two hypotheses compete for explanation. The hyperactivity could be viewed in the framework of the attachment hypothesis or it could be the result of withdrawal of regulatory processes which had been acting to reduce levels of activity while the social relationship was intact. It is difficult to rule out an attachment–separation–distress model, but several predictions of that model failed to be substantiated (Hofer, 1975a). First, the hyperactivity was slow to develop (between 4 and 8 hours after separation), which is not characteristic of the hyperactivity of separation distress. Second, the maternally separated infants in these experiments were housed in the familiar home cage and with seven littermates. Placing them alone and in an unfamiliar environment failed to accentuate the hyperactivity, as predicted by the attachment hypothesis, even at 8 hr when hyperactivity was not near its maximum. Thus, this form of behavioral hyperactivity did not meet our criteria for separation distress. Furthermore, we have recently shown that *immediately* following isolation, infant rats of this age show vocalization and hyperactivity, alleviated by the presence of a familiar social companion. This earlier response, then, is classical separation distress, but not the later changes we have been considering here (see below, Section 5).

What sort of stimulation could serve to reduce or inhibit behavioral arousal levels during the normal mother–infant interaction? Tactile, auditory, olfactory, and vestibular stimulation are the leading candidates, the eyes not yet being open at this age. Since the mother normally licks, noses, scratches, picks up, steps on, rubs, and lies on her infants, tactile stimulation seemed a strong possibility. In order to test this possibility, stimulation was provided for approximately 15 min out of every hour, in accordance with the timing of the visits and absences of the mother at this age. Using mild electric current (0.05 mA constant amperage, just enough to elicit activity), levels of behavior were reduced in the same pattern and to the same degree as occurs after normal mothering during this period. A similar, though not so powerful, effect was obtained by placing infants in a slowly rotating drum. (Because tactile stimulation causes bladder emptying in young pups, we tested whether unstimulated pups were hyperactive because they had full bladders. Separated pups were hyperactive whether bladder emptying had been elicited during separation or not).

What do these results mean? They show that when tactile stimulation, which has the immediate effect of eliciting activity, is repeated, it has the cumulative long-term effect of reducing the pups' level of behavioral reactivity. The results can be said to be consistent with the hypothesis that levels of behavioral arousal of the infant are regulated in part by the levels of tactile stimulation delivered by the mother. A long-term quieting effect from tactile and other forms of stimulation provided to the human infant has been described by Brackbill (1971). Korner and Thoman (1970) have pointed out how effective vestibular stimulation is in calming a crying baby. Cultures differ in how human babies are stimulated, but there appears to be a widespread intuitive understanding of the necessity for regular stimulation in order to avoid fussiness in infants. Such processes may generalize between rats and man.

4.2.3. Vestibular Stimulation

In the course of most early parent–infant interaction, the infant receives intense episodic vestibular stimulation by being picked up and carried and by position changes during nursing, clinging, and following. For example, the infant rat nurses while upside down; at any other time, such a position elicits vigorous righting responses. The infant monkey receives much vestibular stimulation by clinging to a highly mobile, gymnastic mother.

We did not find that either rocking or being turned repeatedly upside down served to prevent or attenuate the hyperactivity of separation in young rats (Hofer, 1975a). However, Thoman and Korner (1971) have shown that vestibular stimulation of infant rats in a rotation apparatus for 10 min each day, had both the acute effect of quieting infants and the long-term effect of increasing exploratory behavior and body weight at 20 days of age, in comparison with similarly handled and swaddled infants which were not rotated.

The most dramatic example of the impact of vestibular stimulation on early development has been reported by Mason and Berkson (1974). One of the consequences of maternal deprivation beginning in the neonatal period in rhesus monkeys is the appearance of stereotyped body rocking, reaching a peak at about 6 months of age, and persisting at high levels thereafter. Similar body rocking is characteristic of human infants who are severely retarded, blind, or autistic and also occurs in some otherwise normal children (Kravitz and Boehm, 1971). This behavior is generally considered to reflect affective arousal and distress in the infant. Among nonhuman primates, macaques, chimpanzees, baboons, geladas, and gibbons apparently show this behavior abnormality and most often as a consequence of early maternal deprivation, whether reared with an artificial surrogate mother or not. But these artificial surrogates were not mobile. Mason and Berkson gave rhesus infants a standard terry cloth surrogate suspended on a wire so it could swing when jumped on by the infant and furthermore gave the surrogate independent mobility by having it moved on the end of its wire in a circle within the cage at irregular intervals during most of the day. The infants spent the same amount of time clinging to this mobile surrogate as to the standard stationary one, but the mobility of the surrogate completely prevented all self-rocking in these infants. Two other self-directed behaviors, self-clasping and sucking, were *not* affected by the mobility of the surrogate, nor was locomotion or distress vocalization. Infants of mobile surrogates spent slightly less total time in contact with the surrogates after the first 3 months, but spent much more time in rough and tumble play with their surrogates than infants with stationary surrogates. Even after the surrogates were permanently removed at 1 year of age, no rocking appeared in the infants previously housed with mobile surrogates. In later tests with novel environments and strange intruder animals, the monkeys raised on mobile surrogates reacted with less timidity, distress vocalization, or extremes of locomotor activity, and with less self-biting and self-rocking. At 4–5 years of age, in tests of social and sexual responsiveness, mobile surrogate-reared monkeys were less emotionally aroused and more responsive to their partners, and benefited more from socializing experience than did stationary surrogate-reared monkeys.

The implication of these results is that vestibular stimulation of the infant by the mother in the course of their social relationship has effects on the development of a variety of behavior patterns. This form of stimulation, as delivered in the normal mother–infant interaction, may function to channel the development of motor behavior into other patterns than self-rocking. In the abscence of vestibular stimulation and the regulatory effects it has on motor behavior, the behavior of self-rocking emerges as early as 1 month postnatally and becomes characteristic and habitual in the juvenile and young adult. In this example, the regulatory function of the mother's stimulation is upon *qualitative* features of the infant's behavior and upon the direction of development, in addition to effects upon levels of behavior as in the previous examples in this section.

4.3. Regulation of Rhythmic Functions of the Infant

Rhythm is such an obvious quality of early social relationships that it is surprising to realize how little we know of its influence on the developing young. In part, this is due to the prevailing view that biological clocks, as the word suggests, are autonomous regulatory mechanisms and to the implication that their maturation is exclusively determined by genes. Of course, one of the major characteristics of circadian and ultradian systems in adults is their entrainment by environmental events, including temperature change, nutrient intake, and social interactions, as well as the more familiar role of light. Entrainment, however, involves only a fine tuning or setting of the biological clocks: other characteristics such as amplitude, main direction, pattern, and period can only be altered slightly if at all by the various entraining stimuli.

We know very little about the extent to which events which occur during very early development may shape major characteristics of the infants' biological clocks, rather than simply acting to trigger or to set them running. The synchronous dancelike interactions of human infants and their mothers were mentioned in the introduction. It is not yet clear who is controlling whom in these play sequences, but Stern's finding (1974) that it is the infant rather than the mother who breaks eye-to-eye contact most of the time (94%) indicates how we cannot assume the infant to be merely a passive recipient of maternal influences. Stern has found that the infant does not alter overall looking rate, but seems to change the distribution of looks during half-hour or hourly segments of time in response to cues from the mother. Similarly, the daily amount of wakefulness seems to be the same under different caretaking (Sander *et al.*, 1972) or different feeding schedules (Gaensbauer and Emde, 1973), but the distribution (e.g., day–night, prefeeding–postfeeding) is regulated by the caretaker through picking the infant up, playing, etc., or by feeding.

These studies with humans could not disclose the nature of the regulation and are only correlative. Studies in rats by Levin and Stern (1975) identified a pattern of predominantly diurnal weight gain in the young rat up to about day 17 which depended upon (1) a mother with an intact visual system and (2) access by the mother to food during the night. Young of blinded mothers had no clear day–night periodicity in weight gain, and young of mothers fed only during the day had reversed cycles (e.g., gained more weight at night). After day 19, normal infants shifted to solid food as the major source of nutrient and to the normal adult rhythm of nocturnal feeding, even in the absence of the mother. Blind infants shifted to a nocturnal pattern at about the same time but only if their mothers were sighted and had a similar rhythm. As soon as the mother was removed from the cage, they reverted to an eating pattern without clear day–night differentiation. Thus, the mother is the source of the diurnal pattern of weight gain in the infant rat prior to 17 days but only continues to be a major determinant of the pattern of food intake of her young after 19 days if the young lack their own visual sensory capacity.

We have little data on long-range developmental consequences of perturbations in early rhythmic functions of the parent in the early social relationship. Sander *et al.* (1972) did find a later effect of rearing human neonates in a hospital nursery for 10 days, an experience which did not allow establishment of the normal day–night activity rhythm during that time. During days 11–25, the direction of the effect of the first 10-day experience on activity rhythm was opposite in females to that in males. There is some evidence in animal work that increased stimulation in early development hastens the development of day–night circadian differences in adrenocortical secretion (Ader, 1969) and that prenatal handling of pregnant female rats hastens development of the infants' day–night activity cycle (Ader and Deitchman, 1970). This latter evidence only indirectly supports our notion, however, since the intermediary process has never been positively identified as originating in the parent–infant social relationship itself, rather than being a delayed result of hormonal change in the pregnant female, acting transplacentally.

More recently, Condon and Sander (1974) have discovered an unexpected relationship between parents' speech rhythms and the rhythm of neonatal limb movements, by using frame-by-frame microanalytic techniques. If an infant was in an awake state, and if it heard human speech, even in a foreign language, the points of change of its limb movements were found to synchronize closely with sound segments of speech. Use of a tape recorder ruled out a role of the adult in promoting synchrony, and the failure of infants to synchronize to isolated vowel or tapping sounds indicated a specificity to the natural rhythms of human speech. Close synchrony was observed as early as 2 days of age. It is worth noting that this relationship was revealed by separating the infant from its mother and presenting key stimuli artificially in her absence, as in some of the experiments above.

The implications of this study are far-reaching, not only for the origins of language, but also for the development of nonverbal communication and subtler aspects of human communication such as empathy. The regulating action of the parents' speech on infant motor rhythms was hidden from ordinary observation and was revealed only by new "microscopic" techniques for behavior analysis and by analytic experiments with distorted and artificial stimuli.

The rhythmic nature of the interactions taking place during the early parent–infant relationship remains almost unexplored territory. It would therefore appear to be a particularly fruitful area for further research on the interactions of parental care with infant development.

4.4. Regulation of the Onset of Puberty

Factors controlling the rate of sexual maturation have generally included nutritional, climatic, genetic, and social factors. Recent evidence, however, suggests a specific role for fathers and like-sexed siblings as well as mothers in the determination of the onset of puberty in young female mice. Vandenbergh (1969) found that the presence of adult males during the preweaning and early juvenile periods hastened the development of sexual maturity as assessed by vaginal opening, uterine weights, and onset of estrus. The degree of effect was not trivial, since maturation was shifted more than 8 days earlier than its usual time of occurrence at 34 days after birth. Conversely, if female mice were raised after weaning in groups composed only of female siblings or adult females, their sexual maturation was delayed more than 30 days later than singly housed females (Drickamer, 1974).

In both cases, these effects were traced to the operation of a pheromone from the urine of the animals with which the developing prepubertal female was in contact. Vandenbergh *et al.* (1976) has isolated and partially characterized the male acceleratory pheromone which he found to be androgen-dependent. It does not appear to be a steroid metabolite of testosterone, however, but rather a polypeptide, and he found that the pheromone itself releases luteinizing hormone from the female pituitary. The resulting early surge of estrogen production by the juvenile female's ovary appears to be responsible for early vaginal opening and onset of estrus. The finding that the pheromone is a polypeptide with these physiological actions suggests that it may indeed be identical to luteinizing-hormone-releasing hormone (LHRH), a hypothalamic hormone which regulates the pituitary output of luteinizing hormone. The pheromone would thus function as an ectohormone, being absorbed via the olfactory mucosa of the young female. This is of course speculation, but not inconsistent with results so far.

The pheromone produced by groups of females living exclusively together which delays sexual maturation has been found by Drickamer to be produced by prepubertal as well as adult females, is contained in their urine, and is added to it during micturition by urethal glands, since it is not present in bladder urine. Indeed, urine painted daily on the nose of a juvenile female will do as well as the presence of females. But the urine must be from females which are allowed full behavioral interaction with each other. Groups of females separated from each other by wire mesh partitions do not produce the pheromone.

The point that emerges from these studies is that the father and the siblings as well as the mother can exert a continuing long-term regulatory effect on the maturation of physiological and structural systems underlying the timing of puberty in the mouse. This is a particularly satisfying example of developmental regulation by members of the infant's family because the processes responsible for the effect have been worked out. Some evidence exists that similar effects occur in species other than rodents, but a clear understanding of these kinds of developmental influences in other mammals is yet to be gained.

4.5. Regulation of Infant Autonomic Function

The infant's behavior and its hormonal state have been shown to be regulated by relatively specific aspects of parental interaction in the foregoing examples. This section will present evidence that autonomic regulation of the infant's cardiovascular system is under the control of the mother–infant interaction through the level of milk she supplies to the infant's stomach. Here we have a form of regulation exerted by the mother over the infant's interoceptive system acting to regulate the neural balance of internal homeostatic control.

It would not be surprising to find that a mother with insufficient milk production caused physiological changes in her offspring. But once sufficient milk is provided to allow weight gain, it is not generally supposed that the level of physiological functions are delicately regulated by the amount of milk provided, within a range of normally occurring weight gains. In this role, the mother would function as an external physiological regulatory agent controlling the behavior of internal systems in her offspring. Such a regulatory function would be more interesting if the nutrient acted through the central and autonomic nervous systems rather than through alterations in the circulatory supply of metabolic substrate. And this is what we found.

The demonstration of this relationship depended upon separating the mother and the 2-week-old infant rat, finding that there was a 40% decrease in cardiac rate and that this, in turn, was the result of a marked reduction in sympathetic cardiac tone (Hofer and Weiner, 1971). When this reduction in rate persisted unchanged, despite maintenance of normal body temperature and return of the infants to nonlactating "maternal" females, our attention turned to a nutritional mechanism. At first it might appear to be a nonspecific debilitative effect of the marked weight loss sustained by these infants, except that their hearts were capable of beating at normal rates if the animals were simply stimulated by tail pinching. Then we found that the heart rate decline persisted, even if they were fed every 4 hr by gastric intubation so that they gained a small amount of weight over the 24 hr the mother was gone (Hofer, 1970). This showed that the low heart rates were not the result of starvation. But the mother provides milk even more frequently (every 1–2 hr) at this age. A series of systematic studies with graded amounts of feeding by stomach tube demonstrated that cardiac rate was delicately tuned to the amount of milk given within the normal range of weight gain for infant rats (Hofer, 1973b). If enough milk was given by tube to produce as much weight gain as a group of mothered infants, then heart rates of separated infants remained at the level of the mothered infants.

Variations ordinarily occur in the amount of weight gained each day by mothered infants. If the mother is disturbed or the litter size is increased, weight gain does not occur or is reversed. Other factors, such as reduction of litter size, seem to promote weight gain. Heart rates follow these fluctuations in the weight gain of mothered infants and illustrate the cumulative action of repeated nursing bouts resulting in long-term regulation of the infant's cardiac rate by the amount of milk supplied over time by the mother.

A series of physiological studies (Hofer and Weiner, 1975) showed that this effect of nutrient on heart rate was most likely mediated by spinal sympathetic pathways and the β-adrenergic receptors on the efferent side. Since lactose and amino acids were only effective if administered intragastrically and not intravenously, the receptor of the afferent mechanism appeared to originate in the gut wall. Simple gastric distention, various gastrointestinal hormones, and the afferent vagus have been ruled out, but afferent mesenteric sympathetic nerves (Sharma and Nasset, 1962) remain a possible pathway by which the brain is informed of the amount of nutrient in the gut.

The heart rates of infant rats (and humans) follow a developmental course characterized by a small rise in the first days (or months), followed by a plateau of high heart rates during midinfancy, which gives way to a slow decline during preadolescence. The high range of heart rates during infancy are the result of an initial high sympathetic tone and the subsequent decline, the result of the gradual establishment of predominant vagal restraint (Hofer, 1974). These developmental stages and transitions had been assumed to be the result of maturation of central neural homeostatic systems, and the set point at any age assumed to be an intrinsic neural function, probably genetically programmed, and heavily buffered from environmental influences. In our studies, we find that the age-characteristic level of heart rate is the result of the infant's nutritional relationship with its mother and can be delicately tuned, over periods of a few hours, by variations in the amount of milk the mother provides. An hypothesis outlining the possible adaptive value of this arrangement for physiological growth processes has been proposed (Hofer, 1978).

This unexpected regulatory phenomenon of the early social relationship in the rat was revealed by the use of mother–infant separation, followed by analytic studies based on a concept of the relationship as a regulator, rather than on the concept of attachment and social bond formation. To have inferred that the low heart rates were a reflection of an emotional state precipitated by disruption of the attachment bond would have been wrong and could have obscured other processes from view.

4.6. Regulation of Brain Enzymes and Growth

Ordinarily we think of the parents as facilitating the physical growth of their children through the food they provide. But clinical experience (Powell *et al.*, 1967) has demonstrated that certain infants "fail to thrive" and are well below growth norms while at home and even when their feeding is supervised in the hospital. Pediatricians have found that if a nurse is assigned to special duty with that infant and spends a good deal of time playing with it, holding it, and generally establishing a relationship, the child will grow dramatically. Often the home environment in these cases is chaotic and the mother withdrawn and ineffective. Abnormal growth hormone responses have been found in these infants, but it has been thought that these were secondary to the malnutrition, not a primary cause of the retarded growth.

In a series of recent papers, Schanberg and co-workers (Butler *et al.*, 1978; Kuhn *et al.*, 1978) have shown that growth hormone levels in infant rats drop rapidly after separation from the mother, as do levels of a brain enzyme essential for the synthesis of brain proteins, ornithine decarboxylase (ODC). To their surprise, they found that decreased ODC levels were not related to nutritional or body temperature changes and could be prevented by supplying the pups with a mother incapable of supplying milk because of mammary duct ligation. Furthermore, the fall in ODC was not prevented by a passive (anesthetized) mother, even though she supplied milk. A drug which blocks release of growth hormone in the infant rat, cyproheptadine, produced falls in ODC similar to those following separation, and administration of growth hormone, intracisternally, increased ODC. The fall in growth hormone and ODC following removal of the mother was very rapid (15–20 min), indicating a sensitive, time-related process.

Apparently some aspect of the pups' interaction with an active mother maintains the infants' growth hormone at their normal high levels. Growth hormone, in turn, induces the enzyme ODC, which determines the levels of synthesis of brain proteins. When the mother is absent, even for a few minutes, growth hormone and ODC levels fall. This sensitive regulation of brain and body growth processes by active interaction with the mother provides an understanding of the process which may underlie so-called emotional dwarfism in human infants and other examples of the dependence of maturation of brain and behavior on a sustained, effective parent–infant relationship. Clearly such a conclusion must be tentative at present, and we need to know more. Particularly interesting will be the search for the vital ingredient of the active interaction with the mother which is necessary to keep up the growth hormone levels of the pup.

Since an active interaction with the mother also seems to be the essential factor in the regulation of infant behavioral arousal described in the first part of this section (4.2), the processes involved in the regulation of growth hormone, the brain enzyme ODC, and behavioral arousal may share some common features.

5. Implications for Aberrant Development

The emphasis throughout this chapter has been on how the elements of the parental relationship influence particular aspects of the infant's behavior and specific physiological or biochemical systems. The influences have been termed regulatory because they seem to control the level, the rhythm, and the patterning of developing systems in the infant. Examples have been given of pronounced long-term effects resulting from alterations in parental behavior, or in the capacity of infants to perceive and respond to a particular modality of stimulation. But studies on "process" in infancy and on "outcome" in adulthood have tended to remain separate. There are not yet any research accounts which begin with a careful and systematic analysis of a single parent–infant regulatory system and show how a given specific aberration in this regulation becomes translated, throughout later developmental experience, into a particular long-term effect on the adult's interaction with its environment.

Another hiatus in our understanding concerns the compensatory or self-correcting influences which tend to return development to a species-specific "normal" course after it has been displaced by some distortion or disruption of the parent–infant interaction at an earlier stage. An understanding of these processes and the environmental factors necessary to support them are of obvious importance if we are to derive therapeutic implications from this line of work.

One of the most-studied parental effects on the young has been the infant's response to loss of this relationship, because of its immediate and dramatic effects. Generally the infant's response has been viewed in terms of stress psychophysiology. The realization that the parent provides important forms of regulation to the infant suggests a revision of how we regard the event of parental loss in early life, and the attention to self-correcting and environmental compensatory influences following separation may help explain the variability in long-term outcome following early parent loss. A discussion of these two topics forms the body of this section.

5.1. Maternal Separation Revisited

The assumption that early mother–infant separation is stressful and traumatic, the profound abnormalities in behavioral development which have been found to result from prolonged maternal deprivation (Spitz, 1945), and the recent findings that mothers are also adversely affected by separation from their infants (Leifer *et al.*, 1972) have directed concepts of separation into the framework of stress psychophysiology. This focus on the *response to separation* has distracted us from considering what may be going on *during* the *previous mother–infant relationship* and from realizing that knowledge about this relationship may help explain some of the responses to separation.

At present, the behavioral and physiological responses to mother–infant separation are generally viewed as related to attachment and to the inferred stress of disrupting such a strong social "bond." An emotional distress response is supposed to ensue with behavioral and physiological components as in classical psychophysiological responses to imposed threat. But (in Section 4) I have described physiological and behavioral responses to separation which do not fit this model, and would now like to examine the concept of attachment and the response to separation from a different point of view (for further discussion of attachment see Gubernick, this volume).

In altricial mammals, the growth and formation of attachment appears to take place slowly, throughout a rather prolonged sensitive period. Initially consisting only of primitive biological approach tendencies (e.g., thermotaxis), the attraction becomes more and more specifically directed. Coincidently, there develops the reaction to separation, which usually consists of behaviors characteristic of emotional distress, as if the animal were in physical pain or threatened with harm. The intensity of this separation response is often taken as a measure of the strength of attachment. These events have been difficult to explain in terms of learning theory, since attachment does not depend on standard reinforcing agents (Bowlby, 1969, 1973; Harlow, 1958; Scott, 1962). Attachment is then viewed as a primary drive, and words are used which convey the sense of an unusual process of stamping in (e.g., "imprinting") or metaphors are used, such as "the bond," which convey a sense of the enduring character of some social attachments and of the traumatic impact of separation.

The formation of attachment develops in the young coincidently with the developmental regulatory processes described in Section 4 above. May not the formation of attachment and the expression of separation distress be related to these regulatory processes? The concept of long-term regulatory effects produced by repeated discrete stimuli is not commonly used in discussing behavioral interactions (for an exception, see Schleidt, 1973) but is, after all, the fundamental process by which neurons communicate and may be the mechanism for much of what the brain does (Kuffler and Nichols, 1976). Trains of repeated nerve impulses along the axon of one neuron impinge on another, altering its membrane characteristics so that it is more or less likely to be fired by an impulse from a third neuron. Aggregates of neurons show facilitation, recruitment, and inhibitory states as a result of repeated stimuli along certain pathways. Trains of stimuli deplete neurotransmitter stores, increase transmitter synthesis rates, and even alter the position and number of receptor sites, trans-synaptically. Such cellular processes presumably underlie the formation of attachment. The analogy for separation arises when some part of this neural traffic suddenly stops: the neurons which had previously received this repetitive input change functionally, due to the unbalanced force of all other regulatory inputs. A transient unopposed activation or inhibition is the usual result. Depending on the mechanisms involved, the shift to stable levels and patterns of activity may take seconds, hours, or days to be completed. Cellular processes known as "disuse supersensitivity" and "disuse potentiation" have been described (Sharpless, 1975) which offer a potential model at the cellular level for the behavioral phenomena following sudden maternal separation. We can thus see that basic neural processes known to operate in relatively simple neural systems have characteristics which are strikingly similar to the characteristics of the changes we have observed following abrupt termination of early social relationships. This analogy does not of course imply identity of underlying mechanisms. Other processes such as the induction of enzymes and the gradual buildup of hormonal levels provide other possible mechanisms for tonic regulatory action and withdrawal responses following social separation.

It is not yet clear exactly how the stimulation provided by the mother promotes the formation of attachment, but the following steps can be outlined. In birds which show imprinting, motion of the object or flashing of a light appears to be sufficient to elicit following. In altricial animals, elicitation of approach, clinging, and following may depend on a series of stimuli in a developmental progression depending on maturation of sensory capacities (Rosenblatt, 1971). These stimuli become gradually combined into a highly specific complex or gestalt. Stimulation from these sources appears to be reinforcing at appropriate ages, eliciting approach and maintaining proximity. This stimulation also has the long-term cumulative effect of lowering emotional arousal and the behavior associated with it. Removal of the mother withdraws the regulatory effect of the repetitive stimulation on the emotional state of the young, and the classical separation distress ensues. Thus, one of the enduring effects of early repetitive sensory stimulation by the parent is to establish and maintain the goal-directed system of behavior which we call attachment. The regulatory nature of attachment has been described by

Bowlby (1969, 1973), and a theoretical model for the elicitation of distress behavior upon withdrawal of the attachment object has been put forward by Hoffman and Solomon (1974), supported by Hoffman's data on imprinting in ducklings.

If we now regard attachment as a regulatory process, established and governed by repetitive specific stimulation, then the classical separation responses of vocalization, locomotor hyperactivity, even stereotyped abnormal behavior, can be viewed as "withdrawal" or release phenomena, analogous to the withdrawal response of narcotic addicts after separation from their repeated drug injections. By this line of reasoning, the results of separation of an infant (of any species) from its mother (at any developmental age) will be in part a function of the tonic or cumulative effects of the stimulation the infant has received from the interaction with its mother.

Thus, the younger the infant, the more completely the responses to separation from the mother are due to withdrawal of the tonic physiological and behavioral regulatory processes inherent in the early mother–infant relationship (see Section 4 above). The older the offspring and the more independent its existence from the parents, the more likely separation responses are to be due to psychological processes such as attachment, operating at higher integrative levels and involving specific perceptual schema, differentiated emotional states, and the learned capacity to respond to signals or symbols for events. The only way to distinguish the kind of process involved in a given change following separation is by testing discrete forms of stimulation, delivered repeatedly after maternal separation, for their capacity to prevent and/or reverse a specific aspect of the response, as in the experiments described in Section 4 above. If these are not effective, if the infant can be shown to be responding with an integrated pattern of responses, and if the response pattern can be prevented only by the mother's proximity, it is likely to be the more complex form of process.

An example which illustrates the need for this distinction in interpretation is the work of Reite *et al.* (1974). Using physiological telemetry, these investigators found that the 6-month-old pig-tailed macaque infant sustained marked decreases in heart rate and body temperature following the initial agitation after their mothers were removed from the colony pen. They also showed a profound loss of REM sleep. The low heart rates and body temperatures occurred both during slow-wave sleep and while the infant was awake in the characteristic immobile hunched posture which has given the infant's response the label "depression." These physiological findings in infant monkeys are identical to those we had previously found in infant rats. But are they the result of similar processes? If heat were supplied, would the immobility and the hunched posture (so typical of behavioral thermal conservation) persist? Could the low heart rates be selectively reversed by providing normal quantities of nutrient? (Weight loss and food intake data were not collected in these studies.) We had also observed a marked decrease in REM sleep in the separated infant rat, but have not yet uncovered its mechanism. It has been argued that the 6-month-old infant monkey, which shows clear-cut, specific attachment to its mother, has matured from the "biological" into the "psychological" phase of its relationship with its mother. Therefore the response to separation should be understood as an integrated psychobiological response termed "depression" or "conservation withdrawal." But it is probably best not to rest with an assumption on this point but to test for the extent to which the complex of responses can be explained by simpler biological processes.

That both kinds of processes can occur in the same individual, if the infant is of an intermediate age, is shown by some recent work in our laboratory (Hofer and Shair, 1978). We recorded the (ultrasonic) vocalizations of 2-week-old rat pups in their home cages and found that they were quiet unless physically disturbed or separated from their littermates and mother. However, even if a pup was apparently asleep in its home cage nest, removal of its mother and littermates resulted in almost immediate awakening and a vocal "outcry" consisting of short, ultrasonic pulses averaging 12/min and lasting for as long as 30 min. When the familiar environment was also absent, a single pup vocalized at about 25 "cries" per minute. But if a single familiar littermate, or the mother, was placed with them in the unfamiliar test box, the test pups' vocalizations were reduced 80–90%. Contact was maintained and vocalizations reduced almost as well with a passive (anesthetized) littermate. The sensory properties of the familiar companion (e.g., texture, odor, contour) were found to be *additive,* not dominated by one sensory modality, and the pup was not at all "comforted" (reduced ultrasound, maintained body contact) by a warm, plastic, suasage-shaped model, or the mere odor of the home cage shavings.

The behaviors of these pups in these experiments have all the hallmarks of the kind of separation response which is dependent on the prior formation of some higher integrative process which we call attachment. The vocalization is immediate, persistent, cannot be prevented by providing one or another sensory modality from the previous relationship, and seems only to respond completely to the presence of a familiar companion from the home cage. The response does not appear to be specific to the mother in this species at this age. This is not too surprising, since the mother rat leaves her young 15–20 times in the course of each day for an average of 40–50 min at a time. The littermates have come to act as parent substitutes, and equal attachment seems to have been formed to them.

The experiments described in Section 4 above, which uncovered separation responses due to withdrawal of maternal regulatory processes, were done on pups of the same age, 2 postnatal weeks. Thus, separation responses based on both regulatory process withdrawal and attachment disruption can be shown to occur in young at the same age. In terms of stages in the developing parent–infant relationship, such an age occupies an intermediate position between the early postnatal stage of biological dominance and the later stage, when psychological processes predominate (see Galef, this volume, for other age-related changes).

It is important to realize that both the simpler biological processes and the more complex psychological ones can be viewed as regulatory: the former involving separate, parallel regulation of several independent biological and behavioral systems, and the latter involving a single integrated biological and behavioral state regulated by a combination and pattern of sensory cues. In the later stages of development, we call the integrated response pattern "emotional" and we can think of the relationship as regulating the emotional state of the infant (and of the parents).

The advantage of this formulation is not only that it allows separation experiments to be used to learn more about the effects of parent–infant interaction, but also that it permits us to seek some understanding of the slower-developing effects of maternal separation (Reite *et al.*, 1974; Kaufman and Rosenblum, 1969; Seay *et al.*, 1962; Spitz, 1946) as the result of withdrawal of specific regulatory actions previously provided by the mother, rather than as some form of prolonged, complex stress response. It allows us to separate out each developmental effect of separation in relation to its specific mechanism, rather than viewing the phenomenon globally as an "emotional stress," "anaclitic depression," or "grief" response, terms which are convenient but do not lead to further questions or new levels of understanding.

5.2. Self-Correcting Tendencies

The cumulative evidence of all the examples given above might lead one to view the development of the infant as entirely at the mercy of the vagaries of the parent–infant relationship. But there appear to be "backup" systems which take over when the parent is deficient, or the infant's sensory systems are impaired, or the mother–infant interaction is altered in some way. For example, the offspring of mouse-killing mother rats (see above, Section 2.2), although showing little mouse killing as adolescents after being raised by non-mouse-killing foster mothers, began to kill mice when they reached young adulthood. The genes were beginning to show. There is a strong element of the determination of developmental outcome which expresses itself by tending to return the course of development to a given trajectory. Waddington (1957), writing of embryological structural development, used the term "creode" to convey the sense of a structure of change over time within the interaction of many units. The self-stabilizing nature of the creode tends to restore the course of development toward a given outcome following disturbances, and provides alternative developmental pathways within the host of available possibilities.

Other examples of the self-correcting nature of aberrant development are given in a later publication from Harlow's lab (Ruppenthal *et al.*, 1976) reporting that the abusive and neglectful behavior of the original motherless females toward their newborn infants was slowly corrected following subsequent pregnancies and experience as mothers until, by the fourth time they were mothers, they were virtually indistinguishable from normals. Contact with peers during adolescence also greatly reduced the probability of inadequate maternal care. Likewise, the profound social deficits of maternally deprived young monkeys were permanent only in an ordinary social group setting. If these previously isolated juveniles were exposed to much younger infants which were not yet fearful and still had strong tendencies to establish close physical contact with anything furry, they could be socially rehabilitated (Suomi and Harlow, 1976). Such encounters are ordinarily prevented by adults in the social group. Additionally, age-mates could not interact with the isolates because of the isolates' inappropriate behavior. The less sophisticated younger monkeys were less specific in their requirements, were much less likely to respond aggressively, and provided the interaction necessary to enable socialization of the isolates.

For the infant rats which showed profound changes in heart rate, and failed to eat after separation from the mother (see Section 4 above), if they were only kept warm, after 3 or 4 days they showed a dramatic recovery and physiological measures returned rapidly to normal without return of the mother (Hofer, 1975b). Heart rate had indeed been regulated by the milk supplied by the mother, but the infant was capable of regulating its own heart rate, first by the food it ate for itself and then later by the development of intrinsic central regulation of the autonomic system. The infant's own self-correcting systems had reestablished physiological control at or near normal levels for juvenile rats. The only residual effect was a slightly higher than normal resting heart rate. By 30 days of age it was only after restraint stress that a thermoregulatory defect (Ackerman et al., 1978), an abnormal sleep pattern and the increased susceptibility to gastric erosions (Ackerman et al., 1975) became evident.

Thus, as in the experiment with short separations of infant rhesus monkeys by Hinde (Section 2.4 above), the effects of the aberrant early parenting became imperceptible a few weeks after reunion under ordinary conditions of groups living. Only in the situation involving exposure to an unfamiliar and stressful environment or unfamiliar companions did the effects of the early separation become evident.

The conclusion, then, is that there are many different routes to the same developmental outcome, that some routes differ from others in that they endow the animal with a latent difference in the form of an altered response to superimposed adaptive challenges. At present, the precise nature of this latent or potential difference must be determined separately for each system and for each early developmental route as well as for each adaptive challenge. No principles with general predictive power are available as yet.

It is clear that even severe deficits can be ameliorated by the proper reparative experience. This should be an encouragement to therapists. However, the most intractable deficits are those in which an early aberration is compounded by a series of later experiences in which the self-correcting potential of the system is thwarted. For example, in the case of the blind infant (see above, Section 4.1), secondary effects sometimes follow from the lack of facial expressiveness. Some mothers respond to the lack of "feedback" by reducing attempts to interact or by neglecting or even abusing the infant (Fraiberg, 1974). Naturally, these altered interactions in turn compound the original effects by depriving the infant of needed sensory stimulation in other modalities, and of the opportunity for compensatory avenues of communication, as well as by the imposition of physical and nutritional abuse. Unless some other communicative system develops to compensate (hand signals, spoken language), the original deficit becomes compounded as one problem leads to another and the experience in the developing child becomes abnormal in different ways at each subsequent stage in development. The resulting severe defects in adulthood, then, have multiple causes, which were, however, set in motion by a deficit in a single sensory system.

With optimal parental flexibility, on the other hand, the development of other communicative pathways allows normal development of all systems, except those depending directly on vision, and encourages the development of heightened skills in other systems which allow the adult to live a normal life.

6. Conclusion

The examples given in this chapter indicate the far-reaching developmental effects which can be demonstrated to result from the action of some of the regulatory processes within early social relationships. These formative influences begin earlier than the learning processes ordinarily described as the parent's major contribution to a child's development. They involve processes which are distinct from learning and at the same time go considerably beyond the mere provision of sufficient nutrient and warmth to permit maturation. They include some of the processes underlying Harlow's classic demonstration (1958) of the vital role played by social interactions in the development of "affectional" systems, but go beyond a single behavior category and specify a number of regulatory processes which affect physiological as well as behavioral development.

Apparently nature, in the form of evolutionary processes, has seen fit to utilize the early parent–infant interaction as a powerful regulator of early development. Mounting evidence that the basic regulators of development are not all within the infant organism is consistent with the intuitive belief that the early environment of the infant is somehow critical for determining outcome. The survival value of this evolutionary development is likely to be the additional adaptability that is possible with this opportunity for cross-generational transmission of information. Events affecting parents prior to conception, for example, in response to a changing ecology, are theoretically capable of affecting basic physiological and behavioral traits of the young, through modification of the parent's role in the early social relationship.

Finally, the existence of these early extrinsic regulatory processes may shed light on the biological function of the intense attachment so characteristic of infants and parents. The arrangement of having important regulators of development within the parent-infant interaction makes it imperative that the infant and parent are highly motivated to maintain a close relationship during this phase of development. The maladaptive consequences of lack of this early developmental regulation may have been an important selection pressure leading to the evolution of attachment and separation distress, in addition to the protection from predators conveyed by social bonds at all ages.

ACKNOWLEDGMENTS. The author's research was supported by a Research Scientist Award (MH-38632) and a project grant (MH-16929) from the National Institutes of Mental Health.

References

Ackerman, S. H., Hofer, M. A., and Weiner, H., 1975, Age at maternal separation and gastric erosion susceptibility in the rat, *Psychosom. Med.* **37**:180–184.

Ackerman, S. H., Hofer, M. A., and Weiner, H., 1978a, Early maternal separation increases gastric ulcer risk in rats by producing a latent thermoregulatory disturbance, *Science* **201**:373–376.

Ackerman, S. H., Hofer, M. A., and Weiner, H., 1978b, The predisposition to gastric erosions in the rat: Behavioral and nutritional effects of early maternal separation. *Gastroenterology* **75**:649–654.

Ader, R. 1969, Early experiences accelerate maturation of the 24 hour adrenocortical rhythm, *Science* **163**:1225–1226.

Ader, R., and Deitchman, R., 1970, Effects of prenatal maternal handling on the maturation of rhythmic processes, *J. Comp. Physiol. Psychol.* **71**:492–496.

Akiskal, H. S., and McKinney, W. T., 1973, Depression disorders: Toward a unified hypothesis, *Science* **182**:20–29.

Barcroft, J., 1938, *The Brain and Its Environment,* Yale University Press, New Haven, Conn.

Becker, G., and Kowall, M., 1977, Crucial role of the postnatal maternal environment in the expression of prenatal stress effects in the male rat, *J. Comp. Physiol. Psychol.* **91**:1432–1446.

Bowlby, J., 1969, *Attachment and Loss,* Vol. 1, *Attachment,* Basic Books, New York.

Bowlby, J., 1973, *Attachment and Loss,* Vol. 2, *Separation,* Basic Books, New York.

Brackbill, Y., 1971, Effects of continuous stimulation on arousal levels in infants, *Child Devp.* **41**:17-26.

Bradley, R. M., and Mistretta, C. M., 1975, Fetal sensory receptors, *Physiol. Rev.* **55**:352.-382.

Broadhurst, P. L., 1961, Analysis of maternal effects in the inheritance of behavior, *Anim. Behav.* **9**:129–141.

Butler, S. R., Suskind, M. R., and Schanberg, S. M., 1968, Maternal behavior as a regulator of polyamine biosynthesis in brain and heart of the developing rat pup, *Science* **199**:445–447.

Condon, W. S., and Sander, L. W., 1974, Neonate movement is synchronized with adult speech: Interactional participation and language acquisition, *Science* **183**:99–101.

Dahlhof, L.-G., Hard, E., and Larsson, K., 1978, Sexual differentiation of offspring of mothers treated with cortisone during pregnancy, *Physiol. Behav.* **21**:673–674.

deFries, J. C., 1964, Prenatal maternal stress in mice: Differential effects on behavior, *J. Hered.* **55**:289–295.

Denenberg, V. H., and Whimbey, A. E., 1963, Behavior of adult rats is modified by the experiences their mothers had as infants, *Science* **142**:1192–1193.

Denenberg, V. H., Hudgens, G. A., and Zarrow, M. Z., 1964, Mice reared with rats: Modification of behavior by early experience with another species, *Science* **143**: 380–381.

Denenberg, V. H., Paschke, R., Zarrow, M. X., and Rosenberg, K. M., 1969*a*, Mice reared with rats: Elimination of odors, vision, and audition as significant stimulus sources, *Dev. Psychobiol.* **2**:26–28.

Denenberg, V. H., Rosenberg, K. M., and Zarrow, M. X., 1969*b*, Mice reared with rat aunts: Effects in adulthood upon plasma corticosterone and open-field activity, *Physiol. Behav.* **4**:705–707.

Drickamer, L. C., 1974, Sexual maturation of female house mice: Social inhibition, *Dev. Psychobiol.* **7**:257–265.

Erway, L., Hurley, L. S., and Fraser, A., 1966, Neurological defect: Manganese in phenocopy and prevention of a genetic abnormality of inner ear, *Science* **152**:1766–1769.

Fairfield, J., 1948, Effects of cold on infant rats: Body temperatures, oxygen consumption, electrocardiograms, *Am. J. Physiol.* **155**:355–365.

Flandera, V., and Novakova, V., 1974, Effect of mother on the development of aggressive behavior in rats, *Dev. Psychobiol.* **8**:49–54.

Fraiberg, S., 1974, Blind infants and their mothers: An examination of the sign system, in: *The Effect of the Infant on Its Care Giver* (M. Lewis and L. A. Rosenblum, eds.), *Origins of Behavior,* Vol. 1, pp. 215–232, Wiley, New York.

Gaensbauer, T. J., and Emde, R. N., 1973, Wakefulness and feeding in human newborns, *Arch. Gen. Psychiatry* **28**:894–897.

Gesell, A., and Amatruda, C. S., 1945, *The Embryology of Behavior,* Harper, New York.

Ginsberg, B., and Allee, W. C., 1942, Some effects of conditioning on social dominance and subordination in inbred strains of mice, *Physiol. Zool.* **25**:485–506.

Gruenwald, P., (ed.), 1975, *The Placenta and Its Maternal Supply Line,* University Park Press, Baltimore.

Harlow, H. F., 1958, The nature of love, *Am. Psychol.* **12**:673–685.

Hinde, R. A., and Spencer-Booth, Y., 1971, Effects of brief separations from mother on Rhesus monkeys, *Science* **173**:111–118.

Hofer, M. A., 1970, Physiological responses of infant rats to separation from their mothers, *Science* **168**:871–873.

Hofer, M. A., 1973*a*, The effects of brief maternal separations on behavior and heart rate of two week old rat pups, *Physiol. Behav.* **10**:423–427.

Hofer, M. A., 1973*b*, The role of nutrition in the physiological and behavioral effects of early maternal separation on infant rats, *Psychosom. Med.* **35**:350–359.

Hofer, M. A. 1973*c*, Maternal separation affects infant rats' behavior, *Behav. Biol.* **9**:629–633.

Hofer, M. A., 1974, The role of early experience in the development of autonomic regulation, in: *The Limbic and Autonomic Nervous System: Advances in Research* (L. DiCara, ed.), pp. 195–221, Plenum Press, New York.

Hofer, M. A., 1975*a*, Studies on how early maternal separation produces behavioral change in young rats, *Psychosom. Med.* **37**:245–264.

Hofer, M. A., 1975*b*, Survival and recovery of physiologic functions after early maternal separation in rats, *Physiol. Behav.* **15**:475–480.

Hofer, M. A., 1978, Hidden regulatory processes in early social relationships, in: *Perspectives in Ethology*, Vol. 3 (P. P. G. Bateson and P. H. Klopfer, eds.), pp. 135–163, Plenum Press, New York.

Hofer, M. A., and Weiner, H., 1971, The development and mechanisms of cardiorespiratory responses to maternal deprivation in rat pups, *Psychosom. Med.* **33**:353–362.

Hofer, M. A., and Weiner, H., 1975, Physiological mechanisms for cardiac control by nutritional intake after early maternal separation in the young rat, *Psychosom. Med.* **37**:8–24.

Hofer, M. A., and Shair, H., 1978, Ultrasonic vocalization during social interaction and isolation in 2 week old rats, *Dev. Psychobiol.* **11**:495–504.

Hoffman, H. S., and Solomon, R. L., 1974, An opponent-process theory of motivation. III. Some affective dynamics in imprinting, *Learn. Motiv.* **5**:149–164.

Hutchings, D. E., 1978, Behavioral teratology: Embryopathic and behavioral effects of drugs during pregnancy, in: *Early Influences* Studies on the Development of Behavior, Vol. 4, (G. Gottlieb, ed.), pp. 7–34, Academic Press, New York.

Kaufman, I. C., and Rosenblum, L. A., 1969, Effects of separation from mother on the emotional behavior of infant monkeys, *Ann. N.Y. Acad.Sci.* **159**:681–695.

Kaufman, I. C., and Stynes, A. J., 1978, Depression can be induced in a Bonnet Macaque infant, *Psychosom. Med.* **40**:71–75.

Korner, A. F., and Thoman, E. B., 1970, Visual alertness in neonates as evoked by maternal care, *J. Exp. Child Psychol.* **10**:67–68.

Kravitz, H., and Boehm, J. J., 1971, The effects of institutionalization on development of stereotyped and social behaviors in mental defectives, *Child Dev.* **42**:399–413.

Kuffler, S., and Nichols, J. G., 1976, *From Neuron to Brain*, Sinaver, Sunderland, Mass.

Kuhn, C. M., Butler, S. R., and Schanberg, S. M., 1978, Selective depression of serum growth hormone during maternal deprivation in rat pups, *Science* **201**:1034–1036.

Lee, M. H. S., and Williams, D. I., 1974, Longterm changes in nest condition and pup grouping following handling of rat litters, *Dev. Psychobiol.* **8**:91–95.

Leifer, A., Leiderman, P. H., Barnett, C., and Williams, J., 1972, Effects of mother–infant separation on maternal attachment behavior, *Child. Dev.* **43**:1203–1218.

Levin, R., and Stern, J. M., 1975, Maternal influences on ontogeny of suckling and feeding rhythms in the rat, *J. Comp. Physiol. Psychol.* **89**:711–721.

Levine, S., and Thoman, E. B., 1969, Physiological and behavioral consequences of postnatal maternal stress in rats, *Physiol. Behav.* **4**:139–142.

Lewis, J. K., McKinney, W. T., Young, L. D., and Kraemer, G. W., 1976, Mother infant separation in rhesus monkeys as a model of human depression—A reconsideration. *Arch. Gen. Psychiatry* **33**:699–705.

Lieberman, M., 1963, Early developmental stress and later behavior, *Science* **141**:824–825.

Liley, A. W., 1972, Disorders of amniotic fluid, in: *Fetal Placental Disorders* (W. S. Assali and C. R. Brinkman, eds.), Pathophysiology of Gestation, Vol. 2, pp. 157–206, Academic Press, New York.

Lund, R. D., 1978, *Development and Plasticity of the Brain*, Oxford University Press, New York.

Malson, L., 1972, *Wolf Children and the Problem of Human Nature*, Monthly Review Press, New York.

Mason, W. A., and Berkson, G., 1974, Effects of maternal mobility on the development of rocking and other behaviors in Rhesus monkeys: A study with artificial mothers, *Dev. Psychobiol.* **8**:197–211.

Money, J., and Ehrhardt, A. A., 1972, *Man and Woman—Boy and Girl,* Johns Hopkins University Press, Baltimore.

Neal, M. V., 1968, Vestibular stimulation and developmental behavior of the small premature infant, *Nurs. Res. Rep.* **3**:1–5.

Powell, G. F., Brasel, J. A., and Blizzard, R. M., 1967, Emotional deprivation and growth retardation simulating idiopathic hypopituitarism, *N. Engl. J. Med.* **276**:1271–1283.

Reading, A. J., 1966, Effect of maternal environment on the behavior of inbred mice. *J. Comp. Physiol. Psychol.* **62**:437–440.

Reite, M., Kaufman, I. C., Pauley, J. D., and Stynes, A. J., 1974, Depression in infant monkeys: Physiological correlates, *Psychosom. Med.* **36**:363–367.

Ressler, R. H., 1963, Genotype correlated parental influences in two strains of mice, *J. Comp. Physiol. Psychol.* **56**:882–886.

Rosenblatt, J. S., 1971, Suckling and home orientation in the kitten: A comparative developmental study, in: *The Biopsychology of Development* (E. Tobach, L. Aronson, and E. Shaw, eds.), pp. 345–410, Academic Press, New York.

Ruppenthal, G. C., Arling, G. L., Harlow, H. F., Sackett, G. P., and Suomi, S. J., 1976, A 10-year perspective of motherless-mother monkey behavior, *J. Abnorm. Psychol.* **85**:341–349.

Russell, P. A., 1971, "Infantile stimulation" in rodents: A consideration of possible mechanisms, *Psychol. Bull.* **75**:192–202.

Salk, L., 1962, Mother's heartbeat as an imprinting stimulus, *Trans. N.Y. Acad. Sci.* **24**:753.

Sander, L. W., Julia, H. L., Stechler, G., and Burns, P., 1972, Continuous 24 hour interactional monitoring of infants reared in two caretaking environments, *Psychosom. Med.* **34**:270–282.

Schleidt, W. M., 1973, Tonic communication: Continual effects of discrete signs in animal communication systems, *J. Theor. Biol.* **42**:359–386.

Scott, J. P., 1962, Critical periods in behavioral development, *Science* **183**:949–958.

Scott, J. P., and Fuller, J. L., 1965, *Genetics and Social Behavior of the Dog,* University of Chicago Press, Chicago.

Seay, B., Hansen, E. W., and Harlow, H. F., 1962, Mother–infant separation in monkeys, *Child Psychol. Psychiatry* **3**:123–132.

Seitz, P. F. D., 1959, Infantile experience and adult behavior in animal subjects. II. Age of separation from the mother and adult behavior in the cat, *Psychosom. Med.* **21**:353–378.

Sharma, K. N., and Nasset, E. S., 1962, Electrical activity in mesenteric nerves after perfusion of gut lumen, *Am. J. Physiol.* **202**:725–730.

Sharpless, S. K., 1975, Disuse supersensitivity, in: *The Developmental Neuropsychology of Sensory Deprivation* (A. H. Riesen, ed.), pp. 125–146, Academic Press, New York.

Skolnick, N. J., Ackerman, S. H., Hofer, M. A., and Weiner, H., 1980, Vertical transmission of acquired ulcer susceptibility in the rat, *Science* **208**:1161–1162.

Smith, D. J., Joffe, J. M., Heseltine, G. F. D., 1975, Modification of prenatal stress effects in rats by adrenalectomy dexamethasone and chlorpromazine, *Physiol. Behav.* **15**:461–469.

Solkoff, N., Yaffe, S., Weintraub, D., and Blase, B., 1969, Effects of handling on the subsequent development of premature infants, *Dev. Psychol.* **1**:765–768.

Sontag, L. W., 1966, Implications of fetal behavior and environment for adult personalities. *Ann. N.Y. Acad. Sci.* **132**:782–786.

Southwick, C. H., 1968, Effect of maternal environment on aggressive behavior in inbred mice, *Comm. Behav. Biol.* **1**:129–132.

Spitz, R. A., 1945, Hospitalism: An enquiry into psychiatric conditions in early childhood, *Psychoanal. Study Child* **1**:53–80.

Spitz, R. N., 1946, Anaclitic depression, *Psychoanal. Study Child* **2**:313–347.

Sterman, M. B., 1972, The basic rest activity cycle and sleep, in: *Sleep and the Maturing Nervous System* (D. Clemente, D. P. Purpura, and F. E. Mayer, eds.), pp. 175–195, Academic Press, New York.

Stern, D. A., 1974, Mother and infant at play: The dyadic interaction involving facial, vocal and gaze behavior, in: *Origins of Behavior,* Vol. 1, *The Effect of the Infant on Its Care Giver* (M. Lewis and L. Rosenblum, eds.), pp. 187–214, Wiley, New York.

Stone, E., Bonnet, K., and Hofer, M. A., 1976, Survival and development of maternally deprived rats: Role of body temperature, *Psychosom. Med.* **38**:242–249.

Suomi, S. J., and Harlow, H. F., 1976, Social rehabilitation of separation-induced depressive disorders in monkeys, *Am. J. Psychiatry* **133**:1279–1285.

Thoman, E. B., and Arnold, W. J., 1968, Effects of incubator rearing with social deprivation on maternal behavior in rats, *J. Comp. Physiol. Psychol.* **65**:441–446.

Thoman, E. B., and Korner, A. F., 1971, Effects of vestibular stimulation on the behavior and development of infant rats, *Dev. Psychol.* **5**:92–98.

Thompson, W. R., Watson, J., and Charlesworth, W. R., 1962, The effects of prenatal maternal stress on offspring behavior in rats, *Psychol. Monogr.* **76**:1–26.

Vandenbergh, J. C., 1969, Male odor accelerates female sexual maturation in mice, *Endocrinology* **84**:658–660.

Vandenbergh, J. C., Finlayson, J. S., Dobrogosz, W. J., Dills, S. S., and Kost, T. A., 1976, Chromatographic separation of puberty accelerating pheromone from male mouse urine, *Biol. Reprod.* **15**:260–265.

Waddington, C. H., 1957, *The Strategy of the Genes,* Macmillan, New York.

Walker, D., Grimwade, J., and Wood, C., 1971, Intrauterine noise: A component of the fetal environment, *Am. J. Obstet. Gynecol.* **109**:91–95.

Ward, I. L., 1972, Prenatal stress feminizes and demasculinizes the behavior of males, *Science* **175**:82–84.

Wehmer, F., Porter, R. H., and Scales, B., 1970, Prenatal stress influences the behavior of subsequent generations. *Comm. Behav. Biol.* **5**:1–4.

4

Offspring Effects upon Parents

Lawrence V. Harper

1. Introduction

The idea that offspring affect the behavior of their caregivers is not new. Early concepts of instinct portrayed the activities of animals as mechanical reactions to external stimulation—including stimuli emanating from the young; the Darwinian revolution simply provided a secular explanation for their origins.

Frequent allusions have been made in the literature to the "adaptiveness" of mutual stimulus exchanges between parent and offspring, but the bulk of experimental research relevant to understanding the ways in which mammalian young affect caregiver behavior has focused on isolating the stimulus–response relationships involved and/or or the physiological mechanisms underlying them. The findings are generally what one would expect from the broad outlines of evolutionary theory. However, until recently, there were few links between models of natural selection and the dynamics of caregiver responsiveness.

2. Phylogenesis: Inclusive Fitness and Parental Investment

Although the ideas relevant to our concerns were first considered by Fisher (1930) and Haldane (1932; cited by Dobzhansky, 1970), theoretical models of parent–offspring relations have been elaborated only within the last decade (see Dawkins, 1976, for review).

2.1. Inclusive Fitness and the Cost of Parental Investment

Current conceptions of selection posit that fitness is determined by the relative proportion of one's genes which appear in subsequent generations. Direct trans-

LAWRENCE V. HARPER • Department of Applied Behavioral Sciences, University of California, Davis, Davis, California 95616.

mission is not necessary; it can pay to facilitate the survival of others who share the same genes (Hamilton, 1964). So long as an animal behaves in ways that ensure its (shared) genes are most efficiently perpetuated, it is enhancing its "inclusive fitness." Trivers (1974) has postulated that an individual has limited (metabolic, etc.) resources available to devote to perpetuating copies of its genes, and that each expenditure has its "cost"—it reduces the organism's remaining capital. Thus, animals should strive to "invest" their resources as efficiently as possible.

Although the costs cannot yet be defined or measured in exact terms, parenting does have its price. The increased metabolic demands created by lactation or the need to provide food for still-dependent weanlings require more intensive and extensive foraging effort, and attempts to meet these demands frequently are accompanied by increased risk of debilitation or predation. It seems that there may be a general trade-off between longevity and fecundity in severe habitats: Nursing mother marmots are relatively less able to accumulate sufficient fat stores to tide them over long periods of hibernation (Armitage and Downhower, 1974), and among wild Himalayan sheep and goats, barren females or those who lose their offspring early are more likely to survive periods of extreme weather (Schaller, 1977).

Reductions in an animal's ability to provide for subsequent young can be illustrated by studies of lactation in domestic cows. If two of the four quarters of the cow's mammary gland are milked regularly through a pregnancy, they will produce less milk than will the unused quarters during the subsequent, postpartum lactation (Smith *et al.,* 1967). Moreover, whereas animals with low milk yield utilize only dietary energy, high-yield cows also convert body tissue in milk production (Hart *et al.,* 1978). As yet, however, there are no generally accepted metrics for comparing the cost of parenting across situations or species (see also Kleiman and Malcom, this volume).

2.1.1. Identifying Beneficiaries

Selection should favor tendencies to restrict investment to those who share genes with their benefactors; however, because not all genes can lead to distinctive phenotypes, individuals should use indirect indices for recognition. The most obvious index of shared genes is kinship, and the clearest evidence of kin is maternity (Alexander and Borgia, 1978; Bertram, 1976; Dawkins, 1976). A prime example of such recognition would be the mutual attachment between mother and young. Insofar as it is usually the mother who becomes attached first, the data are in accord with theoretical expectation.

Field studies indicate that among primates (Hrdy, 1976; Drickamer, 1976), ungulates (Lent, 1974), and elephants (I. Douglas-Hamilton and O. Douglas-Hamilton, 1975) enduring bonds of kinship exist within the maternal lineage. Moreover, naturalistic observations suggest that in some primate groups there even may be a general recognition of parenthood, or at least, "rights of owner-ship": In a captive group of Patas monkeys, a low-ranking mother was able to retrieve her infant from even the alpha female, and when her infant's playmate became too rough, the mother would threaten the offender's *mother* (Hall and Mayer, 1967). Among captive guereza monkeys who "shared" young, a mother was most successful in retrieving her infant (Horwich and Wurman, 1978); and in a group of free-ranging rhesus monkeys, several instances were recorded in which an animal returned an infant to its mother, despite having acquired it from another individual (Breuggeman, 1973). Even among lactating laboratory rats, who are generally responsive to any young, mothers can discriminate their own from the offspring of other females (Beach and Jaynes, 1956a).

Assessing paternity is much more difficult (Trivers, 1972). However, among animals who live in stable groups, males who could have sired offspring should be solicitous of group young. Such seems to be the case in a number of species (see Kleiman and Malcom, this volume).

In short, although selection favors responsiveness to young, the display of solicitude should be discriminating. The model renders intelligible the apparently "maladaptive" tendency of females to attack and/or devour conspecific offspring from alien groups (Gandelman and Davis, 1973; Kruuk, 1972; Rowell, 1960; Schaller, 1972) and even their own groups (Goodall, 1977).

2.1.2. Maximizing Returns

The economic model indicates that animals should invest efficiently; they should cut their losses if short-run costs are so great that continued care of present young threaten to reduce their aggregate reproduction. Thus, it also accounts for what might seem to be paradoxial cases of parental neglect (e.g., Schaller, 1972).

There comes a time when, as Dawkins (1976) put it, "caring and bearing" must be balanced against one another if the maximum number of replicates is to be produced. While a minimum investment is necessary to ensure the survival and reproduction of current offspring, a point of diminishing returns will be encoun-tered. When this happens, parents can enhance fitness by redirecting their remain-ing resources toward others, usually additional offspring.

2.1.2a. Conflict. According to Trivers (1974), reinvestment of resources can be a source of parent–offspring conflict. He argues that, because young share only about one-half of their genes with (either of) their parent(s), it will be to their advantage to extract more investment from caregivers than the latter "should" provide. Selection would favor offspring who most efficiently demand more than their share and parents who can effectively estimate when they have done just enough (as judged in terms of their remaining reproductive potential). Because there exist a variety of ways in which parental care may benefit offspring, and since the needs of the young often change at different rates, there may be a number of periods of conflict within any one caregiving cycle.

The idea that conflict occurs when parents choose to redirect their caregiving energies accords fairly well with existing data. In a number of species, weaning seems to begin later than one would expect in terms of the offspring's abilities to obtain and utilize the adult diet (e.g., Langman, 1977). The "weaning tantrums" of primate young (e.g., Jay, 1963), and the mutual exchanges of "threat-faces" and "swats" between mother lions and their weaning-age cubs (Schaller, 1972) also lend credence to this conception. In general, it seems that young mammals will nurse, if given the opportunity, well after normal weaning age (I. Douglas-Hamilton and O. Douglas-Hamilton, 1975; Lent, 1974; Nash, 1978; Tyler, 1972); that they will accept any opportunity to suck—including others' dams (Autenreith and Fichter 1975; LeBoeuf et al., 1972; Le Neindre et al., 1978); and that older offspring may interfere with the nursing of younger siblings (Kleiman, 1972; Lancaster, 1971).

In this connection, the theory also can accommodate observations of juvenile harassment when mother chimpanzees (van Lawick-Goodall, 1971) and langurs (Hrdy, 1977) are mating. By preventing their mothers from conceiving, the juveniles might be attempting to garner more maternal investment than they could if they had to compete with a younger sibling.

2.1.2b.   Exploitation.   On the other hand, Alexander (1974) has argued that parental care evolved to benefit the *parent's* fitness: Genes which lead to intersibling conflicts detrimental to the mother's overall fitness should not spread, whether or not they enhance the survival of individual offspring, because these same genes will appear in the "conflictor's" own progeny, rendering such "selfish" individuals less successful than their (altruistic) siblings. Furthermore, because parents are more powerful and control resources that their offspring need, they are in a position to enforce offspring compliance.

Alexander does acknowledge that, in sexually reproducing species, parents and individual offspring will not have identical (reproductive) interests, but he feels that, if anything, offspring should evolve to allow parents to "win." He agrees with Trivers (1974) that parents should be responsive to, and dependent upon, offspring signals when the young are in a better position to judge their own needs, but differs with Trivers by implying that selection should favor diminution of either offspring signals or parental responsiveness whenever further investment would be detrimental to the parents' fitness. Rather than conflict, in the sense of a mutual coercion or competition, Alexander emphasizes exploitation of offspring in favor of parental fitness.

There are examples of adults benefiting from the behavior of young. Menzel (1966) described how the curiosity of young Japanese monkeys could provide adults with information about the potential danger posed by unfamiliar objects, and others have documented instances in which primate young have come directly to the aid of their mothers during disputes (Marsden, 1968; van Lawick-Goodall, 1971; Wilson and Vessey, 1968), or when danger threatened (Bramblett, 1973).

Within the confines of caregiving, Alexander (1974) outlined several ways in which parents could serve their own fitness at the expense of some of their young.

One such tactic was to "ration" care to maximize the number of surviving offspring. Recently, P. Klopfer and M. Klopfer (1977) have shown how mother goats respond to the differences in activity level in first- and second-born twin kids so that the stronger, more active firstborn neonate does not get more than its "fair share" of her milk. In accord with the exploitation viewpoint, the firstborn's signals *deny* it an advantage.

Another example of limiting investment to maximize returns might be suckling-induced delayed implantation or development of conceptuses (Banks, 1967; Beniest-Noirot, 1958; Elwood and Broom, 1978; Renfree, 1979; Sharman, 1967). Presumably, births are spaced to allow mothers a chance to "prepare" for the next litter. In gerbils (Elwood and Broom, 1978) and rats (Veomett and Daniel, 1975), females who mated postpartum and whose current litters were nursing showed fewer implantation sites than mothers whose litters were removed. In postpartum-mated rats, if the nursing demands of the current litter were intensified (by increasing litter size) within the first 8 or 9 days after postpartum estrus, implanted embryos were resorbed (Veomett and Daniel, 1975). Insofar as these phenomena depend upon physiological mechanisms *in the mother,* the burden of proof seems to lie with those seeking to explain them as examples of sibling competition, contrary to maternal interest.

One might also consider the mother's tendency to cease nursing a current offspring prior to the birth of subsequent young as an example of rationing resources—especially when mothers who do not give birth continue to nurse yearlings (Lent, 1974; Tyler, 1972). Similarly, the prolongation of maternal protection when no younger siblings are present (Pola and Snowdon, 1975) suggests a conditional investment. Although disputes between mothers and offspring at the time of weaning could be indications of opposing interests, another interpretation is possible: Offspring typically attempt to nurse indiscriminately, and whenever they are indulged by nonrelated or distantly related kin, they and their mother will benefit. It is thus in the mother's interest for her offspring to get "extra rations" from foster parents whenever, and *as long as,* possible. If so, what appears at first glance as a purely selfish offspring strategy could be a side effect of tactics that benefit mothers.

A second exploitative tactic suggested by Alexander was for the mother, under conditions of scarcity, to kill and eat some of her own young, or to feed them to littermates. Recent investigations of cannibalism in mice (Gandelman and Simon, 1977) and hamsters (Day and Galef, 1977) indicate that such behavior keeps litters at or below a "typical" size for that mother. Presumably, the number of pups that a mother will tolerate is approximately the maximum that she can rear so that, on average, the greatest number will survive to reproduce.

Alexander's final tactic calls for parents to have their offspring help care for additional young. Among primates, it is quite common for young members of a troop to show interest in and solicitousness for infants. Where genealogies are known, siblings frequently are the most attentive to infants (Breuggeman, 1973; Ingram, 1977; Klopfer and Dugard, 1976; Schessler and Nash, 1977; van Lawick-

Goodall, 1968). Sibling care also occurs in jackals (H. and J. van Lawick-Goodall, 1971), mongooses (Rasa, 1977), and elephants (I. Douglas-Hamilton and O. Douglas-Hamilton, 1975). Sometimes, juvenile siblings account for a substantial amount of the care received by infants (Box, 1975, 1977; Epple, 1975; Vogt *et al.*, 1978; Bekoff, this volume).

Observations of a captive group of long-tailed macaques indicate yet another way in which young may be exploited for parental gain. A subordinate mother may use her older, less physically vulnerable offspring as "scapegoats" in order to protect her other youngsters from dominant animals' moderate aggression (Chance *et al.*, 1977*b*).

2.1.3. Current Status

At this writing, it is difficult to predict when or where one should expect conflict. On the one hand, although mothers of most species may wean their elder offspring prior to giving birth to subsequent young, weanlings sometimes attempt to "sneak" milk during a subsequent lactation, even to the detriment of their younger siblings (Harper, 1976; Kleiman, 1972; P. H. Klopfer and M. S. Klopfer, 1970; Schaller, 1972). On the other hand, young rats may "wean themselves": A developmental shift in offspring metabolism favors a spontaneous transition from mother's milk to adult diet at about the same time that the mother would give birth to another litter (Yeh and Moog, 1974).

Most existing data do not allow for unambiguous interpretations. Some primate juveniles harass older infant siblings; although this behavior could be "jealousy," or an attempt to gain more maternal attention, it also may increase maternal fitness: To the extent that sibling interference is limited to weaning-age infants, and insofar as mothers remain anovulatory during lactation, speeding up the weaning process hastens the advent of additional young. Such behavior thus could increase the mother's reproductive output and free her from the full burden of discouraging the weanling's nursing attempts. At the same time, because the mother's reproductive output is increased, the interfering sibling would increase its own inclusive fitness (Kurland, 1977).

Theory posits that under certain circumstances, it would be advantageous for the parent to abandon weaker offspring—and even for siblings to sacrifice themselves! The focal offspring should be the least likely of a litter to mature and reproduce (O'Connor, 1978). As yet, we have no unambiguous evidence concerning the cues that lead a parent with several offspring to systematically favor certain individuals. Although Gandelman and Simon (1977) did find that the majority of mouse pups who were cannibalized were below median litter weight, Day and Galef (1977) could not identify weight or apparent vigor as cues distinguishing pups who were chosen as victims for cannibalism among hamsters. Similarly, Deets and Harlow (1974) found that although captive, lactating rhesus monkey mothers (who normally give birth to singletons) were initially responsive to both of two fostered infants one was eventually favored. They too could not determine what characteristics set apart infants who were accepted from those who were rejected.

Both Alexander (1974) and Trivers (1974) agree that when parents are dependent upon offspring cues to signal the need for care they should evolve ways of detecting "cheaters" who attempt to exploit caregiver sensitivities. The unresolved issue is how they do it, and what sorts of outcomes ensue. Although both suggest that offspring should be willing to accept milk from any source so long as they do not endanger themselves, neither predicts that young ground squirrels would behave in such a way (submissively) that unrelated, potential foster mothers reject them (Michener, 1974). At present, applications of theory to parent–offspring relations are the subject of considerable debate—one that will become even more complex when attempts are made to accommodate the fact that where homozygosity is not uncommon, the degree of relatedness of kin can be asymmetrical (Flesness, 1978).

2.2. Sex Differences in Parental Responsiveness

Because investment by male mammals in sperm is relatively insignificant compared with that of females in eggs and parental care, the major concern of males, besides obtaining as many mates as possible, should be to avoid being cuckolded. For their part, females should accept only males who would sire fit offspring and, whenever possible, provide caregiving in return for sexual favors (Trivers, 1972). Corresponding to these fundamental differences in strategy, theory predicts differences in parental responsiveness to, and the effects on caregiving of, offspring stimuli.

2.2.1. Male Infanticide and Defense

A male can be exploited by his mate if he allows himself to invest in offspring sired by other males. Furthermore, in many mammals, females who are actively caring for offspring do not become sexually receptive as soon as those who have lost or weaned their young. Thus, it can be to a male's advantage to exterminate others' offspring so that he can mate sooner and, in addition, not waste energy in protecting unrelated youngsters. Although there is debate concerning the frequency of such behavior (Makwana, 1979), male infanticide has been observed in lions (Bertram, 1976), and changes in male leadership in langurs have been followed by the loss of all young, apparently due to male infanticide (Hrdy, 1977; Wolf and Fleagle, 1977). Additional studies of guereza monkeys (Horwich and Wurman, 1978; Oates, 1977), red-tailed monkeys (Struhsaker, 1977), and chimpanzees (Nishida *et al.*, 1979) suggest that male infanticide may also occur in these species. As for rodents, Mallory and Brooks (1978) showed how "stranger" male collared lemmings may gain a reproductive advantage by killing litters sired by other males.

On the other hand, there is also clear evidence that stimulus control of male defensive behavior can be exerted by infant-specific stimuli. In a field study of vervet, brazza, and sykes monkeys, Booth (1962) sometimes carried hand-reared infants into the field with her. When she was accompanied by young whose coat color was typical of newborns, but not when she had older infants with adultlike pelage, mature animals threatened at close range. Indeed, in one instance, a male vervet approached within arm's reach. Impromptu experiments with stuffed hides indicated that the defense reaction could be released only by the natal coat—when in motion.

Not only do primate males defend young (and their mothers) against predators, in some species they also tend to assist mother-infant pairs within the group, e.g., captive colobus monkeys (Emerson, 1973) and feral baboons (DeVore, 1963). In theory, given the uncertainties surrounding parturition, and with females in varying phases of pregnancy and lactation, the optimal strategy for the male would be to invest most where a live birth (of presumably his own offspring) has already occurred rather than risk resources in defense of barren or gravid females (Judge and Rodman, 1976). The infants' presence can thus cue males' proximity seeking and defending behavior. However, as yet, we know relatively little about the situational and other stimulus conditions that determine whether a male will display solicitude or hostility toward young.

2.2.2. Sharing Duties

Males of a number of species actively participate in caring for their own, and sometimes even foster, young. Among laboratory-reared gerbils (Elwood, 1975), rats (Rosenblatt, 1967; Wiesner and Sheard, 1933), and mice (Beniest-Noirot, 1958) males can be induced to become responsive by (continued) exposure to "maximally attractive" pups. The range of paternal behaviors in these species largely overlaps—if it does not completely mimic—the full complement of maternal activities (see also Kleiman and Malcolm, this volume).

A similarly broad spectrum of parental responsiveness has been seen in some old-world monkeys. Although males regularly display solicitude, it often is restricted to only a few members of the group and/or appears under specific conditions (Alexander, 1970; Breuggeman, 1973; Itani, 1959; Taylor et al., 1978). In a free-ranging troop of stump-tailed macaques, infants under 6 months of age (who are distinctively colored) were particularly singled out for care by males. Moreover, juvenile males paid more attention to their infant siblings than to the newborns of other mothers, indicating that responsiveness not only depended upon general, age-related stimuli, but also may have been moderated by more subtle cues facilitating individual recognition (Estrada and Sandoval, 1977).

One must be cautious in assuming that the bases for caregiving behavior are the same for both sexes. Among captive pairs of Mongolian gerbils, males did display the "same" responses as females; however, the relationships between caregiving behaviors and the number and age of offspring differed according to caregiver sex (Elwood and Broom, 1978). A study of parental solicitude of mated pairs of house mice also revealed sex differences in the degree to which the parents reacted to experimentally controlled pup ultrasounds, and the changes in their responsiveness over days (Bell and Little, 1978). Thus, despite morphologically similar motor patterns, there may be important sex differences in offspring control of caregiving behavior.

3. Function: Social Consequences and Ecological Correlates of Parenting

Analyses of the possible evolutionary origins of species characteristics are plagued by at least two basic uncertainties : The first and most obvious is that one can never be certain of events that occurred in the past. The second, related uncertainty stems from the fact that once a characteristic has evolved it may come to serve new functions bearing little or no relationship to the kinds of selective pressures which originally shaped it. Species characteristics represent the resultant of diverse selective forces; rather than being unambiguously beneficial, many adaptations represent trade-offs between the advantages and disadvantages of behaving in a particular way (Klopfer, this volume). Moreover, across species, behaviors, or other characters may appear similar and yet serve very different ends (Clutton-Brock and Harvey, 1976).

3.1. Kin, the "Con," and Contractual Obligations

One way to evaluate the functional significance of a behavior is to consider who benefits from it.

3.1.1. Kin

Theory predicts that adults (and older siblings) should usually be responsive to signals indicating need emitted by their own or closely related offspring. Although there are occasional instances in which such solicitude can "go overboard" (Taylor *et al.*, 1978), phenomena such as adoptions of infants and juveniles within troops of primates sometimes do benefit parent–offspring or sibling pairs (Hrdy, 1976).

Among those species in which parenting is exclusive to own offspring, benefits are clearly in accord with theoretical expectations. The most obvious uncertainties involve the apparent paradoxes of abandonment or infanticide. Like abortion and fetal resorption, these phenomena may be explained as attempts to cut losses and/or to optimize future investment possibilities (Bertram, 1976; Dawkins, 1976). However, in the absence of a clear metric for quantifying investment, such reasoning must remain speculative.

More difficult problems arise in the interpretation of pup-controlled tendencies for mother mice and rats to attack unfamiliar male intruders (Erskine *et al.*, 1978; Svare and Gandleman, 1973; see Svare, this volume). In the case of rats, even when the intruder can confine himself to a separate compartment unfamiliar to the female, and away from the litter, there is no correlation between maternal attack frequency or intensity and the amount of time the intruder spends in the nest, or even the frequency of pup killing (Erskine *et al.*, 1978). In order for us to understand the functional significance of behavior such as this, our interpretive context will have to integrate factors relating to the social organization and habitat typical of the species (see Section 3.2).

3.1.1a. Siblings, Aunts, and Grannies.   Individuals other than the mother provide "care" for young in one or more representatives of 7 of the 11 orders of mammals (Spencer-Booth, 1970). Among primates, nulliparous females predominate in this extramaternal solicitousness. In accord with the concept of inclusive fitness, in both primates (Hrdy, 1976) and elephants (I. Douglas-Hamilton and O. Douglas-Hamilton, 1975) where genealogies are reasonably well known, close relatives are involved most frequently in caring for another female's offspring. Instances have been reported in which grandmothers (Breuggeman, 1973; I. Douglas-Hamilton and O. Douglas-Hamilton, 1975), as well as sisters and siblings, have assumed care for orphaned young.

Weanlings may occasionally benefit from foster care by older siblings (van Lawick-Goodall, 1968), but sibling care is not an unmixed blessing. Although sibling interest sometimes may be exploited to reduce the parents' burden (e.g., Box, 1975; Section 2.2.3b above), or simply tolerated (Cheney, 1978; Trollope and Blurton-Jones, 1975), among some species (Hunt *et al.*, 1978; Murray and Murdoch, 1977), mothers appear to actively "protect" their neonates from being handled by their older offspring. Juvenile overtures toward newborns in these species are rejected by mothers until the infants are able to help themselves. Even among colobus monkeys, who are unusually relaxed about allowing their very young offspring to be handled by others, mothers tend to monitor their infants closely and retrieve them if they signal distress (Horwich and Manski, 1975; Oates, 1977).

In some cases, maternal vigilance is amply justified. Despite their mothers' tolerance, juvenile guereza monkey females often seem "inept" when they attempt to carry their infant siblings (Horwich and Manski, 1975). Oates (1977) observed one instance in which a guereza infant was dropped from a tree to the forest floor and reported another, fatal case, observed by Struhsaker, in which an infant fell from a handler's body and was not retrieved. These data indicate not only that the attractiveness of infants can cause a mother to be pestered by inquisitive offspring, but that it can be disadvantageous to the extent that juvenile clumsiness and/or distractability, and nonmaternal adults' inattentiveness endanger the life of the young (see also McKenna, this volume). Indeed, the bulk of alloparenting among some Japanese macaques is performed by non-kin (Kurland, 1977).

Nevertheless, a number of benefits could accrue to mothers who allow their offspring to be handled by other females (Hrdy 1976). First, there is the possibility the young will receive care from an "aunt" should the mother become incapacitated. The evidence for this is mixed; on the one hand, Oates (1977) did not observe any fostering in over 1500 hr of observation of guerezas, and Coe *et al.* (1978) found that laboratory-reared infant squirrel monkeys' distress was not markedly reduced by the presence of an aunt during the absence of their mothers. On the other hand, among captive animals, Breuggeman (1973) did report one case in which a lactating grandmother rhesus adopted and reared an infant after its mother (her daughter) had died, and Bernstein (1975) observed another, similar adoption in geladas, although the relationship between the infant and the foster parent was not reported.

A second benefit that might be derived from baby sharing among monkeys is an increase in aunts' tendencies to defend young against infanticidal males. In this connection, it may be significant that both offspring transfer and infanticide during male takeovers have been observed in *Colobus* species (Hrdy, 1977; Oates, 1977). To the extent that interacting with an attractive infant can "induce" (cf. Section 5.2.2c) protectiveness in nonlactating females, this derivative might outweigh the possibility of accidental injury (Oates, 1977; Struhsaker, 1977). Hrdy (1976) also suggests that infant transfers can facilitate maternal mobility. For example, an arboreal primate mother's chances of avoiding a terrestrial predator may be enhanced if she can descend to a water hole unencumbered by a clinging infant. Although lemurs (Richard, 1976) and vervets (Lancaster, 1971) apparently do leave their young with "babysitters" while foraging, we have no basis for estimating how much this increases their ability to flee danger or secure food.

The most unambiguous evidence for the beneficial effects of juvenile care comes from a study of the effects of sibling "helpers" on litter survival in black-backed jackals. In this species, as in most group-living canids, all members of the pack help to feed the lactating mother and her pups. A comparison of the number of pups surviving to weaning with the number of adults at the den yields a very strong, positive correlation. Each helper (in addition to the parents) adds the average equivalent of 1.5 surviving pups to the litter. Because mated pairs only succeed in rearing one pup on average, the siblings' inclusive fitness is maximized by assisting their parents (Moehlman, 1979).

In sum, although the benefits of extramaternal care by kin are not always clear, in some species, attractiveness of offspring to relatives may benefit inclusive fitness.

3.1.2. The "Con"

One of the more frequently cited ways in which young affect adults within their group is a reduction in hostile or aggressive relationships. Schenkel (1966) states that the atmosphere within prides of lions in the Nairobi National Park becomes more tolerant, if not outright "friendly," while young cubs are in the group. Even among the somewhat less solicitous lions of the Serengeti (Schaller, 1972), young cubs are treated more gently than adults, and especially when several lionesses are nursing, cubs act as a "cohesive force" within the pride. Similarly, in some primates, infants generally receive few threats or hostile responses (Chalmers, 1968) and mothers with infants are seldom attacked or otherwise menaced (Chalmers, 1968; A. Estrada and R. Estrada, 1977; Hall and Mayer, 1967; Kummer, 1968; Seyfarth, 1976; but see Dittus, 1977, for a contrasting view). One outcome of this ability of offspring to reduce the probability of attack is that mothers with small infants often enjoy the equivalent of higher dominance status (Schaller, 1963; Seyfarth, 1976) and sometimes new consort relations with other members of the group.*

   3.1.2a.   Group Cohesion.   Because the presence of offspring tends to reduce aggressiveness and the young are generally attractive, many students of primate behavior have suggested that infants play a major role in forming or maintaining intragroup relationships (Cheney, 1978; DeVore, 1963; Horwich and Wurman, 1978; Jay, 1963; Jolly, 1966). This need not be restricted to primates; Wickler (1972) points out that solicitude-inducing responses typical of immatures are commonly exploited by the evolutionary process to facilitate amicable behavior among adults.

   3.1.2b.   Passports and Protection.   Those who otherwise would be recipients of aggression often seem to capitalize upon the fact that adults are less likely to attack individuals who are caring for infants. Among several primates, low-ranking individuals adopt a caregiving or playmate role with infants prior to attempting to establish "friendly relationships" with dominant females (A. Estrada and R. Estrada, 1977; Kurland, 1977) or males (Deag and Crook, 1971; Merz, 1978) before trying to enter the core area of high-ranking individuals (Itani, 1959), or even in attempts to gain admission to a foreign troop (Poirier, 1969). In others, exploitation is blatant; males may pick up infants when conflict is likely (Deag and Crook, 1971; Gouzoules, 1975; Nash, 1978; T. W. Ransom and B. S. Ransom, 1971) and even in order to "extort" grooming (T. W. Ransom and B. S. Ransom, 1971).

   Baboons (S. A. Altmann and J. Altmann, 1970), macaques (Merz, 1978; Trollope and Blurton-Jones, 1975), gelada monkeys (Bernstein, 1975), and chimpanzees (van Lawick-Goodall, 1971) all carry dead infants. This could be in response to the care-eliciting qualities of the young, and to the extent that the infant carried is a member of the matriline (Merz, 1978), the notion of caregiver–young bonding may be applicable. However, the additional facts that the cadavers

*While captive female orangutans (Nadler and Tilford, 1977) and rhesus monkeys (Hinde and Spencer-Booth, 1967) still seem to be attracted to one another's offspring, dominance relations nevertheless are clearly maintained, and extend to the privilege of handling another's young. In captive stump-tailed macaques, increases in dominance of mothers seem to be mediated through the existence or formation of alliances with other adults (Weisbard and Goy, 1976).

often are carried "casually" (van Lawick-Goodall, 1971) and treated more roughly than live infants (Merz, 1978), and that they are carried about until badly decayed, suggest that caregiving is not the central focus. As Hrdy (1976) points out, nonparents often abandon "their" infants (even live ones) after a potential conflict has been avoided, or an alliance formed, suggesting that they can be used as a means to an end—as a shield against hostility. What we seem to be witnessing is an attempt by vulnerable individuals to capitalize upon very strong offspring effects.

3.1.2c. Limitations on Gullibility. Although offspring must be able to evoke solicitude from their kin, unrelated adults must still be selective. R. E. Lickliter (personal communication, 1979) points out that, when monotocous individuals share the same limited resources, the stimuli presented by alien young can also evoke hostile behavior. For example, insofar as sneak suckling can adversely affect a female elephant seal's ability to care for her own offspring, a direct attack on intruding alien pups may be advantageous (Le Boeuf and Briggs, 1977). Furthermore, when population pressure taxes available resources it is conceivable that females may benefit their own offsprings' chances of survival and reproduction by even attacking the young of other group members (e.g., Goodall, 1977; Kurland, 1977).

3.1.3. Contractual Obligations: Reciprocal Altruism

It can also pay to expend energy to benefit an unrelated individual if the expenditure is trivial relative to the beneficiary's gain, and if it is likely that the benefactor will be repaid. Theoretically, investment in nonkin should occur when life expectancy is long, where there are low rates of dispersal, when individuals depend upon one another for protection or other kinds of assistance, and when the existence of dominance relationships does not make the cost–benefit ratios too skewed in favor of only one party (Trivers, 1971). These conditions are often met in social mammals.

3.1.3a. Harem Formation. In primates who live in polygynous groups dominated by a single male, "bachelor" males may defend juveniles and become foci for immature playgroups. To the extent that such solicitous behavior is the basis for developing future reproductive bonds with females, (cf. Kummer, 1968), an obvious reproductive benefit accrues. Often it is not clear whether the benefactors are responding to their protegees' infantile qualities out of parentlike solicitude, or the urge to develop an alliance. In the case of gelada monkeys, the fact that males play with juveniles before and after, but not during, their tenure as harem leaders suggests a mixed motive (Bernstein, 1975).

3.1.3b. Communal Care. Several instances have been recorded in which apparently unrelated females have merged their offspring in a joint rearing effort (e.g., Porter and Doane, 1976). Interpretations of the "functions" of such behavior are hampered by the fact that mammalian social groups often center about maternal lineages (Clutton-Brock and Harvey, 1976), so that individuals who are most likely to meet one another often are related. Nevertheless, available data indicate that reciprocal altruism may play a role in the evolutionary maintenance of communal care.

Among pallid bats, roosting groups are moderately large and clusters form with immature individuals in the center. Although the young clearly are not equally related to all the adults who help shield them from heat loss, their benefit, relative to the adults' expenditure, is great. Presumably, the nonrelated, altruistic adults gain reciprocally to the extent that their own offspring receive similar benefits from other adults. Furthermore, adults can gain, as the unrelated young they benefit may provide similar services for their own grand-offspring in the future (Trune and Slobodchikoff, 1978). An extreme case of this kind seems to be found among communally roosting bats that inhabit very large caves where large numbers are required to maintain adequate warmth. In such cases, there sometimes is no individual recognition of young and females act as a "milk herd," nursing any infant (Humphrey, 1975).

Nursing of both own and unrelated young in rodents (Sayler and Salmon, 1971; Zimmerman, 1974) and carnivores (Schaller, 1972) apparently is facilitated by the young's being of roughly the same age. Although the lionesses in a pride usually are members of the same matriline, reciprocal altruism is suggested because the cubs' sires may not be related and mothers still recognize and sometimes favor their own offspring. The survival rate of the offspring who are nursed communally exceeds that expected for single litters (Bertram, 1976; Sayler and Salmon, 1971; Zimmerman, 1974). Similar advantages of communal suckling have been reported for swine (Lent, 1974).* Among gelada monkeys, females in one-male harems also occasionally nursed one another's infants (Bernstein, 1975). To the extent that these mothers were not closely related, reciprocal altruism might account for such behavior. Given fairly long tenure of the harem males' dominance and at least equal temporal stability of the harem itself, the conditions for reciprocal altruism would be met.

3.1.4. Concluding Comment

The behaviors discussed above may depend upon the uniquely distinctive stimulus features of the young. Although Schaller (1972) attributed the high mortality of Serengeti lion cubs in part to their mothers' greater attraction to the other adults in the pride, and Vaitl (1977) felt that what appeared to be aunt relationships among squirrel monkeys in fact were no more than instances of already-formed bonds that endured despite the otherwise annoying presence of young, other data are consistent with the view that selection favored offspring characteristics which generally elicit solicitous behavior from conspecific adults.

*Additional long-term benefits for lions include the advantages that the cubs gain from growing up together as hunting companions, and for males, in competition to take over new prides (Bertram, 1976).

3.2. Ecological and Social Correlates

At several junctures in the foregoing discussion, statements concerning the ways offspring affect parents were qualified by references to features of social organization and/or the physical setting. Unfortunately, these qualifications are often simply tenative hypotheses to account for otherwise conflicting observations. As mentioned in the introduction to this section, the adaptations of each species represent the outcome of a complex interplay between the genetic heritage of the lineage and the myriad features of its organic and inorganic surroundings. Because the analysis of ecological–behavioral relations is still in its infancy, what follows is presented more in order to raise questions than to resolve them.

3.2.1. The Distribution of Resources

Current evidence suggests that the seasonal and/or spatial distribution of food, water, and places of refuge can influence the social organization of animals (Altmann, 1974; Geist, 1974), and thus the ways in which offspring are likely to affect caregivers.

3.2.1a. Food and Water. The distribution and abundance of food seem to affect mating systems by determining population dispersal. Monogamy, highly correlated with paternal care, is typically associated with sparsely but evenly distributed food supplies, and/or large litter size, which make provisioning of the young costly (Clutton-Brock and Harvey, 1978; Emlen and Oring, 1977). Stable environmental conditions, coupled with high intraspecific competition for resources also appear to be particularly favorable for long-term pair bonding; thus, one would expect male animals in such habitats to be solicitous of young (cf. Kleiman and Malcolm, this volume).

The distribution of food resources also may affect the quality of the female's responsiveness to offspring. In times of scarcity, solicitude for young may wane (Dittus, 1977; Schaller, 1972). Unpredictable food supplies are associated with moderately large litter size and either runts, or abandonment, or cannibalism (Alexander, 1974; Dawkins, 1976) and juvenile dispersal, especially among larger forms (Geist, 1974). Schaller (1972) suggests that communal suckling of young among lionesses is an adaptation allowing the group to be more mobile in pursuit of prey. It is tempting to speculate that the canid pattern in which all members of the pack regurgitate food for young and the mother (Moehlman, 1979; Murie, 1963; H. and J. van Lawick-Goodall, 1971) serves a similar end. However, the fact that social organization within a single population can vary with the source and abundance of food (D. Owens and M. Owens, 1979) indicates that one must be skeptical of simple, broadly applicable "rules" governing social behavior. At a very minimum, we should be prepared to take into account the possible existence of alternative "strategies" whereby a species' behavior can shift markedly in response to environmental conditions.

In a number of species the presence of young is associated with increased female intolerance of conspecifics—especially males (Ewer, 1968; Svare and Gandelman, 1976; Svare, this volume). Insofar as such intolerance affords the female exclusive use of a home range, as in short-tailed shrews (Platt, 1976), or leads to increases in her range, as in chipmunks (Martinsen, 1968), intolerance may function as a means to guarantee sufficient food to support lactation, or intensified postemergence foraging.

On the other hand, some cases of maternal territorial defense may also be related to a need to consolidate the mother–offspring bond, to the tendency of intruders to indulge in cannibalism (Mallory and Brooks, 1978), or even as an attempt at minimizing the conspicuousness of the nest site (see below).

3.2.1b. Shelter.   In several primate species, it seems that social organization is more "relaxed" in troops which inhabit forested areas than among animals whose home ranges are more open (e.g., Rowell, 1966a). Among baboons, the degree to which maternal solicitude is restricted to infants may vary inversely with the availability of cover (T. W. Ransom and B. S. Ransom, 1971). The kinds of signals that are used in communication (see Section 6.2) also may depend upon the nature of available cover (Geist, 1974).

Where weather extremes are encountered, the location of shelter or burrow sites can be an important determinant of social relationships. Perhaps because she depends upon the foodstuffs gathered from adjacent areas for sustenance during lactation and post lactation fat deposition in preparation for the coming winter, the female marmot actively excludes all intruders from the burrow until the young emerge (Barash, 1974). Here, as in bats (Humphrey, 1975), there seems to be an interplay between food, shelter (the thermoregulatory requirements of mother and young), and social organization.

Pinnipeds give birth and breed on land in sheltered coastal areas. How much this represents an attempt to avoid acquatic predators or to accommodate physiological limitations of mother and pups (cf. Peterson, 1968) is uncertain. Whatever the purpose, the fact that these animals tend to return to the same sites each year has important implications for breeding structure and parental care. Insofar as traditional sites are limited in number and size, they become crowded with females—a condition favoring polygyny, and minimal paternal care (LeBoeuf, 1974; Orr and Poulter, 1967).

Hrdy (1976, 1977) has proposed that the degree to which infantile pelage is distinctive (see Section 6.2.4) may be related to the amount of cover the species' habitat affords against predators. Thus, not only can the setting affect the influence of young by affecting social structure, but predation pressure and the nature of cover may influence the evolution of offspring stimulus qualties *per se.*

3.2.1c. Predation.   As indicated in the foregoing section, predation pressure is a factor that appears to influence the form of parental care.

Among the larger species, such as elephants (I. Douglas-Hamilton and O. Douglas-Hamilton, 1975) or rhinoceroses (Schaller, 1972), parental defense may be active. In some highly precocial ungulates, the young may be capable of, and depend upon, flight shortly after birth (Geist, 1971; Lent, 1974). Other ungulate young evade predators by their cryptic coloration and a tendency to remain relatively motionless. In some cases this tactic is combined with active defense or distraction displays by the mother, depending upon the danger posed by the predator (Lent, 1974; Walther, 1969). In others, concealment alone seems to be the primary defense. Among these species, offspring often remain alone for substantial periods of time. Typically, the mother changes locations with each hiding bout so that bed sites are used only once. In animals employing this strategy, the young often actively select the spot in which they "lie out" (Lent, 1974).

Smaller animals who inhabit stable burrow systems, such as European rabbits (Lockley, 1961) or ground squirrels (McLean, 1978), may dig new nest sites away from the main colony in order to raise their young. Preparturient Columbian ground squirrels apparently take great pains to conceal these burrows. They open the entrance hole from underground, leaving no telltale mound at the entrance, and even collect nest material from a different location from the one they use to furnish their regular, colony nest. Then they leave the burrow closed up for about 4 days before littering (McLean, 1978). Even among more solitary forms, females still frequently select new, different nest sites with each litter (Madison, 1978).

A tactic often associated with the strategy of hiding offspring, whether in burrows or in above-ground cover, is for the mother to nurse infrequently for very brief periods (Autenreith and Fichter, 1975; Kruuk, 1972; Lincoln, 1974; Martin, 1966; Zarrow et al., 1965). The relative infrequency of mother–offspring contact presumably minimizes the conspicuousness of the young,* but it also limits the kind of stimulus control that offspring can exert on both the timing of nursing bouts and the course of lactation.

In some gregarious animals, the existence of a brief birth season can be regarded as more than simply a mechanism whereby the energetic demands of lactation are timed to coincide with the availability of good-quality forage: It may be a reciprocally altruistic strategy whereby predators are "swamped" with prey, reducing the probability of any one offspring being taken (Daly and Wilson, 1978).†

It is also possible that predation pressures are exerted by conspecifics; some of the more elaborate defensive or nest-hiding behaviors may be directed at non-resident males (McLean, 1978). The foregoing suggestion must be taken with caution, however, because maternal hostility to congeners does not necessarily mean that cannibalism does occur (Spencer-Booth, 1970). Indeed, among many mammals, especially precocial forms, the neoparturient mother's self-seclusion

---

*Another benefit may be that lying-out minimizes the fluid and food requirements of the young (Langman, 1977).
†In migratory species such as barren-ground caribou, selection for synchronous calving may be intensified by the fact that females who give birth late in the season may be left behind by the herd and thereby denied the protection afforded by group membership (Lent, 1966).

and hostility to conspecifics may represent an attempt to establish a stable, exclusive bond with her infant rather than defense against conspecific infanticide (Autenreith and Fichter, 1975; Collias, 1956; Kleiman, 1972; LeBoeuf *et al.,* 1972; Lent, 1974; Gubernick, this volume).

3.2.2. Offspring Precocity and the Duration of Dependence

The degree and duration of offspring dependence exerts profound influences upon the way that the demands of the habitat combine with other aspects of the species' adaptations to shape parental practices. It is obvious that the longer young depend upon parental care, the greater the interbirth interval is likely to be. Moreover, predation pressure and offspring immaturity may affect the likelihood that nests or burrows will be involved in the care of young (Crook *et al.,* 1976).

3.2.2a. Precocity. Trivers (1974) suggests that natural selection favors greater sensitivity to offspring cues among species bearing immature young who require rapid, extensive, and finely adjusted responsiveness. Therefore, altricial offspring should be in more continuous contact with parents than precocial young. Although probably true on the average, these co-relations must be viewed against the overall context of habitat utilization and defenses against predators. For example, although rabbit pups are relatively altricial, the antipredator tactics of concealed nest sites and brief, infrequent nursing visits are associated with the ability of offspring to develop within the shelter of the burrow without continuous, contingent maternal care (Zarrow *et al.,* 1965).

Whereas the intermittent nursing of (altricial) rabbits and (precocial) ungulates seems to be primarily facets of defense against predators, similar feeding regimens can be observed among (altricial) carnivores, apparently in response to the availability of food. Both hyena (Kruuk, 1972) and leopard (Seidensticker, 1977) cubs must be able to go for periods in excess of 36 hr without nursing while their mothers hunt. For apparently different reasons, probably having more to do with male infanticide and the gregariousness of their mothers, lion cubs who are hidden away from the pride also must be able to go for a day or more without nursing (Schaller, 1972). In these cases there may be little advantage (indeed, often danger) in offspring signaling their momentary needs or in the mother developing a finely tuned sensitivity to them.

The quality of parental responsiveness is likewise associated with developmental rate and habitat utilization. While most lemur offspring are precocial and cling to their mothers, the young of the more highly arboreal variety, *Lemur variegatus,* are relatively immature at birth and are deposited in nests (P. H. Klopfer and M. S. Klopfer, 1970; Klopfer and Dugard, 1976). Presumably, the rich canopy inhabited by this species permits the utilization of a small range, which in turn allows the mother to meet her metabolic needs despite being restricted to a fixed radius of activity defined by the nest site. One would expect mothers whose offspring are altricial and reared in nests to be more likely to retrieve inactive infants than mothers of more precocial species (e.g., Rosenson, 1977).

Despite the obvious exceptions of the extremely immature, single young of some marsupials (Sharman *et al.,* 1966), and the highly precocious multiple off-spring of some hystricomorph rodents (Kleiman, 1974), a general correlate of off-spring precocity (and the duration of gestation) is offspring number. On the average, the larger the litter, the shorter the gestation period (Bulmer, 1970). Among primates, multiple births seem to be associated with either nesting, as in lemurs (Klopfer and Dugard, 1976), or paternal care, as in the case of marmosets (Epple, 1975). The extensiveness of maternal solicitude tends to be limited by the expected number of young. Neoparturient rhesus monkey mothers who normally bear single young will ultimately reject one or the other of two foster infants (Deets and Harlow, 1974), and even among langurs, a species in which there is extensive "infant sharing," mothers remain solicitous of only their own (singleton) offspring despite interest in other females' newborns (Hrdy, 1977).

   3.2.2b. Duration of Dependency.   For many species, limitations on available food, and shelter against the elements or predators interact with developmental parameters to affect the nature and timing of the termination of caregiving (Barash, 1973; Brown, 1966; Eaton, 1974; King, 1955; Smith, 1968). A point can be reached at which the carrying capacity of the habitat is taxed. If the species is one in which reproduction depends upon an individuals' maintaining more-or-less exclusive access to the resources provided by a fixed territory, one might expect that adults will be favored who become insensitive to or even hostile toward their offspring when the latter are sufficiently mature to survive (and reproduce) without parental guidance or protection.

   The relations among duration of dependency and other factors are complex, however. Despite their marked precocity at birth, the *female* offspring of sheep (Grubb and Jewell, 1966) and elephants (I. Douglas-Hamilton and O. Douglas-Hamilton, 1975) remain with their mothers and apparently derive the benefits of protection and guidance well past reproductive maturity. In some "solitary" carnivores, (altricial) cubs of either sex associate with their mothers until they become proficient hunters (e.g., Eaton, 1974; Schaller, 1967). These differences probably are determined by the quality and availability of food, predation pressures, and the development of offspring ability to be self-sustaining. Comparing these species with, say, rodents, it is obvious that the kinds of offspring cues effective in promoting dispersal are likely to be quite different.

   In summary, offspring precocity and the duration of dependency are not perfectly correlated; the kinds of stimulus exchanges which are likely to govern the frequency, quality, or duration of parent–offspring interactions also will depend upon habitat utilization.

3.2.3. Social Organization and Investment

   While it is clear that social organization will depend upon the way in which individuals exploit their habitat and respond to predation, the precise combinations of strategies employed can vary widely (Crook *et al.,* 1976). The effects that offspring exert upon their parents also should differ according to the ways in which *adult* members of their species relate to one another.

Because animals possess no mechanisms to determine exactly the number of shared genes, tactics used to maximize the likelihood of confining care to those whose survival will benefit inclusive fitness should vary according to social organization. For example, when there is a high probability of encountering alien young, selection should favor the ability to discriminate one's own offspring (Daly and Wilson, 1978).

Thus, on the assumption that each species inhabits a single, stable setting or that its social behavior does not vary markedly according to changing environmental conditions, we could generate sets of hypothetical "rules for parenting" that would be amenable to empirical test. However, we still are relatively ignorant of the range and variability of social organization among many species (e.g., marmosets, Neyman, 1978). Thus, given the fact that the nature of social organization can vary according to the availability of cover (Rowell, 1966a) or food (D. Owens and M. Owens, 1979), it seems that we need more extensive studies of animals in their natural habitats before such undertakings can be expected to be fruitful.

4. Changes during the Life Cycle

The degree to which an animal can enhance its (inclusive) fitness by displaying responsiveness to immatures is not constant.

4.1. Age-Related Changes

The idea that each individual has a relatively fixed amount of energetic capital to invest in offspring suggests that females who are capable of repeatedly breeding and rearing young should limit investment in any one offspring or litter, and as an individual matures, its tactics should vary (Daly and Wilson, 1978). The optimal amount to invest in young before becoming reproductively mature may depend upon social organization; early care for siblings or other young should occur only in stable, somewhat closed groups where the beneficiaries are likely to be kin or capable of reciprocation (but see Section 4.2, below). After puberty, early in an individual's reproductive career, when future reproductive potential is the greatest, it may pay to abandon, cannibalize, or resorb young if resource scarcity or danger is so severe that attempting to care for current offspring would materially jeopardize prospects for producing future young. As one becomes older, the balance shifts in favor of current offspring (e.g., Nash, 1978), and at some point in long-lived social species, the risks or uncertainties in becoming a parent may be so great as to favor foregoing further investment in new offspring and facilitating the reproductive success of previous young or grand "children" (Dawkins, 1976; see also Fagen, 1976).

Since senescing individuals have progressively declining prospects for reproducing, they also have less to lose by aiding others. Thus, the likelihood of reciprocal altruism, under favorable conditions (e.g., Section 3.1.3), should increase as
individuals grow older (Trivers, 1971).

Male tactics should also change with age. In free-ranging, provisioned rhesus
monkeys (Breuggeman, 1973) and captive sooty mangabeys (Bernstein, 1976),
with the decline of their chances for successfully competing for females, older
males appear to spend more time safeguarding their kin and offspring than in
attempts to beget future young. (It is not clear whether caring for young could
also serve as an alternative route for developing sexual consort relationships with
females). This strategy of increasing male investment in direct offspring care with
age seems to apply in multimale organizations. However, in one-male groups of
langurs (Hrdy, 1977), guerezas (Oates, 1977), and redtail monkeys (Struhsaker,
1977), male care often takes the form of troop defense. Theoretically, in such cases,
older harem males should be willing to take greater physical risks in order to
defend their females and offspring against takeovers by infanticidal invaders
(Hrdy, 1977).

It is possible that age-related changes in female tactics will vary according to
differences in social structure and the tendency for males to engage in infanticide.
Almost to the end of their reproductive lives, many macaque females compete for
dominance, in terms of priority of access to desired commodities (including proximity to high-ranking males), and once established (Chickazawa *et al.,* 1979), the
entire matriline tends to "share" in the status of the matriarch (e.g., Kurland,
1977). In contrast, among one-male langur troops, female dominance may vary
more directly with reproductive potential. The younger (and smaller) adult
females who have the greatest future prospects tend to vie among themselves for
priority; the older (often larger and stronger) females seem to bow out of the competition. Whereas the younger females are less likely to risk serious injury in
defense of their young during male takeovers or external threats, older females
put up vigorous resistance against attacks on their offspring or other group young.
Older females who have infants are also more likely than younger ones to leave
the troop during takeovers, presumably in order to save their offspring (Hrdy,
1977; Rudran, 1973).

The quality and degree of responsiveness to young thus seem to depend upon
social organization and kinship. Among langurs, the younger females are likely to
be the daughters or the nieces of the elder females. Theoretically, as their reproductive careers reach an end, the "grandmothers" have more to gain by attempting
to protect their existing investments in offspring and grand-offspring. Their adult
young should not take too many risks, because their reproductive future lies ahead
of them (Hrdy, 1977). In contrast, within the larger multimale assemblages typical of macaque monkeys, the several matrilines are probably less closely related
and alliances are formed among close relatives in order to compete within the
group (Drickamer, 1974; Kurland, 1977).

4.2. Prior Experience

Early studies of the effects of isolation upon the parental behavior of rhesus monkeys suggested that prior experience with young might be a necessary condition for the development of parental responsiveness. However, subsequent investigations indicate that isolation-reared females could care adequately for second offspring despite prior aberrant behavior (Harlow *et al.*, 1966). Similarly, observations of parental responsiveness in hand-reared guinea pigs (Stern and Hoffman, 1970), rats (Thoman and Arnold, 1968), and tamarins (Epple, 1975) show that prior experience with conspecifics is not a prerequisite for adequate parenting. Thus, available evidence provides little support for the notion that parenting *per se* must be learned in the traditional sense of the term (Harper, 1970).

On the other hand, there is abundant evidence that parenting may be affected by prior caregiving—effects that are largely, if not entirely, independent of the age-related changes in strategy discussed in Section 4.1. These modifications include both decreases and increases in responsiveness to cues provided by the young, and in greater efficiency of performance (Bell and Harper, 1977; Harper, 1970; Stewart, 1977).

There is general agreement that parents will benefit when their offspring emit signals indicating need for care, and prior experience with offspring can be a valuable guide as to the degree of solicitude that may be required. According to Trivers (1974), experienced parents should be less likely than naive ones to overinvest in their offspring. In mice (Beniest-Noirot, 1958; Noirot, 1964*a*), rats (Carlier and Noirot, 1965), and rhesus monkeys (Seay, 1966) controlled observations indicate that maternal solicitude diminishes to become more closely attuned to the actual needs of offspring as mothers gain experience. These data confirm field impressions that new mothers of several species appear to be "overconcerned" about the welfare of their offspring (e.g., Hrdy, 1977; I. Douglas-Hamilton and O. Douglas-Hamilton, 1975; Horwich and Manski, 1975).

It has been suggested that aunting behavior among primates serves an educative function. Because mothers tend to allow their own (female) kin preferential access to offspring, some of the potential costs of inexperience could be offset by gains in inclusive fitness accruing from increased competence (Hrdy, 1976; Kurland, 1977; Lancaster, 1971). Consistent with this viewpoint, infant and early juvenile langurs are likely to drop, drag, and desert infants when allomothering, whereas older juveniles are much more solicitous and effective in keeping infants comfortable (Hrdy, 1977). Moreover, insofar as later-borns of several species are more likely to survive (Hrdy, 1976; Drickamer, 1974) and inexperienced young do not carry offspring "correctly" (Pola and Snowdon, 1975), the data could support such an interpretation.

On the other hand, despite the fact that young, nulliparous females of a number of other primate species (e.g., Cheney, 1978; Stewart, 1977; Struhsaker, 1967) seem to be most interested in infants, experience gained from allomothering has not been proven to be an important facet in the development of maternal skill. For example, even though the infant's protests at rough handling and the mother's tendency to retrieve offspring who are emitting distress signals could provide feedback facilitating associative or trial-and-error learning, younger animals, who are most likely to cause infants to sound the alarm, are also the *least* likely to be granted access to them. Moreover, despite the fact that older juveniles are more competent than younger ones, experienced adult allomothers also tend to neglect their charges or fail to adjust their behavior to accommodate them (Hrdy, 1977). Although the latter may be explicable in terms of the more experienced females' reduced need for information, it suggests that other factors may be involved (see McKenna, this volume).

According to Struhsaker (1967) the quality of allomothering among vervet monkeys is as much a function of the infant as the prior experience of the aunt. Similarly, Oates (1977) could find little direct evidence for increases in the competence of guereza monkey females as a result of being allowed to handle or care for other females' infants. Thus, while maternal effectiveness seems to increase across offspring, available evidence does not provide unambiguous support for the notion that "play mothering" is important for the development of caregiving skills.

5. Mechanisms: Effects of Offspring on Parental Responsiveness

Analyses of the ways in which offspring affect parents have provided some of the most telling arguments against classifying behavioral change as due to simply heredity or experience (Lehrman, 1953; 1956). Whereas solicitude for young was once thought to depend simply upon hormonal state, it is now clear that, although hormones modify responsiveness to offspring, parental behavior depends upon several factors (Noirot, 1964a; Rosenblatt and Siegel, this volume).

5.1. The Setting as a Determinant of Offspring Effects

Even when an individual is in the appropriate "state" to respond to young, its behavior still depends upon the physical and social setting.

5.1.1. The Physical Setting

It is almost a truism that laboratory or other "artificial" environments may mask or distort "natural" behavior patterns (e.g., Hediger, 1950). Theoretically, among captive animals, failure to care for young or cannibalism represent responses to cues which in the natural habitat provide reliable indices that further investment in offspring would be disadvantageous. For example, in his analysis of the effects of denying rats prior experience in transporting objects on subsequent maternal nest building, Eibl-Eibesfeldt (1961) demonstrated that providing a secluded corner in the animals' cages sufficed for the display of such behavior. Undirected picking up of pups and apparently disorganized parental care probably represented the absence of a sheltered or familiar nest site, a condition inimical to successful reproduction in the wild. Similarly, when lactating mice are presented with pups in an unfamiliar cage they do not engage in "purposeful" transport. However, when their home nest is placed in one corner of the test cage, scattered pups are retrieved promptly once the female has entered the nest (Fraser and Barnet, 1975).

Variations in the setting may lead to changes in the quality of response to the same offspring stimuli even when a caregiving relationship already exists. Whereas the vocalizations of young kittens from afield elicit maternal departure from the nest and investigation of the source, such calls emitted while the mother cat is in the nest with her litter elicit nosing or postural adjustments (Haskins, 1977).

In the natural habitat, exchanges between parent and offspring may thus depend upon the presence of specific environmental features. Among ungulate species in which the young find secluded hiding spots at the end of nursing bouts, the search for, and availability of, suitable locations determine the timing of mother–offspring contacts (Lent, 1974).

Physical surroundings also can modify the overall course of the caregiving relationship. Young macaques reared in bare cages without climbing equipment (Jensen and Bobbitt, 1968) or in smaller cages (Castell and Wilson, 1971) receive more maternal "punishment," sooner, than animals in cages which are liberally supplied with trapeezes, etc., or are simply larger. The differences seem to be due to the former youngsters' tendencies to use their mothers as props for their locomotor as well as their social play.

The availability of food is another feature of the environment that obviously can affect parental responsiveness. Undernourished Himalayan sheep (Schaller, 1977) and laboratory rats (Smart, 1976) appear to be less solicitous of offspring than their more well-nourished counterparts. In rats, not only is general responsiveness reduced by undernutrition, but the tendency to discriminate undernourished from well-nourished pups appears to be reduced (Wiener *et al.,* 1976).

5.1.2. The Social Environment

The social environment exerts similar constraints upon the kinds of effects that offspring can have upon parents, at least in captive monkey colonies. Sensitivity to stimuli typical of offspring sometimes depends upon the actual presence of young in the group. Among marmosets, group members approach tape-recorded infantile "loss of contact" calls only when there have been offspring in the group (Epple, 1968). Mothers' behavior also can be affected by the responses of other individuals. In rhesus (Hinde and Spencer-Booth, 1967) and stump-tailed (Rhine and Hendy-Neely, 1978) macaques, females' interest in infants of low-ranking mothers cause the latter to become restrictive and agitated when their offspring attempt exploratory sorties.* On the other hand, when compared with isolated mother–infant pairs, pig-tailed macaque (Castell and Wilson, 1971) and squirrel monkey (Kaplan, 1972) mothers housed within groups are much less "punitive" or "rejecting" toward their infants.

In free-ranging groups of macaques, the mother's presence may also inhibit the approach tendencies of males toward infants (Estrada and Sandoval, 1977). In a seminatural environment, while squirrel monkey mothers were together with their infants, they threatened and chased males whenever the latter approached them. Because nullipara spontaneously approached the males, and aunts were less antagonistic toward them than mothers, Strayer *et al.* (1975) concluded that the presence of offspring caused the mothers to attempt to exclude males.

The kinds of relationships young have with other members of their group also can affect what parents will do. Many primate parents will defend their young against threats from within as well as from without the troop (Kurland, 1977). Among captive marmosets, nursing bouts are sometimes initiated when siblings who are carrying young attempt to dislodge their burdens; the infants' cries cause mothers to approach and take their offspring, who then typically seek the nipples (Ingram, 1977). Extraparental care may also be affected by the nature of parent–offspring relationships. Among free-ranging rhesus monkeys in Nepal (Taylor *et al.*, 1978) and provisioned Japanese macaques (Itani, 1959), male solicitude for infants seems to be associated with mother–offspring conflicts over weaning, or care for a younger sibling.

When several young are present in a group, the activities of one mother–infant dyad may affect the behavior of others. When mother rats were exposed to the sounds of another female nursing her offspring, their litters gained more weight than the (same-sized) litters of mothers who were deaf but otherwise comparably treated (Deis, 1968). Similarly, among captive Java macaques, nursing bouts lasted longer when both of two mothers were suckling their young as compared to bouts occurring when either mother nursed her infant alone (Chance *et al.*, 1977*a*). Autenreith and Fichter (1975) also suggested that there was "social facilitation" of nursing among wild pronghorn antelope fawns.

5.2. Effects of Offspring upon Parental State

Insofar as parenting in females involves "repeated returns to previous physiological conditions" (Daly and Wilson, 1978), it is of considerable importance to know the degree to which offspring influence these changes.

---

*These upsets also may be due to the lack of kin to aid them in disputes over retrieving their young (White and Hinde, 1975).

5.2.1. Pre- and Perinatal Effects

Although there is uncertainty concerning the mechanisms involved (Dey *et al.,* 1979; Haour and Saxena, 1974; Noonan *et al.,* 1979; Sundaram *et al.,* 1975), most investigators agree that the blastocyst must act upon the uterus in order to suppress maternal immune responses and to affect implantation. Then, the placenta develops from an interaction of fetal and uterine tissues.

5.2.1a. Preterm Changes in Responsiveness. In mice, the number of fetuses carried by the female is positively correlated with the amount of nesting material used from the 4th through the 18th days of pregnancy (Gandelman, 1975). In gravid rats, the degree of prenatal responsiveness to offspring as well seems to be related to the number of fetuses. When pregnancy was terminated by ovariectomy/hysterectomy between the 10th and 21st days of gestation and responsiveness "induced" by exposure to foster litters, animals who had been carrying more than six pups for the first 16 days were more likely to retrieve 1- to 8-day-old pups than were females who had been carrying smaller litters (Mackinnon and Stern, 1977). Even without manipulation of pregnancy, as parturition approaches, females of a number of species become highly responsive to offspring.*

Cross-transfusion experiments with rats (Terkel and Rosenblatt, 1972) indicate that blood-borne factors are involved in the onset of perinatal solicitude, and administration of exogenous hormones suggests that they are probably of endocrine origin (Moltz *et al.,* 1970). Because the fetal–placental unit is a source of hormones, some of these changes may be considered offspring effects. It is possible that distention of the uterus by the fetuses also acts synergistically with the pregnancy-induced hormonal state of the mother rat to increase postpartum responsiveness to offspring (Graber and Kristal, 1977).

Thus, "offspring" are likely to be determinants of some of the physiological and behavioral changes accompanying pregnancy. From the little evidence currently available, it seems probable that although endocrine changes influence maternal responsiveness, the specific hormones involved, their sources, and the behaviors affected vary across species. For example, whereas nulliparous adult hamsters (Rowell, 1961) and guinea pigs (Harper, 1976) are generally unresponsive to young, virgin mice of either sex spontaneously display parental care upon exposure to pups (Beniest-Noirot, 1958).

5.2.1b. Parturition. Because they are in a better position to determine their readiness to survive in the extrauterine environment, the young often initiate the processes leading up to parturition. Clinical observations suggest that delayed

---

*Mice (Noirot and Goyens, 1971), hamsters (Siegel and Greenwald, 1975), rats (Slotnick *et al.,* 1973; Rosenblatt and and Siegel, 1975), sheep and goats (Collias, 1956), white-tailed deer (Townsend and Bailey, 1975), and langurs (Hrdy, 1977).

onset of labor is associated with fetal genotype (Holm, 1967) and experiments have implicated fetal neuroendocrine function in terminating pregnancy (Anderson and Turnbull, 1973). More recent studies demonstrate differences among species in the mechanisms involved.

In goats, the signal from the kid involves fetal adrenocortical activation, which mediates changes in maternal corpus luteum output. At the same time, fetal adrenal hormones hasten the maturation of the kid's lungs. In addition to preparing the uterus for the coordinated contractions necessary for parturition, these changes also facilitate placental separation and delivery, and the initiation of lactation (Currie and Thorburn, 1977a). It appears that the mechanisms are so nicely adjusted that the kid "prepares itself" to be born while preparing its mother to care for it, although the precise timing of birth is not simply under fetal control (Thorburn et al., 1977). During parturition itself, distention of the cervix caused by the expulsion of the fetus causes a reflex release of oxytocin from the maternal pituitary which further facilitates the birth process (Currie and Thorburn, 1977b).

Despite their close taxonomic relationship to goats, the processes underlying fetal control of the onset of labor in sheep are somewhat different. Maintenance of pregnancy appears to be more dependent upon placental as opposed to ovarian hormone output; the adrenocortical output of the fetal lamb alters placental activity, and thereby terminates pregnancy. Thus, in sheep, the fetal pituitary–adrenal system seems to be involved in the initiation of parturition (Thorburn et al., 1977).

5.2.1c.   Puerperal Responsiveness.   Although nulliparous mice are usually responsive to young, hormones associated with lactation can further increase their retrieving performance (Voci and Carlson, 1973). In rats, events correlated with the termination of pregnancy per se also seem to be associated with an increase in responsiveness to young (Rosenblatt and Siegel, this volume). In a number of species, the postnatal period is characterized by very high degrees of parental solicitude.* Physiological priming, presumably resulting from the neuroendocrine events accompanying pregnancy and delivery, may account for an apparently indiscriminate responsiveness among neoparturient females. In some forms, neopara will retrieve young of alien species (Wiesner and Sheard, 1933), and even rear them (Bleich and Schwartz, 1974; Denenberg et al., 1964; Hersher et al., 1963).

Other, less obviously offspring-related responses may also be influenced by maternal state in this period. The tendency of new mothers to be unusually "wary" (Clutton-Brock and Guiness, 1975; Izawa, 1978), to withdraw from their

---

*Even among "hider" species, there is usually a brief period of continuous contact and ad libitum nursing (Autenreith and Fichter, 1975). In flying squirrels (Muul, 1970), as in rats (Wiesner and Sheard, 1933), retrieving behavior is apparently insatiable during the first few days after birth. Similarly, just after birth, mother bush babies are "compulsive" groomers (Sauer, 1967) and squirrel monkeys are "nonselectively" responsive to infants—alive or dead (Kaplan, 1973).

group (Kruuk, 1972), or to be intolerant of conspecifics in the vicinity of their offspring (Collias, 1956) probably represents transient-state-mediated responses to the presence of their young.

Perhaps the best-studied aspect of the puerperal phase is the attachment process. Among precocial ungulates (and probably other animals who give birth in large aggregations, such as seals and sea lions) the postpartum period seems to be particularly important for the formation of an exclusive bond to offspring (Collias, 1956; Hersher *et al.*, 1963; LeBoeuf *et al.*, 1972; Le Neindre and Garel, 1976; P. H. Klopfer and M. S. Klopfer, 1968; Smith *et al.*, 1966). It appears that if the mother does not learn to distinguish her own offspring from other infant conspecifics, she may become indiscriminately solicitous (see Gubernick, this volume, for further discussion).

Bonding is not the only process occurring in the puerperium. The mother hamster's systematic litter reduction by cannibalizing pups is largely limited to the first 3 days after delivery. If litter size is manipulated after that point, the mothers do not cannibalize "excess" young so readily (Day and Galef, 1977).

5.2.2. Postnatal Effects on the Parent

While the physiological changes associated with pregnancy and parturition may "prime" mothers to react to offspring, this state is transitory. If it were not, bereaved mothers might care for unrelated offspring [perhaps another reason why parturient females of so many species seek seclusion (cf. Lickliter, 1979)]. Since loss of young due to predation or other unpredictable causes is a relatively common event, one way to avoid squandering resources is to require further input from offpsring in order to remain responsive.

5.2.2a. Consolidation. As indicated above, and in other chapters, the period immediately after birth often is critical for the establishment of the mother–offspring bond. Among species in which responsiveness to young is not so exclusive, the postnatal period can still be crucial for the maintenance of parental solicitude. For example, rat mothers deprived of their litters for the first 4 days after giving birth, subsequently will be much less responsive to pups than if a comparable period of separation occurs later on in the cycle (Rosenblatt and Lehrman, 1963; Smotherman *et al.*, 1978). Denying rat mothers any contact with their litters from the moment of birth leaves them about as solicitous of pups as nulliparous virgins when tested 25 days postpartum. However, allowing primipara as little as 4–6 hr contact with pups reduces their latency to become responsive to the levels of females who have reared a litter to weaning. These effects do not depend upon parturition itself, the initiation or maintenance of lactation, nor the presence of ovarian secretions. They do, however, depend upon an interaction of the neoparturient state and the presence of pups, because virgins who become responsive to pups as a result of continued exposure still take longer than parous, pup-exposed females to return to a responsive state after 25 days' isolation from young (Bridges, 1975, 1977).

A striking example of the interaction of maternal state and offspring stimuli is provided by experiments with red kangaroos (Sharman, 1967). In these animals, the hormonal correlates of the later phases of the estrous cycle mimic the early postpartum condition. If an infant is put on the nipple of a nulliparous female at about 33 days after she was last sexually receptive, she will begin to give milk.

Responses to other conspecifics also may depend upon early mother–offspring interactions. Neoparturient mother mice are hostile toward intruders; attacks on unfamiliar conspecifics are more prolonged and ferocious in the first few days after giving birth. This intolerance is affected by suckling stimulation from the litter: If the mother's nipples are surgically removed prior to parturition, or if her litter is removed at birth, she will not attack intruders. However, if nipple removal is performed after 2 days of postnatal suckling, mothers will continue to assault intruders up to 1 week later (Svare and Gandelman, 1976; Svare, this volume).

5.2.2b. Sensitization and Induction. Hormonal priming is not always necessary for responsiveness; simply exposing animals to young often affects the probability of caregiving (Rosenblatt and Siegel, this volume). In mice, very brief sessions with 1- to 2-day-old pups hidden in a perforated container increase the level of solicitude shown to "sub-optimal" stimuli (Noirot, 1964c, 1969). Even virgin female hamsters, who usually attack unfamiliar pups, become more likely to build nests and carry subsequently presented young after having been presented with pups—whether or not they attacked them upon their initial exposure (Noirot and Richards, 1966).

Wiesner and Sheard (1933) and Rosenblatt (1967) have shown that continuous exposure to pups can induce the full range of caregiving behaviors in both female and male rats. If caregiving is induced and maintained by continuous exposure to pups, testosterone injections (which otherwise lead to pup killing) do not reduce the solicitousness of virgin female mice (Gandelman and Davis, 1973). Even "nonspecific" stimulation, such as peripherally induced pain (tail pinching) in the presence of pups, can hasten the onset of caregiving behavior in pup-exposed rats (Szechtman *et al.*, 1977).

Among ungulates, once a mother has formed an attachment to her offspring, she usually rejects nursing attempts by other conspecific young. However, in sheep (Collias, 1956), goats (Hersher *et al.*, 1963), and domestic cattle (Hafez and Lineweaver, 1968; Le Neindre *et al.*, 1978), if lactating females are physically restrained so that they cannot prevent foster young from nursing, they will eventually accept them. In fact, sheep have been induced to accept kids and goats to accept lambs by enforced suckling (Hersher *et al.*, 1963). The hormonal patterns of adrenal responsiveness typical of lactating rat mothers can even be induced in virgin, nonlactating females (Korányi et al., 1977).

Despite the fact that young can induce caregiving behavior in previously unresponsive individuals, we cannot dismiss the facilitating role of maternal state. In both mice (Noirot and Goyens, 1971) and rats (Bridges et al., 1972), the responsiveness of lactating females exceeds that of "induced" animals even when the possibility of nipple attachment has been eliminated (Stern and Mackinnon, 1976). At this writing, it seems that we are dealing with a complex interaction of pre- and postnatal events all of which are to some degree influenced by the young, and which lead to heightened maternal solicitude so timed as to maximize the likelihood that a caregiving relationship will be established and consolidated with own offspring.

5.2.2c. Differentiation.   As the young develop there are changes in the degree and kind of parental care required. Maintenance of the lineage demands that some offspring establish reproductive relationships of their own; thus some of their interests will coincide with those of their parents. Similarly, although there may be "disagreements" concerning when it must occur, the fact that offspring can expect to outlive their parents also puts a premium upon their becoming independent.

The first phase of the parent–offspring relationship, however brief, is characterized by intensive mother–young interactions for which the parent is primarily responsible; then the initiative usually shifts to the offspring (e.g., Bell and Harper, 1977; Ingram, 1977; Lent, 1974; Rheingold, 1963). Part of the change in parental attentiveness depends upon cues emitted by the young (e.g., Kleiman, 1972; Noirot, 1964a; Reisbick et al., 1975). If the rate of pup development in rats is slowed by undernutrition, maternal responsiveness wanes more slowly than if the pups are growing at a normal rate (Lynch, 1976). However, as one would expect from the idea that selection favors internal mechanisms preventing over-investment, the decline in maternal solicitude also depends on habituation-like phenomena or hormonal changes. In mice, repeated exposures to pups causes apparent habituation of nest-building and licking behavior (Noirot, 1965). In rats, the decreasing amount of time that the mother spends in the nest with her litter after the first week postpartum also involves endocrine-regulated vulnerability to changes in body temperature (Leon et al., 1978).

In some rodents, the appearance of offspring locomotor skills allows the young to take the initiative in interactions, thus offsetting declining maternal solicitude and yielding fairly stable levels of licking and nursing (e.g., Smotherman et al., 1977a). However, these same developments cause a decline in retrieving behavior (Michener, 1971; Noirot, 1964a, 1972a).

In rats (Leon and Moltz, 1972) and perhaps also in house mice (Breen and Leshner, 1977) and the spiny mouse (Porter and Doane, 1976), the developing offspring stimulate production of a pheromone which attracts newly mobile pups to the mother and nest area. In rats, the production of this olfactory attractant is largely a response to pup age (Holinka and Carlson, 1976; Moltz and Leon, 1973) and can be induced in virgin females by continued exposure to young of the appropriate age (Leidahl and Moltz, 1975). Other changes related to pup age have been observed in nest building and in grooming of young in both mice (Noirot, 1964a,b) and rats (Reisbick et al., 1975).

Just as stimulation from offspring seems to be required to consolidate maternal responsiveness, the maintenance of milk production should continue only so long as there are young to feed: Suckling is a necessary condition for sustaining lactation. Moreover, as the young grow in size, so do their appetites. Among species bearing more than one offspring at a time, a mother must be sensitive to both the variable number and changing needs of her litter. She can adjust by increasing total yield (Fuchs and Wagner, 1963; Kumaresan et al., 1967); or she can adjust by nursing longer, or more frequently, or both (Stern and Bronner, 1970). In some ungulates who have multiple young, a mother will not release milk until all her offspring are sucking (Lent, 1974). In red kangaroos, not only can a mother nurse a "pouch young" and a "young at foot" simultaneously, but the composition of the milk yielded by the two teats will be different (Sharman, 1967).

Among species in which periods of extended mother–offspring separation are common, such as "hider" ungulates (Lent, 1974), pinnipeds (Peterson, 1968), or felids (Schaller, 1972; Seidensticker, 1977), the impetus for initiating a nursing obviously originates in the mother. Even among species in which contact is relatively continuous, there seem to be cycles of readiness to permit young to nurse, as indicated by differences in the duration of bouts according to whether or not they are maternally initiated (Ingram,1977; Schaller, 1977). After the early postpartum period (when the duration of sucking is still largely determined by the young), nursing bouts often are terminated by the mother (Autenreith and Fichter, 1975; Doyle et al., 1969; Fuchs and Wagner, 1963; Geist, 1971; Lent, 1974).

Rabbits nurse their young only once or twice per day. When lactating does' teats are anesthetized rendering them insensitive to tactual stimulation provided by the litters' sucking, nursing durations extend up to three times longer than those of untreated animals (Findlay, 1968). Although teat emptying does not seem to be involved in bout termination in rabbits (Lincoln, 1974), mammary engorgement may be a stimulus for the initiation of nursing bouts (Findlay, 1968). In contrast, rats, who nurse more frequently, suckle their offspring despite artificially reduced milk output (Lu et al., 1976).

5.2.2d.   Termination.   To maximize lifetime reproductive output, parents must be able to "decide" when to cease providing for offspring. The most obvious instance in which responsiveness should cease is when offspring are lost after lactation has been established. In rats, the mechanisms underlying control of milk production are complex. After lactation has been established, 8 days of nursing litters of 6 pups only once per day yields the same decline in mammary development and milk production as complete weaning. Although mothers who are allowed to suckle (foster) litters three times daily produce enough milk for litter survival, only those nursing four times per day maintain mammary development comparable to animals who are with their litters continuously (Tucker *et al.,* 1967).

There also seems to be a mechanism whereby short-term "conservation of resources" can occur in response to transient interruptions in the nursing rhythm without total cessation of lactation in rats. The rate of complete refilling of the mammary glands of rats is reduced if nursing does not occur regularly, but after a delay of up to 16 hr, prolactin will still be released if the mother is allowed to rejoin her litter and permitted to nurse for 30 min. If the delay is longer, say, 24 hr, up to 60 min nursing does not lead to "normal" prolactin release (Grosvenor *et al.,* 1970; Mena *et al.,* 1976). However, even if separation lasts for as long as 9 days after lactation has been established, 7–10 days' exposure to stimulation by vigorous pups can reinduce milk production (Ota and Yokoyama, 1965). There seems to be a gradual decay of responsiveness to suckling; if induction is undertaken after 90 days, no milk will be produced (Fleming, 1976).

As indicated in Section 2.2, mothers should be capable of determining how long to continue focusing investment upon current offspring. In many species nursing blocks ovulation, but under normal conditions, milk production eventually ends and the mother resumes estrous cycling. In rats, this eventual buildup of refractoriness to suckling stimulation is based upon exteroceptive cues in addition to the proximal, tactile stimulation involved in nursing (Grosvenor and Mena, 1973). Not only is the number of functional mammary glands related to the size of the litter (Mena and Grosvenor, 1968), but the duration of lactation diestrus is determined by the number of suckling pups (Tucker and Thatcher, 1968), although not simply as a result of release of prolactin (Lu *et al.,* 1976).

In bonnet macaques tactual stimulation must be combined with high levels of prolactin in order to suppress gonadotropin release (Maneckjee *et al.,* 1976), and in rhesus macaques it is possible that sucking may act in some way to inhibit pituitary synthesis of luteinizing hormone (Weiss *et al.,* 1976). Whatever the hormonal mechanisms, the presence of an (active) infant reduces the sexual receptivity or fertility of a number of species, including ungulates (Fletcher, 1971; Kann and Matinet, 1975). In rats, mammectomized mothers who are provided with foster litters remain anestrous for up to 2 weeks (Moltz *et al.,* 1967), and among yellow baboons (Altmann *et al.,* 1978) and lions (Schaller, 1972), even when mothers have weaned their offspring, the presence of their young is associated with delayed conception.

In some species, sexual activity or changes associated with the rut *per se* are incompatible with caregiving (Autenreith and Fichter, 1975); in others, the caregiving relationship may terminate with the birth of the next offspring (e.g., Lent, 1974; Mackinnon, 1974). Although the (impending) birth of subsequent offspring does not always herald the end of nursing or care for previous young (I. Douglas-

Hamilton and O. Douglas-Hamilton, 1975; T. W. Ransom and B. S. Ransom, 1971), many ungulate mothers separate from their yearlings just prior to giving birth (Langman, 1977; Schaller, 1977). If they do not conceive during a postpartum estrous, rabbits (Lincoln, 1974) and green acouchis (Kleiman, 1972) will continue to nurse for extended periods.

In some primates, weaning of an older offspring is often final with the birth of a new infant (Mackinnon, 1974; Trollope and Blurton-Jones, 1975), but mother–offspring ties may continue, depending upon the species and/or the social/ecological setting (T. W. Ransom and B. S. Ransom, 1971; Yoshiba, 1968).

5.3. Comment

From the evidence reviewed here it is clear that the cues emanating from the young, in conjunction with the nature of the setting and (often offspring–induced) internal state, may provide parents with information that regulates their future reproductive output.

Although I have frequently referred to parental "responsiveness" or "solicitude" for convenience, this should not be taken to mean that parental care is an entity of some sort. Caregiving involves many different activities, each of which varies more or less independently. Maintenance and termination of each of these classes of activity often is elicited by a different kind of offspring signal (Bell and Harper, 1977; Galef and Clark, 1976; Haskins, 1977).

6. Stimulus Effects of Offspring

The theoretical model considered here posits that caregiver responsiveness to offspring evolved as a means to maximize parental fitness and that selection initially favored young who emitted signals which were informative, allowing parents to adjust their behavior in accordance with evaluations of both long-term and momentary requirements. (However, see Chapter 6, this volume, for a contrasting view.) Because parents are in a better position to exert coercive control of young, offspring control of parental behavior should have evolved to influence caregiving primarily by "psychological" means (Alexander, 1974; Dawkins, 1976; Trivers, 1974).

6.1. Age- and Sex-Related Signals Controlling Parenting

The concept of parental investment implies that the "value" of offspring increases with the amount of resources already expended. However, selection also favors reallocating investment from older offspring to younger, more dependent ones so long as the former are better able to withstand mild deprivation (Dawkins, 1976; Trivers, 1974). In addition, because the probabilities of future reproductive payoffs of male and female offspring can differ (depending upon environmental circumstances), parental returns may be maximized by favoring one sex over the other (Trivers, 1972). Therefore, it would be to the parents' advantage to respond differentially according to the age and sex of young.

6.1.1. Responsiveness as a Function of Age

Although the exact stimulus qualities animals use to distinguish among age classes have not been identified, adult discrimination of offspring age is clearly evidenced by the existence of specific responses to immatures.

6.1.1a. Distinctive Patterns of Response. Allogrooming among guinea pigs (Harper, 1976), domestic dogs (Scott and Fuller, 1965), and mountain gorillas (Schaller, 1963) occurs seldom, if at all, yet is a common feature of at least certain stages of the caregiver–offspring relationship in these species; in coatis, the method of grooming young itself is distinctive (Kaufmann, 1962). In feral vervet monkeys (Struhsaker, 1967) and captive rhesus macaques (Hinde *et al.,* 1964) adults play only with juveniles, not with adults. The regurgitation of food among canids is primarily a response to offspring—although it may extend to the mother in the early stages of lactation—and pups are allowed preferred access to kills (H. and J. van Lawick-Goodall, 1971). Among ungulates, mothers may use highly distinctive motor patterns to induce their young to follow them (Autenreith and Fichter, 1975; Lent, 1974) and to threaten immatures (Lent, 1966). Similarly, among feral rats (Calhoun, 1962), prairie dogs (King, 1955), captive marmosets (Stevenson and Poole, 1976), and baboons (Rowell, 1966b), the kinds of agonistic/ dominance responses directed toward young differ qualitatively from those made to adults.

6.1.1b. Failure to Respond. Just as the young may provide the necessary stimuli for eliciting certain classes of response, some as yet unidentified features of juvenile appearance or behavior also seem to inhibit adult responses to "communicative' or provocative behaviors. The alarm displays of elk calves (Altmann, 1960) and antelope fawns (Autenreith and Fichter, 1975) appear to be ignored. So are their adultlike threat signals and those of squirrel monkeys (Ploog, 1966). We have already mentioned the fact that immatures often seem to possess characteristics that inhibit the display of hostile behavior (Section 3.1.2). Additional examples would include the male lion's tolerance of young at the kill (Schaller, 1972) and the tendency of adults of several primates to permit offspring to take food from them (Carpenter, 1965; A. Estrada and R. Estrada, 1977; Hampton *et al.,* 1966; Rowell, 1966a).

Again, the data indicate a unique capacity of the young to influence the behavior of adults. In this case, they may do so by blocking or inhibiting the kinds of reactions that normally would be directed toward mature conspecifics.

6.1.1c. Avoidance. Although stimuli peculiar to offspring must evoke solicitude if the young are to survive, the advantages accruing from selective responsiveness seem to have led to yet another paradoxical effect. Naive female rodents (Noirot, 1972a) and primiparous ungulates (Lent, 1974) often appear to be repulsed or frightened by neonates. Similarly, in captive Japanese macaques, males actively avoid the initial approaches of newborns (Alexander, 1970) and members of both sexes deliberately skirt mothers with new infants (Murray and Murdoch, 1977).

Perhaps this "frightening" aspect of young can account for the apparently extreme negativism shown by primiparous, isolation-reared rhesus monkeys (Seay et al., 1964) and their "rehabilitation" upon giving birth for a second time (Harlow et al., 1966) and, partly, for the lower rate of survival of firstborn infants (Drickamer, 1974). It seems that for some animals, neonates take a little "getting used to" even though their mothers are hormonally primed to be responsive. In this connection, however, it may be worthy of note that where such avoidance seems most marked, the animals concerned were laboratory bred (Noirot, 1972a; Seay et al., 1964) or species in which mothers terminate caregiving prior to the annual birth season (Lent, 1974). Whether the apparently aversive qualities of neonates are artifactual, or function to guarantee the exclusiveness of the early mother–young bond, or restrict solicitude to hormonally primed individuals, or whether they provide an additional explanation for the benefits of aunting, remains to be elucidated.

6.1.1d. *Age-Graded Attractiveness.* Despite the apparently aversive qualities of neonates, the very young are generally attractive to adults in a number of species. In their pioneering work on maternal behavior in rats, Wiesner and Sheard (1933) showed that the probability of caregiving solicitude varied according to pup age. Similar findings have been reported for laboratory mice (Beniest-Noirot, 1958). Indeed, the entire, changing course of responsiveness to offspring can be demonstrated in induced virgin female mice (Noirot, 1964a,b) and rats (Reisbick et al., 1975) by varying the age of foster pups.

Rat pups between the ages of 4 and 10 days seem to be optimal stimuli for eliciting most components of caregiving behavior; by repeatedly fostering new litters, lactation can be extended for at least twice the normal duration (Bruce, 1961; Nicoll and Meites, 1959; Wiesner and Sheard, 1933). In addition, the duration of lactation diestrus can be extended by repeated litter replacement, although not for as long as lactation (Bruce, 1961). Milk production in mice (Selye and McKeown, 1934) and opossums (Reynolds, 1952) also can be maintained for extended periods by litter replacement.

Even among hamsters, who are less easily sensitized to offspring, 7- to 10-day-old pups are more "potent" stimuli than 1- or 5-day-old young (Richards, 1966; Rowell, 1961). In mice, the optimal age of pups for inducing caregiving is younger. When 1- to 4-day-old pups are pitted against 7- to 11-day-old young, the former group elicit more picking-up and retrieving than the latter (Fraser and Barnet, 1975), and the tendencies for mothers to combine litters for communal nursing (Section 3.1.3b) is greater when pups are under 5 days of age (Sayler and

Salmon, 1971). Among flying squirrels, mothers will retrieve young under the age of 40 days, and if presented with several pups of varying ages, they will return the youngest to the nest first (Muul, 1970). Younger Japanese macaques are defended by their mothers whenever they are in disputes with older siblings, regardless of the cause of the problem (Kurland, 1977).

As implied by the findings that the full cycle of caregiving can be mimicked by presenting animals with progressively older test litters, solicitude often wanes as a function of offspring age. Whereas mother ground squirrels tend to be attentive to any young under the age of 24 days, older alien pups often are treated aggressively (Michener, 1974; and just as the production of the maternal pheromone in rats can be induced by providing mothers with stimulation from growing pups between the ages of 8 and 16 days, it can be inhibited early by fostering older pups (Holinka and Carlson, 1976; Moltz and Leon, 1973). Thus, age-related offspring qualities also can account for parental refractoriness.

6.1.2. Responsiveness According to Offspring Gender

Unfortunately, available data do not permit systematic testing of Trivers' (1972) hypotheses concerning the conditions under which one sex should receive more parental investment than the other. Nevertheless, adults of many species have been observed to respond to the gender of even quite young offspring.

In pronghorn antelopes (Autenreith and Fichter, 1975) and lowland gorillas (Hess, 1973) differences in postpartum maternal genital grooming of young indicate an early reaction to offspring sex. A similar preoccupation with inspecting the genitals of very young infants (Hrdy, 1977; Gouzoules, 1975; Rosenblum and Kaufman, 1967) and differential transport patterns (Ingram, 1977) show that monkeys also respond to the gender of offspring at an early age.

In several primates, many of the responses to gender of maturing young seem to be related to the greater activity levels of males (Jensen and Bobbitt, 1968; Mitchell and Schroers, 1973). The high activity levels and sex play of young ungulate males may contribute to their early weaning and to the intolerance of adult males during the rut (Lent, 1974). Among monogamous primates, adults are likely to become intolerant of like-sexed offspring as they approach sexual maturity (Carpenter, 1940; Vogt et al., 1978).

6.2. Analysis According to Sensory Modality

If an advantage accrues to individuals who invest only in appropriate beneficiaries and monitor their expenditures, potential caregivers should use all the information they can. Parenting ought to be possible even under conditions of limited sensory handicap. Although available data are primarily derived from laboratory studies in rats, they are consistent with these expectations (Bell and Harper, 1977). At this stage, absence of evidence for the involvement of more than one sensory avenue is better regarded as an indication of gaps in our knowledge than as a species-specific attribute.

6.2.1. Thermal/Tactile Stimulation

The role of sucking stimulation in the consolidation and maintenance of lactation has been indicated above. Studies showing that the degree of nipple innervation is related to the continuation of milk production (Edwardson and Eayrs, 1967) and that the amount of milk released depends upon the number of pups nursing (Fuchs and Wagner, 1963) confirm the importance of tactual stimulation of the nipples. Other responses typical of lactating females also seem to be dependent upon nipple stimulation. The aggressive response to intruder males shown by lactating or induced virgin female mice varies according to whether or not pups can suckle (Svare and Gandleman, 1976; Svare, this volume) and the attenuated pituitary–adrenal stress response of mother rats requires stimulation of the nipples by the litter (Smotherman et al., 1976).

Similarly, assumption and maintenance of the nursing posture often depend upon tactual stimulation. In rabbits, a doe will not assume the nursing posture until all her offspring are on the teats (Fuchs and Wagner, 1963), and in cats, the duration of nursing bouts depends on the number of kittens sucking (Schneirla et al., 1963). In domestic swine, the posture and vocalizations typical of lactating sows can be induced by rubbing the anterior end of the udder, mimicking the kneading stimulation normally provided by the litter (Fraser, 1976). In guinea pigs (Harper, 1976) and bush babies the young elicit nursing by crawling under their dams. In bush babies, this stimulation tends to inhibit all motor activity— even retrieving (Doyle et al., 1969). Similarly, among captive stump-tailed macaques, when an infant crawls upon the alpha male, he may "freeze" until the youngster is removed—although simple approach by an infant might elicit a "cuff" (Gouzoules, 1975).

In virgin female laboratory rats, the latency to become responsive when caged with foster litters varies inversely with the probability of making (physical) contact with them. The smaller the cage (from 200 through 3000 cm^2), the more rapidly the females become solicitous (Terkel and Rosenblatt, 1971).

The amount of time a mother rat spends with her litter in the postpartum period depends in part upon her (hormonally mediated) sensitivity to changes in body temperature. Presumably, contact with neonates acts to reduce the mother's temperature (which is increased due to the hormonal changes induced by pup suckling). As her pups' thermoregulation matures, they no longer reduce her core temperature and she thus leaves the nest sooner (Leon et al., 1978). Since the course of nest building in pregnant mice nicely follows the curve for changed progesterone levels (Lynch and Possidente, 1978), a similar phenomenon may be operating in this species. Sensitivity to temperature change may thus account for some of the tendency for mother rats (Seitz, 1958; Grota, 1973) and mice (Priestnall, 1972) to spend less time in their nests with larger litters.

In their study of postpartum maternal aggression against unfamiliar mice, Svare and Gandelman (1973) found that, whereas "normal" 14-day-old intruders were attacked vigorously, hairless intruders of the same age were less often attacked and more frequently "mothered." Although some of the difference might be due to increased distress calling of the nude, and thereby cooler, test pups, this suggests that tactual qualities could distinguish immatures from other conspecifics.

6.2.2. Olfactory/Gustatory Stimuli

An apparently prenatally mediated olfactory signal seems to be emitted by pregnant Mongolian gerbils. Whereas males cannibalize infants while caged with nonpregnant females, those caged with females who are at least through the first fifth of gestation become much more solicitous of neonatal test pups. The female's state may be communicated via her urine or increased scent marking late in pregnancy (Elwood, 1977)—there is a strong association between lactation and scent marking among gerbils (Yahr, 1976).

In the majority of mammals studied, sniffing, licking, and/or consuming the birth fluids, membranes, and placenta are among the most frequently encountered perinatal activities. Insofar as fluids associated with birth seem to be highly attractive, they may contribute to a smooth transition from self-grooming to cleaning the young (Autenreith and Fichter, 1975; Lent, 1974; Nadler, 1974; Schneirla *et al.*, 1963; Rosenblatt and Siegel, this volume).

6.2.2a. Identification of Own Offspring. Vigorous licking, etc., directed toward her newborns may provide the mother with cues by which she discriminates them from other conspecific young. Field observations of a number of species in which the mother–offspring bond is exclusive suggest that olfaction plays a major role in recognition (Burton *et al.*, 1975; Gould, 1975; Lent, 1974).

It has been proposed that selective solicitude for own offspring and hostility toward aliens stem from habituation to familiar odors (Lynds, 1976). This hypothesis is consistent with the finding that 20 min *or more* spent licking a lamb is associated with subsequent rejection of alien young (Smith *et al.*, 1966). Interference with olfactory receptors in sheep (Baldwin and Shillito, 1974) and goats (Klopfer and Gamble, 1966) indicates that exclusive bonding in these species depends upon early olfactory stimulation; however, contrary to the habituation hypothesis, mothers denied such input tend to be indiscriminately responsive to conspecific young. Among Corsican wild sheep, covering the anus of lambs with cream or plastic pants also blocks mothers' ability to identify their own lambs (Pfeffer, 1967).

Even among generally responsive animals, there often is discrimination of own young. Lactating laboratory rats accept and care for fostered pups, but they tend to retrieve their own offspring faster than another female's young. If animals are rendered anosmic by removal of the olfactory bulbs during the first postpartum week, they cease to respond preferentially to their own pups, while intact females in the same stage of lactation continue to discriminate (Beach and Jaynes, 1956a).

In captive spiny mice, females of the same colony give no indication of discriminating their own from others' offspring; communal acceptance and nursing of pups is common (Porter and Doane, 1976). However, mothers are likely to cannibalize pups from a different colony. This discrimination seems to depend upon diet; so long as females are given the same kind of feed, they will accept one another's offspring (Doane and Porter, 1978). Apparently, something similar may be occurring in goats; early maternal licking and/or nursing may be the mother's way of labeling their infants (Gubernick, 1980, and this volume). In tree shrews, the mother must scent-mark her offspring in order to avoid their being cannibalized by other members of the group (Autrum and von Holst, 1968). The tendency of nursling hedgehogs to "anoint" themselves with their own saliva when placed in a strange situation may serve a similar end (Brockie, 1976).

6.2.2b.   *Inhibiting or Facilitating Responsiveness.*   Whereas olfactory bulb removal frequently leads inexperienced female rats to cannibalize pups, bulbectomized animals in whom responsiveness is induced become "maternal" unusually rapidly (Fleming and Rosenblatt, 1974a). When reversible olfactory blockage is achieved by means of intranasal injections of zinc sulfate, the incidence of cannibalism is reduced sharply and the latency for induced responsiveness remains very brief (Fleming and Rosenblatt, 1974b).

Prior induction of intact animals reduces the latency for subsequent caregiving, yet anosmic rats are actually slower to reinduce after recovery. Olfactory cues associated with infant rat pups thus may be the basis of their aversive qualities. [Recent separation of the effects of lesions in the vomeronasal and olfactory systems indicate that pup odors are probably complex and exert a variety of response-specific effects (Fleming et al., 1979)]. Presumably, hormonal priming during pregnancy acts in part to affect the prospective mother's responses to pup odors (Fleming and Rosenblatt, 1974b; Mayer and Rosenblatt, 1975).

Whereas olfactory blockage in rats seems to facilitate responsiveness, postnatal ablation of the olfactory bulbs in mice results in cannibalization of almost 80% of young pups. Even if females are bulbectomized prior to mating and produce several litters, the tendency to cannibalize young remains strong. Older pups, about 14 days of age, are not cannibalized, however, indicating that ablation does not disturb the response mechanisms *per se* (Gandelman et al., 1971, 1972). However, bulbectomy in rats also is associated with high rates of cannibalism (Fleming and Rosenblatt, 1974a; but see Orbach and Kling, 1966); therefore, one cannot dismiss the possibility that such cannibalism is an artifact of ablation rather than a simple correlate of anosmia.

Deafened neoparous rats respond to hidden pups with sustained adrenocortical responses like those of hearing females (Zarrow et al., 1972); and so long as their olfactory apparatus remains intact, lactating primipara nurse their young despite multiple sensory impairment (Mena, 1971). In a Y-maze situation, rat and mouse mothers are more likely to initiate retrieving bouts in response to the ultrasonic cries of pups when they are associated with olfactory cues than when sound alone is presented. Indeed, odor alone can be more effective than sound under some conditions (Smotherman et al., 1974, 1978).

Offspring may emit distinctive odors diagnostic not only of their age but also of their sex. If extracts from the tarsal scent glands, the antorbital sacs, or the foreheads of black-tailed deer are presented to other deer, the responses of the latter, especially males, differentiate between sex and age classes (Müller-Schwartze, 1971; Volkman *et al.*, 1978).

In sum, olfactory/gustatory cues play a major role in the onset and mainte-nance of parental responsiveness in mammals. Some of these stimuli are produced by the young themselves, as in deer; others may be mediated by maternal diet, or may even be directly applied to offspring by the mother. In any case, taste and/or odor can be diagnostic of offspring age and sex and provide cues for individual recognition, all of which are important kinds of information for animals whose reproductive success depends upon judicious and discriminating allocation of parental solicitude.

6.2.3. Auditory Stimuli

Theoretically, the degree of offspring maturity might also be signaled through auditory cues to aid parents in gauging when to cease or reduce their investment. Although no systematic data are available on the effects of age-related changes in offspring vocal output, several such changes can be identified. Simply due to the size of the vocal apparatus of the young, their calls often are distinctive (e.g., Bell and Harper, 1977; see also Lenhardt, 1977; Matsumura, 1979; Sand-gren *et al.*, 1973). In several species, age can be determined from the pattern as well as the pitch of the infant's cries (Berryman, 1976; Channing and Rowe-Rowe, 1977; M. S. Hafner and D. J. Hafner, 1979; Matsumura, 1979; McCarley, 1975). In addition to the relationship between age and pitch of calls, in goats there exist sex differences in the maximum energy frequency of kids' calls (Lenhardt, 1977). In many of the same animals, such calls serve to orient and/or elicit maternal approach or retrieving (Bell and Harper, 1977; Haskins, 1977; Pettijohn, 1977; Pola and Snowdon, 1975), although they may not be necessary to retrieving in mice (Busnell and Lehman, 1977) and rats (Smotherman *et al.*, 1974, 1978).

Among rodents and primates, adults' auditory thresholds are lowest in the frequency ranges characteristic of offspring "distress" calls (Nelson 1965; Noirot, 1972*b*; Struhsaker, 1967). Tests using electronically produced stimuli showed that pulses in the center of the frequency range of pup calls evoked maximal behavioral responsiveness from adult mice (Smith, 1976).

It has been argued that the vocalizations of infant rodents only affect mater-nal "arousal" (Bell, 1974). However, the facts that lactating females' responsive-ness is not simply a function of rate of pup calling (Bell and Little, 1978) and that these (ultrasonic) cries vary qualitatively with the kinds of treatment given to young (Noirot, 1972*b*), indicate that some mammalian offspring probably do use auditory channels to signal their needs. In fact, Noirot (1974) showed that the quality of nests built by naive female mice varies according to pups' calls.

The vocalizations of young may elicit other parental responses. Defensive behavior has been observed in response to alarm calls in several species, including ungulates (e.g., Autenrieth and Fichter, 1975; Walther, 1969) and primates (e.g., Bernstein, 1976; Hall and Mayer, 1967). The frequency of nursing bouts in swine (Lent, 1974) and milk ejection in lactating rats (Deis, 1968) also can be increased by exposure to the sounds of suckling young.

In addition to indicating offspring location, state, and activity, the vocalizations of young provide stimuli which can serve as the basis for individual identification. Maternal recognition of own offspring's tape-recorded calls has been demonstrated in reindeer (Espmark, 1971) and elephant seals (Petrinovitch, 1974). Furthermore, under conditions that minimize the influence of other sensory avenues, the calls of own offspring are preferred by squirrel monkeys (Kaplan *et al.*, 1978), galagos (Klopfer, 1970), sheep (Poindron and Carrick, 1976), and perhaps goats (Lenhardt, 1977).

Thus the vocalizations of mammalian offspring can orient their parents to their location, trigger defending behavior, signal needs for warmth or tactual comfort, facilitate milk ejection, and provide cues as to their identity. In addition, these cries have the potential for conveying information concerning the developmental status and gender of the young.

6.2.4. Visual Stimuli

One of the most obvious visual cues that distinguish young from other age classes is size,. According to Geist (1971), this is an important cue for adult dominance among mountain sheep. However, the effects of size on parenting have not yet been investigated systematically.

Another distinguishing feature is body conformation—relatively large head, shorter (longer) legs, shorter neck, etc. Again, the effects of these stimuli on conspecific parenting in animals have yet to receive experimental attention; all we can state at this point is that *humans* react to stimuli typical of the conformation of offspring of most species.

Among many, if not most, mammals the pigmentation or quality of infants' pelage sets them apart from other members of their species. In several primates (e.g., Hendy-Neely and Rhine, 1977) and carnivores (H. and J. van Lawick-Goodall, 1971) there seems to be a relationship between the coloration of the young and the degree of attention that they elicit from others. In some primates coloration of offspring may undergo several changes before the adult hue is finally attained (Hall and DeVore, 1965). The importance of the stimulation associated with infantile coloration in these species can be judged by the ways in which dead infants are used in social interactions (Section 3.1.2) and from Booth's (1962) demonstration that even a stuffed infant released defensive behavior from adults (Section 2.3.1).

In addition to their anatomical peculiarities and coloration, the young also exhibit behavior that provides visual cues concerning age and state. The gait of very young infants, even among relatively precocial mammals, is unsteady and distinctive (Rose, 1977). An early, "extreme" avoidance of infants by captive adult male stump-tailed macaques may be related to the relatively uncoordinated locomotion of neonates (Gouzoules, 1975).

In other forms, the young appear to use postural signals to solicit care. Autenreith and Fichter (1975) have described a stereotyped posture in pronghorn antelope kids that precedes maternal anogenital grooming. Similarly, the tail wagging of sheep and goats (Collias, 1956) and white-tailed deer (Townsend and Bailey, 1975) often are antecedents of maternal perineal licking. Among ungulates, a signal of intent to nurse is to cross directly in front of the mother (Collias, 1956; Lent, 1966, 1974; Schaller, 1977; Tyler, 1972). In addition, the angle from which the young approach their mothers also affects the likelihood that they will be allowed to suck, or the duration of the nursing bout (Autenreith and Fichter, 1975; Lent, 1966).

Other visually distinctive features of offspring behavior, such as the tail posture of juvenile patas monkeys (Hall, 1965), may have some signal value. The locomotor play of Himalayan sheep and goats sometimes induces similar behavior from adults (Schaller, 1977), and the "play faces" or invitation postures of primates and canids serve as signals for adults as well as peers (Bekoff, 1972). The latter behavior may explain why adults appear to ignore otherwise well-executed threat (Altmann, 1960; Ploog, 1966) and alarm (Altmann, 1960; Jay, 1965) gestures of juveniles.

Especially during the puerperium, the amount of offspring activity *per se* may be a determinant of parental response. For many ungulates, the physical activity of neonates seems to be a condition favoring maternal responsiveness (Dagg and Foster, 1976; P. Klopfer and M. Klopfer, 1977; Lent, 1974). Adult male guereza monkeys' interest in infants (Horwich and Manski, 1975), baboon aunts' solicitude for infants (Rowell *et al.*, 1968), and the grooming and retrieving behavior of bush babies (Doyle *et al.*, 1969) seem to depend upon the level of activity of the young. Similarly, rat mothers are more likely to retrieve active than anesthetized pups (Smith and Berkson, 1973). In caribou (Lent, 1966) and lions (Schaller, 1972), young who are stillborn or too weak to walk are abandoned.

6.2.5.   Quantitative and Qualitative Summation

The number of young is an important determinant of caregiving. We have already seen that the timing of implantation and the course of pregnancy can be influenced by the number of young *in utero*, or sucking. Whether mediated by thermal cues alone, or partly by some sort of stimulus satiation or habituation, the litter size also affects the postnatal level of maternal responsiveness in rats. Experiments in which the rate of pup calling is systematically varied show that the intensity of exteroceptive stimulation alone can influence the amount of maternal behavior displayed by rats (Brown *et al.*, 1977).

Thus there may be lower and upper limits for frequency or intensity of offspring stimulation that, on the average, serve as reliable boundaries, below or above which it would be uneconomical to invest at all, or at current levels.

In addition, as we have seen from the interaction of olfactory and auditory stimuli or, from within one modality, of color and motion, several attributes may combine to enhance responsiveness to a fixed number of young. In their studies of sensory control of retrieving in rats, Beach and Jaynes (1956b) have shown that, whereas no one modality is essential for performance, when receptor function is blocked after the first week, increased deafferentiation is associated with impaired functioning. As indicated in Section 6.2.2, the issue of whether olfaction is essential for the consolidation and maintenance of caregiving in mice and rats is currently unclear. In mice, olfaction *may* be highly important (Gandelman *et al.,* 1971, 1972); in rats, the evidence is contradictory, but insofar as Beach and Jaynes' (1956b) work has been replicated with primiparous caesarean-delivered rats whose sensory systems were blocked at or before delivery (Herrenkohl and Rosenberg, 1972), the general concept of polysensory control of maternal behavior seems to be substantiated. It would appear that in order to cause a complete failure of parenting at least two channels of input must be blocked (e.g., Mena, 1971), although loss of a single avenue may reduce effectiveness (Schwartz and Rowe, 1976) or discrimination (Baldwin and Shillito, 1974; Klopfer and Gamble, 1966).

In summary, offspring can and do affect the behavior of their parents via all of the senses. Corresponding to species differences in the overall strategies for caring for offspring, we can expect that there will be differences in the relative importance of discrete input channels. Moreover, as implied by Hrdy's (1976) analysis of species differences in the natal coloration of monkeys, there may also exist systematic correlations between a species' habitat and the kinds of stimuli that are most important in guiding parental behavior.

7. Filial Behavior

Although the reciprocal influences of parent and offspring on one another have been acknowledged for some time, systematic study of filial behavior is literally in its infancy. It is obvious that selection acts upon genes controlling the timing of the appearance of physical and behavioral characteristics (Gould, 1977; Wilson, 1975). Yet, although few would deny the fact that the behavior of the young must be matched to that of their parents, fewer study it.

In this and other chapters, many examples of developmental synchronies between parent and offspring have been discussed. Interactions start almost from the moment of fertilization. Subsequent physiological development of the conceptus obviously must be adapted to the uterine environment, and, at least near term, these adaptations must be controlled by fetal regulatory systems (e.g., Silman *et al.,* 1978).

Postnatally, the neonate mammal must not only be capable of locating the mothers' teats and sucking, but also be able to do so in the habitat peculiar to the species. The developmental timing of approach-and-flight responses in ungulates (Lent, 1974) or clinging and rooting, etc., in monkeys (King *et al.*, 1974; Suomi, 1977) must be synchronized to fit with the kinds of parental solicitude typical of each postnatal phase of the caregiving relationship and, in addition, to the wider ecological setting. For example, the infant hippopotamus must be able to suck even while totally submerged (Laws and Clough, 1966), and the calls of neonate rat pups are controlled not just by temperature but also by the kinds of odors that are present in the environment: Calling is inhibited under conditions that might, on average, lead to cannibalism (Conley and Bell, 1978).

There is no lack of evidence for the ways in which the anatomy and behavior of mother and young are adapted to each other. One only needs to consider the difficulties involved in attempts to foster young from one species to mothers from another in order to realize the intricacies involved (e.g., Beach, 1939; Hersher *et al.*, 1963). Our problem seems to be in appreciating them fully.

Theory suggests that parental care should be influenced by offspring signals, especially whenever it would be difficult for parents to directly assess offspring needs (Alexander, 1974; Trivers, 1974). This means that some facets of the behavior of the young must be under more complex control than that implied by the notion of simple "reflex" responses. As indicated by the rat pup's selective emission of ultrasonic calls (Conley and Bell, 1978), when modulating such behavior may enhance survival, selection should favor the development of situationally sensitive regulation. Given that the behavior of young varies according to internal state (e.g., Hall *et al.*, 1977) and irrespective of whether their signals are "epiphenomena" of other responses to that state (Verhaeghe and Noirot, 1978), when signaling can be inhibited according to the situation (Harper, 1972) we should expect to find evidence of complex control mechanisms.

Whether the differences in the organization of filial behavior of altricial (e.g., Rosenblatt, 1976) and precocial (e.g., Lent, 1974) young will be as marked as the contrast in their developmental status remains to be seen. However, if we are ever to fully untangle the complex fabric of the parent–offspring relationship in any species, ultimately we will have to have a far better understanding of the determinants as well as the effects of the behavior of the young.

ACKNOWLEDGMENTS. The writing of this chapter was partially supported by the Agriculture Experimental Station, University of California.

References

Alexander, B. K., 1970, Parental behavior of adult male Japanese monkeys, *Behaviour* **36**:270–285.

Alexander, R. D. 1974, The evolution of social behavior, *Ann. Rev. Ecol. Syst.* **5**:325–383.

Alexander, R. D., and Borgia, G., 1978, Group selection, altruism and the levels of organization of life, *Ann. Rev. Ecol. Syst.* **9**:449–474.

Altmann, J., Altmann, S. A., and Hausfater, G., 1978, Primate infant's effects on mother's future reproduction, *Science* **201**:1028–1030.

Altmann, M., 1960, The role of juvenile elk and moose in the social dynamics of their species, *Zoologica* **45**:35–39.

Altmann, S. A., 1974, Baboons, space, time and energy. *Am. Zool.* **14**:221–248.

Altmann, S. A., and Altmann, J., 1970, *Baboon Ecology*, University of Chicago Press, Chicago.

Anderson, A. B. M., and Turnbull, A. C., 1973, Comparative aspects of factors involved in the onset of labour in ovine and human pregnancy, *Mem. Soc. Endocrinol.* **20**:144–162.

Armitage, K. B., and Downhower, J. F., 1974, Demography of yellow-bellied marmot populations, *Ecology* **55**:1233–1245.

Autenreith, R. E., and Fichter, E., 1975, On the behavior and socialization of pronghorn fawns, *Wildl. Monogr.* No. 42, pp. 1–111.

Autrum, H., and von Holst, D., 1968, Socialer "stress" bei Tupias *(Tupaia glis)* un seine Wirkung auf Wachstum Korpergewicht und Fortplanzung, *Z. Vergl. Physiol.* **58**:347–355.

Baldwin, B. A., and Shillito, E. E., 1974, The effects of ablation of the olfactory bulbs on parturition and maternal behaviour of soay sheep, *Anim. Behav.* **22**:220–223.

Barash, D. P., 1973, The social biology of the Olympic marmot, *Anim. Behav. Monogr.* **6**:171–245.

Barash, D. P., 1974, The social behaviour of the hoary marmot *(Marmota caligata)*, *Anim. Behav.* **22**:256–261.

Beach, F. A., 1939, Maternal behavior of the pouchless marsupial *Marmosa cinerea, J. Mammal.* **20**:315–321.

Beach, F. A., and Jaynes, J., 1956a, Studies of maternal retrieving in rats. I. Recognition of own young, *J. Mammal.* **37**:177–180.

Beach, F. A., and Jaynes, J., 1956b, Studies of maternal retrieving in rats. III. Sensory cues involved in the lactating female's response to her young, *Behaviour* **10**:104–125.

Bell, R. Q., and Harper, L. V., 1977, *Child Effects on Adults,* Erlbaum, Hillsdale, N.J.

Bell, R. W., 1974, Ultrasounds in small rodents: Arousal-produced and arousal-producing, *Dev. Psychobiol.* **7**:39–42.

Bell, R. W., and Little, J., 1978, Effects of differential early experience upon parental behavior in *Mus musculus, Dev. Psychobiol.* **11**:199–203.

Beniest-Noirot, E., 1958, Analyse du comportement dit 'maternel' chez la souris, *Monogr. Fr. Psychol.* **1** (1).

Bernstein, I. S., 1975, Activity patterns in a gelada monkey group, *Folia Primatol.* **23**:50–71.

Berryman, J. C., 1976, Guinea pig vocalizations: Their structure, causation, and function, *Z. Tierpsychol.* **41**:80–106.

Bertram, B. C. R., 1976, Kin selection in lions and in evolution, *in: Growing Points in Ethology* (P. P. G. Bateson and R. A. Hinde, eds.), pp. 281–301, Cambridge University Press, Cambridge.

Bleich, V. C., and Schwartz, O. A., 1974, Interspecific and intergeneric maternal care in woodrats *(Neotoma)*, *Mammalia* **38**:381–387.

Booth, C., 1962, Some observations on *Cercopithecus* monkeys, *Ann. N.Y. Acad. Sci.* **102**:477–487.

Box, H. O., 1975, A social developmental study of young monkeys *(Callithrix jacchus)* within a captive family group, *Primates* **16**:419–435.

Box, H. O., 1977, Quantitative data on the carrying of young by captive monkeys *(Callithrix jacchus)* by other members of their family groups, *Primates* **18**:475–484.

Bramblett, C. A., 1973, Social organization as an expression of role behavior among Old World monkeys, *Primates* **14**:101–112.

Breen, M. F., and Leshner, A. I., 1977, Maternal pheromone: A demonstration of its existence in the mouse *(Mus musculus)*, *Physiol. Behav.* **18**:527–529.

Breuggeman, J. A., 1973, Parental care in a group of free-ranging rhesus monkeys *(Macaca mulatta)*, *Folia Primatol.* **20**:178–210.

Bridges, R. S., 1975, Long-term effects of pregnancy and parturition upon maternal responsiveness in the rat, *Physiol. Behav.* **14**:245–249.

Bridges, R. S., 1977, Parturition: Its role in the long term retention of maternal behavior in the rat, *Physiol. Behav.* **18**:487–490.

Bridges, R., Zarrow, M. X., Gandelman, R., and Denenberg, V. H., 1972, Differences in maternal responsiveness between lactating and sensitized rats, *Dev. Psychobiol.* **5**:123–127.

Brockie, R., 1976, Self-anointing by wild hedgehogs, *Erinaceus europaeus,* in New Zealand, *Anim. Behav.* **24**:68–71.

Brown, C. P., Smotherman, W. P., and Levine, S., 1977, Interaction-induced reduction in differential maternal responsiveness: An effect of cue-reduction or behavior?, *Dev. Psychobiol.* **10**:273–280.

Brown, L. E., 1966, Home range and movement of small mammals, *Symp. Zool. Soc. London* **18**:111–142.

Bruce, H. M., 1961, Observations on the suckling stimulus and lactation in the rat, *J. Reprod. Fertil.* **2**:17–34.

Bulmer, M. G., 1970, *The Biology of Twinning in Man,* Oxford University Press, London.

Burton, R. W., Anderson, S. S., and Summers, C. F., 1975, Perinatal activities in the grey seal *(Halichoerus grypus)*, *J. Zool.* **177**:197–201.

Busnell, R. G., and Lehmann, A., 1977, Acoustic signals in mouse maternal behavior: Retrieving and cannibalism, *Z. Tierpsychol.* **45**:321–324.

Calhoun, J. B., 1962, *The Ecology and Sociology of the Norway Rat,* U.S. Public Health Serv.Publ. No. 1008, Bethesda, Md.

Carlier, C., and Noirot, E., 1965, Effects of previous experience on maternal retrieving in rats, *Anim. Behav.* **13**:423–426.

Carpenter, C. R., 1940, A field study in Siam of the behavior and social relations of the gibbon, *Comp. Psychol. Monogr.* **16**(5):1–212.

Carpenter, C. R., 1965, The howlers of Barro Colorado Island, *in: Primate Behavior* (I. DeVore, ed.), pp. 250–291, Holt, Rinehart and Winston, New York.

Castell, R., and Wilson, C., 1971, Influence of spatial environment on development of mother-infant interaction in pigtail monkeys, *Behaviour* **39**:202–211.

Chalmers, N. R., 1968, The social behaviour of free living mangabeys in Uganda, *Folia Primatol.* **10**:263–281.

Chance, M. R. A., Jones, E., and Shostak, S., 1977a, Factors influencing nursing in *Macaca fascicularis, Folia Primatol.* **28**:259–282.

Chance, M. R. A., Emory, G. R., and Payne, R. G., 1977b, Status referents in long-tailed macaques *(Macaca fascicularis)*: Precursors and effects of a female rebellion, *Primates* **18**:611–632.

Channing, A., and Rowe-Rowe, D. T., 1977, Vocalizations of South African mustelines, *Z. Tierpsychol.* **44**:283–293.

Cheney, D. L., 1978, Interactions of immature male and female baboons with adult females, *Anim. Behav.* **26**:389–408.

Chickazawa, D., Gordon, T. P., Bean, C. A., and Bernstein, I. S., 1979, Mother-daughter dominance reversals in rhesus monkeys *(Macaca mulatta)*, *Primates* **20**:301–306.

Clutton-Brock, T. H., and Guiness, F. E., 1975, Behaviour of red deer *(Cervus elaphus L.)* at calving time, *Behaviour* **55**:286–299.

Clutton-Brock, T. H., and Harvey, P. H., 1976, Evolutionary rules and primate societies *in: Growing Points in Ethology* (P. P. G. Bateson and R. A. Hinde, eds.), pp. 195–237, Cambridge University Press, Cambridge.

Clutton-Brock T. H., and Harvey, P. H., 1978, Mammals, resources and reproductive strategies, *Nature (London)* **273**:191–195.

Coe, C. M., Mendoza, S. P. Smotherman, W. P., and Levine, S., 1978, Mother-infant attachment in the squirrel monkey: Adrenal response to separation, *Behav. Biol.* **22**:256–263.

Collias, N. E., 1956, The analysis of socialization in sheep and goats, *Ecology* **37**:228–239.

Conley, L., and Bell, R. W., 1978, Neonatal ultrasounds elicited by odor cues, *Dev. Psychobiol.* **11**:193–197.

Crook, J. H., Ellis, J. E., and Goss-Custard, J. D., 1976, Mammalian social systems: Structure and function, *Anim. Behav.* **24**:261–274.

Currie, W. B., and Thorburn, G. D., 1977a, The fetal role in timing the initiation of parturition in the goat, *Ciba Found. Symp.* **47**: 49–72.

Currie, W. B., and Thorburn, G. D., 1977b, Parturition in goats: Studies of the inter-actions between the foetus, placenta, prostaglandin F and progesterone before par-turition, at term or at parturition induced prematurely by corticotrophin infusion of the foetus, *J. Endocrinol.* **73**:263–278.

Dagg, A. I., and Foster, J. B., 1976, *The Giraffe: Its Biology, Behavior and Ecology*, Van Nostrand Reinhold, New York.

Daly, M., and Wilson, M., 1978, *Sex, Evolution and Behavior*, Duxbury Press, North Scituate, Mass.

Dawkins, R., 1976, *The Selfish Gene*, Oxford University Press, New York.

Day, C. S. D., and Galef, B. G. Jr., 1977, Pup cannibalism: One aspect of maternal behavior in golden hamsters, *J. Comp. Physiol. Psychol.* **91**:1179–1189.

Deag, J. M., and Crook, J. H., 1971, Social behaviour and 'agonistic buffering' in the wild Barbary macaque *Macaca sylvana* L., *Folia Primatol.* **15**:183–200.

Deets, A. C., and Harlow, H. F., 1974, Adoption of single and multiple infants by rhesus monkey mothers, *Primates* **15**:193–203.

Deis, R. P., 1968, The effect of an exteroceptive stimulus on milk ejection in lactating rats,*J. Physiol.* **197**:37–46.

Denenberg, V. H., Hudgens, G. A., and Zarrow, M. X., 1964, Mice reared with rats:

Modification of behavior by early experience with another species, *Science* **143**:380–381.

DeVore, I., 1963, Mother-infant relations in free-ranging baboons, in: *Maternal Behavior in Mammals* (H. L. Rheingold, ed.) pp. 305–335, John Wiley, New York.

Dittus, W. P. J., 1977, The social regulation of population density and age-sex distribution in the toque monkey, *Behaviour* **63**:281–322.

Doane, H. M., and Porter, R. H., 1978, The role of diet in mother-infant reciprocity in the spiny mouse, *Dev. Psychobiol.* **11**:271–277.

Dobzhansky, T., 1970, *Genetics of the Evolutionary Process*, Columbia University Press, New York.

Douglas-Hamilton, I., and Douglas-Hamilton, O., 1975, *Among the Elephants*, Viking, New York.

Doyle, G. A. Anderson, A., and Bearder, S. K., 1969, Maternal behavior in the lesser bushbaby *(Galago senegalensis moholi)* under semi-natural conditions, *Folia Primatol.* **11**:215–238.

Drickamer, L. C., 1976, Quantitative observations of grooming behavior in free-ranging *Macaca mulatta*, *Primates* **17**:323–335.

Eaton, R. L., 1974, *The Cheetah: The Biology, Ecology , and Behavior of an Endangered Species*, Van Nostrand Reinhold, New York.

Edwardson, J. A., and Eayrs, J. T., 1967, Neural factors in the maintenance of lactation in the rat, *J. Endocrinol.* **38**:51–59.

Eibl-Eibesfeldt, I., 1961, The interactions of unlearned behaviour patterns and learning in mammals, in: *Brain Mechanisms and Learning* (J. F. Delafresnaye, ed.), pp. 53–73, Thomas, Springfield, Ill.

Elwood, R. W., 1975, Paternal and maternal behaviour in the Mongolian gerbil, *Anim. Behav.* **23**:766–772.

Elwood, R. W., 1977, Changes in the responses of male and female gerbils *(Meriones unguiculatus)* towards test pups during the pregnancy of the female, *Anim. Behav.* **25**:46–51.

Elwood, R. W., and Broom, D. M., 1978, The influence of litter size and parental behaviour on the development of Mongolian gerbil pups, *Anim. Behav.* **26**:438–454.

Emerson, S. B., 1973, Observations on infant sharing in captive *Colobus polykomos*, *Primates* **14**:93–100.

Emlen, S. T., and Oring, L. W., 1977, Ecology, sexual selection and the evolution of mating systems, *Science* **197**:215–223.

Epple, G., 1968, Comparative studies on vocalization in marmoset monkeys *(Hapalidae)*, *Folia Primatol.* **8**:1–40.

Epple, G., 1975, Parental behavior in *Saguinus fuscicollis* spp. (Callithricidae), *Folia Primatol.* **24**:221–238.

Erskine, M. S., Denenberg, V. H., and Goldman, B. D., 1978, Aggression in the lactating rat: Effects of intruder age and test arena, *Behav. Biol.* **23**:52–66.

Espmark, Y., 1971, Individual recognition by voice in reindeer mother-young relationship. Field observations and playback experiments. *Behaviour* **40**:295–301.

Estrada, A., and Estrada, R., 1977, Patterns of predation in a free-ranging troop of stumptail macaques *(Macaca arctoides)*: Relations to ecology II, *Primates* **18**:633–646.

Estrada, A., and Sandoval, J. M., 1977, Social relations in a free-ranging troop of stump-tail macaques *(Macaca arctoides)*: Male care behaviour. I, *Primates* **18**:793–813.

Ewer, R. F., 1968, A preliminary survey of the behaviour in captivity of the Dasyurid marsupial *Sminthopsis crassicaudata* (Gould), *Z. Tierpsychol.* **25**:319–365.

Fagen, R. M., 1976, Three-generation family conflict, *Anim. Behav.* **24**:874–879.

Findlay, A. L. R., 1968, The effect of teat anaesthesia on the milk-ejection reflex in the rabbit, *J. Endocrinol.* **40**:127–128.

Fisher, R. A., 1930, *The Genetical Theory of Natural Selection,* Clarendon Press, Oxford.

Fleming, A. S., 1976, Control of food intake in the lactating rat: Role of suckling and hormones, *Physiol. Behav.* **17**:841–848.

Fleming, A. S., and Rosenblatt, J. S., 1974a, Olfactory regulation of maternal behavior in rats: I. Effects of olfactory bulb removal in experienced and inexperienced lactating and cycling females, *J. Comp. Physiol. Psychol.* **86**:221–232.

Fleming, A. S., and Rosenblatt, J. S., 1974b, Olfactory regulation of maternal behavior in rats. II: Effects of peripherally induced anosmia and lesions of the lateral olfactory tract in pup-induced virgins, *J. Comp. Physiol. Psychol.* **86**:233–246.

Fleming, A. S., Vaccarine, F., Tambosso, L., and Chee, P., 1979, Vomeronasal and olfactory system modulation of maternal behavior in the rat, *Science* **203**:372–374.

Flesness, N. R., 1978, Kinship asymmetry in diploids, *Nature (London)* **276**:495–496.

Fletcher, I. C., 1971, Relationship between frequency of suckling, lamb growth and postpartum oestrus behaviour in ewes, *Anim. Behav.* **19**:108–111.

Fraser, D., 1976, The nursing posture of domestic sows and related behaviour, *Behaviour* **57**:51–63.

Fraser, G. D., and Barnet, S. A., 1975, Pup-carrying by laboratory mice in an unfamiliar environment, *Behav. Biol.* **14**:353–360.

Fuchs, A. R., and Wagner, G., 1963, Quantitative aspects of release of oxytocin by suckling in unaesthetized rabbits, *Acta Endocrinol* **44**:581–592.

Galef. B. G., and Clark, M. M., 1976, Non-nurturant functions of mother-young interaction in the agouti *(Dasyprocta punctata)*, *Behav. Biol* **17**:255–262.

Gandelman, R., 1975, Maternal nest-building performance and fetal number in Rockland-Swiss albine mice, *J. Endocrinol.* **44**:551–554.

Gandelman, R., and Davis, P. G., 1973, Spontaneous and testosterone-induced pup-killing in female Rockland-Swiss mice: The effect of lactation and the presence of young, *Dev. Psychobiol.* **6**:251–257.

Gandelman, R., and Simon, N. G., 1977, Spontaneous pup-killing by mice in response to large litters, *Dev. Psychobiol.* **11**:235–241.

Gandelman, R. Zarrow, M. X., and Denenberg, V. H., 1971, Stimulus control of cannibalism and maternal behavior in anosmic mice, *Physiol. Behav.* **7**:583–586.

Gandelman, R., Zarrow, M. S., and Denenberg, V. H., 1972, Reproductive and maternal performance in the mouse following removal of the olfactory bulb, *J. Reprod. Fertil.* **28**:435–456.

Geist, V., 1971, *Mountain Sheep,* University of Chicago Press, Chicago.

Geist, V., 1974, On the relationship of social evolution and ecology in ungulates, *Am. Zool.* **14**:205–220.

Goodall, J., 1977, Infant killing and cannibalism in free-living chimpanzees, *Folia Primatol.* **28**:259–282.

Gould, E., 1975, Neonatal vocalizations in bats of eight genera, *J. Mammal.* **56**:15–29.

Gould, S. J., 1977, *Ontogeny and Phylogeny,* Harvard University Press, Belknap Press, Cambridge, Mass.

Gouzoules, H., 1975, Maternal rank and early social interactions of infant stumptail macaques, *Macaca arctoides, Primates* **16**:405–418.

Graber, G. C., and Kristal, M. B., 1977, Uterine distention facilitates the onset of maternal behavior in pseudopregnant but not cycling rats, *Physiol. Behav.* **19**:133–137.

Grosvenor, C. E., and Mena, F., 1973, Evidence that suckling pups, through an exteroceptive mechanism, inhibit the milk stimulatory effects of prolactin in the rat during late lactation, *Horm. Behav.* **4**:209–222.

Grosvenor, C. E., Maiweg, H., and Mena, F., 1970, Effect of nonsuckling interval on ability of prolactin to stimulate milk secretion in rats, *Am. J. Physiol.* **219**:403–408.

Grota, L. J., 1973, Effects of litter size, age of young, and parity on foster mother behaviour in *Rattus norvegicus, Anim. Behav.* **21**:78–82.

Grubb, P., and Jewell, P. A., 1966, Social grouping and home range in feral Soay sheep, *Symp. Zool. Soc. London.* **18**:179–210.

Gubernick, D. J., 1980, Maternal 'imprinting' or maternal 'labeling' in goats? *Anim. Behav.* **28**:124–129.

Hafez, E. S. E., and Lineweaver, J. A., 1968, Suckling behaviour in natural and artificially fed neonate claves, *Z. Tierpsychol.* **25**:187–198.

Hafner, M. S., and Hafner, D. J., 1979, Vocalizations of grasshopper mice (genus *Onychomis*), *J. Mammal.* **60**:85–94.

Haldane, J. B. S., 1932, *The Causes of Evolution,* Harper, New York.

Hall, K. R. L., and DeVore, I., 1965, Baboon social behavior, in: Primate Behavior (I. DeVore, ed.), pp. 53–110, Holt, Rinehart and Winston, New York.

Hall, K. R. L. and Mayer, B., 1967, Social interactions in a group of captive patas monkeys *(Erythrocebus patas),* Folia Primatol. **5**:213–236.

Hall, W. G., Cramer, C. P., and Blass, E. M., 1977, Ontogeny of suckling in rats: Transitions toward adult ingestion, *J. Comp. Physiol. Psychol.* **91**:1141–1155.

Hamilton, W. D., 1964, The genetical evolution of social behaviour, Parts I and II, *J. Theor. Biol.* **7**:1–52.

Hampton, J. K. Jr., Hampton, S. H., and Landwher, B. T., 1966, Observations on a successful breeding colony of the marmoset, *Oesipomidas oedipus, Folia Primatol.* **4**:265–287.

Haour, F., and Saxena, B. B., 1974, Detection of a gonadotropin in rabbit blastocyst before implantation, *Science* **185**:444–445.

Harlow, H. F., Harlow, M. K., Dodsworth, R. O., and Arling, G. L., 1966, Maternal behavior of rhesus monkeys deprived of mothering and peer associations in infancy, *Proc. Am. Philos. Soc.* **110**:58–66.

Harper, L. V., 1970, Ontogenetic and phylogenetic functions of the parent-offspring relationship in mammals, *Adv. Study Behav.* **3**:75–117.

Harper, L. V., 1972, The transition from filial to reproductive function of coitus-related responses in young guinea pigs, *Dev. Psychobiol.* **5**:21–34.

Harper, L. V., 1976, Behavior, in: *The Biology of the Guinea Pig* (J. E. Wagner and P. J. Manning, eds.), pp. 31–51, Academic Press, New York.

Hart, I. C., Bines, J. A., Morant, S. V., and Ridley, J. L., 1978, Endocrine control of energy metabolism in the cow: Comparison of the levels of hormones (prolactin, growth hormone, insulin and thyroxine) and metabolites in the plasma of high- and low-yielding cattle at various stages of lactation, *J. Endocrinol.* **77**:333–345.

Haskins, R., 1977, Effect of kitten vocalizations on maternal behavior, *J. Comp. Physiol. Psychol.* **91**:830–838.

Hediger, H., 1950, *Wild Animals in Captivity,* Dover, New York.

Hendy-Neely, H., and Rhine, R. J., 1977, Social development of stumptail macaques *(Macaca arctoides):* Momentary touching and other interactions with adult males during the infant's first 60 days of life, *Primates* **18**:589–600.

Herrenkohl, L. R., and Rosenberg, P. A., 1972, Exteroceptive stimulation of maternal behavior in the naive rat, *Physiol. Behav.* **8**:595–598.

Hersher, L., Richmond, J. B., and Moore, A. U., 1963, Maternal behavior in sheep and goats, in: *Maternal Behavior in Mammals* (H. L. Rheingold, ed.), pp. 203–232, Wiley, New York.

Hess, J. P., 1973, Some observations on the sexual behavior of captive lowland gorillas, *Gorilla g. gorilla* (Savage and Wyman), in: *Comparative Ecology and Behaviour of Primates* (R. P. Michael and J. H. Crook, eds.), pp. 507–581, Academic Press, New York.

Hinde, R. A., and Spencer-Booth, Y., 1967, The effect of social companions on mother-infant relations in rhesus monkeys, in: *Primate Ethology* (D. Morris, ed.), pp. 267–286, Aldine Press, Chicago.

Hinde, R. A., Rowell, T. E., and Spencer-Booth, Y., 1964, Behaviour of socially living rhesus monkeys in their first six months, *Proc. Zool. Soc. London.* **143**:609–649.

Holinka, C. F., and Carlson, A. D., 1976, Pup attraction to lactating Sprague-Dawley rats, *Behav. Biol.* **16**:489–505.

Holm, L. W., 1967, Prolonged pregnancy, *Adv. Vet. Sci.* **11**:195–205.

Horwich, R. H., and Manski, D., 1975, Maternal care and infant transfer in two species of *Colobus* monkeys, *Primates* **16**:49–73.

Horwich, R. H., and Wurman, C., 1978, Socio-maternal behaviors in response to an infant birth in *Colobus guereza, Primates* **19**:693–713.

Hrdy, S. B., 1976, Care and exploitation of nonhuman primate infants by conspecifics other than the mother, *Adv. Study Behav.* **6**:101–158.

Hrdy, S. B., 1977, *The Langurs of Abu,* Harvard University Press, Cambridge, Mass.

Humphrey, S. R., 1975, Nursery roosts and community diversity of nearctic bats, *J. Mammal.* **56**:321–346.

Hunt, S. M., Gamache, K. M., and Lockard, J. S., 1978, Babysitting behavior by age/sex classification in squirrel monkeys *(Saimairi sciureus), Primates* **19**:179–186.

Ingram, J. C., 1977, Interactions between parents and infants, and the development of independence in the common marmoset *(Callithrix jacchus), Anim. Behav.* **25**:811–827.

Itani, J., 1959, Paternal care in the wild Japanese monkey *Macaca fuscata fuscata, Primates* **2**:61–93.

Izawa, K., 1978, A field study of the ecology and behavior of the black-mantle tamarin *(Saguinus nigricollis), Primates* **19**:241–274.

Jacobs, B. B., and Uphoff, D. E., 1974, Immulogic modification: A basic survival mechanism, *Science* **185**:582–586.

Jay, P., 1963, Mother-infant relations in langurs, *in: Maternal Behavior in Mammals* (H. Rheingold, ed.), pp. 282–304, Wiley, New York.

Jay, P., 1965, The common langur in north India, in: *Primate Behavior* (I. DeVore, ed.), pp. 197–249, Holt, Rinehart and Winston, New York.

Jensen, G. D., and Bobbitt, R. A., 1968, Sex differences in the development of independence of infant monkeys, *Behaviour* **30**:1–14.

Jolly, A., 1966, *Lemur Behaviour,* University of Chicago Press, Chicago.

Judge, D. S., and Rodman, P. S., 1976, *Macaca radiata:* Intragroup relations and reproductive status of females, *Primates* **17**:535–539.

Kann, G., and Martinet, J., 1975, Prolactin levels and duration of *postpartum* anoestrus in lactating ewes, *Nature (London)* **257**:63–64.

Kaplan, J., 1972, Differences in the mother-infant relations of squirrel monkeys housed in social and restricted environments, *Dev. Psychobiol.* **5**:43–52.

Kaplan, J., 1973, Responses of mother squirrel monkeys to dead infants, *Primates* **14**:89–91.

Kaplan, J. N., Winship-Ball, A., and Sim, L., 1978, Maternal discrimination of infant vocalizations in squirrel monkeys, *Primates* **19**:187–193.

Kaufmann, J. H., 1962, Ecology and social behavior of the coati, *Univ. Calif. Berkeley Publ. Zool.* **60**:95–222.

King, J. A., 1955, Social behavior, social organization, and population dynamics in a black-tailed prairiedog town in the Black Hills of South Dakota, *Contrib. Lab. Vertebr. Biol. Univ. Mich.* No. 67, pp. 1–123.

King, J. E., Fobes, J. T., and Fobes, J. L., 1974, Development of early behaviors in neonatal squirrel monkeys and cotton-top tamarins, *Dev. Psychobiol.* **7**:97–107.

Kleiman, D., 1972, Maternal behaviour of the green acouchi (*Myoprocta pratti* Pocock), a South American caviomorph rodent, *Behaviour* **63**:48–84.

Kleiman, D. G., 1974, Patterns of behavior in hystricomorph rodents, *Symp. Zool. Soc. London.* **34**:171–209.

Klopfer, P. H., 1970, Discrimination of young in galagos, *Folia Primatol.* **13**:137–143.

Klopfer, P. H., and Dugard, J.,1976, Patterns of maternal care in lemurs: III, *Lemur variegatus, Z. Tierpsychol.* **40**:210–220.

Klopfer, P. H., and Gamble, J., 1966, Maternal "imprinting" in goats: The role of chemical senses, *Z. Tierpsychol.* **23**:588–592.

Klopfer, P. H., and Klopfer, M. S., 1968, Maternal 'imprinting' in goats: Fostering of alien young, *Z. Tierpsychol.* **25**:862–866.

Klopfer, P. H., and Klopfer, M. S., 1970, Patterns of maternal care in lemurs: I, Normative description, *Z. Tierpsychol.* **27**:984–996.

Klopfer, P., and Klopfer, M., 1977, Compensatory responses of goat mothers to their impaired young, *Anim. Behav.* **25**:286–291.

Korányi, L., Phelps, C. P., and Sawyer, C. H., 1977, Changes in serum prolactin and corticosterone in induced maternal behavior in rats, *Physiol. Behav.* **18**:287–292.

Kruuk, H., 1972, *The Spotted Hyena: A Study of Predation and Social Behavior,* University of Chicago Press, Chicago.

Kumaresan, P., Anderson, R. R., and Turner, C. W., 1967, Effect of litter size upon milk yield and litter weight gain in rats, *Proc. Soc. Exp. Biol. Med.***126**:41–45.

Kummer, H., 1968, *Social Organization of Hamadryas Baboons,* University of Chicago Press, Chicago.

Kurland, J. A.,1977, Kin selection in the Japanese monkey, *Contrib. Primatol.* **12**:1–145.

Lancaster, J. B., 1971, Play-mothering: The relations between juvenile females and young infants among free-ranging vervet monkeys, *Folia Primatol.* **15**:161–182.

Langman, V. A., 1977, Cow–calf relationships in giraffe *(Giraffa camelopardalis giraffa)*, *Z. Tierpsychol.* **43**:264–286.

Laws, R. M., and Clough, G., 1966, Observations on reproduction in the hippopotamus, *Hippopotamus amphibius* Linn., *Symp. Zool. Soc. London.* **15**:117–140.

LeBoeuf, B., Whiting, R. J., and Gantt, R. F., 1972, Perinatal behavior of northern elephant seal females and their young, *Behaviour* **43**:121–156.

LeBoeuf, B. J., 1974, Male-male competition and reproductive success in elephant seals, *Am. Zool.* **14**:163–176.

LeBoeuf, B. J., and Briggs, K.T., 1977, The cost of living in a seal harem, *Mammalia* **41**:167–195.

Lehrman, D. S., 1953, Problems raised by instinct theories, *Quart. Rev. Biol.* **28**:337–365.

Lehrman, D. S., 1956, On the organization of maternal behavior and the problem of instinct, in: *L'Instinct* (P. P. Grassé, ed.), pp. 475–514, Masson, Paris.

Leidahl, L. C., and Moltz, H., 1975, Emission of the maternal pheromone in the nulliparous female and failure of emission in the adult male, *Physiol. Behav.* **14**:421–424.

Le Neindre, P., and Garel, J. P., 1976, Existence d'une periode sensible pour l'etablissement du comportement maternel de la vache après la mise-bas, *Biol. Behav.* **1**:217–221.

Le Neindre, P., Petit, M., and Garel, J. P., 1978, Allaitment de deux veaus par des vaches de race Salers II, Étude de l'adoption, *Ann. Zootech.* **27**:553–569.

Lenhardt, M. L., 1977, Vocal contour cues in maternal recognition of goat kids, *Appl. Anim. Ethol.* **3**:211–219.

Lent, P. C., 1966, Calving and related social behavior in the barren-ground caribou, *Z. Tierpsychol.* **23**:701–756.

Lent, P. C., 1974, Mother-infant relationships in ungulates, in: *The Behavior of Ungulates and Its Relation to Management,* Vol. 1 (V. Geist and F. Walther, eds.), pp. 14–55, *IUCN*, New Series No. 24.

Leon, M., and Moltz, H., 1972, The development of the pheromonal bond in the albino rat, *Physiol. Behav.* **8**:683–686.

Leon, M., Croskerry, P. G., and Smith, G. K., 1978, Thermal control of mother-young contacts in rats, *Physiol. Behav.* **21**:793–811.

Lickliter, R., 1979, On the significance of maternal imprinting in the domestic goat *(Capra Hircus)*. Paper presented at the annual meeting of the Animal Behavior Society, June, 1979, New Orleans.

Lincoln, D. W., 1974, Suckling: A time-constant in the nursing behaviour of the rabbit, *Physiol. Behav.* **13**:711–714.

Lockley, R. M., 1961, Social structure and stress in the rabbit warren, *J. Anim. Ecol.* **30**:385–423.

Lu, K. H., Chen, H. T., Huang, H. H., Grandison, L., Marshall, S., and Meites, J., 1976, Relation between prolactin and gonadotrophin secretion in post-partum lactating rats, *J. Endocrinol.* **68**:241–250.

Lynch, A., 1976, Postnatal undernutrition: An alternative method, *Dev. Psychobiol.* **9**:39–48.

Lynch, C. B., and Possidente, B. P., Jr., 1978, Relationships of maternal nesting to thermoregulatory nesting by house mice *(Mus musculus)* at warm and cold temperatures, *Anim. Behav.* **26**:1136–1143.

Lynds, P. G., 1976, Olfactory control of aggression in lactating female housemice, *Physiol. Behav.* **17**:157–159.

Mackinnon, D. A., and Stern, J. J., 1977, Pregnancy duration and fetal numbers: Effects on maternal behavior in rats, *Physiol. Behav.* **18**:793–797.

Mackinnon, J., 1974, The behaviour and ecology of wild orang-utans *(Pongo pygmaeus)*, *Anim. Behav.* **22**:3–74.

Madison, D. M., 1978, Movement indicators of reproductive events among female meadow voles as revealed by radiotelemetry, *J. Mammal.* **59**:835–843.

Makwana, S. C., 1979, Infanticide and social change in two groups of the Hanuman langur, *Presbytis entellus,* at Jodhpur, *Primates* **20**:293–300.

Mallory, F. F., and Brooks, R. J., 1978, Infanticide and other reproductive strategies in the collared lemming, *Dicrostonyx groenlandicus, Nature (London)* **273**:144–146.

Maneckjee, R., Srinath, B. R., and Moudgal, N. R., 1976, Prolactin suppresses release of luteinising hormone during lactation in the monkey, *Nature (London)* **262**:507–508.

Marsen, H. M., 1968, Agonistic behaviour of young rhesus monkeys after changes induced in social rank of their mothers, *Anim. Behav.* **16**:38–44.

Martin, R. D., 1966, Tree shrews: Unique reproductive mechanism of systematic importance, *Science* **152**:1402–1404.

Martinsen, D. L., 1968, Temporal patterns in the home ranges of chipmunks *(Eutamis)*, *J. Mammal.* **49**:83–91.

Matsumura, S., 1979, Mother-infant communication in a horseshoe bat *(Rhinolophus ferrumequinum nipponi)*: Development of vocalization, *J. Mammal.* **60**:76–84.

Mayer, A. D., and Rosenblatt, J. S., 1975, Olfactory basis for the delayed onset of maternal behavior in virgin female rats: Experiential effects, *J. Comp. Physiol. Psychol.* **89**:701–710.

McCarley, H., 1975, Long-distance vocalizations of coyotes *(Canis latrans)*, *J. Mammal.* **56**:847–856.

McLean, I. G., 1978, Plugging of nest burrows by female *Spermophilus columbianus, J. Mammal.* **59**:437–439.

Mena, F., 1971, Release of prolactin in rats by exteroceptive stimulation: Sensory stimuli involved, *Horm. Behav.* **2**:107–116.

Mena, F., and Grosvenor, C. E., 1968, Effect of number of pups upon suckling-induced fall in pituitary prolactin concentration and milk ejection in the rat, *Endocrinology* **82**:623–626.

Mena, F., Enjalbert, A., Carbonell, L., Priam, M., and Kordan, C., 1976, Effect of suckling on plasma prolactin and hypothalamic monomine levels in the rat, *Endocrinology* **99**:445–451.

Menzel, E. W., Jr., 1966, Responsiveness to objects in free-ranging Japanese monkeys, *Behaviour* **26**:130–150.

Merz, E., 1978, Male-male interactions with dead infants in *Macaca sylvanus, Primates* **19**:749–754.

Michener, G. R., 1974, Development of adult–young identification in Richardson's ground squirrel, *Dev. Psychobiol.* **7**:375–384.

Mitchell, G., and Schroers, L., 1973, Birth order and parental experience in monkeys and man, *Adv. Child Dev. Behav.* **8**:159–184.

Moehlman, P. D., 1979, Jackal helpers and pup survival, *Nature (London)* **277**:382–383.

Moltz, H., and Leon, M., 1973, Stimulus control of the maternal pheromone in the lactating rat, *Physiol. Behav.* **10**:69–71.

Moltz, H., Geller, D., and Levin, R., 1967, Maternal behavior in the totally mammectomized rat, *J. Comp. Physiol. Psychol.* **64**:225–229.

Moltz. H., Lubin, M., Leon, M., and Numan, M., 1970, Hormonal induction of maternal behavior in the ovariectomized nulliparous rat, *Physiol. Behav.* **5**:1373–1377.

Müller-Schwartze, D., 1971, Pheromones in black-tailed deer *(Odocoileus hemionus columbianus)*, *Anim. Behav.* **19**:141–152.

Murie, A., 1963, *A Naturalist in Alaska,* Anchor Books, Garden City, N.Y.

Murray, R. D., and Murdoch, K. M., 1977, Mother-infant dyad behavior in the Oregon troop of Japanese macaques, *Primates* **18**:815–824.

Muul, I., 1970, Intra and interfamilial behavior of *Glaucomys volans* (Rodentia) following parturition, *Anim. Behav.* **18**:20–25.

Nadler, R. D., 1974, Preparturitional behavior of a primiparous lowland gorilla, *Primates* **15**:55–73.

Nadler, R. D., and Tilford, B., 1977, Agonistic interactions of captive female orang-utans with infants, *Folia Primatol.* **28**:298–305.

Nash, L. T., 1978, The development of the mother–infant relationship in wild baboons *(Papio anubis)*, *Anim. Behav.* **26**:746–759.

Nelson, J. E., 1965, Behaviour of Australian *Pteropodidae (Megachiroptera)*, *Anim. Behav.* **13**:544–557.

Neyman, P. F., 1978, Aspects of the ecology of free-ranging cotton-top tamarins *(Saguinus O. oedipus)* and the conservation state of the subspecies, in: *Biology and Conservation of the Callithrichidae* (D. Kleiman, ed.), pp. 39–71, Smithsonian Institution, Washington, D.C.

Nicoll, C. S., and Meites, J., 1959, Prolongation of lactation in the rat by litter replacement, *Proc. Soc. Exp. Biol. Med.* **101**:81–82.

Nishida, T., Uehara, S., and Nyundo, R., 1979, Predatory behavior among wild chimpanzees of the Mahale mountains, *Primates* **20**:1–20.

Noirot, E., 1964*a*, Changes in responsiveness to young in the adult mouse. I. The problematical effect of hormones, *Anim. Behav.* **12**:52–58.

Noirot, E., 1964*b*, Changes in responsiveness to young in the adult mouse: The effect of external stimuli, *J. Comp. Physiol. Psychol.* **57**:97–99.

Noirot, E., 1964*c*, Changes in responsiveness to young in the adult mouse. IV. The effects of an initial contact with a strong stimulus, *Anim. Behav.* **12**:442–448.

Noirot, E., 1965, Changes in responsiveness to young in the adult mouse. III. The effect of immediately preceding performances, *Behaviour* **24**:318–325.

Noirot, E., 1969, Changes in responsiveness to young in the adult mouse. V. Priming, *Anim. Behav.* **17**:542–546.

Noirot, E., 1972*a*, The onset of maternal behavior in rats, hamsters and mice, A selective review, *Adv. Study Behav.* **4**:107–145.

Noirot, E., 1972b, Ultrasounds and maternal behavior in small rodents, *Dev. Psychobiol.* **5**:371–387.

Noirot, E., 1974, Nest-building by the virgin female mouse exposed to ultrasound from inaccessible pups, *Anim. Behav.* **22**:410–420.

Noirot, E., and Goyens, J., 1971, Changes in maternal behavior during gestation in the mouse, *Horm. Behav.* **2**:207–215.

Noirot, E., and Richards, M. P. M., 1966, Maternal behaviour in virgin female golden hamsters: Changes consequent upon initial contact with pups, *Anim. Behav.* **14**:7–10.

Noonan, F. P., Halliday, W. J., Morton, H., and Clunie, G. J. A., 1979, Early pregnancy factor is immunosuppressive, *Nature (London)* **278**:649–651.

Oates, J. F., 1977, The social life of a black-and-white Colobus monkey *Colobus guereza*, *Z. Tierpsychol.* **45**:1–60.

O'Connor, R. J., 1978, Brood reduction in birds: Selection for fratricide, infanticide and suicide? *Anim. Behav.* **26**:79–96.

Orbach, J., and Kling, A., 1966, Effect of sensory deficit on onset of puberty, mating, fertility and gonodal weignts in rats, *Brain Res.* **3**:141–149.

Orr, R. T., and Poulter, T. C., 1967, Some observations on reproduction, growth and social behavior in the Steller sea lion, *Proc. Calif. Acad. Sci.* **35**:193–226.

Ota, K., and Yokoyama, A., 1965, Resumption of lactation by suckling in lactating rats after removal of litters, *J. Endocrinol.* **33**:185–194.

Owens, D., and Owens, M., 1979, Notes on the social organization and behavior in brown hyenas *(Hyaena brunnea)*, *J. Mammal.* **60**:405–408.

Peterson, R. S., 1968, Social behavior in pinnipeds with particular reference to the northern fur seal, in: *The Behavior and Physiology of Pinnipeds* (R. J. Harrison, R. C. Hubbard, R. S. Peterson, C. E. Rice, and R. J. Schusterman, eds.), pp. 3–53, Appleton-Century-Crofts, New York.

Petrinovitch, L., 1974, Individual recognition of pup vocalizations by northern elephant seal mothers, *Z. Tierpsychol.* **34**:308–312.

Pettijohn, T. F., 1977, Reaction of parents to recorded guinea pig distress vocalization, *Behav. Biol.* **21**:438–442.

Pfeffer, P., 1967, Le mouflon de Corse (*Ovis ammon musimon* Schreber, 1782). Position systématique, écologie et éthologie comparées, *Mammalia* **31**(Suppl).

Platt, W. J., 1976, The social organization and territoriality of short-tailed shrew *(Blarina brevicauda)* populations in old-field habitats, *Anim. Behav.* **24**:305–318.

Ploog, D. W., 1966, Biological bases for instinct and behavior: Studies on the development of social behavior in squirrel monkeys in: *Recent Advances in Biological Psychiatry*, Vol. 8 (J. Wortis, ed.), pp. 199–223, Plenum Press, New York.

Poindron, P., and Carrick, M. J., 1976, Hearing recognition of the lamb by its mother, Anim. Behav. **24**:600–602.

Pola, Y. V., and Snowdon, C. T., 1975, The vocalizations of pygmy marmosets *(Cebuella pygmaea)*, Anim. Behav. **23**:826–842.

Porter, R. H., and Doane, H. M., 1976, Maternal pheromone in the spiny mouse *(Acomys cahirinus)*, *Physiol. Behav.* **16**:75–78.

Priestnall, R., 1972, Effects of litter size on the behaviour of lactating female mice *(Mus musculus)*, Anim. Behav. **20**:386–394.

Ransom, T. W., and Ransom, B. S., 1971, Adult male-infant relations among baboons *(Papio anubis)*, *Folia Primatol.* **16**:179–195.

Rasa, O. A. E., 1977, The ethology and sociology of the dwarf mongoose *(Helogale undulata rufula)*, *Z. Tierpsychol.* **43**:337–406.

Reisbick, S., Rosenblatt, J. E., and Mayer, A. D., 1975, Decline of maternal behavior in the virgin and lactating rat, *J. Comp. Physiol. Psychol.* **89**:722–732.

Renfries, M. B., 1979, Initiation of development of a diapausing embryo by mammary denervation during lactation in a marsupial, *Nature (London)* **278**:549–550.

Reynolds, H. C., 1952, Studies on reproduction in the opossum *(Didelphis virginiana virginiana)*, *Univ. Calif. Publ. Zool.* **52**:223–283.

Rheingold, H. L., 1963, *Maternal Behavior in Mammals,* Wiley, New York.

Rhine, R. J., and Hendy-Neely, H., 1978, Social development of stumptail macaques *(Macaca arctoides)*: Synchrony of changes in mother-infant interactions and individual behaviors during the first 60 days of life, *Primates* **19**:681–692.

Richard, A. F., 1976, Preliminary observations on the birth and development of *Propithecus verrauxi* to the age of six months, *Primates* **17**:357–366.

Richards, M. P. M., 1966, Maternal behaviour in the golden hamster: Responsiveness to young in virgin, pregnant, and lactating females, *Anim. Behav.* **14**:210–313.

Rose, M. D., 1977, Positional behaviour of olive baboons *(Papio anubis)* and its relationship to maintenance and social activites, *Primates* **18**:59–116.

Rosenblatt, J. S., 1967, Nonhormonal basis of maternal behavior in the rat, *Science* **156**:1512–1513.

Rosenblatt, J. S., 1976, Stages in the early development of altricial young of selected species of non-primate mammals, in: *Growing Points in Ethology* (P. P. G. Bateson, and R. A. Hinde, eds.), pp. 345–383, Cambridge University Press, Cambridge.

Rosenblatt, J. S., and Lehrman, D. S., 1963, Maternal behavior of the laboratory rat, in: *Maternal Behavior in Mammals* (H. Rheingold, ed.), pp. 8–57, Wiley, New York.

Rosenblatt, J. S., and Siegel, H. I., 1975, Hysterestomy-induced maternal behavior during pregnancy in the rat, *J. Comp. Physiol. Psychol.* **89**:685–700.

Rosenblum, L. A., and Kaufman, I. C., 1967, Laboratory observations of early mother-infant relations in pigtail and bonnet macaques in: *Social Communication among Primates* (S. A. Altmann, ed.), pp. 33–41, University of Chicago Press, Chicago.

Rosenson, L. M., 1977, The responses of some prosimian primate mothers to their own anesthetized infants, *Primates* **18**:579–588.

Rowell, T. E., 1960, On the retrieving of young and other behaviour in lactating golden hamsters, *Proc. Zool. Socl. London* **135**:265–282.

Rowell, T. E., 1961, Maternal behaviour in non-maternal golden hamsters, *Anim. Behav.* **9**:11–15.

Rowell, T. E., 1966a, Forest living baboons in Uganda, *J. Zool.* **149**:344–364.

Rowell, T. E., 1966b, Hierarchy in the organization of a captive baboon group, *Anim. Behav.* **14**:430–433.

Rowell, T. E., Din, N. A., and Omar, A., 1968, The social development of baboons in their first three months, *J. Zool.* **155**:461–483.

Rudran, R., 1973, Adult male replacement in one-male troops of purple-faced langurs *(Presbytis senex senex)* and its effect on population structure, *Folia Primatol.* **19**:166–192.

Sandgren, F. E., Chu, E. W., and Vandevere, J. E., 1973, Maternal behavior of California sea otter, *J. Mammal.* **54**:668–679.

Sauer, E. G. T., 1967, Mother-infant relationship in galagos and the oral child-transport among primates, *Folia Primatol.* **7**:127–149.

Sayler, A., and Salmon, M., 1971, An ethological analysis of communical nursing by the house mouse *(Mus musculus)*, *Behaviour* **40**:61–85.

Schaller, G. B., 1963, *The Mountain Gorilla,* University of Chicago Press, Chicago.

Schaller, G. B., 1967, *The Deer and the Tiger,* University of Chicago Press, Chicago.

Schaller, G. B., 1972, *The Serengeti Lion,* University of Chicago Press, Chicago.

Schaller, G. B., 1977 *Mountain Monarchs,* University of Chicago Press, Chicago.

Schenkel, R., 1966, Play, exploration and territoriality in the wild lion, *Symp. Zool. Soc. London.* **18**:11–22.

Schessler, T., and Nash, L. T., 1977, Food sharing among captive gibbons *(Hylobates lar)*, *Primates* **18**:677–689.

Schneirla, T. C., Rosenblatt, J. S., and Tobach, E., 1963, Maternal behavior in the cat in: *Maternal Behavior in Mammals* (H. L. Rheingold, ed.), pp. 122–168, Wiley, New York.

Schwartz, E., and Rowe, F. A., 1976, Olfactory bulbectomy: Influences on maternal behavior in primiparous and multiparous rats, *Physiol. Behav.* **17**:879–883.

Scott, J. P., and Fuller, J. L., 1965, *Genetics and the Social Behavior of the Dog,* University of Chicago Press, Chicago.

Seay, B., 1966, Maternal behavior in primiparous and multiparous monkeys, *Folia Primatol.* **4**:146–168.

Seay, B., Alexander, B. K., and Harlow, H. F., 1964, Maternal behavior of socially deprived rhesus monkeys, *J. Abnorm. Soc. Psychol.* **69**:345–354.

Seidensticker, J., 1977, Notes on early maternal behavior of the leopard, *Mammalia* **41**:111–113.

Seitz, P. F. D., 1958, The maternal instinct in animal subjects. I. *Psychosom, Med.* **20**:215–226.

Selye, H., and McKeown, T., 1934, Further studies on the influence of suckling, *Anat. Rec.* **60**:323–332.

Seyfarth, R. M., 1976, Social relationships among adult female baboons, *Anim. Behav.* **24**:917–938.

Sharman, G. B., 1967, The red kangaroo, *Sci. J.,* March, pp. 53–60.

Sharman, G. B., Calaby, J. H., and Poole, W. E., 1966, Patterns of reproduction in female diprotodont marsupials, *Symp. Zool. Soc. London* **15**:205–232.

Siegel, H. I., and Greenwald, G. S., 1975, Prepartum onset of maternal behavior in hamsters and the effects of estrogen and progesterone, *Horm. Behav.* **6**:237–245.

Silman, R. E., Holland, D., Chard, T., Lowry, P. J., and Hope, J., 1978, The ACTH 'family tree' of the rhesus monkey changes with development, *Nature (London)* **276**:526–528.

Slotnick, B. M., Carpenter, M. L., and Fusco, R., 1973, Initiation of maternal behavior in pregnant nulliparous rats, *Horm. Behav.* **4**:53–59.

Smart, J. L., 1976, Maternal behaviour of undernourished mother rats towards well fed and underfed young, *Physiol. Behav.* **16**: 147–149.

Smith, C. C., 1968, The adaptive nature of social organization in the genus of tree squirrels *Tamiasciurus, Ecol. Monogr.* **38**:31–63.

Smith, F. V., Van-Toller, C., and Boyes, T., 1966, The 'critical period' in the attachment of lambs and ewes, *Anim. Behav.* **14**:120–125.

Smith, L., and Berkson, G., 1973, Litter stimulus factors in maternal retrieval *(Rattus rattus)*, *Anim. Behav.* **21**:620–623.

Smith, J. C., 1976, Responses of adult mice to models of infant calls, *J. Comp. Physiol. Psychol.* **90**:1105–1115.

Smith, A., Wheelock, J. V., and Dodd, F. H., 1967, The effect of milking throughout pregnancy on milk secretion in the succeeding lactation, *J. Dairy Res.* **34**:145–150.

Smotherman, W. P., Bell, R. W. Starzek, J., Elias, J., and Zachman, T., 1974, Maternal responses to infant vocalizations and olfactory cues in rats and mice, *Behav. Biol.* **12**:55–66.

Smotherman, W. P., Wiener, S. G., Mendoza, S. P., and Levine, S., 1976, Pituitary-adrenal responsiveness of rat mothers to noxious stimuli and stimuli produced by pups, *Ciba Found. Symp.* **45**:5–25.

Smotherman, W. P., Brown, C. P., and Levine, S., 1977, Maternal responsiveness following differential pup treatment and mother-pup interactions, *Horm. Behav.* **8**:242–253.

Smotherman, W. P., Bell, R. W., Hershberger, W. A., and Coover, G. D., 1978, Orientation to rat pup cues: Effects of maternal experiential history, *Anim. Behav.* **26**:265–273.

Spencer-Booth, Y., 1970, The relationships between mammalian young and conspecifics other than mothers and peers: A review, *Adv. Study Behav.* **3**:119–194.

Stern, J. J., and Bronner, G., 1970, Effect of litter size on nursing time and weight of the young in guinea pigs, *Psychon. Sci.* **21**:171–172.

Stern, J. J., and Hoffman, B., 1970, Effects of social isolation until adulthood on maternal behavior in guinea pigs, *Psychon. Sci.* **21**:15–16.

Stern, J. M., and Mackinnon, D. A., 1976, Postpartum, hormonal, and nonhormonal induction of maternal behavior in rats: Effects on T-maze retrieval of pups, *Horm. Behav.* **7**:305–316.

Stewart, K. J., 1977, The birth of a wild mountain gorilla *(Gorilla gorilla beringei)*, *Primates* **18**:965–976.

Strayer, F. F., Taylor, M., and Yanciw, P., 1975, Group composition effects on social behavior of captive squirrel monkeys *(Saimiri sciureus)*, *Primates* **16**:253–260.

Struhsaker, T. T., 1967, Behavior of vervet monkeys *(Cercopithecus aethops)*, *Univ. Calif. Berkeley Publ. Zool.* **82**:1–64.

Struhsaker, T. T., 1977, Infanticide and social organization in the redtail monkey *(Cercopithecus ascanius schmidti)* in the Kibale Forest, Uganda, *Z. Tierpsychol.* **45**:75–84.

Sundaram, K., Connell, K. G., and Passantino, T., 1975, Implication of absence of HCG-like gonadotrophin in the blastocyst for control of corpus luteum function in pregnant rabbit, *Nature (London)* **256**:739–740.

Suomi, S. J., 1977, Development of attachment and other social behaviors in rhesus monkeys, *Adv. Study Comm. Affect* **3**:197–224.

Svare, B., and Gandelman, R., 1973, Postpartum aggression in mice: Experiential and environmental factors, *Horm. Behav.* **4**:323–334.

Svare, B., and Gandelman, R., 1976, Postpartum aggression in mice: The influence of suckling stimulation, *Horm. Behav.* **7**:407–416.

Szechtman, H., Siegel, H. I., Rosenblatt, J. S., and Komisaruk, B. R., 1977, Tail-pinch facilitates onset of maternal behavior in rats, *Physiol. Behav.* **19**:807–809.

Taylor H., Teas, J., Richie, T., Southwick, C., and Shrestha, R., 1978, Social interactions between adult male and infant rhesus monkeys in Nepal, *Primates* **19**:343–351.

Terkel, J., and Rosenblatt, J. S., 1971, Aspects of nonhormonal maternal behavior in the rat, *Horm. Behav.* **2**:161–171.

Terkel, J., and Rosenblatt, J. S., 1972, Humeral factors underlying maternal behavior at parturition: Cross transfusion between freely moving rats, *J. Comp. Physiol. Psychol.* **80**:365–371.

Thoman, E. B.,and Arnold, W. J., 1968, Effects of incubator rearing with social deprivation on maternal behavior in rats, *J. Comp. Physiol. Psychol.* **65**:441–446.

Thorburn, G. D., Challis, J. R. C., and Currie, W. B., 1977, Control of parturition in domestic animals, *Biol. Reprod.* **16**:18–27.

Townsend, T. W., and Bailey, E. D., 1975, Parturitional, early maternal, and neonatal behavior in penned white-tailed deer, *J. Mammal.* **56**:347–362.

Trivers, R. L., 1971, The evolution of reciprocal altruism, *Quart. Rev. Biol.* **46**:35–57.

Trivers, R. L., 1972, Parental investment and sexual selection in: *Sexual Selection and the Descent of Man* (B. Campbell, ed.), pp. 136–179, Aldine, Chicago.

Trivers, R. L., 1974, Parent-offspring conflict, *Am. Zool.* **14**:249–264.

Trollope, J., and Blurton-Jones, N. G., 1975, Aspects of reproduction and reproductive behaviour in *Macaca arctoides, Primates* **16**:191–205.

Trune, D. R., and Slobodchikoff, C. N., 1978, Position of immatures in pallid bat clusters: A case of reciprocal altruism? *J. Mammal.* **59**:193–195.

Tucker, H. A., and Thatcher, W. W., 1968, Pituitary growth hormone and luteinizing hormone content after various nursing intensities, *Proc. Soc. Exp. Biol. Med.* **129**:578–580.

Tucker, H. A., Paape, M. J., Sinha, Y. A., Pritchard, D. E., and Thatcher, W. W., 1967, Relationship among nursing frequency, lactation, pituitary prolactin and adrenocorticotropic hormone content in rats, *Proc. Soc. Exp. Biol. Med.* **126**:100–103.

Tyler, S. J., 1972, The behaviour and social organization of the New Forest ponies, *Anim. Behav. Monogr.* **5**:85–196.

Vaitl, E. A., 1977, Social context as a structuring mechanism in captive groups of squirrel monkeys *(Saimiri sciureus)*, *Primates* **18**:861–874.

van Lawick-Goodall, H. and J., 1971, *Innocent Killers*, Houghton Mifflin, Boston.

van Lawick-Goodall, J., 1968, The behaviour of free-living chimpanzees in the Gombe Stream reserve, *Anim. Behav. Monogr.* **1**:165–311.

van Lawick-Goodall, J., 1971, *In the Shadow of Man*, Houghton Mifflin, Boston.

Veomett, M. J., and Daniel, J. C. Jr., 1975, Termination of pregnancy after accelerated lactation in the rat. II. Relationship to number of young, day of pregnancy and length of nursing, *J. Reprod. Fertil.* **44**:513–517.

Verhaeghe, A., and Noirot, E., 1978, Ultrasounds by mouse pups from deaf and hearing strains, *Dev. Psychobiol.* **11**:117–124.

Voci, V. E., and Carlson, N. E., 1973, Enhancement of maternal behavior and nest building following systemic and diencephalic administration of prolactin and progesterone in the mouse, *J. Comp.Psychol.* **83**:388–393.

Vogt, J. L., Carlson, H., and Menzel, E., 1978, Social behaviour of a marmoset
 (Saguinus fuscicollis) group. I. Parental care and infant development, *Primates*
 19:715–726.
Volkman, N. J., Zemanek, K. F., and Muller-Schwartze, D., 1978, Antorbital and fore-
 head secretions of black-tailed deer *(Odocoileus hemionus columbianus):* Their role
 in age-class recognition, *Anim. Behav.* **26**:1098–1106.
Walther, F. R., 1969, Flight behaviour and avoidance of predators in Thompson's gazelle
 (Gazella thompsonii Guenther, 1884), *Behaviour* **34**:184–221.
Weisbard, C., and Goy, R. W., 1976, Effect of parturition and group composition on
 competitive drinking order in stumptail macaques *(Macaca arctoides)*, *Folia Pri-
 matol.* **25**:95–121.
Weiss, G., Bulter, W. R., Dierschke, D. J., and Knobil, E., 1976, Influence of suckling
 on gonadotropin secretion in the postpartum rhesus monkey, *Proc. Soc. Exp. Biol.
 Med.* **153**:330–331.
White, L. E., and Hinde, R. A., 1975, Some factors affecting mother-infant relations in
 rhesus monkeys, *Anim. Behav.* **23**:527–542.
Wickler, W., 1972, *The Sexual Code,* Doubleday, Garden City, N.Y.
Wiener, S. G., Smotherman, W. P., and Levine, S., 1976, Influence of maternal malnu-
 trition on pituitary-adrenal responsiveness of offspring, *Physiol. Behav.* **17**:897–901.
Wiesner, B. P., and Sheard, N. M., 1933, *Maternal Behaviour in the Rat,* Oliver and
 Boyd, Edinburgh.
Wilson, A. P., and Vessey, S. N., 1968, Behavior of free-ranging castrated rhesus mon-
 keys, *Folia Primatol.* **9**:1–14.
Wilson, E. O., 1975, *Sociobiology: The New Synthesis,* Harvard University Press, Bel-
 knap Press, Cambridge, Mass.
Wolf, K. E., and Fleagle, J. G., 1977, Adult male replacement in a group of silvered leaf
 monkeys *(Presbytis cristata)* at Kuala Selangor, Malaysia, *Primates* **18**:949–955.
Yahr, P., 1976, Effects of hormones and lactation on gerbils that seldom scent mark spontaneously, *Physiol. Behav.* **16**:395–399.
Yeh, K., and Moog, F., 1974, Intestinal lactase *d* activity in the suckling rat: Influence
 of hypophysectomy, *Science* **183**:77–79.
Yoshiba, K., 1968, Local and intertroop variability in ecology and social behavior of common Indian langurs, in: *Primates: Studies in Adaptation and Variability* (P. Jay,
 ed.), pp. 217–242, Holt, Rinehart and Winston, New York.
Zarrow, M. X., Denenberg , V. H., and Anderson, C. O., 1965, Rabbit: Frequency of
 suckling in the pup, *Science* **150**:1835.
Zarrow, M. X., Schlein, P. A., Denenberg, V. H., and Cohen, H. A., 1972, Sustained
 corticosterone release in lactating rats following olfactory stimulation from the pups,
 Endocrinology **91**:191–196.
Zimmerman, G. D., 1974, Cooperative nursing behavior observed in *Spermophilus tridecemlineatus* (Mitchell), *J. Mammal.* **55**:680–681.

5

Maternal Aggression in Mammals

Bruce B. Svare

1. Introduction

The mammalian female exhibits a number of complex behaviors during pregnancy, parturition, and lactation that summate to ensure the survival of her young. The behaviors that have been selected for study have traditionally been those exhibited by the adult toward the young. The literature concerning the psychobiology of nursing, retrieving, and other pup-directed maternal activities is characterized by an impressive accumulation of research findings and theoretical principles (see Rosenblatt and Seigel, this volume). However, a less frequently studied dimension of maternal care is the dramatic change in female social behavior that occurs with pregnancy and subsequently with parturition and lactation. Instead of displaying passivity toward conspecifics, the pregnant and lactating mammal exhibits intense aggressive behavior, referred to as maternal aggression. This behavior most likely serves to protect the young but also may be involved in the regulation of social organization and population dynamics. We will speculate on both of these functions later on in the chapter.

Until recently there has been little systematic research concerning the environmental and physiological determinants of maternal aggression. This lack of information is especially ironic in view of the numerous informal and often anecdotal accounts attesting to the aggressiveness of a wide variety of pregnant and lactating mammalian females. Aside from the fact that psychobiologists have concentrated their efforts on the analysis of pup-directed maternal activities, the neglect of maternal aggressive behavior also may stem from the rather chauvinistic assumption that the male is innately more aggressive than the female of most species. We have been repeatedly reminded that the male is spontaneously aggressive whereas the female normally is passive. A cursory examination of the scientific

BRUCE B. SVARE • Department of Psychology, State University of New York at Albany, Albany, New York 12222.

literature would lead one to believe that the female is simply incapable of showing aggressive behavior unless she is either administered male hormones, chronically isolated, lesioned in the hypothalamus, peripherally shocked, or given electrical brain stimulation! As a result, the study of spontaneous female aggression, especially as it may occur during the natural reproductive states of pregnancy and lactation, has been overlooked while much research has been devoted to elucidating the factors that are responsible for male aggression.

The psychobiological basis of maternal aggression, like many emerging research areas, lacks unifying concepts and comparative data. As we shall see, the tendency to exhibit aggression during pregnancy and lactation is characteristic of many mammalian species, but with the exception of recent findings in rodents, virtually nothing is known about the environmental and physiological basis of this behavior. The purpose of the present chapter is fourfold. First, we will review the findings concerning laboratory and naturalistic studies of maternal aggression. Second, we will present a tentative model describing the psychobiological events controlling maternal aggression in rodents. Third, we will examine the role of this behavior in preserving the integrity of the individuals and populations. Fourth, we will from time to time make suggestions for future research. The above approach should serve as a source of information as well as a heuristic tool for future investigators in this area.

A topic not addressed in the present chapter is paternal aggressive behavior. Because we are not aware of any research examining changes in male aggression as it may relate to the care of young, a discussion of paternal aggressive behavior is precluded. Thus, we can begin our suggestions for future research by saying that this is a topic badly in need of systematic analysis.

2. Some Naturalistic Observations of Maternal Aggression

It is not known whether maternal defense is a universal characteristic of all mammals, but the ethological literature indicates that the behavior has been observed in a wide variety of vertebrates.

There are numerous accounts of maternal aggression in females of freely growing rodent populations. Brown (1953) observed that late-pregnant female mice are so aggressive that they frequently kill strange male conspecifics that are unable to escape from the nest site. In his classic studies of population changes and rat social behavior, Barnett (1969) noted that lactating females chase intruders away from the burrows when unweaned young are present. However, some of the most interesting and detailed accounts of maternal aggression come from ethological descriptions of the behavior in higher mammals. For example, Altman (1963) described one very persistent moose cow with her calf that chased a strange male horse for a considerable distance. The horse was chased into the water, where it was ferociously attacked. The moose cow did not give up until the horse was driven to another island! Altman (1963) described another case of maternal aggression in the moose cow in which the postparturient female attacked and badly wounded a bear attempting to encroach on her nest area and carry off her calf. Maternal defense has been observed in numerous other mammals, including cats (Schneirla *et al.*, 1963), squirrels (Taylor, 1966), baboons (Devore, 1963), rabbits (Ross *et al.*, 1963), and sheep (Hersher *et al.*, 1963).

3. Laboratory Studies of Maternal Aggression

Laboratory studies of maternal aggression have been restricted to mice, rats, rabbits, and hamsters. These studies have examined both intra- and interspecific aggressive behavior.

Some reports of maternal aggression have examined the behavior of pregnant and lactating animals toward members of other species. For the most part, however, these studies lacked replicability and generality. Denenberg, *et al.* (1958, 1959) noted increased maternal aggression toward a human handler in lactating rabbits. The behavior was strain specific and varied from merely laying the ears back to foot stamping and vocal objection, to actual attack with the front feet, and finally to biting. Endroczi *et al.* (1958) found that frog killing by rats increased during lactation, but their report could not be replicated by Revlis and Moyer (1969). Flandera and Novakova (1971) observed that mouse killing by rats increased during lactation, but their findings were reported only in Wistar rats maintained under specific-pathogen-free conditions. Recently, Rowley and Christian (1976a, 1977) demonstrated that lactating female mice will vigorously attack juvenile and adult voles.

Studies of maternal aggression toward intruders of the same species generally have been more informative. All of this research has been conducted with rodents and is discussed below.

3.1. Topography of Maternal Aggression

The topography of maternal aggression in mice consists of immediate attacks (latencies usually are less than 10 sec) with biting directed toward the flanks and neck of the intruder (Svare and Gandelman, 1973). The duration of most attacks is about 5–10 sec. Intense attacks continue throughout the first 3–5 min of the test period, but extended observations indicate that the behavior diminishes to a very low point after 30 min of continuous exposure to an intruder (Green, 1978). The behavior of intruder male and female conspecifics during the initial stages of fighting usually is characterized by immediate submission; they rarely initiate attacks or fight back in response to being attacked (Svare and Gandelman, 1973). However, male intruders do show counterattacks after continuous exposure to the female (Green, 1978). Lynds (1976a) compared maternal and intermale aggression in wild house mice and reported that the two types of aggression were indistinguishable with respect to grooming, investigating, boxing, biting, chasing, and tail rattling. However, maternal and intermale aggression in mice differ with respect to the latency to engage in fighting behavior. As previously stated, maternal aggression is characterized by very short latencies, while intermale encounters are typified by longer latencies averaging several minutes in duration.

Maternal aggression in female rats is identical to that of intermale aggression in the same species (Erskine *et al.*, 1978 *a,b*; Price and Belanger, 1977). Attacks are initiated after several minutes of mutual exploration and consist of rapid lunges toward the neck or back region of the opponent followed by kicking, biting, and scuffling. In addition, during a typical bout, the female orients herself to the side of the intruder and may exhibit hip throws which are thrusts of one rear leg toward the intruder. The intensity of the behavior is highest during the first 15 min of an encounter and diminishes rapidly thereafter. Intruder rats, like mice, are almost immediately submissive and rarely fight back in response to being attacked.

In the hamster, the topography of maternal aggressive behavior is similar to what is seen in the rat (Wise, 1974). The average latency to engage in fighting for the pregnant hamster is about 4 min, while the average latency in the lactating female is shorter, averaging about 1 min. Fighting consists of distinct episodes of chasing and pouncing in which the female rapidly pursues the intruder, jumps on its back with its front paws, and bites the intruder in the back. These behaviors continue at a high intensity for at least 10 min, but extended observations have not been conducted to examine how long the behavior persists. It should be noted that the female hamster, unlike most other mammalian females, is more aggressive than the male and is spontaneously aggressive during all reproductive states. By comparison, the female rat and mouse rarely exhibit aggression except during pregnancy and/or lactation.

3.2. Stimuli Eliciting Maternal Aggression

Recent studies indicate that the display of maternal aggressive behavior is dependent upon the degree of familiarity with the intruder. Svare and Gandelman (1973) reported that lactating Rockland–Swiss (RS) albino mice were nonaggressive toward males that had been housed in the females' cages and separated by a wire mesh partition between days 2 and 10 of lactation; on the other hand, strange males were vigorously attacked. Recently, Green (1978) reported that lactating female ICR mice that were cohabited with males throughout pregnancy and early lactation were nonaggressive toward *both* strange and familiar intruders. The lack of differential responsiveness found by Green may be due to a strain difference but most likely reflects the extensive degree of contact between the male and female in their study. Regardless, an olfactory-mediated mechanism may control the recognition of strange and familiar intruders by lactating female mice, since opponents doused with "strange" conspecific urine are attacked more vigorously than opponents coated with "familiar" urine (Lynds, 1976b).

Several studies indicate that the age of the intruder dramatically influences the intensity of fighting behavior in lactating rodents. Erskine et al. (1978a) confronted lactating female rats with unfamiliar male intruders of varying ages. The highest level of fighting behavior was elicited by 35- and 45-day-old intruders, while less fighting was directed toward 55- and 110-day-old animals. As Erskine et al. (1978a) suggest, a number of factors may contribute to the observed differential responsiveness, including the body size and the behavior of the opponents as well as pheromonal stimuli emitted by the intruder. In an earlier study using mice, Svare and Gandelman (1973) demonstrated that 1- and 10-day-old intruders were seldom attacked by lactating mice, while intense aggression was directed against 14- and 20-day-old intruders. Moreover, they demonstrated that 14-day-old intruders whose hair was removed were rarely attacked. Although these findings suggest that the presence of body hair may be one stimulus involved in evoking aggression in the parturient animal, other factors such as body size and olfactory cues may be equally important. More research is necessary in order to elucidate which age-dependent characteristics of the intruder are responsible for the elicitation of attack.

There is some disagreement concerning the influence on maternal aggression of the sex and hormonal status of the intruder. Lactating RS mice attack adult male and female intruders with equal intensity (Svare and Gandelman, 1973). However, CD-1 parturient female mice attack intact males more vigorously than virgin or lactating females (Rosenson and Asheroff, 1975). Moreover, fewer gonadectomized virgin animals of either sex are attacked by CD-1 lactating females as compared to intact animals. Late-lactating (postpartum days 16–20) female intruders are attacked more frequently than early-lactating (postpartum days 3–8) animals. The findings of Rosenson and Asheroff (1975) suggest, of course, that the hormonal condition of the intruder is an important factor in the elicitation of maternal aggression in some strains of mice (see Table 1).

3.3. Experiential Influences on Maternal Aggression

3.3.1. Changes during Pregnancy

Both between and within species differences have been noted in the incidence of maternal aggression during pregnancy. The behavior is limited to the lactation period in RS, BALB, DBA, and HS mice (St. John and Corning, 1973; Svare and Gandelman, 1976a. However, Noirot and her colleagues (Beniest-Noirot, 1958; Noirot et al., 1975) consistently have found an increase in the incidence and intensity of the behavior in albino mice as gestation advances. Impulsive attacks and the duration of attacks toward male intruders dramatically increase over the first 5 days of pregnancy (normally 18 days in duration) and remain high through-

Table 1. The Influence of the Sex and Hormonal Status of the Intruder on the Maternal Aggression Exhibited by Lactating CD-1 Mice[a]

| Type of intruder | N | The number and percentage of lactating females that attacked | | | |
| | | Counting only trials with at least 10 sec fighting | | Counting trials with more than one fight | |
		Number	%	Number	%
Intact male	35	21	60	27	77
Intact female	35	13	37	13	37
Castrate male	35	5	14	5	14
Castrate female	35	1	3	2	6
Mother, pups 3–8 days old	35	0	0	0	0
Mother, pups 16–20 days old	35	8	23	8	23

[a]Intruders were placed in the home cages of individually housed lactating females whose litters were between 3 and 8 days of age. The presence or absence of aggression and the amount of time spent fighting was recorded during one 5-min trial. From Rosenson and Asheroff (1975).

out the gestation period. The peak period of aggressive behavior in the pregnant hamster is around the 10th day of gestation (16-day gestation period) (Wise, 1974). Pregnant rats exhibit considerable aggressive behavior between the 18th and 22nd days of gestation [normal gestation period is 22 days (Erskine et al., 1978b)], but the incidence of the behavior at earlier periods of pregnancy has not been reported for this species. The above differences in the occurrence of aggression during pregnancy may be related to the species studied or to methodological differences between different laboratories. Because aggressive behavior in female rodents is not greatly increased by isolation and fighting experience (Erskine et al., 1978b; Noirot et al., 1975), the heightened aggressive behavior seen during pregnancy is most likely related to the profound physiological changes that occur with advancing gestation. We will speculate on the nature of these changes later on in the chapter.

3.3.2. Changes during Lactation

Most authors consistently find that maternal aggression is most intense early in the postpartum period and declines as lactation advances. In the mouse, numerous studies show that aggression is highest between days 3 and 8, declines between days 9 and 14, and is very low between days 15 and 21 of the lactation period (cf.

Gandelman, 1972; Green 1978; Svare and Gandelman, 1973; St. John and Corning, 1973) (see Fig. 1). Peak frequencies of attacks and bites and low latencies to the first attack are noted on day 9 of lactation in the rat (Erskine *et al.*, 1978*b*) and during the first 5 postpartum days in the hamster (Wise, 1974). The reasons for the decline in aggression with advancing lactation are not known at this time, but they are most likely due to a concomitant decline in the frequency and duration of nursing. Support for this hypothesis is evident from findings showing that lactating mice that are fostered newborn young every 4 days (i.e., elicited high levels of nursing) continue to fight longer at high levels than animals left with their own young (Svare, 1977).

Maternal aggression during the lactation period is controlled by the presence of the litter. Five hours, but not 1 hr, of pup removal during early lactation (postpartum day 6) virtually eliminates maternal aggression in the mouse (cf. Gandel-

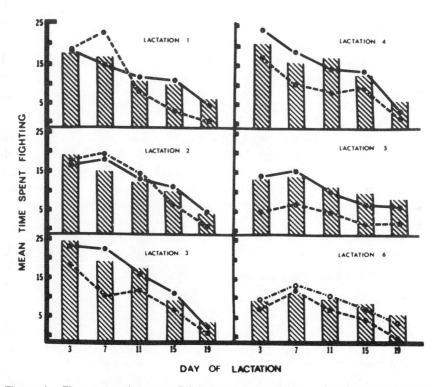

Figure 1. The average time spent fighting across test trials as a function of the lactation period, for Continuous Fighters (exhibiting aggression during each of six lactation periods), Sporadic Fighters (exhibiting aggression on some but not all lactation periods), and Nonrematers (exhibiting aggressive behavior during each of the lactation periods they completed). RS albino mice were given 3-min tests against an adult male intruder on postpartum days 3, 7, 11, 15, and 19. The circles with stars for the sixth lactation period represent the data of animals that were tested for aggression only on the sixth lactation period. From Svare and Gandelman (1976*a*).

man 1972; Svare, 1977, and unpublished data; Svare and Gandelman, 1973). Replacement of the offspring for as little as 5 min following 5 hr of pup removal restores the behavior. In the rat, removal of the litter 4 hr prior to aggression testing on either day 9 or 10 of lactation results in significant decreases in fighting levels (Erskine *et al.,* 1978*b*). As we shall see later, the reduction in aggression with litter removal may be due to profound changes in the dam's reproductive physiology.

Direct physical contact between the dam and her young is not a prerequisite for the short-term maintenance of postpartum aggression in mice. Placement of the dam's entire litter or a single pup behind a double wire mesh partition in the home cage maintains the behavior at a level identical to that of mothers in direct contact with their young. When placed behind the partition, unfamiliar 6-. 13-, and 20-day-old pups, but not 30-day-old mice, maintain the behavior as effectively as the dam's own young (Svare, 1977, and unpublished observations; Svare and Gandelman, 1973). The function of these adaptations would be to ensure the protection and survival of unfamiliar young as well as of pups that inevitably become separated from the nest and their littermates as they grow older. The mechanism responsible for the exteroceptive maintenance of aggression is unknown at the present time, but we shall speculate later in the chapter that olfactory cues from young may sustain the behavior by maintaining the release of circulating hormones in the dam.

In the mouse, changes in postpartum aggression are also related, in part, to experiential factors. Lactating female mice that are repeatedly tested for aggressive behavior from days 2–22 postpartum continue to exhibit aggression at higher levels than different groups of females tested only once at various times during lactation (Green, 1978). Moreover, isolation during pregnancy appears to promote postpartum aggression. Pregnant females paired with a male prior to parturition are less aggressive following delivery of young than females housed in isolation during gestation.

3.3.3. Changes across Several Reproductive Cycles

The only study to examine changes in maternal aggression over several reproductive cycles was one recently reported by Svare and Gandelman (1976*a*). Female RS mice were tested for aggression during six successive pregnancies and lactation periods or until they ceased to remate. They found that maternal aggression was limited to the lactation phase of each reproductive cycle and that most of the animals exhibited the behavior on each lactation period that they completed. Also, the intensity of the behavior was highest during the beginning of each lactation interval and was lowest at the end. Moreover, the intensity of the behavior increased across the first three lactation periods and then declined on later lactation periods to a point well below initial levels. The fighting behavior exhibited by multiparous animals was not due simply to previous fighting experience, since some multiparous animals exhibited postpartum aggression when tested for the first time on the sixth lactation period (see Figs. 1 and 2). Svare and Gandelman

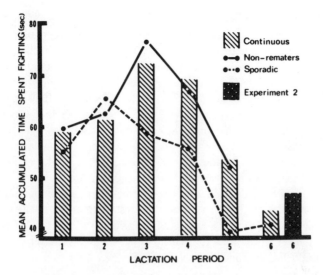

Figure 2. The average accumulated time spent fighting (i.e., the total duration of time spent fighting on five tests for aggression during each lactation period) for Continuous Fighters (exhibiting aggression during each of six lactation periods), Sporadic Fighters (exhibiting aggression on some but not all lactation periods), and Nonrematers (exhibiting aggressive behavior during each of the lactation periods they completed). RS albino mice were given 3-min aggression tests against an adult male intruder on postpartum days 3, 7, 11, 15, and 19. The additional data (experiment 2) for the sixth lactation period represent the average accumulated time spent fighting for animals that were tested for aggression only on the sixth lactation period. From Svare and Gandelman (1976*a*).

(1976*a*) have suggested that the changes in the intensity of aggression across successive lactation periods may be due either to alterations in the display of other maternal behaviors such as retrieving, nursing, or nest building or to age-related alterations in the secretory patterns of hormones. Further experimentation is needed in order to identify which, if any, of the above factors is responsible.

Svare and Gandelman (1976*a*) also noted in their study that animals that exhibited higher overall levels of aggression tended to produce litters with more young than did animals exhibiting lower overall levels of the behavior. These findings may be especially important for our understanding of population dynamics.

3.4. Psychobiological Determinants of Maternal Aggression

3.4.1. The Role of Nipple Development and Suckling Stimulation

We recently reported a series of studies examining the relationship between pregnancy termination, lactation, nipple growth, suckling stimulation, and the postpartum initiation of maternal aggression in albino mice. Let us examine these studies in depth.

3.4.1a. Pregnancy Termination Studies.   Postpartum suckling from young initiates a number of physiological changes in the mother that enable her to secrete milk. Milk secretion and the resulting body weight changes in the pups is commonly referred to as lactation. Our previous work (Gandelman and Svare, 1975; Svare and Gandelman, 1973; Svare and Gandelman, 1976a) showed that lactation and aggression are related, since mice of the RS strain exhibit fighting only during the lactation phase of the reproductive cycle. One strategy that we found to be successful in examining this relationship consisted of attempting to induce lactation and aggression in animals that are not lactating (Gandelman and Svare, 1974). In previous work (Gandelman and Svare, 1975), we determined that lactation, as assessed by weight gain in young, could be induced in mice following hysterectomy on the 12th and subsequent days of pregnancy by maintaining them in the presence of foster young (the normal gestation length of the mouse is 19 days). In view of these findings, we examined whether the artifical termination of pregnancy alone or in combination with the presentation of foster young is a sufficient condition for the induction of maternal aggression at a time earlier than when it would normally appear.

Primiparous female mice were divided into four groups on the morning of the 14th day of pregnancy (the day the copulatory plug was found was called day O). One group of mice was anesthetized with ether and hysterectomized. Hysterectomy in the mouse is a simple surgical procedure which appears to be relatively nontraumatic to the animal. To perform the operation, a midventural incision is made along the abdominal wall and both horns of the uterus are exposed, ligated at each extremity, and removed. A second group of animals was treated similarly except that ligation and uterine removal were not performed. A third group was left intact and was permitted to deliver its young, which normally occurred on day 19 of pregnancy. Immediately following hysterectomy or sham hysterectomy or within 24 hr of parturition, the animals were presented with five preweighed 1-day-old RS mouse pups. In the case of the parturient animals, their offspring were first removed. On the next morning the behavior of the adults toward the young was observed for 15 min following which the pups were removed, weighed and replaced by five preweighed 1-day-olds. We continued this procedure for 2 additional days. A fourth group of mice was also hysterectomized on the 14th day of pregnancy, but the animals of this group were not presented with foster young immediately following hysterectomy or throughout the duration of the three tests for aggression. However, the mice were presented with five preweighed 1-day-olds immediately after the third fighting test. On the following morning, these females were tested for aggression and their pups were weighed.

All of the females were observed to display maternal activities such as nest building, pup retrieval, and the assumption of nursing postures each day regardless of whether they were scored as lactating. This was also true for the mice that were hysterectomized but were not given young until the 4th day following parturition.

An animal was scored as lactating if the average weight of its foster young (i.e., total litter weight/number of pups) equalled or exceeded its prefostering weight (i.e., the weight taken 24 hr previously). We used the average weight of foster young as opposed to the total weight, since a pup occasionally would be found dead the morning following fostering. We noted that when the average weight of the foster young did equal or exceed the prefostering average weight, milk bands were invariably seen in the stomachs of the young. Furthermore, once an animal began to lactate it continued to do so. Mice that experienced parturition, and mice that were presented with young immediately following hysterectomy, lactated, whereas animals only presented with young did not lactate (see Table 2). Also, hysterectomized animals given young on the 4th postoperative day did not lactate during the following 24 hr. The results for fighting behavior were consistent with the lactation findings. Aggression was exhibited by normally lactating animals and animals induced to lactate by hysterectomy and the presentation of young, but intact animals given young and hysterectomized animals not presented with young were not aggressive. In the case of the latter group, aggression was not observed during the additional fighting test held on the morning following the fostering of young on the 4th post-hysterectomy day. The hysterectomized and parturient animals that were fostered young did not differ from each other with respect to the number of tests on which fighting occurred or with respect to the topography of aggression.

Our findings at this point indicated that lactation and maternal aggression may be related in that aggression was displayed by animals induced to lactate by artificially terminating pregnancy and fostering them with young. If lactation is necessary for maternal aggression, we felt that fighting should not occur if pregnancy termination and the presentation of young occur prior to day 12 of gestation, a period during which lactation cannot be induced (Gandelman and Svare, 1975).

To examine this hypothesis, separate groups of pregnant females were hysterectomized on gestation day 10, 11, 12, 13, 14, or 15. Pup fostering and aggression testing took place in a manner identical to that of the previous experiment with the exception that these procedures were repeated for 7 days instead of 4. We found that the number of animals lactating gradually declined the earlier in gestation that hysterectomy and the fostering of young was performed (see Table 2). Furthermore, day 12 of gestation appeared to be a pivotal day with respect to the induction of lactation, since more mice lactated following hysterectomy on day 12 than on day 11, while fewer mice hysterectomized on day 12 lactated as compared to animals hysterectomized on day 13. Like the lactation findings, the number of animals exhibiting aggression gradually declined as hysterectomy and the fostering of young was performed earlier in gestation. We also found that with few exceptions those animals that did engage in fighting behavior also were judged to be lactating. Indeed, of the 33 animals that fought and lactated, only 5 exhibited fighting behavior prior to the day they began to lactate.

Lactation is the end result of a complex process which depends upon the appropriate timing of both physiological and behavioral events. The initiation of this process is, of course, dependent upon the growth of the teats as pregnancy

Table 2. The Role of Nipple Development, Suckling Stimulation, and Lactation in the Maternal Aggression Displayed by Mice[a]

Experiment	Aggression[b]	Suckling[c]	Lactation[d]	Reference
Pregnancy termination [*Hysterectomy (HYST)*]				
HYST (day 14)	Yes	NM[e]	Yes	Gandelmen and Svare (1974)
PART[f]	Yes	NM	Yes	
S-HYST[h] (day 14)	No	NM	No	
HYST (day 14)—No pups	No	NM	No	
HYST (day 15)	Yes	NM	Yes	Gandelmen and Svare (1974)
HYST (day 14)	Yes	NM	Yes	
HYST (day 13)	Yes	NM	Yes	
HYST (day 12)	Few	NM	Few	
HYST (day 11)	Few	NM	Few	
HYST (day 10)	No	NM	No	
HYST (day 15)	Yes	Yes	Yes	Svare (unpublished
HYST (day 13)	Yes	Yes	Few	observations)
HYST (day 11)	Few	Few	No	
HYST (day 9)	No	No	No	
Nipple removal [*Thelectomy (THEL)*]				
Prior to mating	No	No	No	Svare and Gandelman (1976b)
Day 18 of pregnancy	No	No	No	
Postpartum day 5	Yes	No	No	
Following 48 hr of suckling	Yes	No	No	
Following 24 hr of suckling	No	No	No	
Hormone induction[g]				
PART	Yes	Yes	Yes	Svare and Gandelman (1976c)
Virgin female + EB + P	Yes	Yes	No	
Virgin female + oil	No	No	No	
OVX virgin female + oil	No	No	No	
PART-THEL	No	No	No	Svare and Gandelman (1976c)
PART-S-THEL	Yes	Yes	Yes	
Virgin female + EB + P − THEL	No	No	No	
Virgin female + EB + P − S-THEL	Yes	Yes	No	

[a]Unless otherwise noted, the animals in all the experiments received new foster young on a daily basis.
[b]Aggressive behavior was assessed by observing the presence or absence of fighting toward an adult male intruder.
[c]Suckling was assessed by counting the number of red and distended nipples.
[d]Lactation was judged by weight gain in young and/or the presence of milk in the mammary glands at autopsy.
[e]Not measured.
[f]PART, parturient.
[g]Hormone treatment consisted of 19 daily injections of 0.02 μg estradiol benzoate (EB) and 500 μg progesterone (P) for 19 days. S-THEL, sham thelectomy; OVX, ovariectomized.
[h]S-HYST, sham hysterectomy.

advances as well as upon postpartum suckling stimulation from young. Thus, in the previous study, the failure of animals to fight following hysterectomy early in pregnancy most likely was due to inadequate suckling stimulation as a result of poorly developed nipples, instead of to the absence of lactation *per se*. To further examine the relationship between the growth of the nipples, suckling stimulation, and the initiation of aggression, we conducted an experiment similar to the one described above with the exception that several additional measures were noted (Svare, unpublished observations).

Nulliparous R-S female mice were mated with adult male mice and checked daily for vaginal plugs. When a plug was found (day 0), the females were individually housed as described earlier. Separate groups of animals either were allowed to deliver their own young or were hysterectomized under ether anesthesia on day 9, 11, 13, or 15 of gestation. Pup fostering was identical to the previous experiments except that the procedure was repeated every 24 hr for 9 days. The presence or absence of suckling also was assessed daily by counting the number of red and distended nipples (the mouse has five pairs of nipples). Informal observations of maternal behavior were conducted throughout each day. Aggressive behavior was judged every other day (total of four tests) beginning on the 3rd day following parturition or hysterectomy. An unfamiliar adult male was placed into the home cage of the female for 3 min and the number of attacks (sustained periods of biting) was recorded. To assess nipple growth during pregnancy, eight additional animals were placed into each group. They were sacrificed on day 9, 11, 13, 15, or 17 of gestation, and their nipples were removed and measured with dial calipers under a dissecting microscope.

Our informal observations showed that all animals exhibited the complete repertoire of maternal activity, including nest building, the assumption of nursing postures, pup licking, and pup retrieval. Animals hysterectomized and fostered young early in gestation (days 9 and 11) seldom fought, while parturient animals or animals hysterectomized late in gestation frequently exhibited aggression (see Table 2). The incidence of suckling and the number of nipples suckled paralleled the aggression findings in that these measures were significantly lower in animals hysterectomized and fostered young early in gestation (days 9 and 11) as compared to parturient animals or animals hysterectomized late in gestation. Aggression and suckling clearly were related, since only 4 of the 55 animals that fought did not show evidence of suckling. However, aggression and lactation were not related, since animals frequently fought without having exhibited lactation (19 of 55 animals). As expected, nipple length significantly increased with advancing gestation (see Fig. 3). The number of animals lactating, like the suckling and aggression findings, was significantly lower in animals hysterectomized and fostered young early in gestation (days 9 and 11) as compared with parturient animals and animals hysterectomized late in gestation.

Thus, our findings showed that maternal aggression is present in pregnant animals hysterectomized and fostered young at a time when nipple growth is

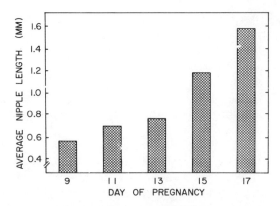

Figure 3. Mean nipple length of pregnant female RS mice as a function of time of gestation. Separate groups of pregnant animals were sacrificed on day 9, 11, 13, 15, or 17 of gestation, and their nipples were removed and measured with dial calipers under a dissecting microscope. Because nipple position did not influence nipple length, the data for each animal represent the average of ten nipples. From Svare (unpublished observations).

adequate for pup attachment to take place. Conversely, our findings showed that the behavior is not evident in pregnant animals hysterectomized and fostered young at a time during gestation when nipple growth is inadequate for pup attachment to take place. Thus, suckling stimulation rather than lactation appears to be a prerequisite for the behavior to be exhibited.

3.4.1.b. *Thelectomy (Nipple Removal) Studies.* As demonstrated by the previous findings, stimulation received from the young is an important factor in the initiation of maternal aggression. The absence of this stimulation renders lactating animals nonaggressive. Since suckling stimulation is the major source of the stimulation received by the lactating animals from the young, we felt that it might prove important to further examine the influence of this stimulation upon aggressive behavior. This was accomplished by observing the fighting behavior of female mice deprived of their nipples (thelectomy) prior to and during pregnancy and lactation while maintained in the presence of foster young (Svare and Gandelman, (1976*b*).

If aggression depends upon suckling stimulation from young, we reasoned that the behavior would probably be disrupted in animals whose nipples are removed prior to parturition. To study this question, we divided virgin females into three separate groups. One group of mice was anesthetized with ether and their nipples removed. The two remaining groups of animals were sham-thelectomized. Sham operations consisted of making two 1-mm incisions along the ventral midline of each animal. Twenty-four hours following surgery the mice were timed-mated and isolated. On the morning of parturition, the offspring of the animals from the thelectomized groups were removed and replaced with five 3- to 8-day-old foster young from donor mothers. On the next morning the behavior of the adults toward the young was observed for 15 min followed by an aggression test. Immediately following the test, the pups were removed and replaced by five 3- to 8-day-old foster young from donor lactating mothers. We repeated this procedure for 7 successive days. The remaining group of sham-thelectomized animals was tested for aggression in the same manner except that they were not presented with foster young after parturition or on any of the test days. Immediately following the last test for aggression, each animal was sacrificed by cervical dislocation and their mammary glands were examined macroscopically for the presence of milk. In addition to a visual inspection for milk, the glands were cut and lightly squeezed with forceps in order to express milk.

Our observations showed that all of the animals from the thelectomized and sham-thelectomized groups fostered pups were scored as being in the nursing posture each day. Other observations at various times throughout each day indicated that the animals from these two groups engaged in other maternal care activities such as nest building, anogenital licking, and pup retrieval. Thelectomy with, and sham thelectomy without, the fostering of young resulted in the absence of maternal aggression (see Table 2). In contrast, a large proportion of sham-thelectomized animals given pups exhibited aggression. The sham-thelectomized animals that were fostered pups tended to fight on all tests. In contrast to these intensely aggressive animals, the thelectomized animals that were fostered pups and the sham-thelectomized animals that were not fostered pups were nonaggressive. Instead, they engaged in anogenital sniffing and frequently allowed the intruder access to the nest and pups. Finally, the autopsies showed that only the mammary glands of the sham-thelectomized animals that were fostered pups contained milk.

Our findings indicated that animals thelectomized prior to mating and given foster young following parturition fail to display aggressive behavior. However, we speculated that the absence of aggression may not have been directly related to the absence of suckling stimulation, since some other changes may have occurred during the long interval between surgery and testing. To rule out this possibility, animals were either thelectomized or sham-thelectomized on day 18 of gestation, fostered young, and tested for aggression in a manner identical to that of the previous experiment. As was found in the previous experiment, prepartum thelectomy dramatically reduced the proportion of animals exhibiting maternal aggression (see Table 2).

We next sought to examine whether thelectomy attenuates aggressive behavior in animals that have had postpartum suckling stimulation. For this experiment, virgin female mice were timed-mated and isolated as previously described. On the day of parturition, the newborn young were removed from each female, examined for milk bands, and replaced by five 3- to 8-day-old foster young from donor mothers. This procedure continued through the morning of postpartum day 5, at which time one group of animals was thelectomized while another group of animals was sham-thelectomized. On the next morning, each animal was tested for aggression and fostered young as previously described. This procedure continued for 7 additional days. After the test the animals were sacrificed and their mammary glands examined for milk.

Postpartum thelectomy did not influence maternal aggressive behavior (see Table 2). In addition, we observed that the pups of both groups of animals had milk in their stomachs when examined each day prior to surgery. However, following surgery, only the pups of sham-thelectomized animals had milk in their stomachs. Furthermore, the autopsies showed that none of the mammary glands of the thelectomized females contained milk, while the mammary glands of all of the sham-thelectomized animals contained milk.

Our findings showed that aggression is not impaired by nipple removal when it is performed on postpartum day 5 and aggression testing commences on the next day. However, we felt that the time period between thelectomy and the initial test (24 hr) may not have been long enough to produce deficits in aggressive behavior as a result of the absence of suckling stimulation. To test this possibility, we thelectomized or sham-thelectomized lactating female mice on day 5 of lactation and fostered replete young as previously described. Unlike the previous experiment, however, aggression testing did not begin until postpartum day 12. Thelectomy did not alter the proportion of animals fighting, thus indicating that suckling is not important for aggression in animals with previous exposure to this stimulus. Our autopsies also showed that none of the mammary glands of thelectomized females contained milk, while the mammary glands of all of the sham-thelectomized animals contained milk.

At this stage, our findings showed that prepartum thelectomy reduces the proportion of animals that exhibit maternal aggression, but postpartum thelectomy in animals with at least 5 days of previous suckling stimulation does not influence subsequent fighting behavior. We next examined whether there is a critical duration of postpartum suckling exposure required to initiate maternal aggression. We examined this question by testing for aggression different groups of parturient females exposed to various durations of suckling prior to thelectomy.

On the day prior to parturition (day 18 of pregnancy), the home cages of isolated timed-mated animals were slightly modified so as to prevent the adults from receiving suckling stimulation during the immediate postpartum interval. A ¼-inch wire mesh screen was placed inside each cage 1½ inches above the floor. Thus, during parturition, the pups dropped through the screen. On the day of parturition, the mesh was taken out of the cage, the pups were removed from the floor and checked for milk bands, and the parturient animal was proffered five 3- to 8-day-old foster young. The fostering procedure was conducted again on the next day, at which time one group of animals was thelectomized while another group of animals was sham-thelectomized. On the following morning each animal was tested for aggression and the pups were removed and replaced by five recently suckled foster young. This procedure was continued for 7 additional days. A third group of animals was thelectomized following 48 hr of suckling exposure. The animals of this group received daily aggression tests beginning 24 hr following thelectomy and were treated in a manner identical to that of the previously described groups. As in the previous experiments, the animals were killed following the last aggression test and their mammary glands were examined for milk.

More animals fought following 48 hr of suckling exposure as compared with animals that only received 24 hr of suckling exposure (see Table 2). Also, animals that received 48 hr of suckling exposure did not differ from animals sham-thelectomized after 24 hr of suckling stimulation. On the day of parturition, no milk was seen in the stomachs of the pups, thus indicating that the wire mesh screen successfully prevented suckling exposure. Prior to thelectomy, or sham thelectomy, the stomachs of the pups of all mothers contained milk. Only the pups of sham-thelectomized mothers contained milk in their stomachs following surgery. Upon autopsy it was found that only the mammary glands of sham-thelectomized animals contained milk. These findings show that 24–48 hr of postpartum suckling stimulation is a prerequisite for the postpartum initiation of maternal aggression. Less than 24 hr of postpartum suckling does not result in the percentage of animals fighting that is normally observed.

To summarize our findings at this point, we found that suckling stimulation is necessary for the initiation of maternal aggression in RS mice but is not essential in animals with previous recent suckling exposure. Females that were thelectomized prior to mating or just before parturition and continuously fostered young following delivery failed to exhibit fighting behavior. In contrast, the fighting behavior of females that were thelectomized following 5 days of postpartum suckling stimulation and maintained in the presence of foster young was comparable to that of normal lactating females regardless of whether the first aggression test took place 1 or 7 days following thelectomy. A critical duration of postpartum suckling exposure for the initiation of aggression was identified from findings showing that 48, but not 24, hr of postpartum suckling exposure prior to thelectomy and that the fostering of young was sufficient to establish aggressive behavior comparable to that of normal lactating animals.

Our results also showed that lactogenesis, as assessed by the presence of milk in the mammary glands of recently suckled mothers, is not always associated with aggression. Animals thelectomized after having lactated for several days did not show galactopoiesis (milk production) after surgery but did exhibit fighting behavior.

3.4.1c. Hormone Induction Studies. Our work clearly showed at this stage that suckling stimulation is an important factor in the initiation of maternal aggression in parturient mice. Animals whose nipples were removed prior to parturition and that were fostered young daily following delivery failed to display aggressive behavior. In addition, we had repeatedly failed over a period of several years to induce aggression in virgin animals continuously fostered newborn young. One reason for this failure may have been that virgin females, owing to the lack of nipple development that normally occurs during pregnancy, did not receive suckling stimulation from the foster young. We hypothesized (Svare and Gandelman, 1976c) that if nipple growth were induced in virgin females, they would be able to receive suckling stimulation and hence would exhibit aggressive behavior comparable to that of parturient animals.

To test the above line of reasoning, virgin female mice were divided into four groups. The animals of one group were bilaterally ovariectomized under ether anesthesia and administered a regimen of hormones that was previously found in pilot work to induce nipple growth. The regimen consisted of 19 daily subcutaneous injections of 0.02 μg estradiol benzoate dissolved in 0.02 ml sesame oil and 500 μg progesterone in 0.02 ml oil. The requirement of both steroids for nipple growth was confirmed by pilot data showing that little or no nipple growth occurred in response to injections of either estrogen or progesterone alone. This synergism between estrogen and progesterone also is evident in many ruminant species (Larson and Smith, 1974; Turner and Bagnara, 1976). Another group of virgin females was ovariectomized and administered oil for 19 days, while a third group of mice was sham-ovariectomized and treated daily for 19 days with oil. A final group was left intact and timed-mated. In order to verify that the hormone regimen did produce nipple growth, six additional animals were placed into each treatment group. They were killed, and a representative nipple from each animal (the right second thoracic) was removed and measured microscopically following either 19 days of hormone or oil treatment or on day 19 of gestation.

The virgin mice were tested for aggression 24 hr following the final hormone or oil injection. The animals permitted to mate were tested on day 19 of gestation. Tests against a male opponent were used, and the presence or absence of aggression was recorded. Immediately following the test, each animal was presented with five 3- to 8-day-old foster young. Pregnant animals were allowed to deliver, and their young were removed prior to presenting them with foster young. On the next morning each animal was observed for maternal behavior for 15 min, after which a test for aggression was conducted. Immediately following the aggression test, the nipples of each animal were examined for color and size (red and distended nipples indicate that the female had received suckling stimulation) and the pups were removed and replaced by five other 3- to 8-day-olds. This procedure was continued for 9 additional days or until an animal exhibited aggression on two successive tests. The animals were killed at the termination of testing in order to determine whether milk was present in the mammary tissue.

We found that every animal exhibited the complete repertoire of maternal behavior, including pup retrieval, licking of the pup's anogenital area, and the assumption of the nursing postures. None of the animals fought when tested prior to exposing them to young (see Table 2). Instead, they sniffed and groomed the intruder. Following exposure to young, hormone-treated virgin females displayed aggressive behavior while oil-treated ovariectomized and oil-treated sham-ovariectomized virgins did not fight. Aggressive behavior typically occurred on the first test following pup exposure, although several animals began to fight on the second or third test. The groups did not differ with respect to the average number of tests on which fighting occurred. The topography of the fighting behavior of the hormone-treated virgins was indistinguishable from that of the parturient animals. The initial attack latencies were always short, and the attacks consisted of bites to the neck and flanks of the intruder male. Tail rattling and anogenitial sniffing were observed during fights but not prior to the initial attack.

Daily inspection of the parturient and hormone-treated animals following pup exposure revealed red and distended nipples. In both groups, the pups attached themselves to the nipples such that they were able to maintain contact with the adult when the latter was lifted by the tail. In contrast, red and distended nipples and attachment were never observed in either of the groups of virgins given oil injections. It also was found that only the mammary glands of the parturient animals contained milk.

The level of nipple development seen in the hormone-treated virgins was less than that observed in pregnant animals, but both groups showed substantial development when compared to virgins treated with oil (see Table 3).

Our findings showed that virgin female mice can be induced to exhibit aggressive behavior comparable to that of parturient animals by producing nipple growth with estrogen and progesterone treatments followed by the presentation of foster young. The hormone treatment did not directly induce the fighting behavior, since the hormone-treated animals never exhibited aggression prior to exposing them to young. Thus, this further confirmed our previous finding that suckling stimulation is essential for the initiation of aggression.

In order to further corroborate the importance of suckling stimulation for aggression we performed two additional experiments. In the first experiment, aggressive behavior was examined in hormone-treated virgins and parturient animals whose nipples were removed. To perform this experiment, virgin female

Table 3. The Hormonal Induction of Nipple Growth in RS Virgin Female Mice[a]

Group	N	Average nipple length (mm)	Average nipple base diameter (mm)
Pregnant	6	$1.40^{b,c} \pm 0.05$	$0.66^{b,c} \pm 0.06$
OVX + EB + P	6	$0.99^{b} \pm 0.07$	$0.55^{b} \pm 0.14$
OVX + oil	6	0.49 ± 0.08	0.43 ± 0.08
S–OVX + oil	6	0.47 ± 0.13	0.41 ± 0.06

[a]The data represent the average nipple length and base diameter of a representative nipple (second right thoracic) for pregnant mice (Pregnant), ovariectomized virgin mice treated with estradiol benzoate and progesterone (0.02 µg and 500 µg a day for 19 days) to induce nipple growth (OVX + EB + P), ovariectomized and treated with oil (OVX + oil), or sham-ovariectomized and given oil (S-OVX + oil). Nipples were removed and measured on the 19th day of pregnancy or following 19 days of sterioid or oil treatment. From Svare and Gandelman (1976c).
[b]Significantly different from group OVX + oil and S-OVX + oil.
[c]Significantly different from group OVX + EB + P.

mice were divided into four groups. Two groups of mice were ovariectomized and treated with the previously described hormone regimen. One group of the hormone-treated mice was subjected to thelectomy, while the other group was sham-thelectomized. The two remaining groups of mice were left intact and mated with RS males. One of these groups of animals was thelectomized on the 19th day of pregnancy, and the other group was sham-thelectomized. The testing and fostering procedures were identical to those previously described with the exception that animals were sacrificed following the final aggression test and examined for the presence of milk in the mammary tissue.

Our observations showed that all animals were comparable with respect to the display of maternal behavior. None of the animals exhibited aggression during the initial test which occurred prior to the exposure of young (see Table 2). Following pup exposure, sham-operated parturient mice and sham-thelectomized hormone-treated virgins displayed aggression, while thelectomized parturient animals and thelectomized hormone-treated virgins rarely exhibited aggression. Daily examination of both groups of sham-operated animals once again revealed red and distended nipples, thus indicating that they had received suckling stimulation. Only the parturient animals were found to have milk in their mammary glands at the postmortem examination.

These findings showed that hormone-treated virgins respond to thelectomy in a manner identical to that of parturient mice, with thelectomy significantly reducing the porportion of animals that exhibit aggressive behavior. The final experiment in this series examined the influence of suckling on aggression by removing and re-presenting foster young.

Virgin female mice either were ovariectomized and administered the estrogen–progesterone hormone regimen or were mated. The animals were tested for aggression on the 3rd day of exposure to foster young. Hormone-treated virgins and parturient animals that fought the intruder had their foster young immediately removed following the fighting test and were retested 24 hr later. If fighting did not occur on the retest, five foster young were placed into the cage, and beginning 24 hr later, daily aggression tests along with the daily fostering of young were performed until fighting reappeared. Those animals that fought 24 hr following the removal of young continued to receive daily aggression tests in the absence of young. As soon as fighting was no longer exhibited, foster young were reintroduced and daily tests were conducted until fighting reappeared. The animals were sacrificed and examined for milk in the mammary glands following the reestablishment of fighting behavior.

We found that all of the animals stopped fighting following pup removal and resumed fighting following pup replacement. The hormone-treated virgins and the parturient animals did not differ from each other in terms of the number of days of pup removal required to terminate fighting or the number of days of pup replacement required to reestablish fighting. The median number of days to stop fighting following pup removal was 1 (range 1–5) for the hormone-treated virgins and 1 (range 1–3) for the parturient animals. All animals resumed fighting by the first test, which occurred 24 hr following the reintroduction of the foster young. Red, distended nipples and nipple attachment were found for every animal following the replacement of the foster young. In addition, only the parturient animals were found to have milk in their mammary glands. These findings showed that the aggressive behavior of both hormone-treated and parturient mice can be eliminated by removing the suckling stimulus and reestablished by reintroducing that stimulus.

The above findings supported our assumption that nipple growth and suckling stimulation are prerequisites for the postpartum initiation of maternal aggressive behavior in mice. They further indicate that lactogenesis is not necessary for the establishment of the behavior.

3.4.2. Additional Biological Determinants

As previously reviewed, Noirot *et al.* (1975) consistently found that their female mice are aggressive during gestation. If the incidence of the behavior is related to the complex physiological changes occurring during this reproductive state, then one would expect to see enhanced aggressive behavior in virgin females made pseudopregnant by exposure to a vasectomized male. Noirot *et al.* (1975) reported that pseudopregnant female mice are as aggressive as pregnant females, suggesting, of course, that ovarian hormones may modulate the appearance of the behavior during gestation. An issue that needs to be addressed in the future is the generality of the above findings to other species as well as the exact nature of the presumptive changes (i.e., estrogen, progesterone, or both).

Recent work suggests that the onset of the behavior during the postparturient period may be related to the pituitary hormone prolactin. Inhibitors of prolactin release such as ergocornine and estrogen have been shown to suppress maternal aggression. Twice-daily injections of 0.5 mg ergocornine for 3 days inhibit the behavior in recently parturient female hamsters, while prolactin replacement therapy reverses this trend in ergocornine-treated animals (Wise and Pryor, 1977) (see Table 4). The daily administration of 0.5 μg estradiol benzoate to lactating mice that previously exhibited aggression reduces the amount of time spent fighting as well as the proportion of animals exhibiting the behavior (Svare and Gandelman, 1975). Because ergocornine and estrogen have a number of other effects on physiology (e.g., alteration of dopaminergic function) in addition to inhibiting prolactin release, it is premature to conclude at this time that prolactin may mediate postparturient aggression.

Finally, large within and between strain differences have been observed in the maternal aggression exhibited by inbred and random-bred mice. The behavior in RS, BALB, DBA, and HS mice is limited to the lactation period (St. John and Corning, 1973; Svare and Gandelman, 1976a) but some agouti and albino strains of mice are aggressive during pregnancy as well (Beniest-Noirot, 1958; Noirot *et*

Table 4. The Percentage of Ovariectomized–
Hysterectomized Lactating Hamsters Exhibiting Maternal
Aggression Following Injections of Ergocornine (Group ERGO),
Ergocornine and Prolactin (Group ERGO–PRL), or Vehicles
(Group NORM–LAC)[a]

| | | Aggressive responses | | | |
Group	N	Pounce	Fight	Roll	Chase
ERGO	6	0	0	16.6	0
ERGO–PRL	7	71.4	42.9	71.4	28.6
NORM–LAC	14	76.9	84.6	69.2	69.2

[a]The lactating hamsters received the different treatments on postpartum days 4–7. The
first injection was administered following surgery on postpartum day 4; then injections
were repeated twice daily on postpartum days 5 and 6, and a single injection was given
3 hr prior to testing on postpartum day 7. Ergocornine hydrogen maleate and ovine
prolactin were administered at a dosage per injection of 0.5 mg and 1.0 mg, respectively. Aggression tests using adult male intruders were conducted in a neutral cage on
postpartum day 7. Each female was tested only once, and each test lasted 10 min. From
Wise and Pryor (1977).

al., 1975). A relatively high percentage (40–80%) of lactating females from RS,
CD-1, BALB, DBA, and HS lines exhibit the behavior, but lactating females
from C57BL/6 and C3H strains rarely show aggression (Rosenson and Asheroff,
1975; St. John and Corning, 1973; Svare and Gandelman, 1973). Interestingly,
St. John and Corning (1973) found that the incidence of intermale aggression in
various inbred and random-bred strains is positively correlated with the incidence
of maternal aggression (see Table 5), suggesting that aggressive behavior in the
two sexes may be similarly affected by the same genes.

4. A Psychobiological Model of Maternal Aggression in the Mouse

The findings reviewed here enable us to propose a model for maternal
aggression in the mouse. The sequence of events involved in maternal aggression
may conveniently be divided into three phases. In order of their appearance, these
phases can be called (1) the substrate preparation phase, (2) the postpartum initiation phase, and (3) the postpartum maintenance phase. We will examine these
phases and the events in each that are important for the appearance of maternal
aggression.

4.1. Substrate Preparation Phase (Conception to Parturition)

During the gestation period, the nipples of the mouse increase in length and
diameter as a result of the action of hormones. Our findings (Svare and Gandel-

man, 1976c) indicate that estrogen and progesterone, two hormones that are at their highest level in the mouse during pregnancy (McCormack and Greenwald, 1974; Murr *et al.*, 1974), synergize to stimulate nipple growth in the virgin mouse. Although this synergism is evident for other species as well (Larson and Smith, 1974; Turner and Bagnara, 1971), it is possible that other hormones (e.g., growth hormone, corticosterone) may also contribute to nipple development. Nevertheless, the substrate preparation phase enables an animal to receive suckling stimulation. Since suckling is necessary for the initiation of aggression during lactation, this stage is a prerequisite for the postpartum exhibition of the behavior.

The fact that estrogen and progesterone will stimulate nipple growth and, hence, allow animals to receive suckling stimulation important for the establishment of aggressive behavior, is evidence that hormones may at least play an indirect role in maternal aggression by preparing peripheral structures for the suckling stimulus. This finding is reminiscent of others in the behavioral endocrinology literature showing that sexual behavior in some species may be mediated by hormonal sensitization of structures used in copulatory acts. Several investigators (Komisaruk *et al.*, 1972; Kow and Pfaff, 1973) have shown that estrogen stimulation influences the female rat's cutaneous sensitivity in the area adjacent to the vagina. Such hormonal preparation may enable her to receive maximal stimulation during the copulatory act.

Other findings suggest that ovarian hormones, in addition to their role in preparing the substrate for maternal aggression may also play a role in the heightened aggressive behavior seen during pregnancy in rats, hamsters, and in some mouse strains. As previously noted, Noirot *et al.* (1975) showed that pseudopregnant female mice are as aggressive as pregnant animals. Because pseudopregnant animals undergo changes in ovarian hormones identical to those of normal pregnant animals, the heightened aggressive behavior seen during pregnancy may be

Table 5. A Comparison of the Incidence of Maternal and Intermale Aggression in Breeding Pairs of Inbred (C57BL, C3H, DBA, and BALB) and Random-Bred (HS) Mouse Strains[a]

	C57BL		C3H		BALB		DBA		HS		Total	
	N	%	N	%	N	%	N	%	N	%	N	%
Female	24	0	24	0	59	53	24	54	67	54	198	40
Male	24	0	24	13	59	44	24	63	67	75	198	46

[a]Aggression tests were conducted on postpartum days 6–8 with the pups present but the mate removed. On each test, a target adult male mouse from the HS line was dangled in the cage by the experimenter for 2 min and the presence or absence of aggression was noted. Each member of the pair was tested once. From St. John and Corning (1973).

due to the synergistic action of estrogen and progesterone. A more direct test of this hypothesis has not, as yet, been performed.

4.2. Postpartum Initiation Phase (Parturition to Postpartum Day 2)

Following parturition, the female mouse gathers her young and permits them to suckle by crouching over them in a nursing posture. However, our findings indicate that suckling stimulation, in addition to providing nourishment, controls the mother's aggressive behavior toward other adult members of her species. This follows from the findings that the presence of this stimulation results in the appearance of aggressive behavior, while the absence of such stimulation either through thelectomy (Svare and Gandelman, 1976b), the absence of young (Gandelman, 1972; Gandelman and Svare, 1974; Svare and Gandelman, 1973), or the absence of nipple growth (Svare and Gandelman 1976b,c) renders parturient and hormone-treated animals nonaggressive.

The substrate preparation and initiation phases must be properly timed for the appearance of the behavior. Animals of the RS mouse strain do not exhibit aggression during pregnancy even in the presence of suckling stimulation (Gandelman and Svare, 1974). Thus, although nipple development has been completed by the final days of pregnancy, something prevents suckling stimulation from inducing aggression until after parturition, at which time 24–48 hr of suckling stimulation is required before aggression is exhibited (Svare and Gandelman, 1976b). The nature of this inhibition is not known at this time, but it may be related to sharply declining levels of progesterone in juxtaposition to increasing estrogen levels (McCormack and Greenwald, 1974; Murr et al., 1974). The necessity of these two phases and their proper timing is further illustrated by our findings that hysterectomy and the fostering of young after day 12 of pregnancy results in normal numbers of animals fighting (Gandelman and Svare, 1974), thus indicating that the substrate preparation phase is sufficiently completed and the initiation phase capable of starting as early as day 12 of gestation. Because nipples are undeveloped prior to day 12 of pregnancy, hysterectomy before that day in combination with the fostering of young does not result in attachment by young or fighting behavior.

The question remains as to how suckling stimulation influences the initiation of maternal aggression. The answer may reside in the numerous endocrinological, biochemical, and neuronal changes that occur when the suckling stimulus is applied to the postparturient female. For example, it is well known that suckling stimulation results in the discharge of a number of lactation-promoting hormones, including oxytocin, vasopressin, prolactin, somatotropin, thyroid-stimulating hormone, adrenocorticotropic hormone, and melanocyte-stimulating hormone (Grosvenor and Mena, 1974). These hormonal changes may of course facilitate aggressive behavior in the parturient animal. As previously reviewed, blockage of prolactin release does produce deficits in hamster and mouse maternal aggressive behavior (Svare and Gandelman, 1975; Wise and Pryor, 1977). Recall, however, that lactogenesis, as measured by milk in the mammary glands, was found not to be necessary for the initiation of aggression in mice, since hormone-primed virgin females that fought following the fostering of young were never found to have milk in their mammary glands (Svare and Gandelman, 1976c). While more sensitive indices of hormone activity are necessary before more conclusive statements can be made, this finding suggests that prolactin and other lactogenic hormones may play only a minor role in the initiation of aggression or that the thresholds for hormonal activation of aggression and lactation may differ.

It is also possible that the initiation of maternal aggression may be independent of hormonal control. The behavior may be mediated instead by suckling-induced alterations in the firing patterns of hypothalamic neurons and/or changes in the synthesis of hypothalamic monoamines. Brooks *et al.* (1966) showed that the paraventricular neurons of the cat hypothalamus respond to peripheral stimulation of the nipple with increases in their rate of discharge, while Voogt and Carr (1974) reported that 20 min of suckling stimulation significantly increased the rate of synthesis of hypothalamic dopamine in lactating rats deprived of suckling stimulation for 5 hr. Future research should examine whether some or all of the above suckling-induced changes contribute to the initiation of maternal aggression.

Regardless of the mechanism responsible for the initiation of maternal aggression, the importance of suckling stimulation for the behavior is especially interesting in light of its nonessential nature for the initiation of maternal behavior in the rat (Moltz *et al.*, 1967; Bridges, 1975). While this may simply reflect a species difference, it is significant to note that previous investigators of maternal behavior in the rat have always chosen to study pup-directed maternal responses such as retrieving, licking, and nursing as opposed to adult-directed maternal responses such as aggression (see, for example, Rosenblatt and Siegel, this volume). This underscores the need for a more thorough analysis of the entire maternal repertoire before we can arrive at valid conclusions concerning the mechanism responsible for the initiation of maternal responding.

4.3. Postpartum Maintenance Phase (Postpartum Days 2–21)

Following a period of at least 24–48 hr of suckling stimulation, maternal aggression is initiated and reaches its highest intensity in the immediate postpartum period. As the lactation period progresses, the intensity of aggression wanes to a point where it is very low by the time pups reach the age of weaning (Svare and Gandelman, 1973). The reasons for the decline in aggression with advancing lactation are not well known at this time. The findings reviewed here suggest that, once initiated, the behavior may be modulated by the freqency with which nursing occurs and the physiological changes associated with high as compared with low frequencies of nursing. Thus, animals that were continuously fostered newborn young (i.e., nursed their young frequently) continued to fight longer at high levels than animals left with their own young (i.e., nurse their young infrequently) (Svare, 1977).

While suckling is clearly involved in the initiation of aggression and perhaps in the maintenance of the intensity of fighting, it is not needed for the presence of fighting behavior during the maintenance phase. This is illustrated by our findings showing that thelectomy in animals with at least 2 days of suckling exposure does not eliminate aggressive behavior (Svare and Gandelman, 1976b) and that the placement of young behind a wire partition in the dam's home cage maintains aggression (Svare and Gandelman, 1973). The mechanism responsible for the maintenance of aggression in the absence of suckling may involve the aforementioned suckling-induced hypothalamic changes, which may be long lasting. However, we feel that an interpretation based upon a hormonal mechanism is a more appealing explanation at this time. Prolactin and corticosterone, hormones related to lactation and aggression, are known to be responsive to olfactory stimulation from young and, therefore, may be involved in maternal aggression. For example, removal of young from rat mothers for 5 hr but not 1 hr leads to a marked reduction in the plasma levels of prolactin and corticosterone (Mena and Grosvenor, 1971; Zarrow et al., 1972). Perhaps the reduced levels of these hormones following the removal of pups is responsible for the attenuation of fighting behavior. Levels of prolactin and corticosterone in lactating rats also are known to increase when young are reintroduced in a manner which precluded direct physical contact with the lactating animal (Mena and Grosvenor, 1971; Zarrow et al., 1972). The increased levels of these hormones following pup reintroduction may explain why aggression is still observed when pups are placed behind a wire partition in the home cage, and why the reintroduction of pups restores aggression in lactating animals whose young are removed for 5 hr (Svare and Gandelman, 1973).

In summary, the events contributing to the display of maternal aggression can be divided into three phases. During the substrate preparation phase, hormones secreted during pregnancy facilitate aggressive behavior and produce growth in the nipples. The growth of the nipples enables the parturient dam to receive suckling stimulation from her young. Suckling stimulation during the first several days following parturition induces changes in the dam that facilitate aggression during the postpartum initiation phase. Suckling-induced pituitary–hypothalamic changes are strongly suspected of playing a major role during this phase. Aggression is controlled during the postpartum maintenance phase by the frequency of suckling and exteroceptive stimuli from young.

Regardless of the mechanism involved in the preparation, initiation, and maintenance of maternal aggression, it should be emphasized that the growth of nipples and suckling stimulation and the physiological changes produced by suckling are necessary conditions for aggression in mice, but they are not sufficient. This follows from our observation that fighting behavior is displayed by only 50–60% of normally parturient, pregnancy-terminated, and hormone-primed females. Thus, it is not likely that fighters and nonfighters differ in terms of the physiological events that accompany suckling. It is more likely that they differ with respect to their sensitivity to the aggression-promoting factors that accompany suckling. The fact that 20–35% of RS mice do not exhibit aggression and that some strains of mice are nonaggressive (St. John and Corning, 1973) suggests that some animals are more responsive to environmental cues that may disrupt aggressive behavior. Thus, it is possible that genetic and within-strain differences may represent unique adaptations that evolved as a result of environmental and ecological constraints.

Finally, we might conceptualize maternal aggression as a form of irritable aggression and not, as Moyer (1976) has suggested, a separate class of aggressive behavior. One primary characteristic of irritable aggression, the nonspecificity of the stimuli which evoke the response, would seem to apply to the aggressive behavior of lactating mice. Postpartum female mice are known to attack young and old, male and female conspecifics (Svare and Gandelman, 1973) as well as adult rats (V. Denenberg, personal communication). On occasion, it has been noted that lactating mice will attack the hand of the experimenter. Future work in this area should examine the possible similarities between maternal aggression and other forms of aggressive behavior, most notably irritable aggression. For example, one area of investigation that may prove profitable concerns the hypothalamus. As previously stated, numerous suckling-induced changes occur in the hypothalamus, and it is well known that this area of the brain is implicated in irritable and other forms of aggressive behavior (cf. Delgado, 1969; MacDonnell and Flynn, 1966; Panksepp, 1971). Perhaps pharmacological blockade, electrical stimulation, or lesions of the hypothalamus of parturient animals will in the future be found to alter maternal aggressive behavior.

5. Relationship to Defense of Young and Population Dynamics

Does maternal aggressive behavior function to protect the young? The answer to this question is uncertain at the present time. However, two recent experiments have some bearing on this issue. Using an extended cage apparatus, Erskine *et al.* (1978a) demonstrated that the lactating female rat is not effective in preventing access of the intruder to the litter. Even though the intruder had the opportunity to avoid contact with the mother and her litter by fleeing to another part of the cage, the majority of the intruders ended up spending more time in the nest than the mothers. Moreover, cannibalization of the litters by the intruder frequently occurred and the number of attacks by the females was not correlated with either the number of times the intruder moved in and out of the nest or the amount of time the intruder spent in the nest. Similar findings have been reported for mice (Green, 1978). These findings apparently do not substantiate the claim that the behavior functions as a response to a perceived threat to the young.

Because forced cohabitation was an unnatural feature of both of the above studies, it would seem that a naturalistic study of the behavior is the only way to adequately test the notion that maternal aggression serves a protective function.

Does maternal aggressive behavior play a role in population regulation? Previous work with house mice indicates that a number of factors are implicated in the decline of population growth at high densities. For example, increased intraspecific competition, altered hormone titers, fetal resorption, inhibition of sexual maturation, increased infant mortality, delayed implantation, and inhibition of estrous cycles are frequently cited as population-limiting mechanisms (cf. Christian, 1971; Christian *et al.*, 1965). Notably absent from this list is the role that maternal aggression and other maternal behaviors may play in population regulation. The neglect is not total, however, as some evidence exists to warrant further investigations of maternal behaviors and population density in rodents. For example, Brown (1953) and Southwick (1955) noted that nursing and nest construction were disrupted by other mice as population density rose. Also, Christian and LeMunyan (1958) demonstrated that the crowding of female mice caused deficient lactation which resulted in the diminished growth of young. The young themselves were affected so that they in turn had smaller progeny apparently as a result of deficient lactation. Other than the few studies mentioned above, little information of a systematic nature is available regarding maternal behavior and its relation to population dynamics. Perhaps maternal aggression breaks down in periods of high population density, thus enabling next encroachment, pup killing by other conspecifics, and, as a result, the reduction of population size. On the other hand, maternal aggression may increase under conditions of high population density and disperse the population into adjacent unoccupied territories (cf. Rowley and Christian, 1976b). Studies of maternal aggression in freely growing populations are needed for a more complete understanding of the potential role of maternal behaviors in population dynamics.

Finally, some findings suggest that reproductive performance may be related to maternal defense. Lloyd and Christian (1969) found that high-ranking female mice were responsible for almost all of the offspring produced in the freely growing populations of *Mus musculus* they studied, and Svare and Gandelman (1976a) found that lactating female mice that exhibited high overall levels of aggression tended to produce litters with more young than did animals exhibiting lower overall levels of aggression. The above findings suggest a biologically adaptive mechanism that favors increased fecundity of highly aggressive dams as opposed to less aggressive dams.

6. General Conclusions

As previously stated, the study of maternal aggressive behavior in mammals is still in its infancy. More research using other species is critically needed before we can arrive at general conclusions concerning the experiential and biological determinants of the behavior. In addition, questions concerning the potential role of the behavior in population dynamics and defense of the young remain unanswered.

In spite of the above limitations, a solid body of research findings has been generated for the mouse, rat, and hamster. These findings have allowed us to pinpoint several factors that contribute to the display of the behavior under laboratory conditions. They ultimately will lay the groundwork for what should be an exciting area of psychobiological investigation in the ensuing years.

References

Altman, M., 1963, Naturalistic studies of maternal care in the moose and elk, in: *Maternal Behavior in Mammals* (H. L. Rheingold, ed.), pp. 233–253, Wiley, New York.

Barnett, S. A., 1969, Grouping and dispersive behavior among wild rats, in: *Aggressive Behavior* (S. Garattini and E. B. Sigg, eds.), pp. 3–14, Wiley, New York.

Beniest-Noirot, E., 1958, Analyse du comportement dit 'maternal' chez la souris, *Monogr. Fr. Psychol. Paris,* No. 1.

Bridges, R. S., 1975, Long-term effects of pregnancy and parturition upon maternal responsiveness in the rat, *Physiol. Behav.* **14**:245–249.

Brooks, C. M., Ishikawa, T., Korzumi, K., and Lu, H., 1966, Activity of the neurones of the paraventricular nucleus of the cat hypothalamus, *J. Physiol.* **182**:217–228.

Brown, R. Z., 1953, Social behavior, reproduction, and population changes in the house mouse. *Ecol. Monogr.* **23**:217–240.

Christian, J. A. and LeMunyan, C. D., 1958, Adverse effects of crowding on reproduction and lactation of mice and two generations of their progeny, *Endocrinology* **63**:517–529.

Christian, J. J., 1971, Population density and reproductive efficiency, *Biol. Reprod.* **4**:248–294.

Christian, J. J., Lloyd, J. A., and Davis, D. E., 1965, The role of the endocrines and the self-regulation of mammalian populations, *Recent Prog. Horm. Res.* **21**:501–568.

Delgado, J. M. R., 1969, Radio stimulation of the brain in primates and man, *J. Int. Anes. Res. Soc.* **48**:529–542.

Denenberg, V. H., Sawin, P. B., Frommer, G. P., and Ross, S., 1958, Genetic, physiological and behavioral background of reproduction in the rabbit: IV. An analysis of maternal behavior at successive parturitions, *Behaviour,* **13**:131–142.

Denenberg, V. H., Petropolus, S. F., Sawin, P. B., and Ross, S., 1959, Genetic, physiological and behavioral background of reproduction in the rabbit: VI. Maternal behavior with reference to scattered and cannibalized newborn and mortality, *Behaviour* **15**:71–76.

Devore, I., 1963, Mother-infant relations in free ranging baboons, in: *Maternal Behavior in Mammals* (H. L. Rheingold, ed.), pp. 305–335, Wiley, New York.

Endroczi, E., Lissak, K., and Telegdy, G., 1958, Influence of sexual and adrenocortical hormones on the maternal aggressivity. *Acta Physiol. Hung.* **15**:353–357.

Erskine, M., Denenberg, V. H., and Goldman. B. D., 1978a, Aggression in the lactating rat: Effects of intruder age and test arena, *Behav. Biol.* **23**:52–66.

Erskine, M., Barfield, R. J., and Goldman, B. D., 1978b, Intraspecific fighting during late pregnancy and lactation in rats and effects of litter removal, *Behav. Biol.* **23**:206–218.

Flandera, V., and Novakova, V., 1971, The development of interspecific aggression in rats towards mice during lactation, *Physiol. Behav.* **6**:161–164.

Gandelman, R., 1972, Mice: Postpartum aggression elicited by the presence of an intruder, *Horm. Behav.* **3**:23–28.

Gandelman, R., and Svare, B., 1974, Pregnancy termination, lactation and aggression. *Horm. Behav.* **5**:397–405.

Gandelman, R., and Svare, B., 1975, Lactation following hysterectomy of pregnant mice, *Biol. Reprod.* **12**:360–367.

Green, J. A., 1978, Experiential determinants of postpartum aggression in mice, *J. Comp. Physiol. Psych.* **92**:1179–1187.

Grosvenor, C. E., and Mena, F., 1974, Neural and hormonal control of milk secretion and milk ejection, in: *Lactation: A Comprehensive Treatise,* Vol. 1 (B. L. Larson and V. R. Smith, eds.), pp. 227–276, Academic Press, New York.

Hersher, L., Richmond, J. B., and Moore, A. U., 1963, Maternal behavior in sheep and goats, in: *Maternal Behavior in Mammals* (H. L. Rheingold, ed.), pp. 203–232, Wiley, New York.

Komisaruk, B. R., Adler, N. T., and Hutchinson, J., 1972, Genital sensory field: Enlargement by estrogen treatment in female rats, *Science* **178**:1295–1298.

Kow, L. M., and Pfaff, D. W., 1973, Effects of estrogen treatment on the size of the receptive field and response thresholds of pudendal nerve in the female rat, *Neuroendocrinology* **15**:419–427.

Larson, B. L., and Smith, V. R., eds., 1974, *Lactation: A Comprehensive Treatise,* Vol. 1, Academic Press, New York.

Lloyd, J. A., and Christian, J. J., 1969, Reproductive activity of individual females in three experimentally freely growing populations of house mice *(Mus musculus), J. Mammal.* **50**:49–59.

Lynds, P. G., 1976a, A comparison of the behavioral components of postpartum and intermale aggression in wild housemice, Paper presented at Nebraska Academy of Sciences Meeting.

Lynds, P. G., 1976b, Olfactory control of aggression in lactating female housemice, *Physiol. Behav.* **17**:157–159.

MacDonnell, M. F., and Flynn, J. P., 1966, Control of sensory fields by stimulation of the hypothalamus, *Science* **152**:1406–1408.

McCormack, J. T., and Greenwald, G. S., 1974, Progesterone and oestradiol 17B-concentrations in the peripheral plasma during pregnancy in the mouse. *J. Endocrinol.* **62**:101–107.

Mena, F., and Grosvenor, C. E., 1971, Release of prolactin in rats by exteroceptive stimulation: Sensory stimuli involved, *Horm. Behav.* **2**:107–116.

Moltz, H., Geller, D., and Levin, R., 1967, Maternal behavior in the totally mammectomized rat, *J. Comp. Physiol. Psychol.* **64**:225–229.

Moyer, K. E., 1976, *The Psychobiology of Aggression,* Harper and Row, New York.

Murr, M. M., Stabenfeldt, G. H., Bradford, G. E., and Geschwind, I. I., 1974, Plasma progesterone during pregnancy in the mouse, *Endocrinology,* **94**:1209–1211.

Noirot, E., Goyens, J., and Buhot, M., 1975, Aggressive behavior of pregnant mice toward males, *Horm. Behav.* **6**:9–17.

Panksepp, J., 1971, Effects of hypothalamic lesions on mouse-killing and shock-induced fighting in rats, *Physiol. Behav.* **6**:311–316.

Price, E. O., and Belanger, P. L., 1977, Maternal behavior of wild and domestic stocks of Norway rats, *Behav. Biol.* **20**:60–69.

Revlis, R., and Moyer, K. E., 1969, Maternal aggression: A failure to replicate. *Psychon. Sci.* **116**:135–136.

Rosenson, L. M., and Asheroff, A. K., 1975, Maternal aggression in CD-1 mice: Influence of the hormonal condition of the intruder, *Behav. Biol.* **15**:219–224.

Ross, S., Sawin, P. B., Zarrow, M. X., and Denenberg, V. H., 1963, Maternal behavior in the rabbit, in: *Maternal Behavior in Mammals* (H. L. Rheingold, ed.), pp. 94–121, Wiley, New York.

Rowley, M. H., and Christian, J. J., 1976a, Interspecific aggression between *Peromyscus* and *Microtus* females: A possible factor in competitive exclusion, *Behav. Biol.* **16**:521–525.

Rowley, M. H., and Christian, J. J., 1976b, Intraspecific aggression of *Peromyscus leucopus, Behav. Biol.* **17**:249–253.

Rowley, M. H., and Christian, J. J., 1977, Competition between lactating *Peromyscus leucopus* and *junvenile Micorosus pennsylvanicus Microtus, Behav. Biol.* **20**:70–80.

Schneirla, T. C., Rosenblatt, J. S., and Tobach, E., 1963, Maternal behavior in the cat, in: *Maternal Behavior in Mammals* (H. L. Rheingold, ed.), pp. 122–168, Wiley, New York.

Southwick, C. H., 1955, Regulatory mechanisms of house mouse populations: Social behavior affecting litter survival, *Ecology* **36**:627–634.

St. John, R. S., and Corning, P. A., 1973, Maternal aggression in mice, *Behav. Biol.* **9**:635–639.

Svare, B., 1977, Maternal aggression in mice: Influence of the young, *Biobehav. Rev.* **1**:151–164.

Svare, B., and Gandelman, R., 1973, Postpartum aggression in mice: Experiential and environmental factors, *Horm. Behav.* **4**:323–334.

Svare, B., and Gandelman, R., 1975, Postpartum aggression in mice: Inhibitory effect of estrogen, *Physiol. Behav.* **14**:31–36.

Svare, B., and Gandelman, R., 1976a, A longitudinal analysis of maternal aggression in mice, *Dev. Psychobiol.* **9**:437–446.

Svare, B. and Gandelman, R., 1976b, Postpartum aggression in mice: The influence of suckling stimulation, *Horm. Behav.* **7**:407–416.

Svare, B., and Gandelman, R., 1976c, Suckling stimulation induces aggression in virgin female mice, *Nature (London)* **260**:606–608.

Taylor, J. C., 1966, Home range and agonistic behavior in the grey squirrel, in: *Play Exploration, and Territory in Mammals* (P. A. Jewell and C. Loizos, eds.), pp. 229–236, Academic Press, New York.

Turner, C. D., and Bagnara, J. T., 1971, *General Endocrinology,* W. B. Saunders, Philadelphia.

Voogt, J. L., and Carr, L. A., 1974, Plasma prolactin levels and hypothalamic catechol-amine synthesis during lactation, *Neuroendocrinology* **16**:108–110.

Wise, D. A., 1974, Aggression in the female golden hamster: Effects of reproductive state and social isolation, *Horm. Behav.* **5**:235–250.

Wise, D. A., and Pryor, T. L., 1977, Effects of ergocornine and prolactin on aggression in the postpartum golden hamster, *Horm Behav.* **8**:30–39.

Zarrow, M. X., Schlein, P. A., Denenberg, V. H., and Cohen, H. A. 1972, Sustained corticosterone release in lactating rats following olfactory stimulation from the pups, *Endocrinology* **21**:191–196.

6

The Ecology of Weaning

Parasitism and the Achievement of Independence by Altricial Mammals

Bennett G. Galef, Jr.

1. An Overview of Weaning

At birth and for some time thereafter altricial mammals are, by definition, almost totally dependent on their parents or other caretakers for provision of many of the necessities of life. Although many mammals are born relatively helpless, none continue to depend on conspecifics for sustenance throughout their life cycles. Prior to reaching reproductive age each altricial mammal must become competent to acquire directly from the environment those goods and services which it previously acquired only indirectly as the result of interaction with conspecific caretakers. This transition from an infantile dependence on others for transduction of environmental resources to an adult mode of independent acquisition of necessities is in two senses a gradual process. First, independence with respect to any single need is rarely achieved suddenly; most often a series of graded transitional stages intervene between total dependence on caretakers for supply of a given resource and independent acquisition of that resource. The developing rat pup, for example, moves successively from *in utero* total dependence for nutrition on mother's blood, to postparturient total dependence on mother's milk, to a mixed diet of mother's milk and solid food, to a diet of solid food transported to the home burrow by adult conspecifics, to feeding trips with adult conspecifics, to independent acquisition of solid food. Second, the juvenile may exhibit adultlike behavior in meeting each of its various needs at different ages. The rat pup, for example,

---

BENNETT G. GALEF, JR. • Department of Psychology, McMaster University, Hamilton, Ontario, Canada L8S 4K1.

initiates endogenous thermoregulatory behavior some time before it initiates independent feeding on solid foods.

Thus, while weaning in the dictionary sense of "accustoming an organism to the loss of mother's milk" frequently occurs during a fairly restricted period during ontogeny, weaning in the broader sense implied above (i.e., achievement of the adult degree of independence of conspecifics in the acquisition of the necessities of life) is an ongoing process extending from birth (after which oxygen is acquired independently) to shortly prior to reproductive maturity, when, for example, independent construction of harborage sites may be initiated. It is the latter broad view of the weaning process with which I will be concerned here.

The extended period during which weaning in this sense occurs has been studied in detail by psychologists with two distinct goals: first, to determine the factors mediating observed changes in the form of mother–young interaction during the weeks following parturition, and second, to elucidate the processes underlying the development of behavior.

In the present paper I will discuss the possibility that consideration of parental investment theory (Trivers, 1974) suggests the necessity of a change in theoretical orientation with respect to the selection pressures responsible for the apparent complementarity between the time course of maternal behavior of the dam and the changing requirements of her developing young. I will further argue that this change in orientation concerning the behavioral synchrony of the mother–young dyad suggests the need for a broadening of the range of factors considered as important in determining the time course of behavioral development in altricial mammals (see Hofer, this volume, for a related view). Last, I will propose that understanding of both behavioral development and mother–young interaction depend in some measure on interpretation of the process of weaning. I will draw an analogy between various stages in the weaning process and various types of parasitism. And, to illustrate the implications of the theoretical position suggested above, I will use this analogy to interpret the time course of behavioral development in rat pups during the weaning period.

1.1. Mother–Young Interaction

Maternal behavior during the period of mother–young interaction has been described as "changing in ways which are correlated with changes in the needs of the developing young" (Rosenblatt and Lehrman, 1963, p. 8). While such a statement of correlation is descriptive and does not imply that adjustments in the behavior of mothers are caused by changes in the need states of their young, it does focus attention on one possible cause of the synchrony in the behavior of mothers and their young (i.e., infant needs) at the expense of others. One could as easily describe the behavior of the young as changing in ways which are correlated with changes in the willingness of the dam to provide them with resources (i.e., parentally regulated). Or, more equivocally, one could describe the behavior of both dam and young as changing so as to correlate changes in the resource contribution of the dam with changes in the needs of the young (i.e., mutually adjusted). The three different statements of correlation suggest three different hypothetical histories of the evolution of the observed synchrony of mother–young interaction as well as three different general models of the proximate causation of changes in the behavior of mothers and young during their period of interaction.

There is little question that each of the three correlational statements mentioned above points to an important feature of the evolution of the behavior of mother and young. Clearly, during evolutionary time, the needs of the young have become adapted to the willingness and ability of their dam to provide them with resources, just as the willingness of the dam to invest in her offspring has become adapted to the needs of the young for energy inputs to assure adequate growth. Given the partial communality of reproductive interests of parent and young, such co-evolution is inevitable.

It has, however, recently been pointed out that the reproductive interests of parent and offspring are not identical (Trivers, 1974). To a certain extent a mother and her young are competing, each seeking to maximize its own inclusive fitness, even if such individual maximization is achieved at the expense of the other member of the parent–offspring pair. To put the matter simply, there is reason to believe that offspring should be selected to attempt to acquire more parental investment* than parents will be selected to give (Trivers, 1974). This is because the young of a given litter show only reproductive profit from parental investment in them, while the dam has to consider both the costs and benefits of continuing to transduce energy to a given litter rather than saving it for investment in future litters.

Trivers' (1972, 1974) discussions of parental investment and parent–offspring conflict have several possible implications for the study of the proximal causation of the behavior of both mother and young during their period of interaction. First, Trivers' model implies that the conflict between mother and young evolves to a situation in which it is the reproductive interests of the mother which set the upper bound on the amount and duration of her investment in any given litter of young. The irreducible needs of the young may set a lower bound on maternal investment, but this lower limit is probably seldom arrived at in a successful reproductive effort. The observation that the willingness of the dam to invest in a given litter sets the upper limit on energy flow to the young suggests that it might be more useful, in discussions of parent–young interaction, to treat the dam as providing the environment to which the young must respond rather than to treat the young as presenting needs which the dam must meet.

Second, in psychological discussions of parent–offspring interactions, as in classical evolutionary theory, the parent is frequently treated as the active partner.

---

*Parental investment in an offspring is defined as "anything done by the parent for the offspring that increases the offspring's chance of surviving while decreasing the parent's ability to invest in other offspring" (Trivers, 1974, p. 249).

The current approach in evolutionary theory, briefly described above, is to view dependent young as relatively equiactive participants in a parent–offspring conflict. Such a theoretical position suggests that at the level of mechanism it might be fruitful to treat the neonate as acting to acquire investment from the dam rather than as a passive recipient of the dam's beneficence (see Bell and Harper, 1977, for a related view).

Third, in both the psychological and biological literature the neonate is often treated as to some extent inefficient, inept, and inadequate. Trivers' position suggests that it might be more valid to treat neonates as exhibiting behaviors and physical characteristics adapted to acquiring parental investment and thus maximizing survival, growth, and, ultimately, inclusive fitness of the young within the environmental niche defined by their parents.

In the present paper I will, therefore, treat dependent young as acting so as to maximize net energy gain within the constraints imposed by their dam's limited willingness to invest in them.

1.2. Development of Behavior

The progressive increase over time in the apparent complexity of the behavior of mammalian neonates is generally viewed as resulting from two factors and their interaction. First, maturation in the complexity and sophistication of the neural and muscular structures underlying behavior is treated as necessary for increased behavioral complexity. Second, experiences of the neonate in interaction with the physical and social environment early in development are treated as necessary conditions for the organization of behavior appearing later in development (Schneirla, 1957).

Lehrman and Rosenblatt's (1971, p. 10) description of the development of feeding behavior in kittens provides an elegant example of this point of view:

> ... although we see the same relationships between sucking and food intake during the whole period through weaning, the internal relationships between the animal's behavior and the needs which it serves are constantly changing through some interaction between the growth of the infant and the experience it has in the interaction with its mother made possible by that growth. At first the sucking behavior is reflexly initiated by stimuli offered by the initiative of the mother, or simply by the proximity of the mother to the infant. The growth of the infant and its experience in the situation described by this interaction lead to the emergence of an internal connection between (a) the sucking behavior by which the infant gets food, (b) the felt need for food, and (c) the external source of food, so that the infant becomes able to perceive the source of food as such, to orient to it at its own initiative, and to learn to distinguish many subtle details of the food source.

There is a second view of the developmental process, complementary to the first, which has been largely ignored by psychologically oriented students of behavior. In discussing the ontogeny of regulatory mechanisms in the rat pup, Adolph (1957, p.131) has clearly described that view.

> We tend to speak as though the regulations with which the adult is endowed are the only adequate ones, superior to those of infants. In doing so, we recognize that they are in use longer than those of infants. Actually we are not justified in believing that adult regulations could successfully be imposed on infants. *As far as we know, each stage of development is functionally complete in its own right,* and the common supposition that the adult stage enjoys special advantages cannot rigorously be sustained, because we have no criterion of advantages and disadvantages other than the frequency of survival in natural circumstances. (Emphasis added)

Williams (1966, p. 71) makes the same point somewhat more succinctly:

> The succession of somata in the life cycle of an organism must provide an adjustment of each stage to the one before and the one after in addition *to an adaptive selection of environmental niches and precise adaptation to each niche.* (Emphasis added)

Although considerable attention has been paid to the development of structures underlying behavior and the role of experience in the achievement of the adult behavioral phenotype, little attention has been paid to the fact that the organism at every point in development is a functional entity, adapted to the environment in which it finds itself. The reason for the inattention paid to this view by comparative psychologists, with at least a secondary interest in using the study of animal development to understand human development, is not difficult to surmise. But there is some reason to question whether the role of experience in the development of adult behavior in nonprimate mammals is nearly so great as in primates. The human juvenile clearly requires specific experiences during the period of dependence on its parents to acquire the language, social, and motor skills necessary to develop a successful adult behavioral phenotype. Nonhuman primates may exhibit an analogous dependence on experience during ontogeny for the development of important aspects of their adult social behavior (H. F. Harlow and M. K. Harlow, 1965).

Recent technical innovations (Hall, 1975) have made it possible to rear rat pups both in isolation from all conspecifics (including their dam) and without the opportunity to experience oral ingestion from days 2–18 postpartum. Pups raised in this fashion exhibit essentially normal patterns of feeding when they reach weaning age and do not exhibit detectable deficits in food recognition, patterns of ingestion, or response to various nutrient stresses at the time of introduction to solid food (Hall, 1975). Thoman and Arnold, (1968) similarly found relatively small effects of hand rearing of rat pups on subsequent reproductive behaviors. While I would most certainly not wish to argue that experiences unique to the juvenile period, such as suckling from a dam, play no role in the development of their adult behavioral analogues, findings such as those of Hall and of Thomas suggest that engagement in some juvenile patterns of behavior are not necessary conditions for the emergence of a normal adult behavioral phenotype. If this is so, then understanding of the function of the behavior of juveniles must to some extent be sought in other factors in addition to their contribution to the development of later behavior.

The argument that the development of behavior must await the development of supporting neural coordinating centers and muscular strength also has obvious validity, but it does not contribute to understanding why neural substrates sufficient for some behaviors (e.g., nipple attachment, suckling, and huddling) are present at birth, while substrates sufficient for the support of other behaviors are absent (Darwin, 1877). Further, it is difficult to determine whether the failure to observe a given behavior pattern in a neonate is due to the absence of the necessary neural and muscular substrate for that behavior. The fact that an organism fails to exhibit some pattern of behavior early in life can surely not be taken as evidence that the underlying structures necessary to support that behavior are absent, or the explanation becomes circular. It is always possible either that the necessary supporting structures are present but their expression is actively inhibited at some points during development (Bower, 1976; Graham *et al.,* 1978) or that the external conditions necessary for the expression of a behavior in the neonate have not been identified. For example, Hall (1979) has recently found that rat pups as young as 3 days of age are capable of feeding independently of the mother by lapping milk, but only if they are both deprived of food and placed in a warm environment, (see also Smith and Spear, 1978).

I would suggest that consideration of the behavioral adaptations required by the environment in which the neonate finds itself at each point in development may provide some insight into the observed course of development. The remainder of the present chapter is, in essence, an examination of the heuristic value of treating the neonate as exhibiting behavioral adaptations necessary for survival and growth within the ecological niche defined by its caretaker at each point in its life cycle.

1.3. Parasitism: An Analogy to the Life Strategy of the Altricial Juvenile Mammal

In attempting to understand the relationship between the juvenile mammal and the environment provided by its dam I was struck by the parallels between the mother–young relationship and the host–parasite relationship. In the present section I will develop the analogy between host–parasite and parent–young interactions, which I will use to interpret the process of weaning and the time course of development in later sections.

Parasitism is usually defined' as an intimate relationship between two heterospecific organisms during which the parasite, usually the smaller of the two partners, is metabolically dependent on the host. I will simply broaden the definition by considering metabolically dependent members of the same species to be parasites on their source of nutrition. The analogy between dependent neonates and parasites is surely appropriate in terms of the somatic relationship of the young to their dam and may even be justified to some extent in terms of their genetic relationship. So long as there is not genetic identity between the investing dam and her dependent offspring (as is always the case in sexually reproducing species), the foreign genes of the young are in fact parasitic on the genes of the investing parent.

A broadening of the definition of parasitism to include members of the same species is not unique to the present discussion. For example, the males of a number of species of angler fish (*Ceratiidae, Linophrynidae,* and *Photocorynidae*) spend much of their lives attached to the much larger females. The males have their own gills and respiratory system, but their food is obtained directly from the bloodstream of the female. They may be completely dependent on the female for nourishment and they often exhibit specializations to their parasitic mode of life, including reduced dentition and a simplified intestinal tract. Such relations between members of the same species have been labeled "sexual parasitism" (Villwock, 1973; Kikkawa and Thorne, 1971) for much the same reasons that I would propose to treat altricial mammals as "reproductive parasites" of their dam.

Parasitic relationships in nature are of many different kinds. The parasite may be internal to its host (an *endoparasite*) or attached to its exterior (an *ectoparasite*). It may be completely dependent on its host metabolically (an *obligate parasite*) or only partially dependent (a *facultative parasite*). The parasite may be *active* in seeking its host or may *passively* await the arrival of its host in a fixed location. It may be continuously (a *constant parasite*) or intermittently (a *periodic parasite*) in contact with its host. Clearly different modes of parasitic interaction with a host require different sensory, motor, somatic, and behavioral adaptations on the part of the parasite if it is to succeed in its parasitic way of life.

The older view of parasitic animals is that they are degenerate because many have lost some if not all of the organs that their nonparasitic relatives have. However, such alterations in phenotype can be, perhaps, more profitably viewed as adaptive specializations for the particular mode of life parasites have adopted and the particular ecological niche they occupy than as the result of degeneracy (LaPage, 1963; Bates, 1961). As Bates (1961, p. 180) has proposed, "The degeneracy of parasitism is specialization involving the loss of structures and functions that are no longer needed." Similarly, altricial mammals are sometimes described as "helpless and poorly developed" (Vaughan, 1972, p. 363). Again there might be something to gain from viewing altricial mammals as adapted to the particular ecological niches which they occupy.

Part of the usefulness of the parasitism analogy is that it makes clear the fact that as the young mammal increases in age, both its means of utilizing the dam as a host and its degree of dependence on the mother as a resource base change. The young mammal moves through a series of ecological niches with movement from niche to niche required by changes in the energy resources provided by the dam. In many invertebrate parasites one sees similar changes in host exploitation during the life cycle or ontogeny of the organism and correlated changes in phenotype associated with the particular niche exploited in each life stage (see footnote on page 236 for an example). Further, the parasitism analogy makes explicit the point that the young are to some extent exploiters of their dam rather than cooperative partners in their interaction with her.*

In terms of the present analogy, prior to parturition the fetal mammal is functionally an obligate, constant endoparasite. At parturition it becomes an ectoparasite, obligate, periodic, and passive. As it increases in age it moves from a passive to an active mode of interaction with its host, and the interaction moves from an obligate to a facultative parasitism. Once milk delivery from the dam ceases, the juvenile may becomes a commensal of its host dam. In succeeding sections I will use this parasitic analogy to discuss the series of ecological niches which the dam provides for her young and the behavioral and phenotypic adaptations which the young exhibit to the succession of niches made available to them by their caretaker.

2. The Rat Pup as Parasite

Energy passes from mother to young in three forms: (1) as calories in milk, (2) as heat conducted from mother to young during periods of contact, and (3) as mechanical energy utilized in physical movement of the young, nest or burrow construction, or transport of food. In the following sections I will describe the willingness of the dam to commit resources of each type to her young as changing over time and the young as exhibiting compensatory adjustments in their physical characteristics and behavior. Such an analysis clearly requires rather detailed knowledge of the energy commitment of the dam and of the development, both somatic and behavioral, of the young. To my knowledge such information is available at the present time only for rats *(Rattus norvegicus)*, and my discussion will therefore be focused on the interaction of rat dams and their litters.

2.1. The Environment Provided by the Host: Parental Investment by the Dam

Because the dam provides the environment to which the young must adapt, it is necessary to first consider the changes over postpartum time in the willingness of the dam to invest in her offspring.

---

*One might well argue that the analogy between the relationship of parasite and host and that of mother and young is inappropriate in that hosts are typically adversely affected by interaction with their parasites while dams benefit from interaction with their offspring. Whether one considers the young as beneficial or harmful to the parent depends on the level of analysis employed. In genetic terms, offspring are clearly beneficial to their dam in that they provide a potential medium for replication of the dam's genetic compliment. In physiological terms the offspring are, of course, a considerable burden to her. Thus, the parasitism analogy, while failing to reflect important positive aspects of the relationship of mother to young, calls attention to the metabolic costs associated with parental investment and is useful in that regard.

2.1.1. Milk Flow from from Mother to Young

Figure 1 presents data on the amount of milk flow from a lactating rat to her litter of 8 pups as measured by rate of ^{85}Sr transfer (Babicky *et al.,* 1970). The important findings (confirmed by Ostadalova *et al.,* 1971) are (1) that there is a gradual increment in milk transfer which reaches a maximum on day 15 post-partum, and (2) that milk transfer declines gradually from day 15 until is ceases on days 27–28 postpartum. Variation in litter size (3, 8, or 15 sucklings) has been found to effect the total amount of milk transferred but not the day of peak transfer or the day of termination of milk delivery (Ostadalova *et al.,* 1971).

2.1.2. Conductance of Heat from Mother to Young

The cost to the mother of maintaining contact with her young, and thus conducting heat to them, is less obvious than the cost to her of provision of milk or mechanical energy (Pond, 1978); the mother exhibits a rise in body temperature at the same time that the pups gain heat from her, because the mother–young huddle reduces the surface to volume ratio of all participants and is energy conserving (Leon *et al.,* 1978). The cost to the mother of prolonged periods of contact with her offspring is probably twofold: First, it reduces her potential foraging time, and second, it results in the dam suffering hyperthermia (Leon *et al.,* 1978; Woodside, 1978).

Figure 2 presents data describing the percentage of the day which the dam spends in contact with her offspring (Grota and Ader, 1969). As contact is both a necessary and sufficient condition for heat transfer, the data provide a rough indication of the mother's contribution of heat to her young during the first three postpartum weeks. The data indicate (1) a steady decrease in contact time from parturition to day 17 postpartum (see also Leon *et al.,* 1978; Woodside, 1978) and (2) a subsequent low but constant level of contact.

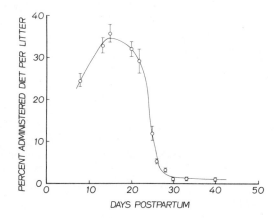

Figure 1. Percentage of labeled SrCl$_2$ injected into the dam appearing in pups 48 hr following injection, as determined by whole-litter radiation counts (adapted from Babicky *et al.,* 1970).

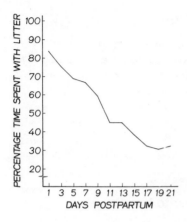

Figure 2. Percentage of time lactating rats spent with their litters during 21 days postpartum (adapted from Grota and Ader, 1969).

2.1.3. Mechanical Energy Provided by the Dam

2.1.3a. Movement of Pups. Figure 3 presents data indicating the probability of dams retrieving entire litters of pups during the 30 min subsequent to the displacement of the pups from the nest site and the probability of the mother transferring her entire litter to a new nest site during the 3 hr following destruction of her original nest (Brewster, unpublished data). The data indicate (1) a high probability of movement of the young to a place of safety from parturition to day 14, (2) followed by a gradual decline in willingness to move pups, (3) ending with a total unwillingness to move pups by day 22. (See also Grota and Ader, 1969; Rosenblatt and Lehrman, 1963.) That such changes in the dam's probability of moving young are due in some measure to changes in her internal condition rather than changes in the movement eliciting properties of the young is strongly indicated by the finding of Rosenblatt and Lehrman (1963) that 4-day

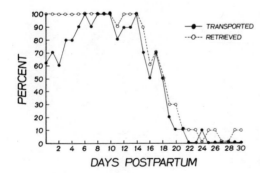

Figure 3. Percentage of females retrieving entire litters or transporting entire litters to new nests during test periods (Brewster, unpublished).

postpartum females will retrieve 19-day-old pups, while 19-day postpartum females will not.

2.1.3b. Nest Building and Burrow Construction. As can be seen in Fig. 4, the willingness of the mother to engage in nest-building behavior (Rosenblatt and Lehrman, 1963) shows a similar pattern to that of pup movement: (1) A period of high levels of nest-building activity, (2) a period of gradual waning of investment beginning on day 13 or 14, and (3) a total cessation of activity on day 17 or 18.

Little is known of maternal investment in burrow construction or food transport to the nest site, but Calhoun (1962, pp. 29, 33–38) reports that just prior to weaning of the young (postpartum day 17 or 18), lactating females enlarge their burrows and form a food cache.

2.1.4. Discussion

There are two general features of the changes in maternal investment over the period of mother–young interaction of particular importance to subsequent discussions. First, the energy transfer from mother to young in each category wanes prior to day 22 postpartum. Second, in each case the energy input by the dam to the young decreases gradually rather than ending abruptly. It cannot be determined from the data presented in Figs. 1–4 whether this gradual waning in energy investment by dams is an individual or group phenomenon, but both published data (Grota and Ader, 1969) and personal communications indicate that at least in the cases of retrieval, transport, nest building, and contact time individual dams exhibit a gradual reduction in investment (Brewster, personal communication; Croskerry, personal communication; Leon, personal communication).

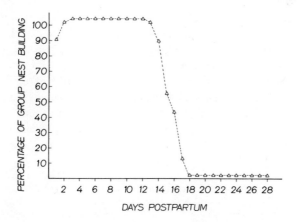

Figure 4. Percentage of mothers building nests on each of 28 days postpartum (adapted from Rosenblatt and Lehrman, 1963).

The fact that investment by the dam changes gradually, suggests that the pups are not exposed to sudden shifts in their environment, and one would therefore expect an absence of abrupt transitions in pup behavior or morphology over the course of development. By contrast, invertebrate parasites often undergo abrupt transitions from one ecological niche to another (i.e., from one host to the next, or from a parasitic to independent mode of existence) and their morphology and behavior exhibit discrete, clearly identifiable stages correlated with the succession of clearly defined niches they occupy. The absence of such abrupt niche shifts in altricial mammals makes the identification of moments of transition difficult and consequently makes it difficult to identify well-defined stages in the development of the young. Therefore, rather than attempting to identify and discuss stages in pup development, I have chosen to describe behavior of the young at three representative points in development: (1) in infancy, when the pup is totally dependent on the dam (day 5), (2) during the transition from dependence to independence (day 18), and (3) when independent existence is a possibility (day 40). The goal of each discussion will be to see if some sense can be made of the pup's succession of phenotypes, when those phenotypes are considered in relation to the successive environments dams provide.

In effect, my approach will be a comparative one, seeking to understand the adaptations of a series of closely related phenotypes in terms of the environment to which each must adapt. In this model the study of the same organism over developmental time is a comparative enterprise, in that although the underlying genotype remains constant, aspects of the genotype expressed in the phenotype change as a function of age.

2.2. The Rat Pup as Obligate, Passive Ectoparasite—Day 5 Postpartum

During the days immediately following birth the altricial mammal has available to it an ecological niche which it is uniquely adapted to exploit and in which competition for resources is at a minimum. The recently parturient dam is both able and willing to provide energy in all forms sufficient for pup growth, and competition for the energy resources of the dam is limited (from the point of view of any single pup) to the dam herself, contemporaneous siblings, and any fetuses the dam may be carrying in a postpartum pregnancy. At this time the juvenile rat exhibits a phenotype highly adapted to the function of exploitation of the dam as an energy source.*

*The functional significance of neonatal phenotypes as releasers of parental investment is clearly illustrated by the young of Estrilid finches and their brood parasites, various species of widow bird. Food-begging behavior by the young is very highly specialized in both the finch young and the mimetic young of the widow-bird parasite. "The Estrilid finches have highly specialized gape patterns. . . . The [food] begging gape presented to the parents is characterized by black spots on the palate, tongue, and lower mandible. In addition, the greatly thickened outgrowths on the upper and

As one might expect, given the total dependence of the 5-day-old pup on resources obtained from the dam and the pup's consequent need to exploit the dam in an efficient fashion, the 5-day-old pup gives every indication of being an optimal eliciter or releaser of maternal investment. Replacing a dam's pups every 5 days with neonates results in the maintenance of the high levels of suckling behavior characteristic of the early stages of mother–young interaction. Replacing 5-day-old pups with those 12 days of age results in a rapid decline in suckling time (Grosvenor and Mena, 1974; see also Grota, 1973).

The efficiency of neonatal rat pups as releasers of nurturent behavior by conspecifics is further exemplified by their ability to cause even nonreproductives exposed to them to behave "maternally" (Rosenblatt, 1967). The high level of specialization of the neonate for the elicitation of nurturance is suggested by the finding that exposure to 1- to 2-day-old pups induces "maternal" behavior in nonreproductive conspecifics both more reliably and more rapidly than exposure to older pups (Stern and MacKinnon, 1978).

As discussed in the preceding section, the pup cannot continue indefinitely to depend totally on the dam for energy inputs. As the pup changes phenotype to begin exploitation of energy sources other than the dam, it apparently must sacrifice some of those characteristics which optimize its exploitation of the dam. The nature of the resulting compromises in the phenotype of the young which result in the deterioration of pups' ability to elicit maternal investment are of considerable interest within the present conceptual framework.

2.2.1. Thermoregulation

The exothermic or poikilothermic characteristic of rat pups and other altricial mammals is often treated as resulting from the poor state of development of the young. However, it is important to keep in mind that homoiothermy or endothermy is not the goal or end point of a progressive phylogenetic evolution. Rather, it is one solution to the challenges posed by sets of environmental conditions. Most of the adult members of existing species are, in fact, successful exotherms, and their ability to become inactive during periods of stress in temporarily unsuppor-

---

lower mandible carry one or two rounded papillae, which are usually white, or blue in some species. . . . The number and position of spots vary [among species], they may be lost from the tongue or the lower beak, or fuse to form arches. . . . The parents pump food only into the right pattern and ignore young with an incompatible pattern which are smuggled into the nest. Such young would rapidly die, and so it is easy to predict the appearance of the beaks of the young widow birds [the brood parasites]. As expected, the nestlings of the different widow-bird species are found to possess gape patterns indistinguishable from those of the host's young. The Estrilid finches exhibit a number of additional important signals contributing to the overall begging signal. The nestlings utter specific begging calls and rotate the head in an unusual manner while begging. Both these features are copied exactly by the widow birds. Finally, the coloration of the plumage of the parasite is identical to that of the host. Distinct differences can only be observed in the adults" (Wickler, 1968, pp. 196–197).

tive environments can be viewed as an advantage rather than a disadvantage (Schall and Pianka, 1978). The neonatal rat exists in an ecological situation totally different from that of the adult of its species, and it is at least possible that the exothermy of the pup represents an adaptation to its niche rather than a failure to achieve the endothermic mechanisms of the adult.

Because of the great ratio of surface area to body weight of the neonatal rat pup the energy expenditure required for homoiothermy would be great (Conklin and Heggeness, 1971). In order to maintain a constant body temperature, a pup would have to use an additional 4 cal/10 g per day for every 1°C it elevated its body temperature above ambient (Hahn et al., 1961). Energy used for thermal homeostasis would be a cost to be subtracted from the energy inputs acquired by the pup from its dam. Thus maximization of net energy gain and, hence, growth by infant rats is facilitated by limiting the energy expended on endogenous heat production (Hopson, 1973) and dependence on exogenous sources of thermal energy.

At day 5 postpartum the dam may spend as much as 80% of the day in contact with her young and during periods of contact simultaneously provides warm milk and, by conduction, an exogenous source of thermal energy. An exothermic pup can reduce its metabolism during periods when it is unable to acquire nutrients (i.e., when its mother is absent) and have its body temperature and metabolic rate raised passively only during periods when the opportunity for food acquisition exists. This ability of the pup to allow its body temperature and, consequently, its metabolism to fall during periods in which the opportunity for food acquisition is absent (see Fig. 5) results in very considerable energy saving. For example, the total caloric requirement of a 1-day-old pup maintained in 33°C ambient is 47 cal/10 g per day, while its requirement at 20°C ambient is only 37 cal/10 g per day (Hahn and Koldovsky, 1960). Hahn et al. (1961) have estimated that if a rat pup kept at 20°C both maintained its body temperature at 33°C by endogenous heat production and continued normal growth it would have to consume more than twice as much milk as an exothermic pup. Clearly exothermy and the consequent ability to allow body temperature and metabolic rate to fall

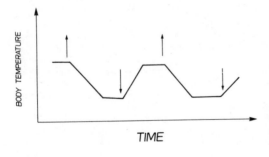

Figure 5. Schematic representation of infant rat body temperature in the presence or absence of the dam. Upward arrows indicate the mother's leaving and downward ones the mother's returning to the nest (adapted from Hahn et al., 1961).

during periods when food acquisition is impossible provide important benefits to the obligate, parasitic neonatal rat.

The minimization of endogenous heat production in the neonate may have an additional important effect. Leon *et al.* (1978) have provided evidence consistent with the view that the duration of maternal nesting bouts is limited by the rate of maternal temperature increment during her periods of contact with the young in the nest. To the extent that pups can decrease their rate of endogenous heat production, they can increase the duration of periods of contact with the dam and, thereby, their access to the energy the dam provides. Blackmore (1972) has reported increasing rates of pup growth from days 2–8 postpartum as a function of increasing hours of pup contact with their dam. While it cannot be determined from Blackmore's data whether increased pup stimulation of the dam results in increased milk letdown or increased pup suckling time results in increased milk delivery from a constant supply, it is clear that pups can increase their net energy gain by maximizing their duration of contact with the dam.* To the extent that pup endothermy would decrease mother–young contact time it would decrease the rate of energy acquisition by the pups.

Increased access to the dam not only increases potential feeding time for the young rat but also increases the amount of stimulation which the pups may provide to the dam. There is strong evidence that stimulation of the dam by the young is necessary both to maintain the rat dam's willingness to invest in them and to prevent her initiation of investment in future litters. Removal of pups from their mother for 2–4 days immediately following parturition results in the gradual disappearance of maternal behavior in the dam (Rosenblatt and Lehrman, 1963). Further, removal of pups from their mother results in her ovulating three days later, permitting her reimpregnation (Van der Schoot *et al.,* 1978).

2.2.2. Feeding Behavior

The neonatal rat pup is totally dependent on the dam to provide nutrients and waits passively in the nest for her arrival. Active pursuit of food is, in normal circumstances, limited to searching, rooting, nipple attachment, and sucking behaviors necessary for milk acquisition. Though different from the feeding behavior of adults, the ingestive behavior of the young (described in detail in Hall and Rosenblatt, 1977) is both highly complex and specifically adapted to the exploitation of the dam. While the neonate lacks the dentition necessary for the mastication of solid food it does have an enzyme, lactase, which permits it to absorb and utilize lactose, a source of carbohydrates and calories in maternal milk.

---

*This statement is an oversimplification. There must be an upper limit to mother–young contact time beyond which additional suckling of the pups by the dam would reduce the time available to her for foraging to the point where she could no longer support herself and her young.

High levels of lactose activity in the small intestine of rat pups are initiated just prior to birth and terminate at 21–22 days postpartum. As Alvarez and Sas (1971, p. 827) have proposed, "the physiological immaturity of the newborn seems to be in this case rather a physiological adaptation (in a broad sense) to a milk diet."

The young depend on their mother not only for the gathering and preliminary processing of food but for regulation of intake as well. Pups exhibit a willingness to attach to their dams' nipples independent of nutritional deprivation state, will remain attached to the teats so long as they are available, and will ingest whatever milk is ejected from the nipples until the pups are on the verge of drowning in milk overflow from their stomachs (Hall and Rosenblatt, 1977). Thus, within broad limits (Friedman, 1975) the food intake of the infant rat is a function of food delivery by the dam.

The regulatory behavior of the rat pup with respect to food intake can, like its thermoregulatory behavior, be viewed as a "primitive analogue of adult behavior" (Hall et al., 1977, p. 1141) or as a sophisticated means of exploiting the host dam. Because the net energy acquisition and growth rate of the pup are determined in the main by the amount of milk it can acquire from a single dam unwilling or unable to supply all the milk the pup could potentially utilize, the pup has nothing to gain from refusing or delaying nipple attachment regardless of its internal state. Because continued investment by the dam is dependent, at least in part, on nipple stimulation by the pups, attachment and suckling may provide important long-term benefits to the young as well as the immediate benefit provided by direct milk acquisition. For example, suckling stimulation provided by the rat pup has been demonstrated to stimulate prolactin release by the dam (Grosvenor et al., 1970). High circulating levels of prolactin in turn stimulate both future lactation and the elevated levels of food intake by the dam which allow nursing rats to feed their young without depleting their own energy reserves (Fleming, 1976, 1977; Tucker, 1974; Cotes and Cross, 1954).

I have placed considerable emphasis in the preceding discussion on the importance of direct pup stimulation of its mother in the elicitation of continued parental investment. My underlying assumption has been that the greater the stimulation by the litter the greater the investment by the dam. There are unfortunately relatively few data to support such an assumption. The present approach to mother–young interaction points to the need for further work on the extent to and means by which the parasite pup may affect the behavior of the host dam and modify it to the pup's own advantage during periods of contact with her.*

---

*The ability of parasites to manipulate the behavior of hosts to their own advantage is certainly not unique to the mother–infant relationship. To take a spectacular example, the lancet fluke (Dicrocoelium dentriticum), which encysts in the suboesophogeal ganglia of host ants, changes the behavior of the ant host so that instead of returning to its nest during the cooler parts of the day the ant clings to the top of a blade of grass. This parasite-induced pattern of behavior keeps the infected ant exposed to accidental ingestion by the crepuscular grazing ungulates which are the next host of the lancet fluke (Holmes and Bethel, 1972).

2.2.3. Locomotory Behavior

The neonatal rat depends on its dam not only for energy in the form of milk and conducted heat and for energy conservation provided by the dam in the form of insulation provided by the nest and burrow, but also for physical transport from one point to another. The locomotory capacity of the pup is limited to the ability to move in an oriented fashion only the very short distances necessary to modify the surface to volume ratio of the huddle for purposes of thermoregulation (Alberts, 1978 *a, b*) or to orient to the nest over distances of a few centimeters (Cornwell-Jones and Sobrian, 1977). In general, rat pups expend little energy on locomotion and, instead, both induce the dam to retrieve them to the nest site if they are displaced more than a short distance from it, or to move them to a new nest site should the original nest lose its insulating properties as a result of flooding or disturbance (Brewster and Leon, 1980a; Sturman-Hulbe and Stone, 1929). Again one could consider the lack of independent locomotory capacity of rat pups as evidence of their poor state of physical development or as an elegant means of reducing energy expenditure.

Although the young depend almost totally on the dam for the mechanical energy required for locomotion, they are capable of increasing the probability that the dam will move them, should they require movement, by emitting ultrasonic vocalizations (Allin and Banks, 1971; Brewster and Leon, 1980, *a*) and are also capable of assuming a special curled posture which facilitates their movement by the mother (Brewster and Leon, 1980 *b*). In one sense, arguing that the locomotory capacity of the pup is primitive in comparison with that of the adult, is like arguing that horseback riding is a primitive form of locomotion in comparison with walking. Although the locomotory behavior of the adult may be behaviorally or psychologically more complex than that of the juvenile, the behavior of each is functionally adaptive within its niche and neither is simple or primitive in that respect. Because human psychological and behavioral functioning is the most complex of that of any species, we tend to attribute superiority to complex behavioral mechanisms, though there is no good reason to do so. It is surely equally reasonable to consider simple solutions to a problem as more elegant and sophisticated than complex ones. In this perspective the neonatal pup is seen as the near-perfect parasite able to exploit its dam for its caloric, thermal, and mechanical energy needs while conserving its own energy for growth.

2.3. The Rat Pup as Facultative, Active Ectoparasite—Day 18 Postpartum

Two events occur on or about day 17 postpartum which require rat pups to abandon once and for all their obligate, passive parasitic niche. First, as can be seen in Fig. 6, adapted from Babicky *et al.* (1973), the energy delivered by the dam via her milk begins to drop precipitously in the face of steadily increasing

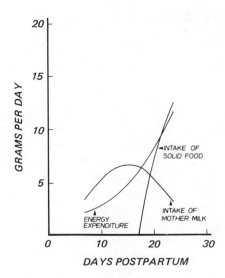

Figure 6. Schematic representation of the energy balance of litters of 8 rat pups (adapted from Babicky *et al.*, 1973).

energy demands by the young. Although the minimal metabolism of the pups at neutral temperature (as measured by oxygen consumption per kilogram per minute) remains essentially constant from birth to maturity (Taylor, 1960), the body weight of each pup and, consequently, its energy requirements grow rapidly. Even if the dam were to maintain her day 15 postpartum peak level of milk production indefinitely, by the time her pups reached 18–19 days of age, they would still either have to abandon growth or seek alternative sources of nutrition. In order to maintain a net positive energy flow the pups are obliged to begin ingesting solid food and to make the consequent transition from obligate to facultative dependence on the dam for nutrition. Second, as indicated in Fig. 7, adapted from Rosenblatt (1965), at about the same time that the dam reduces her nutritional contribution to her young she becomes increasingly unwilling to return to the nest site to initiate nursing bouts and, on occasion, even actively avoids her young when they attempt to suckle from her (Rosenblatt, 1965). The pups are thus forced to become active rather than passive in their exploitation of their dam as an energy source.

Whether the failure of the milk supply of the dam and her unwillingness to suckle play a role as proximal stimuli for initiation of feeding on solid food or whether the pups are endogenously driven to initiate ingestion of solid food is not clear. It is, however, clear that to continue to show a positive net energy flow in the face of a falling commitment by the dam the pup must begin to exploit alternative resources. The active, facultative parasitic niche of the 18-day-old obviously requires a markedly different phenotype for success than the passive, obligate parasitic niche of the 5-day-old, and in fact, the young pup has undergone a major

THE ECOLOGY OF WEANING

metamorphosis in the intervening weeks. It has acquired an adult coat of fur, its ears and eyes have opened, giving it the full range of sensory systems, and its bodily proportions and musculature have altered so as to permit independent locomotion. As discussed below, the behavior of the pup has undergone changes as profound as its morphology.

2.3.1. Thermoregulation

In order either to actively seek the dam outside the nest site or to independently acquire solid foods in the general environment, the pup must be capable of activity even when out of contact with the dam, its major exogenous source of heat. According to Hahn (1956) and Hahn *et al.* (1956), chemical thermoregulation as well as insulation from the fur reach full development when the animals open their eyes at approximately 16 days of age (Croskerry, 1974). Thus, the 18-day-old rat can maintain constant colonic temperature over a fairly broad range of ambient temperatures, though even the 21-day-old pup does not possess the full adult capacity for maintaining a stable body temperature (Conklin and Heggeness, 1971). It should be kept in mind, however, that in the natural environment the rat pup does not leave the thermally protected environment of the burrow for extended periods of time until it reaches approximately 34 days of age (Boice, 1977; Calhoun, 1963). Prior to this time, movement by the pup is confined to explorations within the insulating burrow or brief trips on the surface in the vicinity of the burrow entrance. The microenvironment of the pup, even at 18 days of age, is thus not so thermally stressful as that of the adult.

2.3.2. Feeding Behavior

Perhaps the most marked behavioral changes associated with the change from obligate to facultative and from passive to active dependence on the dam as a

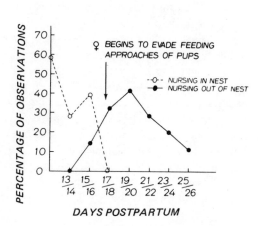

Figure 7. Percentage of observations of rat dams nursing in and out of the nest during the later stages of lactation (adapted from Rosenblatt, 1969).

source of energy are concerned with the control of feeding. Again the observed behavioral phenotype appears appropriate for the expanded niche of the 18-day-old. Whereas the neonate incurs little cost in attempting to feed and has access only to the finite supply of food provided directly by the dam, the 18-day-old must expend energy (in pursuit either of the dam or of solid food) in order to eat and has access to the functionally infinite supply of food available in the real world as well as to the declining energy supply of the dam's milk. Thus, the 18-day-old faces many of the same cost–benefit decisions concerning when and where to eat as do conspecific adults. One would expect, therefore, to see the emergence of adult patterns of control of feeding at about the time when facultative parasitism is imposed on the young by the failure of the milk supply of the dam. It has recently been reported (Hall and Rosenblatt, 1977, 1978) that whereas 15-day-old pups do not show increased ingestion with increasing deprivation, 20-day-old pups do and that nutritive preloads reduce intake of milk in 20-day-olds but not in 10-day-olds. Similarly, Hall *et al.* (1977) have found a dramatic increase in the latency of nondeprived pups to attach to the dam's nipple in 14- to 16-day-old young but not in 11- to 13-day-old young. Further, Hall *et al.* (1977) have found that pups at 15–17 days of age sucking on a dry nipple begin to exhibit a tendency to abandon that nipple and transfer attachment to another. Thus, the rat pup becomes sensitive both to its own internal state and to the nutritive return from its ingestive activity only when it has options other than mother's milk to satisfy its nutritive requirements.

Whether the absence in the obligate, parasitic neonate of the behaviors appropriate for multisource feeding are the result of an absence of appropriate neural structures to support those behaviors or are the result of an active inhibition of existing structures is open to question. The results of a recent series of studies by Martin and Alberts (1979) suggest that in at least one case it is inhibition of a developed neural mechanism rather than a failure of development which is responsible for the absence of a behavior during the obligate, parasitic state which emerges in the facultative, parasitic stage. Martin and Alberts have found that 15- to 16-day-old rat pups receiving flavored milk via a tongue cannula learn to avoid the flavor introduced into the milk only if milk delivery occurs while the pups are not attached to the nipple. Littermates treated identically while attached to the nipple fail to form the aversion. In contrast, nipple attachment fails to inhibit taste aversion learning in pups 21–22 days old at training. I would suggest that obligate, parasitic 15-day-olds cannot afford to learn an aversion to the taste of their dam's milk, experienced while suckling, and that the lack of capacity to learn aversions while attached to the nipple is thus an adaptation to the obligate, parasitic mode of life. Facultative parasites can learn to avoid the flavor of mother's milk and still survive, and such learning may play an important role in inducing pups to abandon mother's milk and find other sources of nutrition.

It might be argued that mother's milk is never toxic to young rats, so that the capacity to form aversions to flavor experienced while on the nipple is never of use. However, the development of lactase deficiency and consequent lactose intolerance in 21- to 22-day-old rats (Alvarez and Sas, 1961), taken together with the phenomenon reported by Martin and Alberts (1979), suggests that the combination of lactase disappearance in the pup and the emergence of the capacity to associate gastrointestinal distress with milk experienced while on the nipple may be an important proximal factor in motivating pups to seek food in the general environment. It has been argued (M. Lieberman and D. Lieberman, 1978; Rozin, 1976) that progressive lactase deficiency in young mammals and consequent experience of gastrointestinal distress in association with milk ingestion may be a mechanism regulating both the time and rate of abandonment of suckling in young mammals.* Of considerable interest with respect to this hypothesis the known temporal relationship in rats between the relative time of onset of (1) lactase deficiency [days 21–22 postpartum (Alvarez and Sas, 1961), (2) the ability to associate illness with cues experienced while suckling [between days 16 and 20 postpartum (Martin and Alberts, 1979)], (3) the failure of the milk supply of the dam to provide energy sufficient for continued pup growth [days 16–18 postpartum (Babicky *et al.,* 1973)], and (4) the onset of ingestion of solid food [day 17 postpartum (Galef, 1979; Babicky *et al.,* 1973)]. The findings presented above suggest that in rats, initiation of ingestion of solid food may result from failure of the dam's milk supply (see Davies, 1978), while the rate of transition to independent feeding may be accelerated by lactase deficiency and consequent learned aversions to mother's milk.

2.3.3. Locomotory Behavior

As mentioned above, the 19-day-old pup, while capable of rapid and efficient independent locomotion, is still an adequate stimulus for the elicitation of retrieval behavior in recently parturient dams (Rosenblatt and Lehrman, 1963). However, the pup's own mother at 19 days postpartum is no longer willing to engage in the physical movement of her young from one location to another. Eighteen-day-old pups must provide the mechanical energy to bring themselves in contact with their dam and other sources of nutrition. Although the mother is not willing to transport the young to resources, she does provide mechanical energy to transport resources to the young. Rat pups in the wild do not leave the natal burrow until 28 days of age, and there is some question as to where the solid foods which the pups could begin to utilize on day 17 (Galef, 1979; Babicky *et al.,* 1973) might come from. Rats are known to hoard food in their burrows (Barnett, 1975; Calhoun, 1962; Pisano and Storer, 1948), and it is likely that the food cache of the dam is the first

---

*The role of lactase in the utilization of milk by neonatal rats and the role of lactase disappearance in the abandonment of mother's milk as a source of nutrition in weanling rats is probably less important than in other species. Whereas rat milk has only 5.9% of its calories in its carbohydrate fraction, the percentage of calories in carbohydrates in the milk of other species is far higher (cow milk, 29.5%; human milk, 45.1%) (Cox and Mueller, 1937).

solid food ingested by her young. While I know of no reports in the literature of laboratory studies of the time course of hoarding during the period of mother–young interaction, Calhoun (1962, p. 29) reports formation of a food cache by a lactating female rat in the wild immediately prior to weaning of her young. This observation suggests that although the dam will no longer physically move her young, and the young no longer elicit maternal locomotion by emitting ultrasonic vocalizations (Allin and Banks, 1970), the dam is still willing to invest mechanical energy in the young by transporting food to them. Investigations of postpartum hoarding by the dam is required to determine the extent of the dam's commitment of mechanical energy in this form.

2.4. The Juvenile Rat as Commensal—Day 40 Postpartum

In a commensal relationship the commensal is not metabolically dependent on its host; to the contrary, it is fully capable of independent survival. The spatial proximity of the participants in a commensal relationship does, however, allow the commensal to share resources (most frequently food and shelter) with its host without obvious detriment or benefit to the host. The success of the commensal, be it a pilot fish *(Naucrates ductor)* accompanying a shark or a rat living in a human dwelling and eating refuse, often depends in considerable measure on the tolerance of the host for the presence of the commensal.

The facultatively parasitic 18-day-old rat pup is already a commensal of its dam, sharing the food which she may transport to the home burrow and benefiting from the protection which that burrow provides against both extremes of temperature and humidity and the intrusion of predators.

On the 28th day postpartum, milk transfer from a rat dam to her young ceases (Ostadalova *et al.,* 1971; see also Fig. 1) and the young, which have been gradually increasing the proportion of energy they acquire from sources other than mother's milk (Babicky *et al.,* 1973; see also Fig. 6), can no longer directly exploit their dam as a source of nutrition. As discussed in the preceding section, a cache of food may be available in the burrow to sustain the pups for some time, but on about the 34th day of postpartum life rat pups (at least in seminatural environments) (1) begin to explore on the surface at some distance from the nest site, (2) begin to acquire food in the extra-burrow environment, and (3) undertake their first harboring with conspecifics other than their own mothers and siblings (Calhoun, 1962, p. 152). Thus, by 40 days of age the juvenile rat pup has extended its commensal relationships, with respect to both food and shelter, from its dam to others of its species. This new ecological situation places new demands on the pup and is reflected in changes in both its behavior and physiognomy.

2.4.1. Thermoregulation

The 40-day-old rat has the full adult capacity to maintain colonic temperature in the face of cold stress at least as severe as 5°C (Adolph, 1957). The final improvement in ability to control internal temperature, probably due to development of vascular mechanisms (Hahn, 1956; Hahn *et al.*, 1961, p. 134), is achieved at about day 30 postpartum, prior to the pup undertaking extended trips away from the thermal protection of its home burrow (Calhoun, 1962).

Although capable of maintaining independent thermal homeostasis, neither the juvenile 40-day-old nor the reproductively capable adult do so under normal circumstances. Both exhibit a tendency to huddle in groups in response to cold stress (Steiniger, 1950), as does the neonate. While the factors important in the control of huddling have been investigated in neonates (Alberts, 1978*a*,*b*), little is known of the factors affecting huddling in adults, though it has been suggested that huddling at later ages serves important, if unspecified, social functions (Barnett, 1975, p. 106) in addition to its thermoregulatory role.

The fact that juvenile rats, even those as old as 40 days, maintain a core temperature several degrees lower than that of adult conspecifics at any of a wide range of ambient temperatures (Adolph, 1957) suggests that the conductance of metabolic heat between fully and partially grown individuals may provide greater benefits to the latter animals than to the former. Full equivalence in heat transfer is not reached until the juvenile achieves adult size.

2.4.2. Feeding Behavior

Although both patterns of ingestion and the caloric regulatory behavior of 40-day-old rats are indistinguishable from those of adult conspecifics, selected aspects of feeding behavior in the juvenile remain distinctive. While adults independently select feeding sites and dietary items (Galef, 1977*a*), immature rats, even those as old as 40 days of age (Galef and Clark, 1971*a*; Galef and Heiber, 1976), are profoundly influenced in their selection of feeding sites and diets by the behavior of adults of their species (Galef, 1977*b*). The juvenile is guided by both visual and olfactory cues from adults in the selection of a location in which to feed and is guided in its selection of a diet by gustatory cues experienced both during ingestion of mother's milk and while feeding on solid food in the nest site (Galef and Sherry, 1973; Galef and Henderson, 1972; Clark and Galef, 1972). Although the 40-day-old rat has the full range of thermoregulatory, locomotory, and ingestive behaviors sufficient to permit independent acquisition of food, it lacks the experience of resource distribution possessed by adult colony members that have a history of food acquisition in the area in which the pup undertakes its first feeding trips outside the burrow. By approaching adults at a distance from the nest site (Galef and Clark, 1971*b*), by following adults as they move about the clan territory (Calhoun, 1962, p. 149), and by using residual olfactory cues deposited by conspecifics (Galef and Heiber, 1976; Telle, 1966) the juvenile rat can both reduce its probability of ingesting toxic substances (Steiniger, 1950) and reduce the time and energy it must spend in locating needed nutrients in the general environment (Galef, 1977*b*). Adult conspecifics, already familiar with the location and safety of the various potential ingesta to be found in their home range, have less need to attend to the behavior of conspecifics in their feeding site selection and, in fact, exhibit relatively little social influence on their feeding behavior (Galef, 1977b; Galef and Clark, 1971a).

The 40-day-old rat, usually treated as already weaned, is thus both dependent on food resources shared with adult conspecifics and on information acquired from conspecifics for its sustenance.

2.4.3. Locomotory Behavior

As mentioned above, the 40-day-old pup is influenced in its movements about its home range by the chemical cues deposited by adult conspecifics as well as by the physical paths which adults create through snow, dense grass, and other barriers to locomotion (Calhoun, 1962, p. 54–84; Telle, 1966). Both of these patterns of behavior, like thermoregulatory huddling, extend into adulthood and therefore cannot be considered behavioral adaptations specific to the juvenile period.

Young rats do, however, exhibit a pattern of locomotory behavior which is observed at its highest frequency during the same period when the ability to share resources both with the dam and other adult conspecifics is of greatest importance (Baenninger, 1967). Such behavior patterns are frequently described as "play behavior," and while no satisfactory definition of such activities exists, their functional significance has been an active area of discussion. It is generally assumed that the benefits of play are delayed rather than immediate (Fagan, 1974), and result from the opportunity to practice, acquire, or develop motor and social skills, thus affording some as yet undemonstrated benefit in later life (Fagan, 1976). Interpretations of the functions of play behaviors are thus congruent with interpretations of the function of many of the other activities of juvenile organisms, which are similarly treated as necessary precursors of the emergence of a successful adult phenotype. It is, of course, possible that the distinctive locomotory patterns of juveniles serve a function at the time of their expression rather than (or in addition to) any delayed beneficial effects which they might produce. As mentioned in the introduction to the present section, the success of the commensal juvenile rat depends in large measure on the tolerance of conspecific adults for its presence at feeding sites and in burrows, in that the young rat is incapable of defending itself successfully against sustained attack by older and, therefore, larger conspecifics. Both informal description (Steiniger, 1950) of the interaction of juveniles and adults at feeding sites and experiments in which foreign juveniles were introduced into established wild rat colonies (Barnett, 1975, p. 121) suggest that adult rats are surprisingly tolerant of the presence of juveniles. Either the phenotype of the commensal juvenile is not such as to elicit aggression from adult conspecifics or the young exhibit types of behavior which actively inhibit aggression. It seems possible that the distinctive gait and "playful" mode of social interaction of the juvenile serve either to conceal from adult conspecifics aggression eliciting patterns of movement (Alberts and Galef, 1973) or to inhibit tendencies to attack, normally elicited by the presence of unfamiliar conspecifics (Galef, 1970; for a similar view see Ghiselin, 1974, p. 261).

In either case, the determination of the properties of juveniles which inhibit aggression during their period of commensalism with adults requires further investigation as it is one of the more important features of the success of juvenile rats.

2.4.4. Sequela

As suggested in the immediately preceding discussions, the dependence of rats on conspecifics for a variety of the necessities of life never ceases. To carry the parasitism analogy to its inevitable and somewhat overdrawn conclusion, most adult rats are in many ways symbiotes of their fellows, exchanging thermal energy by huddling in cold temperatures, exchanging physical energy in the joint construction of burrows and trails and defense of the clan territory, and feeding off one another's food caches (Calhoun, 1962). It is only those individuals which leave the home range of their natal clans to establish new colonies that are faced with the need to acquire independently all the necessities of life.

3. Conclusion

Within the Aristotelian philosophical system each "type" or species was considered to begin life as simple, undifferentiated matter endowed with the "potentiality" or capacity for "realizing" the adult "form" of its species. The process of development was viewed as fundamentally one of achieving this adult form. While contemporary students of development would feel uncomfortable with the teleological implications of the Aristotelian position, the notion that the adult form is in some sense more realized than antecedent somas is an intuitively comfortable one. The existence of a developmental analogue of the *scala naturae* with the fetus or newborn infant at its base and the reproductively active adult at its apex is often assumed in discussions of ontogeny; the adult phenotype is treated as a goal toward which antecedent somas progress. While it is true both that interactions of the environment and organism early in life may affect later development and that interactions of environment and organism late in life cannot affect early development, it cannot be inferred from these observations that the organism's development has evolved with the production of the adult phenotype as its end point. At a genetic level of analysis the adult is as much a precursor of the fetus as the fetus is of the adult.

The intuitively appealing notions that some stages of development are better adapted than others and that the adult phenotype is the end point or goal of the life cycle tend to lose their appeal when one examines the life histories of species which, unlike most vertebrate species, do not move from a protected to an unprotected environment as they mature. The message implicit in such life histories is that the life cycle is, in fact, a cycle without definable end point. While the life history of a species such as the liver fluke, *Fasciola hepatica,** is more easily seen as a series of adapted phenotypes, each meeting the criterion of gene maintenance, than is the life history of a mammalian species such as *R. norvegicus,* the function of the developmental succession of somas is identical in the two cases. The maturing or weaning mammal moving through its life cycle is not *progressing* from a dependent to an independent state; rather, as the individual grows older, it *changes* both its behavior and the ecological niche it occupies so as to maximize its probability of maintaining the genes it temporarily houses.

One can conceive of each point in the life cycle as having two theoretically discriminable functions. First, it has to be adequate to deal with the challenges to survival posed at that point. Second, any given stage (whether fetus or adult) must produce the next succeeding stage, if the genes which it contains are to be maintained (Williams, 1966, p. 44). It is the underlying thesis of the present paper that too little attention has been directed to the study of the first of these functions of the early stages of the life cycle of mammalian organisms and that there has been a consequent overemphasis on the second. The resulting imbalance in perception of the nature of the phenotypes expressed prior to the emergence of the adult form has resulted in a failure within the psychological tradition to explicitly consider the behavioral and somatic adaptation of nonadult phenotypes to their environment.

The two views of development, one stressing the contribution of a given stage to its successors and the other the interaction of each stage with its environment, are clearly complementary. Each has its contribution to make to the understanding of the organism during the nonadult period of its life. The preceding pages are intended to suggest that there may be heuristic value in treating the organism as adapted to the succession of environments in which it must survive and grow in order to insure the protection of its genetic complement for future generations.

There is, of course, a danger inherent in interpretations of behavior and morphology deduced from their hypothetical functions. Such explanations are no more

---

*The liver fluke *(F. hepatica),* for example, develops from a zygote into a first larval stage, the miracidium, that lives independently in an aquatic environment about which it swims by means of a covering of cilia. It has the neuromuscular machinery to identify, approach, and burrow into a particular species of aquatic snail. Inside the snail the miracidium metamorphoses into a second larval stage, the sporocyst, which reproduces by internal budding to produce the third larval stage, the redia. Rediae migrate within the snail and are capable of producing other rediae asexually and of metamorphosing into the last of the larval stages, the cercaria, which are provided like the earlier miracidia with a means of migrating to a new host. The cercaria burrows out of its snail host, swims by wiggling its tail to a blade of grass, attaches itself to the plant, casts off its tail, and encloses itself in a protective membrane. Inside the membrane the cercaria changes into a more or less amorphous mass, the metacercaria, which when ingested by a sheep or other herbivore hatches out as a young fluke. The immature fluke in time develops into a hermaphroditic adult capable of laying fertilized eggs, which continue the cycle (LaPage, 1963, p. 32).

than working hypotheses, and their value is dependent on the interest of the empirical investigations which they may generate.

ACKNOWLEDGMENTS.   Preparation of this manuscript was greatly facilitated by funds provided by the National Research Council of Canada (Grant A0307) and by the McMaster University Research Board. I thank the many friends and colleagues, especially Jeffrey Alberts, JoAnne Brewster, Mertice Clark, Martin Daly, Alison Fleming, Ted Hall, Herb Jenkins, Michael Leon, Jay Rosenblatt, Judy Stern, Paul Rozin, Linda Siegel, David Sherry, W. John Smith, and Barbara Woodside, who supplied preprints, reprints, references, unpublished data, and most helpful advice, criticism, and encouragement during the drafting of this chapter.

References

Adolph, E. F., 1957, Ontogeny of physiological regulations in the rat, *Quart. Rev. Biol.* **32**:89–137.

Alberts, J. R., 1978*a*, Huddling by rat pups: Multisensory control of contact behavior, *J. Comp. Physiol. Psychol.* **92**:220–230.

Alberts, J. R., 1978*b*, Huddling by rat pups: Group behavioral mechanisms of temperature regulation and energy conservation, *J. Comp. Physiol. Psychol.* **92**:231–245.

Alberts, J. R., and Galef, B. G., Jr., 1973, Olfactory cues and movements: Stimuli mediating intraspecific aggression in the wild Norway rat, *J. Comp. Physiol. Psychol.* **85**:233–242.

Allin, J. T., and Banks, E. M., 1970, Effects of temperature on ultrasound production by infant albino rats, *Dev. Psychobiol.* **4**:149–156.

Allin, J. T., and Banks, E. M., 1972, Functional aspects of ultrasound production by infant albino rats *(Rattus norvegicus)*, *Anim. Behav.* **20**:175–185.

Alvarez, A., and Sas, J., 1961, β-Galactosidase changes in the developing intestinal tract of the rat, *Nature (London)* **190**:826–827.

Babicky, A., Ostadalova, I., Parizek, J., Kolar, J., and Bibr, B., 1970, Use of radioisotope techniques for determining the weaning period in experimental animals, *Physiol. Bohemoslov.* **19**:457–467.

Babicky, A., Parizek, J., Ostadalova, I., and Kolar, J., 1973, Initial solid food intake and growth of young rats in nests of different sizes, *Physiol. Bohemoslov.* **22**:557–566.

Baenninger, L., 1967, Comparison of behavioral development in socially isolated and grouped rats, *Anim. Behav.* **15**:312–323.

Barnett, S. A., 1975, *The Rat: A Study in Behavior,* University of Chicago Press, Chicago.

Bates, M., 1961, *The Nature of Natural History,* Scribner's, New York.

Bell, R. Q., and Harper, L. V., 1977, *Child Effects on Adults,* Erlbaum, Hillsdale, N.J.

Blackmore, D., 1972, The effects of limited, moderate, and lengthy daily separation from the mother during the early postnatal period of the rat on concurrent and subsequent growth, and of concurrent oxygen consumption at, as well as below, thermal neutrality, *Biol. Neonate* **21**:268–281.

Boice, R. E., 1977, Burrows of wild and albino rats: Effects of domestication, outdoor raising, age, experience, and maternal state, *J. Comp. Physiol. Psychol.* **91**:649–661.

Bower, T. G. R., 1976, Repetitive processes in child development, *Sci. Am.* **235**:38–74.

Brewster, J., 1978, Transport of young in the Norway rat, Ph.D. dissertation, McMaster University.

Brewster, J., and Leon, M., 1980*a*, Relocation of the site of mother–young contact: Maternal transport behavior in Norway rats: Maternal transport behavior, *J. Comp. Physiol. Psychol.* **94**:69–79.

Brewster, J., and Leon, M., 1980*b*, Facilitation of maternal transport by Norway rat pups, *J. Comp. Physiol. Psychol.* **94**:80–88.

Calhoun, J. B., 1962, *The Ecology and Sociology of the Norway Rat,* U.S. Department of Health, Education and Welfare, Bethesda, Md.

Clark, M. M., and Galef, B. G., Jr., 1972, The effects of forced nest-site feeding on the food preferences of wild rat pups at weaning, *Psychon. Sci.* **28**:173–175.

Conklin, P., and Heggeness, F. W., 1971, Maturation of temperature homeostasis in the rat, *Am. J. Physiol.* **220**:333–336.

Cornwell-Jones, C., and Sobrian, S. K., 1977, Development of odor-guided behavior in Wistar and Sprague-Dawley rat pups, *Physiol. Behav.* **19**:658–688.

Cotes, M. P., and Cross, B. A., 1954, The influence of suckling on food intake and growth of adult female rats, *J. Endocrinol.* **10**:363–367.

Cox, W. M., and Mueller, A. J., 1937, The composition of milk from stock rats and an apparatus for milking small laboratory animals, *J. Nutr.* **13**:249–261.

Croskerry, P. G., 1974, Normal prenatal and postnatal development of the hooded rat and the effects of prenatal treatment with growth hormone, Ph.D. dissertation, McMaster University.

Darwin, C., 1877, A biographical sketch of an infant, *Mind* **2**:285–294.

Davies, N. B., 1978, Parental meanness and offspring independence: An experiment with hand-reared great tits *(Parus major)*, *Ibis* **120**:509–514.

Fagan, R., 1974, Selective and evolutionary aspects of animal play, *Am. Nat.* **108**:850–858.

Fagan, R. M., 1976, Exercise, play, and physical training in animals, in: *Perspectives in Ethology,* Vol. 2 (P. P. G. Bateson and P. M. Klopfer, eds.), pp. 189–219, Plenum Press, New York.

Fleming, A., 1976, Control of food intake in the rat: Role of suckling and hormones, *Physiol. Behav.* **17**:841–848.

Fleming, A., 1977, Effects of estrogen and prolactin on ovariectomy-induced hyperphagia and weight gain in female rats, *Behav. Biol.* **3**:417–423.

Friedman, M. I., 1975, Some determinants of milk ingestion of suckling rats, *J. Comp. Physiol. Psychol.* **89**:636–647.

Galef, B. G., Jr., 1970, Aggression and timidity: Responses to novelty in feral Norway rats, *J. Comp. Physiol. Psychol.* **70**:370–381.

Galef, B. G., Jr., 1977*a*, The social transmission of food preferences: An adaptation for weaning in rats, *J. Comp. Physiol. Psychol.* **91**:1136–1140.

Galef, B. G., Jr., 1977*b*, Mechanisms for the social transmission of food preferences from adult to weanling rats, in: *Learning Mechanisms in Food Selection* (L. M. Barker, M. Best, and M. Domjan, eds.) pp. 123–148, Baylor University Press, Waco, Tex.

Galef, B. G., Jr., 1979, Investigation of the functions of coprophagy in juvenile rats, *J. Comp. Physiol. Psychol.* **93**:295–305.

Galef, B. G., Jr., and Clark, M. M., 1971*a*, Social factors in the poison avoidance and feeding behavior of wild and domesticated rat pups, *J. Comp. Physiol. Psychol.* **75**:341–357.

Galef, B. G., Jr., and Clark, M. M., 1971*b*, Parent-offspring interactions determine the time and place of first ingestion of solid food by wild rat pups, *Psychon. Sci.* **25**:15–16.

Galef, B. G., Jr., and Heiber, L., 1976, The role of residual olfactory cues in the determination of feeding site selection and exploration patterns of domestic rats, *J. Comp. Physiol. Psychol.* **90**:727–739.

Galef, B. G., Jr., and Henderson, P. W., 1972, Mother's milk: A determinant of the feeding preferences of weaning rat pups, *J. Comp. Physiol. Psychol.* **78**:213–219.

Galef, B. G., Jr., and Sherry, D. F., 1973, Mother's milk: A medium for the transmission of cues reflecting the flavor of mother's diet, *J. Comp. Physiol. Psychol.* **83**:374–378.

Ghiselin, M. T., 1974, *The Economy of Nature and the Evolution of Sex,* University of California Press, Berkeley and Los Angeles.

Graham, F. K., Leavitt, L. A., Strock, B. D., and Brown, J. W., 1978, Precocious cardiac orienting in a human anencephalic infant, *Science* **199**:322–324.

Grosvenor, C. E., and Mena, F., 1974, Neural and hormonal control of milk secretion and milk ejection, in: *Lactation: A Comprehensive Treatise,* Vol. 1 (B. L. Larson and V. R. Smith, eds.), pp. 227–276. Academic Press, New York.

Grota, L. J., 1973, Effects of litter size, age of young, and parity on foster mother behaviour in *Rattus norvegicus, Anim. Behav.* **21**:78–82.

Grota, L. J., and Ader, R., 1969, Continuous recording of maternal behavior in *Rattus norvegicus, Anim. Behav.* **17**:722–729.

Hahn, P., 1956, The development of thermoregulation. III. The significance of fur in the development of thermoregulation in rats, *Physiol. Bohemoslov.* **5**:428–431.

Hahn, P., and Koldovsky, O., 1960, The effect of complete starvation on body composition and energy losses in rats of different ages, *Physiol. Bohemoslov* **9**:172.

Hahn, P., Krecek, J., and Kreckova, J., 1956, The development of thermoregulation. I. The development of thermoregulatory mechanisms in young rats, *Physiol. Behemoslov.* **5**:283–289.

Hahn, P., Koldovsky, O., Krecek, J., Martinek, J., and Vacek, Z., 1961, Endocrine and metabolic aspects of the development of homeothermy in the rat, in: *Somatic Stability in the Newly Born* (G. E. W. Wolstenholme and M. O'Conner, eds.), pp. 131–155, Little, Brown, Boston.

Hall, W. G., 1975, Weaning and growth of artificially reared rats, *Science* **190**:1313.

Hall, W. G., 1979, Feeding and behavioral activation in infant rats, *Science* **205**:206–208.

Hall, W. G., and Rosenblatt, J. S., 1977, Suckling behavior and intake control in the developing rat pup, *J. Comp. Physiol. Psychol.* **91**:1232–1247.

Hall, W. G., and Rosenblatt, J. S., 1978, Development of nutritional control of food intake in suckling rat pups, *Behav. Biol.* **24**:413–427.

Hall, W. G. Cramer, C. P., and Blass, E. M., 1977, The ontogeny of suckling in rats: Transitions toward adult ingestion, *J. Comp. Physiol. Psychol.* **91**:1141–1155.

Harlow, H. F. and Harlow, M. K., 1965, The affectional systems, in: *Behavior of Non-human Primates,* Vol. 2 (A. M. Schrier, H. F. Harlow, and F. Stollnitz, eds.), pp. 287–334, Academic Press, New York.

Holmes, J. C., and Bethel, W. M., 1972, Modification of intermediate host behavior by parasites, in: *Behavioral Aspects of Parasite Transmission* (E. U. Canning and C. A. Wright, eds.), Academic Press, New York.

Hopson, J. A., 1973, Endothermy, small size, and the origin of mammalian reproduction, *Am. Nat.* **107**:446–452.

Kikkawa, J., and Thorne, M. J., 1971, *The Behavior of Animals,* New American Library, New York.

LaPage, G., 1963, *Animals Parasitic in Man,* Dover, New York.

Lehrman, D. S., and Rosenblatt, J. S., 1971, The study of behavior development, in: *The Ontogeny of Vertebrate Behavior* (H. Moltz, ed.), pp. 2–29, Academic Press, New York.

Leon, M., Croskerry, P. G., and Smith, G. K., 1978, Thermal control of mother-young contact in rats, *Physiol. Behav.* **27**:793–811.

Lieberman, M., and Lieberman, D., 1978, Lactase deficiency: A genetic mechanism which regulates the time of weaning, *Am. Nat.* **112**:625–627.

Martin, L. T., and Alberts, J. R., 1979, Taste aversions to mother's milk: The age-related role of nursing in acquisition and expression of a learned association, *J. Comp. Physiol. Psychol.* **93**:430–445.

Ostadalova, I., Bibr, B., Babicky, A., Parizek, J., and Kolar, J., 1971, Transfer of ^{85}Sr from lactating rats to sucklings as affected by the size of the litter, *Physiol. Bohemoslov* **20**:397.

Pisano, R. G., and Storer, T. I., 1948, Burrows and feeding of the Norway rat, *J. Mammal.* **29**:374–383.

Pond, C. M., 1978, The significance of lactation in the evolution of mammals, *Evolution* **31**:177–199.

Rosenblatt, J. S., 1965, The basis of synchrony in the behavioral interaction between the mother and her offspring in the laboratory rat, in: *Determinants of Infant Behaviour III* (B. M. Foss, ed.), pp. 3–45, Methuen, London.

Rosenblatt, J. S., 1967, Nonhormonal basis of maternal behavior in the rat, *Science* **156**:1512–1514.

Rosenblatt, J. S., 1969, The development of maternal responsiveness in the rat, *Am. J. Orthopsychiatry* **39**:36–56.

Rosenblatt, J. S., and Lehrman, D. S., 1963, Maternal behavior of the laboratory rat, in: *Maternal Behavior in Mammals* (H. L. Rheingold, ed.), pp. 8–57, Wiley, New York.

Rozin, P., 1976, The selection of foods by rats, humans, and other animals, in: *Advances in the Study of Behavior,* Vol 6 (J. S. Rosenblatt, R. A. Hinde, E. Shaw, and C. Beer, eds.), pp. 21–76, Academic Press, New York.

Schall, J. J., and Pianka, E. R., 1978, Geographical trends in numbers of species, *Science* **201**:679–686.

Schneirla, T. C., 1957, The concept of development in comparative psychology, in: *The Concept of Development* (D. B. Harris, ed.), pp. 78–108, University of Minnesota Press, Minneapolis.

Steiniger, F. von, 1950, Beitrage zur Soziologie und Sonstigen Biologie der Wanderrate, *Z. Tierpsychol.* **7**:356–379.

Stern, J. M., and MacKinnon, D. A., 1978, Sensory regulation of maternal behavior in rats: Effects of pup age, *Dev. Psychobiol.* **11**:579–586.

Taylor, P. M., 1960, Oxygen consumption in new-born rats, *J. Physiol.* **154**:153–168.

Telle, H. J., 1966, Beitrag zur Kenntnis der Verhaltensweise von Ratten, vergleichend dargestellt bei *Rattus norvegicus* und *Rattus rattus, Z. Angew. Zool.* **53**:129–196.

Thoman, E. B., and Arnold, W. J., 1968, Effects of incubator rearing with social deprivation on maternal behavior in rats, *J. Comp. Physiol. Psychol.* **65**:441–446.

Trivers, R. L., 1972, Parental investment and sexual selection, in: *Sexual Selection and the Descent of Man* (B. Campbell, ed.), pp. 136–179, Aldine-Atherton, Chicago.

Trivers, R. L., 1974, Parent-offspring conflict, *Am. Zool.* **14**:249–264.

Tucker, H. A., 1974, General endocrine control of lactation, in: *Lactation: A Comprehensive Treatise* (B. L. Larson and V. R. Smith, eds.), pp. 227–318, Academic Press, New York.

Van der Schoot, P., Lankhorst, R. R., DeRoo, J. A., and DeGreef, W. J., 1978, Suckling stimulus, lactation, and suppression of ovulation in the rat, *Endocrinology* **103**:949–956.

Vaughan, T. A., 1972, *Mammalogy,* Saunders, Philadelphia.

Villwock, W., 1973, Order: Anglerfish, in: *Grzimek's Animal Life Encyclopedia,* Vol. 4 (B. Grzimek, ed.), pp. 398–404, Van Nostrand Reinhold, New York.

Wickler, W., *Mimicry in Plants and Animals,* 1968, World University Library, London.

Williams, G. C., 1966, *Adaptation and Natural Selection,* Princeton University Press, Princeton.

Woodside, B. C., 1978, Thermoendocrine limitation of mother-litter contact in the Norway rat, Ph.D. dissertation, McMaster University.

7

Parent and Infant Attachment in Mammals

David J. Gubernick

1. Introduction

Parental care is one strategy which helps ensure the survival of offspring and thereby enhances the parents' reproductive success (see Klopfer, this volume; Pianka, 1970). In mammals, parental care usually involves behavioral interactions between parents and offspring. One form these interactions can take is that of parent and infant attachment, the subject of this chapter.

The interactions between parents and offspring that have received the widest attention have been those between the mother and her young, while male parental care has been less well studied (cf. Hrdy, 1976; Kleiman and Malcolm, this volume; Mitchell and Brandt, 1972; Spencer-Booth, 1970). The maternal–filial relationship has been considered a special relationship, the disruption of which often has adverse consequences on the developing young (e.g., Bowlby, 1969; Denenberg and Whimby, 1963; Harlow, 1962; Scott, 1963; Seay *et al.*, 1964). A primary focus of research in maternal–filial relationships has been on the attachment ("bond") of the infant to its mother. Relatively little is known of the parent's attachment to its infant.

My emphasis here will be placed on mammals other than humans.* Three general questions will be addressed: what is attachment, where is attachment found, and why are attachments formed?

---

*The interested reader can approach the voluminous literature on human parent and infant attachment by first consulting Ainsworth (1972, 1973, 1977), Ainsworth *et al.* (1978), Alloway *et al.* (1977), Bowlby (1969, 1973), Gewirtz, (1972*a*), Klaus and Kennel (1976), and *Parent-Infant Interaction, Ciba Found. Symp.* (1975), Schaffer (1977).

DAVID J. GUBERNICK • Department of Zoology, Duke University, Durham, North Carolina 27706. Present address: Department of Psychology, Indiana University, Bloomington, Indiana 47405.

2. What Is Attachment?

The term "attachment" has been used in a general sense to refer to some emotional bond between individuals, such as mates, or between individuals and inanimate objects, such as nest site attachment (Lorenz, 1935; Weinraub *et al.*, 1977; Wickler, 1976). It has also been used to refer to a mechanism by which individuals learn the "identity" of their species (cf. Lorenz, 1935; Roy, 1980). However, attachment has been more widely used to refer to certain aspects of the relationship between parents and infants. It is in this latter sense that the term will be exclusively employed here. A discussion of the similarities and differences between parent–infant attachment and the other forms of attachment is beyond the scope of this chapter.

Typically attachment is defined as a special affectional relationship between two individuals that is specific in its focus and endures over time. Further, it is often assumed "that the object of attachment serves a special psychological function for which others cannot substitute and that he elicits affective and social responses that differ from those elicited by other figures" (Cohen, 1974, p. 207).

The important features of such definitions are (1) the formation of a special emotional relationship (i.e., affective bond), (2) with a specific individual (i.e., specificity of the bond), (3) towards whom certain responses are directed rather than toward other individuals (i.e., differential responding). These features assume the recognition and discrimination of the attachment figure from other figures.

Several authors have conceptually distinguished between attachment as an affectional tie or bond and attachment behavior (Ainsworth, 1972; Bowlby, 1969; Cairns, 1972; Lamb, 1974) and have espoused different viewpoints regarding the relationship between the two.

Attachment behaviors are considered to be those behavior patterns which predictably result in proximity to a specific (attachment) figure and whose essential function is assumed to be protection of the young (Ainsworth, 1972, 1973; Bowlby, 1969; King, 1966). Attachment behaviors may include, for example, gazing, vocalization, clinging, and approach, such that as a consequence of these behaviors proximity is achieved or maintained. Further, it is these behaviors through which some consider the attachment bond to be formed, maintained, and mediated (Ainsworth, 1973; Bowlby, 1969; Harlow and Harlow, 1965). The affectional bond itself, however, is not any of these behaviors but rather is inferred from such behavior.

According to one view, attachment behavior may change as a function of the situation and the motivation of the individual, but the affective attachment bond is more or less constant and enduring (Ainsworth, 1972). Accordingly, it is argued that the strength or intensity of an attachment bond cannot be directly measured, although the existence of an affectional bond can be inferred. What can be measured then is the intensity or degree to which attachment behaviors are activated

(Ainsworth, 1973). Additionally, with the development of an attachment bond the infant presumably develops a "sense of security" from the familiar attachment figure. As a result, the infant can use that caregiver as a base for exploring the environment, which involves leaving the proximity of the caregiver and as such is antithetical to attachment behaviors (Ainsworth, 1969, 1973; Harlow, 1958). This "sense of security" is likewise inferred from observing the behavior of the infant in the presence of its caregiver.

Others, however, have used attachment behaviors as indices of an attachment bond and have explicitly or implicitly viewed attachment (as a bond) as definable in terms of these measures (e.g., Harlow, 1958; Mason, 1970; Scott, 1971). That is, the indices themselves reflect the strength or intensity of the attachment bond.

Both viewpoints consider the attachment bond as a construct but differ in how the bond is reflected in behavior. A problem arises, however, when attachment as a construct is used as a motivator or organizer of behavior or as an explanation, often for the very behavior from which it was inferred. For example, the disruption in the young's behavior when separated from its primary caregiver has often been attributed to a disruption of the bond between them (attachment as explanation), yet this bond is often inferred to exist from the very observation that the infant is distressed upon separation from its caregiver.

To illustrate the problem, consider, for example, a worm impaled on a hook, writhing and squirming. Obviously the response was elicited by the entry of the hook, and presumably the response is adaptive in helping the worm escape. Need we, however, ascribe any emotional feelings to the worm in order to account for its behavior? Questions regarding the immediate causation, function, or ontogenetic development of the worm's behavior can still be addressed without reference to some hypothetical construct that energizes or directs its behavior.

A concept certainly may be of heuristic value in generating further questions. However, a difficulty with the use of any concept (e.g., territoriality, aggression, attachment) is that it may give the impression that we are studying a unitary phenomenon in regard to, for example, the mechanisms underlying the behavior, its functions, or its development. Another problem is that the concept may become reified, that is, treated as if it had material existence or an identity independent of the observed behavior from which it was inferred. Further, the concept may then be used in such a way as to give the impression as mentioned earlier that we have explained behavior by referring to the concept. Such, I think, is the case with an attachment bond construct.

Other investigators have rejected the need for an attachment bond construct and view attachment in terms of the stimulus–response contingencies within the parent–infant relationship itself (Cairns, 1966, 1972; Gewirtz, 1972b; Rosenthal, 1973; Weinraub et al., 1977). Their focus is on the interactions (or, if you will, the attachment behaviors) of the infant and parent and the various conditions controlling the expression of these interactions.

I believe that questions regarding the evolution, function, ontogeny, and immediate causation of parent and infant attachment can continue to be profitably addressed without the use of an attachment bond construct.

The term "attachment" as used here will be considered a summary statement or a shorthand descriptive label that refers to the *preferential responding between parents and infants as demonstrated by various operational criteria*. These operational criteria are presented in the following section.

2.1. Criteria of Attachment

Various operational criteria have been commonly used as evidence of attachment, in particular, of the infant's attachment to its mother. These criteria have been primarily developed from research on humans and other primates. Since these criteria are so frequently used it is important to examine the difficulties associated with employing them as indicators of attachment.*

These criteria include:

1. Preference for one individual (the presumed attachment figure) over another.
2. Seeking and maintenance of proximity to that figure; also used as evidence of a preference.
3. Response to brief separation from the presumed attachment figure.
4. Response to extended periods of separation.
5. Response to reunion with the presumed attachment figure.
6. Use of the attachment figure as a secure base from which to explore the world.

Most of the criteria, except perhaps (6) a secure base for exploration, can also be applied to studies of parental attachment even though they were developed originally with the infant in mind. Each of these indices is briefly examined below.

The demonstration of preferential responding towards one individual over another is considered the most important requisite of an operational definition of attachment (Cohen, 1974). Preferential responding can be demonstrated by showing that an infant (or parent) is selectively responding to that presumed attachment figure and not to some general class of stimuli that would also elicit the same behavior if provided by another individual. Evidence for such specificity must be provided whenever any of the criteria are utilized.

Preference. The infant is typically given a choice between its mother and a strange female, who are presented to the infant either simultaneously or sequentially. Recognition of and preference for the mother are usually determined by a difference in some behavior (e.g., sniffing, clinging) directed toward or shown in

---

*For further discussions of characteristics or criteria of attachment see, in addition to the above sources, Ainsworth (1977), Bernal (1974), Bischof (1975), Gewirtz (1972c), Lamb (1976), Rajecki *et al.* (1978), and Sroufe and Waters (1977).

the presence of the mother, compared with the same behavior directed towards or shown in the presence of the stranger.

If the infant can physically interact with both the mother and the stranger, the infant's behavior will be affected by the behavior of the others and the others will also be affected by the behavior of the infant. The infant may, for example, cling more to the mother or spend more time in proximity to her than the stranger because it recognizes and prefers its mother, or because the mother recognizes the infant, or because the stranger avoided or threatened the infant, or all of the above. Based on this procedure we could not discriminate between these interpretations. A similar difficulty would arise if the infant was prevented from physically contacting its mother and the stranger, but the infant and mother or infant and stranger were allowed to see, hear, and smell each other as, for example, from behind a plexiglass or wire-mesh partition.

This problem can be overcome by preventing such interactions either through anesthetizing the females (e.g., Kaplan *et al.,* 1977) or through the use of one-way mirrors (e.g., Rosenblum and Alpert, 1974). A difficulty may still remain, however, from the fact that the infant may spend more time in proximity to its mother (i.e., shows a "preference") as an indirect consequence of its avoidance of the stranger. The infant may simply approach the familiar (mother) and avoid the unfamiliar (Zajonc, 1971). On the other hand, if the infant shows no preference for either (i.e., spends equal amounts of time with each) or even spends more time near the stranger, the infant may in fact be attached to its mother and be more likely to explore the environment in her presence and thereby approach the stranger (e.g., Ainsworth, 1972; Harlow, 1958). Additionally, infants placed alone in a novel environment will approach and maintain proximity to an unfamiliar social stimulus (e.g., Pettijohn *et al.,* 1977; Rosenblum and Alpert, 1974). What may be needed here, and what is typically not done, is to compare the infant's response to its mother with that of its response to a stranger and another familiar individual (Cohen, 1974). Comparing the responses to the combinations of familiar figure and stranger, mother and stranger, and mother and familiar figure should indicate the effects of familiarity and the specifity of the infant's response to its mother. Determining what constitutes a familiar figure is not without its own problems (Weinraub *et al.,* 1977).

The recommendation here is that, where possible, preference tests incorporate another familiar figure and minimize interactions between the infant and the various social stimuli.

Proximity. The absence of proximity seeking or maintenance is not by itself evidence of the lack of a preference for or attachment to another. As the young develop, more proximal forms of maintaining proximity and contact (e.g., clinging) may change to more distal forms (e.g., watching from a distance, vocalizations). The task then becomes one of identifying what constitutes proximity seeking and proximity maintenance during the course of development. Further, the types of behavior that do qualify will vary with motivational state, context, sex, age, etc. Proximity alone is not sufficient to indicate attachment. The difficulties of using proximity measures for determining attachments of freely moving animals to individuals, objects, and areas has already been discussed by Wickler (1976; see also Hinde, 1977).

Separation. The effects of separation on the mother–infant relationship and the subsequent behavior of the infant have been amply documented (e.g., Bowlby, 1969; Harlow and Harlow, 1965, 1969; Hinde and Davies, 1972). Typically, infants show "distress" reactions upon separation from their (presumed) attachment figure. However, such reactions are not inevitable and may be influenced by a variety of factors, such as the social relationship between the mother and infant prior to separation, the social context of the separation environment, who is removed, and the species [cf. Hinde and McGinnis (1977) and Mineka and Suomi (1978) for reviews on primates]. Does the disruption in behavior upon separation indicate the presence of attachment and does the lack of such reactions indicate the absence of an attachment to the primary caregiver? Perhaps. However, the removal of the mother may also remove stimulation (e.g., warmth, motion) that normally regulates the behavior of the infant. Perhaps it is the removal of this stimulation that induces the observed distress reactions [see Hofer (1978, this volume) and Section 3 below]. The mother may also provide cues that have become associated with or conditioned to many of the infant's organized behavior patterns (Cairns, 1966; Gewirtz, 1972b). The removal of these cues may then result in the disruption of organized sequences of behavior.

Separation studies are certainly important for understanding the nature of parent–infant relationships, the effects such separations have on the developing young, and the means for alleviating the traumatic effects such separations may induce. However, by themselves separation studies may not provide sufficient evidence to argue for the presence of filial or parental attachment. Some experimental verification of attachment (e.g., preference tests; see above) independent of the separation study would be needed.

Reunion. Response to reunion with the (presumed) attachment figure can also be influenced by various factors, e.g., duration of separation, the social relationship between mother and infant prior to reunion, and the separation and reunion environment (cf. Mineka and Suomi, 1978, for a review). Upon reunion the infant may make immediate and sustained contact with the mother. In other cases the infant might not even return to its mother if during the separation period the infant adopted another caregiver as has been reported for North Indian langur monkeys (Dolhinow, 1978; McKenna, this volume). We are justified in referring to attachment when the changes in behavior upon reunion with the primary caregiver are specific to that caregiver and are not elicited by another caregiver who shares or provides the same stimulus characteristics as the primary caregiver (see also Section 3 below).

Base for Exploration. Most reports indicate that an infant is more likely to explore a novel environment or stranger in the presence of its caregiver (e.g., Ainsworth, 1973; Harlow, 1958).

In some cases the infant may remain in the proximity of the mother, while in other cases the infant may leave and explore the novel environment or approach a stranger should one be present. Based on the infant's proximity to the mother one might conclude that the infant is attached to the mother in the first case, but not in the second case. However, the same difficulties and comments regarding the use of discrimination tests and proximity measures are applicable here. Again one must be able to demonstrate that infants are responding to the presence of that specific caregiver and not of another familiar figure.

As mentioned earlier, these same criteria can be applied to investigations of parental attachment. Another criterion for assessing parental attachment is defense of the young.

Defense of Young. There are numerous field observations and laboratory studies indicating that caregivers will defend their young against conspecifics and predators (cf. Svare, this volume). It is surprising therefore to find that defense of the young has not been systematically used as an operational measure of attachment. Although the details must be worked out for each case, the caregiver can be presented simultaneously with its own young and another's young, both of which are in the presence of a threatening stimulus. One must determine that the caregiver's response is specific to its own young rather than to just any young, or simply a function of the particular site of the test (e.g., home cage), and that it is not a reaction that will be elicited even in the absence of any young.

The indices of attachment discussed above should not be construed as being rigid criteria. They may change as further research sheds additional light on the behavioral interactions specific to a parent and its offspring. It is obvious, however, that no single index of attachment is completely satisfactory and that multiple criteria need to be used. Further, the view taken here is that attachment is not a unitary phenomenon but rather is multidimensional and that, as such, multiple criteria and convergent approaches would be more useful.

These indices may not, however, be as applicable for some species as for others. As an extreme example, rabbits nurse their young once every 24 hr (Zarrow *et al.*, 1965), while tree shrews do so once every 48 hr (Martin, 1966). The use of separation studies with these species may not reveal much about attachment, since these infants normally undergo frequent and prolonged separations. A similar case can be made for ungulates that tend to hide their young for extended periods of time after birth, periodically returning to them during that time [e.g., elk (Altmann, 1963)]. However, separation studies might be applicable for those ungulates whose young follow and stay with their mother shortly after birth [e.g., wildebeest (Estes and Estes, 1979)].

3. Attachment and Separation

In various species individuals other than the mother may provide care for the young (cf. Hrdy, 1976; Gurski and Scott, in press; McKenna, this volume; Spen-

cer-Booth, 1970; see Section 5 below). In the absence of the mother these alternative caregivers may minimize the stress to the infant associated with the loss of
the mother. Studying how substitute caregiving alleviates this reaction may provide further insight into the problems discussed earlier.

Separation from or loss of the primary caregiver can have profound effects
on the young. Such separations and loss may lead to a debilitation in species-
appropriate social, maternal, sexual, aggressive, and exploratory behavior in primates (Bowlby, 1969; Harlow and Harlow, 1965, 1969; Hinde and Spencer-
Booth, 1971; Seay *et al.*, 1962), dogs (Fuller, 1967; Scott and Marston, 1950;
Scott *et al.*, 1974), cats (Rosenblatt *et al.*, 1962; Schneirla *et al.*, 1963), and rodents
(Beach and Jaynes, 1954; Denenberg and Whimby, 1963; Levine, 1967). These
effects are considered somewhat analogous to human psychopathology (Bowlby,
1969; Harlow and Suomi, 1974) and indicate the importance of studying ways to
alleviate such effects of separation.

The response to separation has been characterized as one of initial "protest"
upon separation followed by "despair" or "depression" (Bowlby, 1969; Seay *et
al.*, 1962). The stage of initial "protest" often consists of hyperactivity and
increased vocalizing, while the "despair" phase is marked by substantial inactivity,
withdrawal from social interactions with others, and self-directed behavior such
as thumb sucking and self-rocking.

These reactions, however, are not always shown, since, as mentioned earlier,
there are a number of variables which affect the young's response to separation
(cf. Mineka and Suomi, 1978). These include the nature of the interactions
between mother and infant prior to separation and the nature of the separation
environment. The importance of the social context in reducing separation distress
has been amply demonstrated in the now classic studies by Kaufman and Rosenblum (1967*a,b*, 1969; see Swartz and Rosenblum, this volume) of bonnet and
pigtail monkeys. Briefly, bonnet monkey infants have early and frequent contact
with other adult females. When their mother is removed from the group, the
infants show initial agitation or protest but no depressive reactions (see above)
while they are adopted by a substitute caregiver. In contrast, pigtail infants have
contact primarily with their mother. They do not receive substitute care from other
adult females. When their mother is removed from the group, pigtail infants show
agitation followed by severe depression (see Swartz and Rosenblum, this volume,
for further elaboration).

Such studies have emphasized the effects of substitute caregiving on reducing
the disruptive consequences of separation. The question raised here is whether
substitute caregivers alleviate separation distress as a result of the infant's attachment to the caregiver or as a result of providing the same stimulation normally
provided by the mother.

One answer derives from Bowlby (1969), who argued from an evolutionary
perspective that the function of infant attachment is protection from predation
through maintaining proximity with the caregiver (see, however, Section 4.1.1).
The infant presumably derives an emotional "sense of security" from an attachment with its caregiver such that if the attachment figure is removed the infant

responds with distress or protest. Although the infant is considered to form its first and principal attachment to a single primary caregiver, later secondary attachments may occur. These secondary attachment figures may help reduce separation distress (presumably through the sense of security they provide), although they are not interchangeable with the primary attachment figure. Thus, in the case of the bonnet infant, alleviation of separation distress is considered a result of the sense of security provided by the substitute caregiver (Kaufman, 1973).

From this perspective we would need to demonstrate two things *prior to separation* of mother and infant. First, has the infant formed an attachment to its mother, preferring her over other familiar substitute caregivers? Second, has the infant developed a secondary attachment to one particular substitute caregiver in preference to another?

Related to this second point are the two studies, one of rhesus monkeys (Sackett *et al.,* 1967), the other of mice (Gurski, 1977), that have systematically manipulated the amount of contact with substitute mothers. The results were quite similar in both cases. Infants given the same amount of contact with their biological mother and several substitute mothers showed no preference for one caregiver over the other, while infants reared only with their own mother preferred her to a stranger.

Multiple attachments may indeed be present prior to removal of the mother. Perhaps an attachment also develops between the infant and the substitute caregiver *during the separation interval.* What is suggestive in this regard are some recent findings on North Indian langur monkeys (Dolhinow, 1978). These langurs show extensive and very early handling and care of the infants by other females. When mothers of 6- to 8-month-old infants were removed from the group for 2 weeks, most of the infants adopted an adult female as a substitute caregiver. The adoptions were initiated by the infant and not by the adult female and resulted in the reduction of separation distress. There was no clear-cut relationship between an infant's choice of a substitute and its earlier experience with that substitute (suggesting perhaps that there was no prior attachment between them).

Separation reactions decreased despite the quality of substitute care (e.g., "punishing," "permissive"), and this was interpreted as indicating that the availability of caregivers was more important than the adequacy of substitute care (Dolhinow, 1978; see below). Following the return of the mother, 5 of 11 of the infants that had adopted a substitute caregiver stayed with that substitute rather than returning to their mothers (McKenna, 1979), suggesting that some sort of attachment had developed. These observations are of course open to other interpretations.

The point is this: we need to examine the relationship between the infant and substitute caregivers more directly to at least first determine whether attachment is present and how substitute caregivers can alleviate an infant's response to separation.

Perhaps substitute caregivers reduce separation distress by providing some general class of stimuli similar to those provided by the mother, and it is the presence of these stimuli that reduces distress and not the specific substitute caregiver.

This suggests a different view of separation distress, one that does not emphasize an attachment bond or multiple attachments. This view, offered by Hofer (1978, this volume), is that the mother interacts with her young through stimulation of various sensory systems (olfactory, visual, vestibular, tactile, etc.). This stimulation plays a role in directing and controlling the behavior and physiology of the infant. When the mother is removed, this stimulation, for example, food, heat, is withdrawn. The withdrawal of this stimulation results in various behavioral and physiological changes. For rats, removal of the mother may lead to a $3°C$ drop in the pup's body temperature. The decreases in locomotion and self-grooming that occur when the mother is removed can be restored to normal by providing sufficient heat. Further, decreases in heart rate can be restored to normal by supplying appropriate amounts of milk (see also Galef, this volume).

Although rat pups do not show a specific attachment to their own mother (Section 6.1), the above view highlights the possibility of such regulatory processes existing in other groups, such as primates, that do exhibit infant attachment. For example, infant rhesus monkeys reared in isolation with a stationary surrogate typically develop a stereotyped self-rocking and heightened "emotional" responsiveness. Vestibular input provided by a moving surrogate (input a mother might provide while carrying her infant) suppressed the expression of these reactions (Mason and Berkson, 1975).

Separation from the mother does have physiological and biochemical effects in primate infants. Reite *et al.* (1974, 1978*a*) found that 4 days of maternal separation in group-living pigtail infants (where there is no aunting) may induce profound physiological changes, for example, changes in heart rate, body temperature, sleep cycle, and amount of REM sleep, among others. These changes may continue and exceed the duration of any observable behavioral stress and may therefore affect subsequent interactions with the mother and others. Similar changes though of lesser magnitude were also found for surrogate-reared pigtail infants (Reite *et al.*, 1978*b*).

Hormonal changes following maternal separation have been reported for rhesus monkey infants living in groups (Smotherman *et al.*, 1979), with only the mother (Meyer *et al.*, 1975) or with a surrogate (Hill *et al.*, 1973; Mason, 1978) and for squirrel monkey infants living in groups (Coe *et al.*, 1978), with mother only (Levine *et al.*, 1978; Mendoza *et al.*, 1978) or a surrogate (Hennessy *et al.*, 1979; Mendoza *et al.*, 1978). In each instance increased levels of plasma cortisol (reflecting arousal) were obtained following separation. In general, however, infants reared with their mother evidenced a more pronounced cortisol response to separation than surrogate-reared infants, suggesting that attachment to the mother is different at least quantitatively from attachment to an inanimate surrogate. This is important since the behavioral response to separation may not differentiate between mother-reared and surrogate-reared infant monkeys (Suomi *et al.*, 1970).

Do these hormonal and physiological changes reflect the removal of some regulatory stimulation provided by another individual or a disruption of an affectional bond with a specific attachment figure? In the case of the group-living squirrel monkey infants, the increased cortisol levels were obtained during a brief 30-min separation despite the fact that the infants were being aunted and were *not* behaviorally agitated (Coe *et al.,* 1978). This could suggest that aunting, at least in squirrel monkeys, does not alleviate the effects of separation. Perhaps, however, more than 30 min of contact with a substitute caregiver is needed to reduce physiological stress, although such brief contact does eliminate behavioral agitation.

This same result has been taken to indicate that the squirrel monkey infants had specific attachments to their mothers, since the arousal induced by 30 min of separation was not ameliorated by contact with a familiar female (Coe *et al.,* 1978). This implies that reunion with the mother following 30 min of separation would reduce the elevated cortisol levels present during separation. However, Levine *et al.* (1978) found that infants continued to show elevated cortisol levels after 30 min of reunion with their mother following 30 min of separation. These results would suggest, contrary to Coe *et al.* (1978), that the infant squirrel monkey does not have a specific attachment to its mother. Alternatively, these results may simply indicate that cortisol levels in infant squirrel monkeys take longer than 30 min to return to preseparation levels (Levine *et al.,* 1978). Longer-term reunions with mother and contact with substitute caregivers will be necessary to further assess the possible hormonal correlates of infant attachment.

Another finding of interest was that 30 min of separation also induced increases in cortisol levels in squirrel monkey mothers (Coe *et al.,* 1978; Levine *et al.,* 1978; Mendoza *et al.,* 1978) and that after 30 min of reunion with their infant, mothers' cortisol levels had returned to base line (Levine *et al.,* 1978). Perhaps such hormonal changes may be used to assess maternal attachment under conditions of brief separation and reunion. However, whether these changes would reflect maternal attachment or the reinstatement of stimulation that modulates maternal behavior and physiology would need to be demonstrated (for ways infants affect the behavior and physiology of their caregivers, see Harper, this volume). Reuniting the mother with a familiar infant would help differentiate between these two possibilities. One further related point this research highlights is that the effects of separation on the caregiver have not received nearly as much attention as have the effects on the infant.

These hormonal and physiological studies are important because they demonstrate that (1) separation has effects on physiological and hormonal systems, (2) behavioral changes induced by separation may disappear while hormonal and physiological changes continue (which could potentially affect subsequent interactions), and (3) such physiological and hormonal changes may differentiate attachment to various attachment figures especially in cases where behavioral differences are not shown. Additionally, they represent the relatively few studies of possible hormonal and physiological correlates of attachment.

Thus the behavioral changes we observe in the infant during separation may not reflect the disruption of some affectional bond. Rather separation distress may reflect the withdrawal of stimulation to the infant that is normally provided by the mother.

Perhaps, then, substitute caregivers may also provide the necessary stimulation and as such reduce separation distress. This gains added support from findings that features such as texture, warmth, and contact reduce separation distress in monkeys (Harlow and Suomi, 1970; Harlow and Zimmerman, 1959), dogs (Pettijohn *et al.*, 1977; Scott *et al.*, 1974), and rodents (Hofer, this volume) as may the behavior of the substitute in response to the infant (Hennessy *et al.*, 1979; Mason, 1978; Pettijohn *et al.*, 1977). Whether substitute caregivers are also capable of reducing physiological stress is still in question. Hofer's (1978, this volume) ideas on separation stress are provocative and deserve detailed study in species showing attachment.

We actually know very little about the nature of the relationship between infants and their substitute caregivers, or about the mechanisms by which substitute caregiving alleviates separation distress.

4. Why Form an Attachment?

I assume that natural selection has operated to promote the formation of attachment. However, parent and infant attachment is not always necessary for survival and reproduction. Attachment is but one strategy parents and infants can employ to enhance their fitness. We need to ask then why parents or infants of some species form attachments. This question can be considered from the four different perspectives (following Tinbergen, 1963) of the (1) evolution, (2) function, (3) ontogeny, and (4) immediate causation of attachment. The function and evolution of attachment are addressed here.

4.1. Function of Attachment

The ultimate function of parental and filial attachment must be assessed in terms of the reproductive success it confers on the parent and the young, that is, how attachment enhances fitness (function in a "strong" sense; Hinde, 1975). To my knowledge there are no data directly addressing this question.

Parental and filial attachment do, however, have obvious beneficial consequences with respect to the immediate survival of the young [function in a "weak" sense (Hinde, 1975)]. The ultimate and proximate consequences (functions) of parental and filial attachment will be discussed separately.

4.1.1. Ultimate Function

The major function of filial attachment (at least in humans and other primates) has in the past been presumed to be that of protection from predation through attachment behaviors which promote and maintain proximity between the parent(s) and its young (Bowlby, 1969). This hypothesis is based on three facts: (1) isolated animals are more likely to be seized by a predator, (2) attachment behavior is easily and intensely elicited in those particularly vulnerable to predators (e.g., young, pregnant females), and (3) attachment behavior is elicited at high intensity in situations of alarm such as in the presence of a predator (Bowlby, 1969). Furthermore, caregivers have also evolved a complementary behavioral system (i.e., parental behavior) which presumably has the same function of protection of the young from predation (Ainsworth *et al.*, 1978; Bowlby, 1969). That predation of the young is responsible for the evolution of infant attachment and the corresponding parental care is, in part, correct; that it is the major or only factor is certainly incorrect for several reasons.

First, according to the above, we should expect to find filial attachment in species where predation of offspring is moderate to high and minimal or no attachment among large predatory species. Infant mortality is indeed high in various mammalian species (cf. Caughley, 1966; French *et al.*, 1975; Sadleir, 1969), but while filial attachment is found in some of these species, e.g., primates and wild sheep and goats, it is not found in others, e.g., mice, rats, and seals, indicating that other factors are involved in the evolution of infant attachment. Filial attachment is found among some predatory species, e.g., spotted hyenas (Kruuk, 1972), lions (Rudnai, 1973), but not others, e.g., cheetah (Schaller, 1972), cape hunting dog (van Lawick and van Lawick-Goodall, 1971; cf. Ewer, 1968, 1973).

Second, the promotion and maintenance of proximity between caregivers and their young may have more to do with offspring obtaining necessary resources, such as food, warmth, and stimulation, provided by the caregiver (see below; Section 3; and Galef, this volume). In other words, proximity between parents and offspring can have other important benefits which probably affected the evolution of such behavior in many mammals, including humans (Blurton-Jones, 1972; DeVore and Konner, 1974). Thus any general statements regarding the effects of predation are certain to be inadequate to account for the evolution of filial (or maternal) attachment.

Selection should favor a parent that provides care for its own young, since it would probably reduce its own fitness to care for the young of others. It should make no difference to the survival of the infant who provides the necessary resources as long as they are provided. We might then expect selection to favor infants that elicit care from other potential caregivers. But if parents restrict caregiving to their own young, such elicitation of care by other young would be unsuccessful and possibly detrimental. Infants should thus be more selective in forming attachments to their own parents.

The ultimate function of parental attachment is viewed as insuring the survival of one's own offspring by providing the necessary resources to them and not to those of another. The ultimate function of filial attachment is viewed here as simply insuring survival by sequestering resources typically provided by the parent. According to this scheme, both parents and infants would be expected to form attachments to each other, with parents because of their advanced sensory–motor development probably forming an attachment earlier than infants.

In any case, if the infant or the caregiver forms an attachment they both ultimately benefit in terms of enhancing their inclusive fitness. However, since these same benefits may be obtained in species that do not show filial or parental attachment, we need to know under what conditions we can expect to find filial and parental attachment (see Section 5 below). These conditions would include, but would not be limited to, ecological factors (e.g., food availability and distribution), social organization, reciprocity of infant care, and the stage of maturation of the young at birth (i.e., precocial, altricial).

For example, if we assume, as above, that the function of parental attachment is that of providing care for only one's own young then attachment would not be essential among species with altricial young whose movements were confined to a nest (e.g., many rodents) since a mother would typically encounter only her own young in the nest. Some mechanism however would be needed for them to recognize and maintain proximity to the nest site (see Section 6.1). In contrast, among precocial species in which the young are mobile shortly after birth (e.g., most ungulates), much more rapid formation of attachment may be necessary.

The stage of maturation of the young at birth does not influence attachment in isolation from other factors such as the social organization of the species. A mother living with her precocial infant but with minimal contact with other conspecifics (e.g., moose) should be under different selection pressures regarding attachment than another mother and her precocial infant who normally live together with other conspecifics (e.g., elk, caribou). A similar argument can be made for species with altricial young. Social organization and the opportunity for reciprocal infant care are correlated with each other, since the opportunity for such care is present in the more gregarious species and is minimal or absent in the more "solitary" species. Thus the evolution of attachment will depend upon a variety of factors acting in concert, such as those discussed above (see Section 5 for further elaboration).

A potential consequence of the formation of attachment may have been to affect the likelihood of the young dispersing or remaining with their caregiver. This may have then influenced the evolution of other social systems from what has been considered the basic mammalian social unit consisting of a female and her offspring (Eisenberg, 1966).

One final hypothesis that must be considered is that perhaps attachment has no function but simply evolved as a consequence of the close association between mothers and offspring. At best this could only be part of an answer, since some species have a close association between mother and young without any attachment.

4.1.2. Proximate Function

The proximate consequences to the infant of either the caregiver forming an attachment to the infant or the infant forming an attachment to the caregiver are essentially the same. These include, among others, obtaining food, warmth, stimulation (e.g., vestibular, tactile; cf. Hofer, this volume), shelter, protection from predators and conspecifics, and integration into the social group (see Swartz and Rosenblum, this volume). However, these same benefits are found in some groups, such as rodents, that do not show either parental or filial attachment (cf. Harper, Hofer, this volume; see Sections 5 and 6).

The proximate consequences to the caregiver of forming an attachment to its own young are less obvious but might include enchancement of social status, as in some primates (Seyfarth, 1976), and access to resources. Such benefits, however, are not essential to the maintenance of parental behavior (Rosenblatt and Siegel, this volume) or parental attachment as, for example, in chiroptera, ungulates, and pinnipeds.

Thus, the proximate consequences for the infant would be similar whether it formed an attachment or not; the same applies to the caregiver. We are again left with the question of why parents and infants form attachments, which can be answered, in part, by knowing in which species and under what conditions we find parental and filial attachments.

5. Where Is Attachment Found?

There is simply not enough information currently available to draw any firm conclusions regarding the conditions which would favor parent and infant attachment in mammals. However, a preliminary classification scheme is presented here which may be of heuristic value in organizing available information about attachment and in generating further research.

This scheme is a relatively simple one, based on only two parameters: (1) mobility of young and (2) social structure. Mobility of young is roughly equivalent to the stage of development of the young at birth (i.e., altricial, semi-altricial, precocial). Mobility is of course a continuum but is treated here as consisting of discrete categories for purposes of classification. Immobile young remain in a relatively fixed location such as a nest and most often include altricial young (e.g., naked, eyes closed) but also some precocial young with limited mobility, such as pinnipeds and some chiropterans. Semimobile or semi-altricial young mammals are usually furred, with eyes open at or shortly after birth and exhibiting strong clinging reflexes. These infants are often carried by their mother, who thus acts as a "moveable nest." Mobile refers to infants that are furred, with eyes open and capable of independent locomotion shortly after birth, and is equivalent to precocial. An infant may pass from immobile to mobile during successive stages of development. In some cases selection then may favor the formation of attachment by the time the young are capable of independent mobility. Mobility of the young will likely act as a constraint on other features of social organization and responses to ecological demands (Crook et al., 1976; Eisenberg, 1966).

Social structure refers to the organization of individuals within a social system (Eisenberg, 1966; Crook et al., 1976). The basic social unit is the mother–infant (Eisenberg, 1966). A male may also invest directly or indirectly in the care of the young (Kleiman and Malcolm, this volume) as may other group members (Bekoff, McKenna, this volume). The presence of others in the groups in addition to a mother–infant unit is probably an important factor influencing attachment.

These two factors, mobility and social structure, are displayed as a matrix in Table 1. Included in the table are predictions regarding the likelihood of finding parent and infant attachment. These predictions are based on the assumption that the function of attachment is to insure that care is provided for one's own young and not "squandered" on those of another (see Section 4). It follows from this assumption that attachment would be favored where the chances were greater for misdirecting such care towards unrelated young as would be the case among more gregarious species and among species where the young are highly mobile (see also Ewer, 1968).

Among gregarious species a third factor, kinship (degree of relatedness), may also influence the evolution of attachment. Relatively high degrees of relatedness between group members (e.g., females) might favor cooperative care of the young if the inclusive fitness of the individuals (the sum total of an individual's fitness plus that of its relatives) was enhanced through such cooperation (Hamilton, 1964*a*,*b;* West Eberhard, 1975). Under such conditions attachment would prob-

Table 1. Predicted Presence of Parent and Infant Attachment as a Function of Social Structure and Mobility of Young

Social structure	Mobility of young		
	Immobile	Semimobile	Mobile
Solitary[a]	No[b]	No	Possibly
Family[c]	No	No	Possibly
Extended family[d]	No? CC[e]	No? CC	Possibly
Matriarchy[f]	Possibly; CC	Possibly; CC	Yes
Harem[g] or allied (Unrelated) females	P.A.[h]— Possibly F.A.[i]—?	Yes	Yes
Multimale/ Multifemale[j]	P.A.—Possibly F.A.—?	Yes	Yes

[a]Solitary: female + recent offspring.
[b]Refers to both parental and filial attachment.
[c]Family: one male + one female + recent litter.
[d]Extended family: a single, reproductive pair + offspring from previous breeding seasons.
[e]CC, cooperative care of the young.
[f]Matriarchy: a female + several daughters or sisters; a male may be associated with them.
[g]Harem: a male + a group of allied (unrelated) females and their young.
[h]P.A., parental attachment.
[i]F.A., filial attachment.
[j]Multimale/Multifemale: cohesive mixed sex groups associated together for long durations.

ably not be favored. In contrast, when females in a group are unrelated, care of another individual's young would be a form of altruism, adaptive only when reciprocated (Trivers, 1971). Under these conditions attachment would probably be more common. Intermediate degrees of relatedness might favor a mixed strategy of cooperative care and attachment. Kinship or degree of relatedness is used here as a broad conceptual tool and does not refer to any discrete numerical values. The following sections consider how well the predictions listed in Table 1 conform to the available information.

5.1. Solitary Social Structure

The association between a mother and her offspring is the only stable social unit in many mammals (Eisenberg, 1966). Individuals of solitary species may defend a territory or their home ranges may overlap. A male and female only briefly associate together to mate (Eisenberg, 1966; Crook, 1977). Individuals are commonly antagonistic towards intruders, especially those of the same sex. One consequence of such behavior is that others are kept away from the female and her young. The maintenance of a territory, as, for example, through scent marking, would also have a similar consequence.

Under such conditions attachment would probably not be necessary prior to weaning, at least in those solitary species with immobile or semimobile young, since only one's own offspring will usually be found at the nest site or be carried by the mother (see Table 1). Additionally, attachment would not be favored if the young disperse shortly after emergence from the nest site, which often coincides with time of weaning.

5.1.1. Immobile Young

Richardson's ground squirrels, *Spermophilus richardsonii,* live in colonies, but each adult maintains a home range containing an area around its burrow which it defends against all other conspecifics except at the time of mating (Michener, 1973; Yeaton, 1972). Members of these colonies do not associate with each other. Females do not recognize their own young prior to the young's emergence at 28 days postpartum, since females will retrieve and care for alien young up to this time in both the laboratory (Michener, 1971) and in the field (Michener, 1972). The infants do not interact with unrelated adults until 2–3 weeks after emergence, when their increasing activity brings them into a neighboring female's area (Michener, 1973). At this time an infant will act submissively in the presence of unfamiliar adults but act amicably in the presence of its mother (Michener, 1973, 1974; Michener and Sheppard, 1972). Likewise, adults act amicably towards their own young and agonistically with unfamiliar young. The adult's and infant's reactions may be based in part on the location and behavior of the infant, since an infant in a stranger's home range acts differently than when it is in its mother's (familiar) home range (Michener, 1973). These results do not prove that after emergence mothers or young identify each other as individuals (Michener, 1973). A mother need only recognize familiar from unfamiliar young, and each infant need only distinguish familiar adults from unfamiliar adults in order to share the mother's home range prior to dispersal (Michener, 1974).

The thirteen-lined ground squirrel, *S. tridecemlineatus,* also lives in colonies, but male and female home ranges overlap (McCarley, 1966). The young emerge from their natal burrow when they are weaned, around 30 days postpartum, and start to disperse 2 weeks later. Females in captivity will nurse alien young prior to the time of emergence (Zimmerman, 1974), suggesting an absence of attachment.

Although a female typically rears only her own young, she may share a communal nest and nurse alien young at higher population densities or in captivity, suggesting an absence of attachment as, for example, in prairie deer mice, *Peromyscus maniculatus,* white-footed mice, *P. leucopus,* and canyon mice, *P. crinitus* (Eisenberg, 1962, 1963, 1966, 1968; Hill, 1977; King, 1963), and house mice, *Mus musculus* (Crowcroft and Rowe, 1963; Sayler and Salmon, 1969), golden hamsters, *Mesocricetus auratus* (Richards, 1966; Rowell, 1961), striped skunks, *Mephitus mephitus* (Verts, 1967), and domestic cats, *Felis domestica* (Ewer, 1961), to name but a few (for other examples, cf. Eisenberg, 1966; Spencer-Booth, 1970).

5.1.2. Semimobile Young

Marsupials provide an interesting group for study, since the young of most species spend their altricial development in a pouch or marsupium and emerge in an advanced stage of development. During the pouch phase a female and her young can be considered "solitary," since the female does not normally encounter other pouched young and her young do not normally encounter other mothers.

The available evidence suggests that mothers do not recognize their young while still in the pouch, since forced transfers of young from one pouch to another were successful in Virginia opossums (Hunsaker and Shupe, 1977) and in kangaroos and wallabies (Merchant and Sharman, 1966). The Virginia opossum, *Didelphis virginiana,* and mouse opossum, *Marmosa robinsini,* are both solitary species, nest in trees, and give birth to large litters. Young of the Virginia opossum are carried on the mother's back after emergence from the pouch (Hunsaker and Shupe, 1977). Although the young may hiss at or retreat from strangers (McManus, 1970), it is not yet known whether maternal or infant attachment occurs after emergence. Infant mouse opossums, *M. robinsini* and *M. cinerea,* initially cling to the mother for the first 30 days, since she has no pouch, and then later are left in a nest (Beach, 1939; Eisenberg and Maliniak, 1967). Females will retrieve alien young during the clinging phase (Beach, 1939; Thrasher *et al.,* 1971), suggesting an absence of maternal attachment. It is not known whether females recognize their young during the nesting phase, although the likelihood of encountering alien young in the nest is minimal.

Tarsiers *(Tarsius syrichta, T. bancanus)* are nocturnal arboreal prosimians, typically solitary or paired, and may defend a territory (Niemitz, 1979). The single infant produced is semi-altricial and clings to the mother, since a nest is not built. Lactating females will retrieve and adopt alien young (Niemitz, 1979), suggesting an absence of maternal and infant attachment prior to weaning. The lesser mouse lemur, *Microcebus murinus,* is also a nocturnal arboreal prosimian. Females maintain relatively exclusive home ranges (Martin, 1973) and are difficult to categorize as strictly solitary, since females that are presumed to be a mother and previous daughters may sleep together in a nest but forage alone and evidence no organized social interactions (Martin, 1973). During the nesting phase a lactating female will accept an alien infant (Schilling, 1979).

Cheetah, *Acinonyx jubatus,* are also considered solitary (Eaton, 1970; Schaller, 1972). Cubs are weaned at about 3 months, at which time a female may not be able to distinguish her own cubs from a strange cub (Schaller, 1972). This is based on the observation that two females with cubs encountered each other at a kill and one cub followed the other female and her two cubs when they moved off. The female now with three cubs reacted to the cubs following her by swatting at and avoiding all three (Schaller, 1972). These examples support the prediction that attachment would not be favored among solitary species with semimobile young.

5.1.3. Mobile Young

Little is known about solitary species with highly mobile young, since these species tend to be associated with "closed" habitats which provide concealment, e.g., forests, rather than with "open" habitats, e.g., savannahs (Eisenberg, 1966; Estes, 1974), and are therefore difficult to observe. Such species can be found among the Artiodactyla, but none have been studied in sufficient detail to warrant any firm conclusions (see Estes, 1974; Jarman, 1974; Leuthold, 1977, for various examples and further discussion).

The solitary moose cow *(Alces alces)* chases her yearling away before the birth of her next calf (Altmann, 1963; Geist, 1963). The cow and new calf remain close together for the first 4–20 days postpartum and have little contact with conspecifics (Altmann, 1958, 1963). From 20–90 days postpartum the female behaves aggressively towards other moose, which serves to keep them away from her calf (Altmann, 1958, 1963; Houston, 1974). During this time the calf may approach and follow other moose (Altmann, 1958, 1963), suggesting a lack of infant attachment.

Muntjacs, *Muntiacus muntjak* and *M. reevesi,* are solitary forest-dwelling deer. Male home ranges overlap those of females, but same-sex individuals are intolerant of each other (Barrette, 1977a). The single fawn remains concealed for up to 4 weeks, and the mother may largely ignore her offspring after the first few weeks (Barrette, 1977a,b). The infant may be weaned by 8 weeks of age, being visited perhaps once every 24 hr up until that time (Yahner, 1978). In a captive muntjac, the 2-week-old fawn followed and attempted to nurse from another female. Females were increasingly agonistic towards other females and scent marked more at 5 weeks postpartum that at other times, which coincided with the peak activity (mobility) of their young (Yahner, 1978). Such increased aggression and scent marking probably keeps others out of the female's area and decreases the likelihood of the infant and mothers encountering other conspecifics.

Thus the available evidence appears to moderately support the prediction that a solitary social structure combined with immobile or semimobile young would not favor the evolution of parent and infant attachment (see Table 1).

5.2. Family

A reproductive unit of a male paired with a single female occurs relatively infrequently throughout the Mammalia but is often found among carnivores, some primates, and ungulates (Kleiman, 1977a; Kleiman and Malcolm, this volume). Such families or monogamous units are typically dispersed from other such units. Thus they are essentially solitary for purposes of this discussion and include only their most recent offspring, in contrast to extended families (see below). The expectations for finding attachment among these units would be similar to those for finding attachment among the kinds of solitary social structures discussed above (see also Table 1).

The California deer mouse, *P. californicus,* is considered monogamous (Eisenberg, 1963; Kleiman, 1977a; McCabe and Blanchard, 1950), as is the prairie vole, *Microtus ochrogaster* (Thomas and Birney, 1979), and both produce altricial young. In captivity female and male *P. californicus* (Gubernick, unpublished observations) and *M. ochrogaster* (Thomas and Birney, 1979) will care for the young of others. Among the ungulates only the Suidae produce large litters of altricial young (Ewer, 1968). The wart hog, *Phacochoerus aethiopicus,* lives in family groups (Frädrich, 1974), and in captivity females occasionally share the same nest and nurse one other's young (Frädrich, 1965, cited in Lent, 1974).

As is the case of solitary species, there is a paucity of information about monogamous pairs with highly mobile young, since these also tend to be confined to "closed" habitats (Eisenberg and Lockhart, 1972; Estes, 1974). The possibility of finding parental and filial attachment in this group remains an open question. For example, among ungulates, Kirk's dik-dik *Madoqua kirki,* live as a more or less permanent pair and in a territory which both individuals maintain through olfactory marking (Hendrichs and Hendrichs, 1971). Only the male actively defends the pair's territory. The single precocial infant is evicted by one or both parents when it reaches maturity at 8–10 months or before the next birth (Hendrichs and Hendrichs, 1971). Other evidence suggests that African bovids of the Cephalophini (duikers) and Neotragini (e.g., dik-dik, klipspringer, steinbuck) tribes may also show a similar social structure (Estes, 1974; Leuthold, 1977). Although the relatively long association of the infant with its parents might favor attachment, it may be unnecessary due to the essentially solitary rearing of the young.

A similar argument can be made for the agouti, *Dasyprocta punctata* (Caviomorpha), which is considered monogamous and territorial and which usually gives birth to a single precocial infant (Kleiman, 1977a; Smythe, 1978). The infant remains in a separate nest for the first 3 weeks postpartum, during which time the female will not allow her mate near the nest (Smythe, 1978). The mother and infant continue to groom each other until the infant disperses at 4–5 months when a new infant is born. If a new litter is not born the infant may remain with its parents (Smythe, 1978). However, in captivity infants are reportedly bitten by conspecifics other than their own mothers (Roth-Kolar, 1957, cited in Spencer-Booth, 1970), indirectly suggesting the possibility of maternal attachment. In contrast, the related green acouchi, *Myoprocta pratti*, which is also considered monogamous and territorial and which bears precocial young, may accept alien young in captivity (Kleiman, 1972, 1977a), suggesting an absence of attachment.

Although the available evidence is scanty, there appears to be some support for the prediction that a family social structure combined with immobile young probably would not favor attachment. Obviously more research on family social structure is sorely needed.

5.3. Extended Family

An extended family consists of a single reproductive pair that does all of the breeding, the recent offspring, and those from previous breeding seasons (Eisenberg, 1966, 1977). Since the same male usually sires all the offspring, members of an extended family are highly related to each other and thus kin selection could favor cooperative care (Hamilton, 1964a,b; West Eberhard, 1975). Because only the founding pair breeds, the necessity for parent and infant attachment may be minimal. Although cooperative care may reflect the absence of parental or filial attachment, caution must be exercised in drawing such a conclusion. Cooperative care does not in itself indicate that an infant or its mother cannot distinguish each other from other individuals or exhibit a preference for one another (Spencer-Booth, 1970). Further, the possibility remains that such cooperative care reflects in part the existence of multiple attachments (see Section 3).

The basic social unit among the Canidae is considered to be a male-female pair. Packs such as found in wolves, cape hunting dogs, and dholes are probably extended families (Kleiman and Eisenberg, 1973). The wolf, *Canis lupis,* and cape hunting dog, *Lycaon pictus,* are similar in that the dominant male and female are typically the only ones to breed and the family members, which are related individuals making up several generations, help provision the mother and her young (van Lawick and van Lawick-Goodall, 1971; Mech, 1970; Rabb *et al.,* 1967; Schaller, 1972).

Coyotes, *Canis latrans,* are found in monogamous pairs and in packs presumed to be extended families (Camenzind, 1978). Coyote packs may defend a territory and have distinct social hierarchies in which the dominant male and female breed, although occasionally several females in a pack may give birth and nurse and provision one another's cubs (Andrews and Boggess, 1978; Camenzind, 1978; Gier, 1975).

Clans or packs may be considered large extended families in which more than one female breeds, as in the Hyaenidae. Brown hyenas, *Hyaena brunnea,* associate together in small clans (8–13) of males and females which defend a common territory and exhibit a distinct social hierarchy (Owens and Owens, 1978, 1979*b*). By 4 months of age, young are brought to a central communal den where, in contrast to spotted hyena (below), females nurse cubs other than their own, even ones of vastly different ages (Owens and Owens, 1979*a,b*). Young are nursed for at least 10 months, and clan females bring food to the den. Clans of brown hyenas may be based on extended families, but unfortunately kinship within these groups remains largely unknown.

The spotted hyena, *Crocuta crocuta,* associates in clans of 30–70 individuals in which the females are dominant over males and both sexes defend the clan's territory (Kruuk, 1972). Although the semimobile young are kept in communal dens, food is not brought to the den, young are not fed by regurgitating, and males do not assist in feeding the cubs (Kruuk, 1972), all in contrast to the pattern common among brown hyena (Owens and Owens, 1979*a,b*) and the social Canidae (Kleiman and Eisenberg, 1973; see above). A mother will nurse only her own young, although young may attempt to suckle from any mother (Kruuk, 1972), suggesting the presence at least of maternal attachment. Possibly clan members are not related (Kruuk, 1972), but this has not yet been confirmed. If clan members' average degree of relatedness is low, then the spotted hyena might be more appropriately considered in the multimale/multifemale social structure whose predictions regarding attachment would coincide with the findings for spotted hyena (see Table 1).

The dwarf mongoose, *Helogale parvula,* is reportedly found in an extended family group in which only the dominant pair breeds and all family members cooperate in caring for the young (Rasa, 1977; Rood, 1978). Within a family of banded mongoose, *Mungos mungo,* several females may give birth synchronously and raise and nurse one another's young in a communal den (Neal, 1970; Rood, 1974).

Among New World monkeys the marmosets and tamarins (Callitrichidae) presumably live in monogamous family groups in which only the founding pair breed and produce twins twice a year (Epple, 1975; Napier and Napier, 1967). All members of the family group help carry and care for the young (Dawson 1977; Ingram, 1977; cf. Kleinman, 1977*b*) Recent field studies indicate that these family groups may actually include unrelated subordinate members that have transferred from another group (Dawson, 1977; Neyman, 1977). Unfortunately there is no information as to the degree of relatedness between these immigrants and the family group members. Nor is it known whether parental or filial attachment is present.

The basic social unit of black-tailed prairie dogs, *Cynomys ludovicianus,* is the coterie (King, 1955; Tileston and Lechleitner, 1966). A coterie usually consists of one adult male, several females, yearlings, and young of the year, which are all probably closely related to each other. The coterie is a closed social group which defends a common territory against other coteries (King, 1955). Interactions between members of a coterie are frequent and amicable with no female dominance relations except during the rearing of young, when a female defends her nesting burrow prior to emergence of the young (King, 1955; Tileston and Lechleitner, 1966). The four to five young of a litter probably have no contact with other coterie members except their mother, prior to emergence. Pups are probably weaned by the time they first appear above ground. After emergence the pups will solicit care and attempt to nurse from other females, although no nursing occurs above ground (King, 1955; Tileston and Lechleitner, 1966). The pups are treated amicably by all members of the coterie, and the pups may sleep with other females. These results suggest that there are no attachments between a mother and her infants, although the young are likely to develop enduring associations to the coterie, as is the case among the adults.

In all but one of the examples cited above, cooperative care of the young is evident. However, in none of these cases is it clear that parental or infant attachment is absent (or present, for that matter). As mentioned earlier, caution must be exercised in drawing any such conclusions in the absence of direct evidence. The allegiance members have to their pack cannot be taken as *prima facie* evidence for attachment, since factors such as the necessity for cooperative hunting will select for continued association with others (Kleiman and Eisenberg, 1973; Bekoff, this volume).

5.4. Matriarchy

A matriarchy consists of a female, several daughters or sisters, and their offspring (Eisenberg, 1966, 1977). One or more adult males may be loosely associated with the matriarchy. Although the females are related, the average degree of relatedness among the females will vary since they may have different fathers (if they all had the same father it would essentially be an extended family). Thus we might expect a strategy of attachment to own offspring with some cooperative care of other young. This strategy should be more likely in cases where the young are immobile or remain in a fixed location (see Table 1).

Most members of the family Felidae are solitary; only the lion, *Panthera leo,* has an advanced social organization (Kleiman and Eisenberg, 1973). The basic social unit of lions consists of a mother and her daughters from previous litters and two or more adult males usually unrelated to the females of the pride (Schaller, 1972). All cubs are communally nursed, but a female sometimes gives her own young preference in suckling by withdrawing with her cubs from the pride or by chasing other young away (Schaller, 1972). An infant may run to its own mother when disturbed even if it was sitting near another adult (Rudnai, 1973). Such a mixed strategy of attachment and care for other young may have been favored in part by kin selection, since females in a lion pride are related on average a litter greater than full cousins (Bertram, 1976).

The coati *Nasua narica,* lives in bands of females and their young of the previous 2 years (Kaufmann, 1962; Smythe, 1970). The bands are loosely organized and frequently divide into smaller groups to forage during the day. The bands do not defend a territory, and a solitary male's home range may overlap with that of a band which he joins temporarily during the breeding season (Kaufmann, 1962). The band temporarily disintegrates when each pregnant female leaves to give birth to three to five young in a tree nest. A female defends her nest against her own young from previous years. When the infants are about 5 weeks old and highly mobile, the females and their young reform the band (Kaufmann, 1962). Adult females may nurse other young in the band until they are weaned at about 8 weeks of age (J. Russell, personal communication). The females in the band may be closely related, but this has not yet been established.

Parent and infant attachment is commonly found among presumed matriarchies with highly mobile young, as, for instance, in red deer, *Cervis elephas* (Darling, 1937) and white-tailed deer, *Odocoileus virginianus* (deVos *et al.,* 1967) (for additional examples see Eisenberg, 1966; Estes, 1974; Geist and Walther, 1974; Leuthold, 1977; Spencer-Booth, 1970). Females of the African elephant, *Loxodonta africana,* and Asiatic elephant, *Elephas maximus,* associate together in an extended matriarchy and give birth to precocial young (Douglas-Hamilton, 1973; Laws and Parker, 1968; McKay, 1973). However, contrary to expectations for a matriarchy with highly mobile young (see Table 1), infant elephants are allowed to nurse from any lactating female in the group. Filial attachment appears to be present since infants follow and remain close to their mothers, although they are in the company of other adult females (McKay, 1973). Perhaps the reasons for these findings are related to the extremely long life span (50–70 years), prolonged lactation (several years), and length of association between females (10–20 years) (Douglas-Hamilton, 1973; Eisenberg and Lockhart, 1972; Laws, 1974).

Thus the evidence for matriarchies appears to generally confirm the prediction for a mixed strategy of attachment and cooperative care when young are mobile and semimobile, and in even one case when the young are highly mobile.

5.5. Harem

A harem refers to the association of a male with a group of females. In contrast to the females in a matriarchy, the females of a harem are generally not related. The male is associated with the females during the rearing of the young (Crook *et al.*, 1976; Eisenberg, 1966). A harem may occur in relative isolation from other harems (e.g., equids) or within larger social groups containing other harems (e.g., some pinnipeds). In either case, selection should favor attachment if the females are unrelated with little or no cooperative care of the young. This is particularly true if the young are semimobile or mobile and possibly so if the young are immobile (see Table 1).

Pinnipeds are adapted to living and feeding in water but generally haul out onto land in large aggregations for short periods once a year to give birth to a single precocial infant and to mate (Matthews, 1971; Peterson, 1968). Although the young are precocial they are poorly adapted for locomotion on land and are relatively immobile. Among the Otariidae (earless seals and sea lions) males may defend a territory in which females aggregate in temporary harems. As a rule, among otariids mothers recognize and nurse only their own young, e.g., Alaskan fur seal, *Callorhinus ursinus* (Bartholomew, 1959; Bartholomew and Hoel, 1953), Cape fur seal, *Arctocephalus pusillus* (Peterson, 1968; Rand, 1955), California sea lion, *Zalophus californicus* (Peterson and Bartholomew, 1967), and Australian sea lion, *Neophoca cinerea* (Marlow, 1975), among others.

In many of the true seals (Phocidae) males from a dominance hierarchy and associate with groups of females in a temporary harem. Alien young are occasionally allowed to nurse, although most females will reject such attempts, e.g., Northern elephant seal, *Mirounga angustirostris* (Fogden, 1968; LeBoeuf *et al.*, 1972), harbor seal, *Phoca vitulina* (Bishop, 1967, cited in LeBoeuf *et al.*, 1972), and grey sea, *Halichoerus grypus* (Fogden, 1968, 1971; Smith, 1968). The Weddell seal, *Leptonychotes weddelli,* nurse only their own young (Tedman and Bryden, 1979) except under extremely crowded conditions, when alien young may occasionally be tolerated (Mansfield, 1958, cited in Tedman and Bryden, 1979).

Most investigators believe that the basis for maternal recognition of young is the vocalizations and odor of the pup. There has only been one clear demonstration that mothers can discriminate their own pup's vocalizations (Petrinovich, 1974). Most reports suggest that pups do not recognize and discriminate their own mother from others, yet there have been no systematic attempts to confirm this. Pinnipeds may then be one of the few groups wherein maternal attachment, but not filial attachment, occurs (see Table 1). Perhaps the combined factors of nursing sedentary young in large aggregations, short lactation periods, and high infant mortality prior to weaning (LeBoeuf and Briggs, 1977) favored the development of maternal attachment. These same factors may also have favored opportunistic nursing by infants.

Female hamadryas baboons, *Papio hamadryas,* associate with other females in one-male units and give birth to a semimobile infant (Kummer, 1968). One-male units are found in bands with other one-male units. However, these one-male harems rarely intermingle and a male mates only with females in his own unit, although females may copulate with young males from outside the unit. Within a harem females with infants interact very little with other mothers. There is no cooperative nursing or handling of infants among females, although the unit male may occasionally carry infants (Kummer, 1968). Maternal and filial attachment is present, since females will protect and retrieve their own young, and infants when disturbed or separated will seek their own mother (Kummer, 1968).

Harems are found in ungulates with highly mobile young such as the Camelidae and Equidae, and in particular the mountain and plains zebra, *Equus zebra* and *E. quagga* (Klingel, 1974), horses, and New Forest ponies, *E. caballus* (Klingel, 1974; Tyler, 1972; Waring *et al.*, 1975). In these equids both maternal and filial attachment is present and there is no cooperative nursing of the young (Klingel, 1974; Tyler, 1972; Waring *et al.*, 1975; Wolski *et al.*, 1980).

The evidence for harems supports the predictions that parental and filial attachment are favored when the young are semimobile or mobile and that the presence of immobile young favors parental attachment but not necessarily filial attachment (see Table 1).

5.6. Allied Females (Unrelated)

Rearing of young within a group of allied (unrelated) females and their young without the presence of an adult male occurs in a number of mammals. The expectations regarding attachment are the same as for a harem (see Table 1).

Bats typically give birth to only one infant, which is carried by the mother for some days postpartum (Bradbury, 1977). Groups of females and their young are often found together during the lactation period, and the young are usually left together in aggregations while their mothers are out foraging. Under such conditions it would be essential for a mother to recognize and nurse her own young upon return to the roost. This is the case for the little brown bat, *Myotis lucifugus* (Gould, 1971); large brown bat, *Eptesicus fuscus* (Davis *et al.*, 1968; Gould, 1971); serotine, *E. serotinus,* noctule, *Nyctalus noctula,* and pipistrelle, *Pipistrellus pipistrellus* (Kleiman, 1969); evening bat, *Nycticeius humeralis* (Jones, 1967); and pallid bat, *Antrozous pallidus* (Brown, 1976). In most cases there is no indication that infants recognize their own mothers except perhaps in the little brown bat (Turner *et al.*, 1972) and the pallid bat (Brown, 1976). In the two reported cases in which communal nursing occurs, females leave their infants in nursery groups of hundreds of thousands and each female upon return to the roost nurses any two infants (Bradbury, 1977; Davis *et al.*, 1962). Increased density of infants was also noted to induce indiscriminate nursing in some pinnipeds (Fogden, 1968; Smith, 1968).

The thick-tailed bushbaby, *Galago crassicaudatus,* is a nocturnal prosimian which nests in trees and is found in a loose association with other individuals that may also include daughters from previous litters (Clark, 1978; Doyle and Bearder, 1977). Adult females are agonistic towards young females that are not their daughters (Clark, 1978). Usually two semi-altricial young are born and remain in the nest for the first 21 days, by which time mothers recognize their own young

(Klopfer, 1970). Infants are then carried and nursed until about 10 weeks of age. Whether these galagos are considered as allied females, a matriarchy or a combination of both, the prediction is the same: that parental attachment and filial attachment are present (see Table 1).

In many ungulates groups of females and their precocial young associate together after birth (see Estes, 1974; Lent, 1974; Leuthold, 1977; Schaller, 1977, for various examples). In almost every reported case there is evidence for both mother and infant attachment and the absence of cooperative nursing, e.g., elk, *Cervus canadensis* (Altmann, 1963); reindeer, *Rangifer tarandus* L. (Espmark, 1971); barren-ground caribou, *R. tarandus groenlandicus* (Lent, 1966; deVos *et al.*, 1967); ibex, *Capra ibex,* and Himalayan tahr, *Hemitragus* (Schaller, 1977); *Ovis* sp. (Geist, 1971; Schaller, 1977); giraffe, *Giraffa camelopardalis* (Langman, 1977); bison, *Bison bison* (McHugh, 1958, 1972); water buffalo, *Bubalus bubalis* (Eisenberg and Lockhart, 1972); and impala, *Aepyceros melanopus* (Jarman, 1976), among others.

The evidence for allied (unrelated) females supports the predictions given in Table 1. Parental and filial attachment are favored when the young are semimobile or mobile. The presence of immobile young favors parental attachment but not necessarily filial attachment.

5.7. Multimale/Multifemale Groups

Multimale/multifemale groups are cohesive social structures wherein several or more males and females and their young are combined together as a unit. Such groups are found in several mammalian orders, most notably among primates, and are considered the most advanced social structure (Crook, 1970, 1977; Crook *et al.*, 1976; Eisenberg, 1966, 1977; Eisenberg *et al.*, 1972; Wilson, 1975). Such groups consist of related and unrelated members. As a consequence, kin selection is probably important in social interactions within such groups (Hamilton, 1964a,b; Wade, 1979; West Eberhard, 1975; Wilson, 1975).

Within multimale/multifemale groups paternity is often uncertain, while matrilineal kinship groups can be more easily determined (Hrdy, 1976; Kleiman and Malsolm, this volume). It is within matrilineal kinship networks that we might expect selection to favor cooperative care of the young. Such a matrilineal network might be considered as a matriarchy within a larger social group. Selection might then favor a mixed strategy of attachment and cooperative care within the matrilineage among species with immobile or semimobile young. Cooperative care may also occur among (presumably) unrelated members of a group, as a form of reciprocal altruism (Trivers, 1971). In either case selection would favor parent and infant attachment.

However, the requisite long-term genealogical information necessary to determine kinship networks is available for only a few species, and those are primates. Such information is available for the chimpanzee, *Pan troglodytes* (van

Lawick-Goodall, 1967, 1968, 1971), rhesus monkey, *Macaca mulatta* (Koford, 1963*a,b;* Sade, 1965), and Japanese macaque, *M. fuscata* (e.g., Itani, 1959; Kawamura, 1958; Yamada, 1963). In such matrifocal groups an infant usually remains associated with its mother after the birth of the next infant. In each of the above species it is usually the mother's kin, especially previous daughters, that help care for and protect the infant (see also Poirier, 1968, for Nilgiri langurs, *Presbytis johnii*). Although other group companions may hold, groom, or carry the infant, both the infant and its mother recognize and prefer each other, e.g., rhesus monkeys (Kaufmann, 1966; Rowell *et al.*, 1964).

Most members of a primate troop are attracted to a newborn and will attempt to touch or groom it. Primates vary in the extent to which infants are handled and cared for by others (Hrdy, 1976; McKenna, this volume). Infant care is provided by others in, for example, the North Indian langur, *Presbytis entellus* (Jay, 1963), Nilgiri langur, *P. johnii* (Poirier, 1968), black-and-white colobus, *Colobus guereza* (Wooldridge, 1971) [but not the red colobus, *C. badius* (Struhsaker, 1975)], vervets, *Cercopithecus aethiops* (Lancaster, 1971), squirrel monkeys, *Saimiri sciureus* (Baldwin, 1971; Rosenblum, 1968), the ring-tailed lemur, *Lemur catta* (Jolly, 1966), and the mountain gorilla, *Gorilla gorilla beringei* (Schaller, 1963), among others. In most of these cases it is the mother's "permissiveness" or lack of stable dominance relationships and not relatedness which may determine access to her infant (e.g., Lancaster, 1971; McKenna, this volume). In other cases, mothers are more restrictive and others provide limited if any care for the young, e.g., olive baboon, *Papio anubis* (DeVore, 1963), yellow baboon, *P. cynocephalus* (Altmann, 1978), pig-tailed monkey, *M. nemestrina* (Rosenblum, 1971), and bonnet monkey, *M. radiata* (Simonds, 1965). Most of these studies and others provide evidence that a mother and her infant recognize each other. Such evidence is in the form of an infant's orienting towards its mother when separated from her, remaining in proximity to her, when distressed running to its mother, or refusing to be transferred or held by another. The evidence for recognition by a mother of her own infant usually consists of protecting or retrieving the infant, responding to the infant's vocalizations, monitoring the infant when away from her, or suckling only her own infant. Thus the available evidence overwhelmingly supports the prediction of parent and infant attachment among species with multimale/multifemale social structures and semimobile young. Additionally, the likelihood of finding coooperative care under such conditions is also supported.

The information is scanty for multimale/multifemale groups with immobile young. Bats can be found in such groups, and for at least one such bat, *Rousettus aegypticus,* the mother recognizes and nurses only her own infant (Bradbury, 1977; Eisenberg, 1966).

The Macropodidae (kangaroos and wallabies) are considered the most socially advanced marsupials (Kaufman,, 1974*b*). The whiptail wallaby, *Macropus parryi,* lives in large groups of mixed ages and sexes, with a male dominance hierarchy and where the young are tolerated or ignored by adults (Kaufmann,

1974*a*). Although permanently out of the pouch by 9–10 months of age, the single young follows the mother but does not at that age appear to visually recognize her, since the infant will still attempt to enter any female's pouch (Kaufmann, 1974*a*). Such attempts, however, are rebuffed, suggesting that mothers can recognize their own young. In contrast, the red kangaroo, *Megaleia rufa,* has a loose, mixed-sex social organization (Kaufmann, 1974*b*) and a mother may sometimes foster alien young-at-foot (Sharman and Calaby, 1964), although usually a female rejects such attempts of alien young entering her pouch (Russel, 1970).

Large, mixed-sex herds are found among ungulates inhabiting "open" habitats, especially those within the subfamily Bovinae (Estes, 1974; Jarman, 1974). These species have highly precocial young. In all reported cases maternal and infant attachment are present while cooperative care is absent. Maternal attachment develops much more rapidly than filial attachment (see also Section 6). For example, female blue wildebeests, *Connochaetus taurinus,* are found in both mixed-sex and female nursery herds as either sedentary or migratory populations (Estes, 1974, 1976). A female gives birth in the herd to an extremely precocial infant, to which it forms a rapid attachment, nursing only its own young (Estes and Estes, 1979). The infant's attachment to its own mother may take several days. The African buffalo, *Syncerus caffer,* gives birth away from the herd, where the single infant remains briefly hidden. Later the infant follows its mother into the herd, where the female recognizes and nurses only her own infant (Sinclair, 1977). The bontebok, *Damaliscus dorcas,* gives birth in the herd and joins a female nursery group within the herd wherein females nurse only their own young (David, 1975).

Thus the evidence for multimale/multifemale groups strongly supports the prediction that both parental and filial attachment would be formed when the young are semimobile or mobile. Further, the likelihood of finding cooperative care of semimobile young and not mobile young is also supported.

5.8. Comment on a Scheme

Most of the examples presented in this section support the qualitative predictions about parent and infant attachment listed in Table 1. As such they support the assumption on which the predictions were made, namely, that the function of attachment is to enhance an individual's fitness by ensuring that parental care is provided one's own offspring and not those of another (see also Section 4). Allowance must be made in those cases where cooperative care of related individuals may improve one's inclusive fitness. More research is obviously needed to further test the generality of these predictions. The classification scheme presented above is based on only two parameters, social structure and mobility of young. The inclusion of other parameters would probably allow a more fine-grained analysis of the conditions favoring attachment.

Please note, however, that this scheme does not imply a causal relationship between these parameters and attachment. Additionally, there is the danger from such schemes of underestimating the complexity involved and the variability. And finally, this scheme is not intended to imply that attachment is a unitary phenomenon (see also Section 2). Attachment will have multiple determinants (e.g., ecological, physiological, hormonal), and although numerous species exhibit attachment, the underlying mechanisms and developmental timing of attachment may differ as discussed in the following section.

6. Mechanisms of Parent and Infant Attachment: Selected Examples

This section discusses in more detail some of the information available on parent and infant attachment in mammals. The discussion will be limited to certain species of rodents, primates, and ungulates. These groups were selected because there is more information about them than other groups and because they represent contrasting strategies of parent and infant attachment. These strategies range from the relative absence of attachment in some rodents to the rapid formation of attachment in ungulates (see Section 5). Attention is placed on the sensory basis and developmental timing of parent and infant attachment in each group.

6.1. Rodents

Rattus norvegicus, the brown or Norway rat, is predominantly nocturnal, lives in colonies, nests in burrows, and produces large litters of altricial young (Barnett, 1963). The domesticated ("laboratory") rat has been used extensively in studies of mother–infant relationships.

It is not known whether rats communally rear young (Gurski and Scott, in press); however, lactating females are often used as foster mothers for pups from their own and other species and will retrieve and show maternal behavior towards pups of other litters (Moltz, 1971). Mothers will retrieve their own and alien pups and in some cases retrieve their own young faster than alien young (Beach and Jaynes, 1956). In other cases, mothers may retrieve foster pups sooner than their own young which were raised together with the foster pups (Misanin *et at,* 1977). Thus, based on behavioral evidence rat mothers do not appear to form a specific attachment to their own young. Furthermore, changes in mothers' plasma corticosteriod levels (reflecting arousal) following separation from their pups were not specific to the removal of the mothers' pups, although mothers remained responsive to pup emitted cues (Smotherman *et al.,* 1977a,b).

Do rat offspring form an attachment to their mother? The altricial rat pup becomes attracted to (approaches) chemical cues (soiled bedding) associated with its mother at about 12–14 days postpartum and prefers these odors to those from virgin females (Leon and Moltz, 1971). However, rat pups showed no preference for odors of their own mother over those of another lactating female (Leon and Moltz, 1971). Pups continue to respond to these odors emitted by lactating females until 27 days postpartum (Leon and Moltz, 1972). The initiation and terminaion of the young's attraction towards these "maternal pheromones" is synchronized with the development and waning of production of the pheromones by the mother (Leon and Moltz, 1972). The pheromone is produced by bacterial action in the cecum of lactating females and is emitted in anal excreta as cecotrophe (Leon, 1974).

Of further interest here is that maternal diet may influence the nature of the pheromone. Pups raised with mothers on one diet preferred the odor of anal excreta of other mothers raised on the same diet as their mother to odors from mothers raised on another diet (Leon, 1975).

Additionally, mothers may "mark" their young through the maternal pheromone. Pups are attracted to odors from their own litter and prefer these odors to odors from litters whose mothers were inhibited from producing maternal pheromones (Leon, 1974). Again, however, pups were equally attracted to odors of litters from other mothers producing caecotrophe. The mechanism of transmission of these marks appears to be through the young's ingesting the anal excreta of their mothers (Leon, 1974), although ingesting mother's milk and eating the same foods as the mother may also be involved (Leon, 1978; see also labeling in goats, below).

The maintenance of maternal behavior seems to be dependent on stimulation from the young (Rosenblatt and Siegel, this volume). As this responsiveness declines, the young modify their behavior in response to changes in the mother (Galef, this volume). In particular the young are attracted to a pheromone emitted by their mother, which likely functions among other things to help keep the young in the proximity of the mother and nest site at a time when the young are mobile but not yet weaned.

These studies of maternal pheromones provide an excellent illustration of the mutual adaptations of parents and infants without the necessity of either one forming an attachment (see Galef, this volume, for other such illustrations).

However, perhaps a word of caution is warranted here, since a recent study suggests that another response measure may be useful for investigating attachment in rodents. Rat pups emit ultrasounds (which are detected by adults), and during the 1st week postpartum these vocalizations were highest when pups were exposed to odors from an unfamiliar lactating female and were suppressed when presented with odors from their own home cage, an unfamiliar adult male, or a virgin female (Conely and Bell, 1978). At this stage these results are difficult to interpret but indicate that studying ultrasonic vocalizations in rodents as a means of assessing infant and maternal attachment may lead to some unsuspected discoveries.

House mice, *Mus musculus,* produce altricial offspring and may show communal nesting and nursing of young in both wild and laboratory populations

(Crowcroft and Rowe, 1963; Sayler and Salmon, 1971). As is the case for rats, pups are attracted to odors of their mother but at 18 days of age show no preference for maternal odors over odors of an unfamiliar lactating female, suggesting that mice also produce a maternal pheromone (Breen and Leshner, 1977). Although these results seem to indicate that neither maternal nor infant attachment is present in house mice, there is some evidence that laboratory mice might show attachment to their mother. Mice raised with their mother for 20 days were tested at 22 days in an open field where their mother and three unfamiliar lactating females were located (Gurski, 1977). All were free to interact. Measures of proximity and following indicated that infants were more often closer to their own mother than to the other females and followed their own mother more often than they followed other females. However, it is not clear from this report whether an infant remained near its mother because it recognized her or because she was the only one who did not avoid or attack the pup (see Section 2.1). Perhaps these results could also be construed to indicate that the mothers recognized their own young. Thus the possibility that infant mice form attachments must await further confirmation.

The spiny mouse, *Acomys cahirinus,* is unique among the murid rodents (i.e., rats and mice) in giving birth to precocial young. Newborn young 2–12 hr old show a rapid olfactory imprinting, since a brief, 1-hr exposure to a chemical cue induced a preference for that artificial odor 24 hr later (Porter and Etscorn, 1974). One-day-old pups show a preference for bedding soiled by their mother over bedding soiled by a nonpregnant female, but they show no preference for bedding soiled by their mother over that of another lactating female (Porter and Doane, 1976; Porter and Ruttle, 1975), This suggests the presence of a maternal pheromone as in rats and house mice (see above). By 25 days of age, the approximate age of weaning, pups are no longer attracted to the maternal pheromone (Porter *et al.,* 1978).

Further, 1- and 3-day old pups preferred chemical cues from a conspecific lactating female fed on the same diet as their own mother over one fed on a different diet (Doane and Porter, 1978; Porter and Doane, 1977). Pups even preferred cues from lactating females of a different species *(M. musculus)* fed on the same diet as their own mother over those of a conspecific lactating female fed on a different diet (Porter and Doane, 1977). Thus mothers produce pup-attractant odors, and the pup's preferences for such cues are diet-dependent, as is the case for rats (see above).

Of further interest is the finding that spiny-mouse mothers presented with two strange pups, one whose mother was raised on the same diet as the test mother and the other on a novel diet, retrieved both pups but retrieved faster the pups on the same diet, thus indicating that chemical cues associated with the pup may also depend upon maternal diet. (Doane and Porter, 1978). Additionally, mothers tested up to 4 days postpartum retrieved their own young as rapidly as they retrieved an alien pup (presumably on the same diet) (Porter and Doane, 1978). Females housed together in pairs after giving birth nursed their own and alien young equally often, with dominant females nursing them more frequently than did subordinate females (Porter and Doane, 1978).

These results, taken together, suggest that mothers do not recognize and prefer their own young and that infants do not recognize and prefer their own mother. Such specificity may not be necessary under natural conditions, since the attractiveness of the chemical cues (which are diet-dependent) associated with the mother and nest site may help ensure that the young remain in the proximity of the nest site. Further, since mothers are not likely to eat the exact same foods, the cues associated with the mother and her young would likely be different from those of other mothers. Unfortunately, there are no field studies of the spiny mouse.

Chemical cues are only one potential basis for individual recognition. Perhaps investigation of other sensory modalities and cues (e.g., audition and vocalizations) may reveal specificity of maternal and infant recognition in spiny mice.

At this stage of our knowledge the evidence indicates that rats, house mice, and spiny mice do not form filial or parental attachments.

Another rodent that has received wide attention is the domestic guinea pig, *Cavia porcellus*, which also produces precocial young. Wild guinea pigs are diurnal surface dwellers that feed together in groups (Rood, 1972). The infants may form attachments to animate and inanimate objects within the 1st week postpartum (Beauchamp and Hess, 1971; Harper, 1970; Sluckin, 1968; Sluckin and Fullerton, 1969) and also show preferences for odors they were exposed to during the first 3 days after birth (Carter and Marr, 1970).

However, when given a choice between their own mother and an unfamiliar lactating female in a two-choice discrimination task, infants from 1 to 23 days of age showed no preference as measured by the amount of time spent in proximity to each (Porter *et al.*, 1973*b*). These infants could see, hear, and smell the mothers but had no physical contact with them. The authors suggested that proximal cues may be necessary for infants to recognize their own mother or that the lack of a preference was due to infant guinea pigs' being attracted to the unfamiliar female in the presence of the mother, since the mother may act as a secure base for exploration of the environment (Porter *et al.*, 1973*a*).

In contrast, mothers tested in a similar manner 36–48 hr postpartum preferred their own litter to a similar-aged but unfamiliar litter, and in later tests also preferred artificial odors that had been associated with their own litter (Porter *et al.*, 1973*b*). Thus, although infant guinea pigs gave no indication of being attached to their mother, mothers formed an attachment within 48 hr of giving birth. However, this maternal attachment does not seem to be very stable, since under seminatural conditions mothers living in large groups may indiscriminately nurse other young (King, 1956; Rood, 1972). Even when only two mothers and their litters are housed together for the first 12 days postpartum mothers will nurse the other infants, although their own young may be nursed more frequently (Fullerton *et al.*, 1974).

Thus it appears that mothers can recognize their own young but form a "loose" attachment to them (Fullerton *et al.*, 1974), while the evidence from these studies for filial attachment is less convincing.

The presence of filial attachment in infant guinea pigs was investigated in two recent studies (Pettijohn, 1979*a,b*). Infant guinea pigs raised with both parents were given a two-choice discrimination test between their mother and father (a familiar social stimulus, thereby eliminating the problem of fear or of exploration of a novel social stimulus; see Section 2.1). Infants had auditory, olfactory, and visual but not physical contact with the parents during the test. Infants at 2 and 4 weeks of age preferred their mother as measured by the amount of time in proximity to each (Pettijohn, 1979*b*). It is during this age span that mothers are still nursing and caring for their young (Pettijohn, 1978) and that mothers, but not fathers, are most responsive to infant vocalizations (Pettijohn, 1977). At 6 weeks of age infants spent similar amounts of time in proximity to each parent during the test. By this time infants are already weaned and are interacting more with the father in the home cage (Pettijohn, 1978).

In another study Pettijohn (1979*a*) investigated the effectiveness of various social stimuli (i.e., mother, father, sib, and an unfamiliar adult female) on reducing separation-induced distress vocalizations of infant guinea pigs. Distress vocalizations of infant guinea pigs upon being separated from their familiar surroundings and being placed alone in an unfamiliar area are highest during the first 4 weeks postpartum and decline thereafter. All the social stimuli reduced such infant distress calls during the first 4 weeks of life. However, the most effective stimulus in reducing distress vocalizations was the presence of the mother, followed by the father, sib, and, least effective of all, the unfamiliar adult female. There is the possibility, however, that since infants in both studies could interact with the social stimuli that the infants' preferences or distress vocalizations could have been affected by the reactions of the others rather than indicating infant attachment (see Section 2.1 above).

It is interesting to note that intrasexual aggression among female domestic guinea pigs, *C. porcellus,* is minimal even to strange females and that females will nurse other young, while among their wild counterpart, *C. aperea,* females are highly aggressive to other females and nurse only their own young (Rood, 1972). Perhaps domestication has had some effect, since F₁ hybrids *(C. aperea* × *C. porcellus)* have unstable female dominance relationships and also nurse indiscriminately (Rood, 1972).

It would be of some interest to use distress vocalizations and discrimination tasks to assess attachment among communally reared domestic and wild guinea pigs. There have yet to be any systematic investigations of the sensory basis (e.g., olfaction, audition, vision) for parent and infant attachment in guinea pigs.

Table 2. Summary of the Sensory Basis and Developmental Timing of Parent and Infant Attachment in Selected Species of Mammals

	Who forms an attachment?		Sensory basis of attachment	Attachment develops by (time postpartum)
Rodents				
Rat[a]				
House mouse[a]	No attachment			
Spiny mouse[b]				
Guinea pig[b]	Parent	Perhaps	?	36–48 hr
	Infant	(Perhaps)	?	?
Primates[c]				
Rhesus monkey	Parent	Yes	?	?
	Infant	Yes	Vision	60 days
Pig-tailed monkey	Parent	Yes	?	3–17 days
	Infant	Yes	Vision	90 days
Bonnet monkey	Parent	Yes	?	?
	Infant	Yes	Vision	90 days
Squirrel monkey	Parent	Yes	?	?
	Infant	Yes	Olfaction; vision	30 days
Galago sp.	Parent	Yes	Olfaction? Audition?	7 days
	Infant	?	?	?
Ungulates				
Sheep[b]	Parent	Yes	Olfaction	30 min
	Infant	Yes	Vision; audition	> 8 days
Goats[b]	Parent	Yes	Olfaction	< 4 hr
			Audition	4 days
	Infant	Yes	Vision? Audition?	4 days
Pigs[a]	Parent	Perhaps	Olfaction	> 2 days
	Infant	No		

[a]Altricial young.
[b]Precocial young.
[c]Semialtricial young.

6.1.1. Rodent Summary

Parental and infant attachment is absent in rats, house mice, and spiny mice. Although there is some evidence of maternal recognition in domestic guinea pigs, there is also indiscriminate nursing. These results are summarized in Table 2.

Rats and house mice give birth to altricial young in nests hidden away from other conspecifics. Neither mothers nor fathers are likely to encounter other young prior to the youngs' emergence from the nest. There may then be little need for individual recognition as long as the young remain near the nest, as a result, for example, of their attraction to a maternal pheromone, and as long as the mother can readily locate her own nest site (see Section 5).

Precocial infant guinea pigs and their mothers have early contact with other infants and adults, and here individual recognition would be expected to be more important especially if there was no reciprocity among mothers for care of the young. Such is the case among wild guinea pigs. However, female domestic guinea pigs do provide care for other young and yet show some indication of recognition of their own young. Perhaps part of the difference between the wild and domestic forms is accounted for by the extent of intrasexual female aggression, since such aggression (as in the wild form) would effectively prevent females from contacting other females and their young. The question then becomes one of accounting for female dominance hierarchies (McKenna, this volume).

6.2. Primates

Some primates give birth in a nest where the young remain until they are mobile [e.g., galago species (Sauer, 1967)]. Most primates, however, do not build nests but carry their young with them, as is the case for the monkeys discussed below. Although most infants are furred and have their eyes opened at birth, their locomotor capability is limited and they depend upon clinging to the mother for transport (Ewer, 1973a). The monkeys discussed below are found in multimale/ multifemale social groups (see Section 5.7).

The rhesus monkey, *M. mulatta,* has been used extensively to study mother–infant relationships and the development of social behavior. However, other than Harlow's early and classic studies on surrogate-reared infant rhesus there has been surprisingly little research on the sensory basis or perceptual aspects of infant attachment in this species.

Infants reared with both cloth-covered and wire surrogates rapidly came to spend more time on the cloth surrogate regardless of which one provided nourishment (Harlow, 1958), thus demonstrating that "contact–comfort" and not food was overwhelmingly important in maintaining the infants' physical closeness to its surrogate "mother." When presented with a fear-provoking stimulus in the home cage, infants would cling tenaciously to the cloth surrogate even when the wire surrogate which supplied food was available to them (Harlow, 1958). The same occurred in open-field tests (Harlow, 1958; Harlow and Zimmerman, 1959). In the presence of the cloth surrogate, infants would explore the fear-inducing stimuli, thereby indicating that the cloth mothers also provided some "emotional reassurance" (Harlow, 1958; Harlow and Zimmerman, 1959).

Food was of some importance in maintaining contact, since infants from 10 to 100 days of age spent more time on the cloth surrogate providing them with milk than on the cloth surrogate that did not give milk (Harlow and Suomi, 1970; Harlow and Zimmerman, 1959).

By 40–50 days of age, infants raised with both cloth and wire surrogates preferred to watch through a window a cloth surrogate or another infant more than a wire surrogate or an empty box (Harlow, 1958), indicating that infants can at least visually discriminate one type of surrogate from another. By 90 days of age infants displayed fear responses to a new face put on their surrogate (Harlow and Suomi, 1970). Additionally, infants spend more time on a rocking surrogate than on a stationary one (Harlow and Suomi, 1970). Further, infants show preference for warm surrogates over cold ones (Harlow and Suomi, 1970; Harlow and Zimmerman, 1959).

These studies clearly demonstrate that there are several stimulus variables, e.g., contact–comfort, food, warmth, movement, that are important for maintaining the infant rhesus monkey's proximity to its surrogate "mother." Additionally, contact and to a lesser extent sight of the surrogate reduced emotional distress. Perhaps odor should also be added to this list. Cloth surrogates probably accumulate the odors of the infant, while wire surrogates are less likely to do so. Perhaps the familiar odor was partly responsible for the maintenance of contact with the cloth surrogate (see squirrel monkeys, below).

The possibility remains, however, that these infants had not developed an attachment to one specific surrogate (e.g., own cloth) over another (e.g., wire) but rather were responding to some general class of stimuli provided by each. If an infant prefers to cling to a cloth surrogate in an open-field test when a familiar wire surrogate is also available, this only indicates that the infant can discriminate cloth from wire. The infant may or may not have previously developed an attachment to that surrogate even though the presence of the cloth surrogate reduces emotional distress (see Section 3 for further elaboration). However, if the infant clings equally to any cloth surrogate that was presented, that is, shows no preference, then the infant has not likely formed an attachment but is responding to some class of stimuli provided by both surrogates. Should the infant prefer its own cloth surrogate over another (e.g., on the basis of odor or visual cues) then the infant can be said to have developed an attachment to that surrogate. Such a choice, however, was not provided. The above evidence indicates that infants can at an early age discriminate between various types of surrogates.

Infant rhesus may visually discriminate familiar from unfamiliar surrogates by approximately 2 months of age (Mason et al., 1974). Infant rhesus emit distress vocalizations when placed along into an unfamiliar or familiar environment. During the 1st month of life distress vocalizations were suppressed when infants could see and clasp their own surrogate or an unfamiliar surrogate, but being able to see and cling to the surrogates was more effective than being able to only see the surrogates (Mason et al., 1974). At this age both surrogates were equally effective in reducing distress vocalizations. This confirms Harlow's earlier results (above) that contact and to a lesser extent visual cues could reduce emotional distress. By 2 months of age infants would readily cling to both surrogates, but the familiar surrogate was more effective in suppressing distress vocalizations and was also more effective (as measured by heart rates) than the unfamiliar surrogate when only visual access was provided.

Rhesus monkey infants 1–4 months of age and reared in a nuclear family prefer their mother and father to both familiar and unfamiliar adult females and males and prefer their mother to their father (Suomi *et al.,* 1973), but the basis for the discrimination was not determined.

The evidence suggests that filial attachment in rhesus monkeys can develop by 2 months of age, at least on the basis of static cues. Infants probably could develop an attachment to a real mother before 2 months [e.g., less than 20 days of age (Harlow and Harlow, 1965)], although there has been no experimental verification of this. What the sensory basis is for maternal attachment in rhesus monkeys or when maternal attachment develops has yet to be determined.

Filial attachment has also been studied in two other macaque species, the bonnet monkey *(M. radiata)* and pig-tailed monkey *(M. nemestrina).* Bonnet adult females tend to remain physically close to each other, often in huddles, while pigtail adults do not usually make such physical contact (Rosenblum and Kaufman, 1967). Soon after giving birth the bonnet mother returns to close contact with her group and allows other females to handle and groom her infant, whereas the pigtail mother remains apart from the group and guards her infant from contact with others (Rosenblum, 1971). In general, bonnet mothers are less restrictive, less punitive, and less likely to retrieve their infant than pigtail mothers. As a result, infant bonnet monkeys have earlier and more sustained contact with other females (who may act as "aunts") than does the pigtail infant. They are also less disturbed by maternal separation than are pigtail infants (Kaufman and Rosenblum, 1967*a,b*; Rosenblum, 1971; Swartz and Rosenblum, this volume).

Group-raised infant bonnet and pig-tail monkeys show a clear perference by 3 months of age for their mothers compared with a strange female when given only visual access to them through a one-way mirror (Rosenblum and Alpert, 1974, 1977). From 3 to 6 months to 1 year of age the preference of the bonnet infants was stronger than that of pigtail infants. This species difference has been attributed to the differences in maternal care in the earlier and frequent contact with others as in the bonnet may result in earlier preferences for mother and wariness of strangers (Rosenblum and Alpert, 1977). In support of this are findings that bonnet infants reared only with their mother showed no preferences during their 1st year of life for their mother or a stranger (Rosenblum and Alpert, 1974, 1977). A similar result has also been found with squirrel monkey infants (Kaplan and Schusterman, 1972). Thus infant bonnet and pig-tailed monkeys develop an attachment to their mothers by 3 months of age on the basis of visual cues. However, such filial attachment may develop earlier since other sensory modalities likely influence filial attachment, but these have not yet been studied.

Maternal attachment has also been studied in pig-tailed monkeys. In one study, two females tested at 5–7 months postpartum with either their own or a strange infant acted maternally only towards their own infant (Jensen and Tolman, 1962). However, since a mother and infant were free to interact with each other, the relative role each played in recognizing the other cannot be ascertained from this study, nor the basis on which discriminations were made. There is some

indication that within 3–17 days postpartum pig-tailed monkey mothers can visually discriminate their own infant from a similarly aged infant (Jensen, 1965).

Pigtail mothers deprived of postpartum contact with their own infants (i.e.,
infant immediately removed at birth) preferred their own infants to strange infants
at 1 day postpartum (Wu and Sackett, personal communication). At 3 and 7 days
postpartum mothers preferred the younger of two infants whether or not it was
their own infant. Nonpregnant females had no preference for either younger or
older infants. Since at 1 day postpartum a mother's own infant was also usually
the younger infant it is not possible to conclude at this stage that mothers had
indeed recognized their own infants, although this possibility cannot be ruled out
(Wu and Sackett, personal communication).

While filial attachment appears to develop in bonnet and pig-tailed monkeys
by 3 months of age, if not sooner, maternal attachment in pig-tailed monkeys may
occur earlier than this. Further research is needed to determine the developmental
timing and sensory basis of both filial and maternal attachment in these two
species.

To date, the perceptual factors involved in the development of filial attachment have been best studied in the New World squirrel monkey, *S. sciureus*.
Infants reared on different colored surrogates, which also accumulated the infant's
odor, were presented with surrogates differing in either odor or color (Kaplan and
Russell, 1974). Infants tested at 4 weeks of age had a clear preference for surrogates containing the infant's own familiar odor to those containing an unfamiliar
infant's odor or no odor, but had no preference for a familiar colored surrogate
over an unfamiliar one. Thus by 4 weeks of age infant squirrel monkeys can recognize familiar odor cues but seem less responsive to static visual cues.

Do normally reared infants recognize their own mother on the basis of olfaction? Infants at 8 weeks of age can discriminate their mother from an unfamilar
lactating female on the basis of odor when given free access to both mothers
(Kaplan, 1977; Kaplan *et al.*, 1977). Mothers were anesthetized (thereby eliminating all behavioral cues) and were either washed (thereby removing odor cues)
or left unwashed and additionally had their heads covered (removing static visual
cues) or uncovered. Whenever the odor cues were available, infants preferred their
own mother, but they showed no preference when odor cues were removed. Visual
cues did not enhance the infants' discriminations. Thus infant squirrel monkeys
at 8 weeks of age can recognize and prefer the natural odor of their own mother.
They may be able to do so even earlier, since as mentioned above, surrogate-reared
infants can discriminate odors at 4 weeks of age. Note that infants at 8 weeks of
age spend 60% of the time riding on their mother's back (Rosenblum, 1968).
Therefore, washing the mother probably also removed the infant's odors. It would
be interesting to know to what extent infants were responding to the absence of
their own odor.

Although static visual cues did not enhance infant squirrel monkeys' recognition of their own mothers (Kaplan *et al.*, 1977), infants at 4–6 weeks of age can

discriminate at a distance between their awake mother and another mother (Rosenblum, 1968). However, since the infants and mothers could interact with one other it is not clear whether the infant recognized its own mother or whether the mother recognized the infant. However, when mothers are presented behind one-way mirrors and auditory and olfactory cues are greatly reduced, 4-week-old infant squirrel monkeys show a preference for their own mothers (Redican and Kaplan, 1978).

In summary, infant squirrel monkeys, by 4–8 weeks of age, discriminate and prefer their own mother on the basis of olfaction and vision.

Although it is not clear yet how early or on what basis maternal attachment develops in squirrel monkeys, mothers can recognize their own infant's vocalizations at 1–7 months postpartum (Kaplan *et al.*, 1978).

A factor that likely affects the development and sensory basis of attachment in all the monkeys discussed above is the amount and type of contact between mothers and infants. For example, a squirrel monkey infant almost immediately after birth climbs onto its mother's back, where it will spend 85–90% of its time during the first 4 weeks postpartum (Rosenblum, 1968). A squirrel monkey mother only minimally handles or explores her infant.

In contrast, infant and mother macaques (i.e., rhesus, bonnet, and pig-tailed) maintain close ventral–ventral orientation (face-to-face) and contact for long periods (Hinde and Spencer-Booth, 1967; Rosenblum, 1968, 1971). Additionally, a macaque infant is often handled, groomed, and explored by its mother.

Because of the close contact infants have with their mothers we would expect that olfaction would provide an early basis for filial attachment. However, visual cues might be initially more important to a macaque infant and its mother (because of their face-to-face orientation) than to a squirrel monkey infant and its mother. Alternatively, auditory cues may initially be more important for maternal attachment in a squirrel monkey because of its infant's "piggyback" orientation.

The relatively few studies of the mechanisms of maternal and filial attachment in primates have been of New and Old World monkeys, while only one such study is available for a prosimian. In a two-choice simultaneous discrimination task, female galagos (bushbabies), which are nocturnal prosimian primates, recognize and prefer their own infants at 1 week postpartum, presumably by olfaction and audition (Klopfer, 1970). We do not yet know whether other primates use olfaction as a basis for maternal attachment, although other mammals do so (see below).

6.2.1. Primate Summary

Most of the research on the primates cited above has been devoted to studying the development of filial attachment, while our knowledge of parental attachment is relatively meager. Filial attachment is present in most of those species and has developed at least by 1–3 months postpartum on the basis of olfaction or vision (see Table 2). Filial attachment may occur earlier, but this has yet to be demonstrated. The evidence of maternal attachment is very limited but suggests that mothers may develop an early attachment to their infants. There appear to be no systematic studies on the development and perceptual basis of maternal attachment in primates.

6.3. Ungulates

The most intensive investigations of maternal attachment have been those of domestic sheep and goats; each are found in allied female groups when rearing young (see Section 5.6).

Domestic sheep, *Ovis aries,* like many of the ungulates living in open habitats, are found in herds and give birth to one or two precocial young. The preparturient ewe usually removes herself from the flock and gives birth (Geist, 1971; Smith, 1965). Mothers will nurse only their own young and reject alien lambs. There appears to be a sensitive period following parturition during which ewes will be responsive to all lambs, since ewes separated from their lambs at birth will still show maternal behavior towards any lamb within 4–8 hr postpartum (Poindron *et al.,* 1979; Smith *et al.,* 1966).

However, after 20–30 min of contact with her own lamb a ewe will reject alien lambs (Poindron *et al.,* 1979; Smith *et al.,* 1966). This maternal attachment is based on olfaction, since ewes made anosmic prior to parturition would accept both their own and alien lambs, while intact ewes would accept their own lamb but reject alien young (Baldwin and Shillito, 1974; Bouissou, 1968; Rosenblatt and Siegel, this volume).

From 1 to 4 days postpartum, ewes can apparently show some indication of visually recognizing their own young at a distance, while auditory recognition is minimal (Alexander, 1977; Shillito and Alexander, 1975). With increasing age of the young, ewes can recognize their lambs from a distance based on both the appearance and vocalizations of the lamb (Alexander, 1977; Lindsay and Fletcher, 1968). However, when in close contact mothers continue to use olfaction as the primary basis for identification (Morgan *et al.,* 1975).

A lamb at 5 days of age may recognize its own mother on the basis of olfaction (Poindron, 1976), but does not seem to recognize her at a distance (Alexander, 1977). However, lambs more than 8 days old can recognize their own mother at a distance on the basis of the mother's appearance and vocalizations (Arnold *et al.,* 1975; Shillito, 1975) with visual cues becoming more important relative to auditory cues as the lambs get older (Alexander, 1977; Poindron and Carrick, 1976).

The domestic goat, *Capra hircus,* is a member of the family Bovidae (as are sheep). Prior to parturition the doe withdraws from the herd, gives birth to one or two precocial kids, and leaves her kids hidden for several days (Rudge, 1970). Although goat kids will initially attempt to approach and nurse from any mother, a mother will accept only her own kid (Collias, 1956; Hersher *et al.*, 1963; Rudge, 1970).

Apparently, only 5–10 min of contact with at least one of her kids immediately after birth is sufficient to establish a maternal bond (Hersher *et al.*, 1958), since a mother will then accept all her kids of that litter even after a 3-hr separation but would reject alien young (Klopfer *et al.*, 1964). Presumably, all the kids of her litter share some litter specific cues for recognition (Klopfer, 1971). The absence of postpartum contact with any young leads to the waning of maternal responsiveness and the rejection of all young even after only a 1-hr separation (Collias, 1956; Klopfer *et al.*, 1964), which is a shorter sensitive period of maternal responsiveness than for sheep. Chemoreception has been implicated in the establishment of maternal attachment, since olfactory (and probably gustatory) impairment at parturition appears to reduce subsequent own–alien-young discriminations (Klopfer and Gamble, 1966). Vision and audition are apparently not involved (Klopfer, 1971). Thus maternal attachment in goats seems to be specific, rapidly formed, and fairly stable, perhaps somewhat analogous to imprinting in birds (Hess, 1973). In fact, such rapid acquisition of maternal attachment is considered one of the few and most clear-cut examples of imprinting in a mammal (Hess, 1973).

Recent findings, however, suggest that maternal attachment in goats may not occur as rapidly as previously reported. Mothers given 5 min postpartum contact with their own kid generally failed to later discriminate between their own and alien young after a 1-hr separation (Gubernick *et al.*, 1979). Alien kids were accepted if they had had no contact with their own mother, but were rejected if they had been kept with their own mother for more than 24 hr (Fig. 1).

We suggested that mothers in contact with their kids "label" them and that such labeled kids are then rejected by other mothers [alien kids used in earlier studies had been kept with their own mother, e.g., Klopfer *et al.*, (1964)]. We further hypothesized that a mother may label her kid directly through licking it

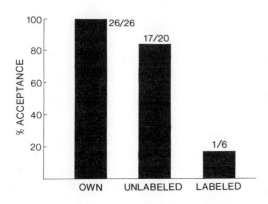

Figure 1. Percentage of mothers accepting unlabeled and labeled kids (data from Gubernich *et al.*, 1979; criteria of acceptance is ≥ 2 min of licking or nursing; only data from mothers accepting their own kid are presented. Numbers are the ratio of mothers accepting to mothers tested).

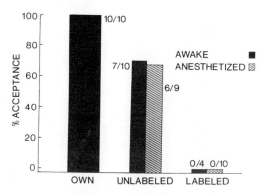

Figure 2. Percentage of mothers accepting unlabeled and labeled kids (data from Gubernick, 1980).

or indirectly through the kid's milk intake. Licking may transfer rumen micro-fauna to the kid's body surface, while ingestion and digestion of milk and subse-quent defecation may influence mouth, body, and anal odors, respectively. Such a labeled kid may then be recognized and accepted by its own mother but be rejected by another mother, while any unlabeled kid might still be acceptable to a mother within a few hours after parturition (see above for rodents "marking" their young).

Neither age alone nor behavior of the alien kids was a factor in acceptance or rejection, since unlabeled alien kids were accepted whether they were anesthe-tized (thus eliminating behavior) or were older (2 days old) and awake, while labeled kids were rejected under both conditions (Gubernick, 1980; see Fig. 2).

Additional experiments were performed to determine whether licking and/ or milk intake provide a label (Gubernick, unpublished results). Mothers were given 5 min of contact with their own kid immediately postpartum. After a 1-hr separation, mothers were presented with their own kid and alien kids in a series of 10-min acceptance–rejection tests. Kids were presented individually and in ran-dom order. All the alien kids had been kept with their own mothers. Some, how-ever, were only licked by their mother, others were nursed on mother's milk but were not licked, and others were both licked and nursed on mother's milk (labeled aliens). Since those kids were kept with their mothers it was possible that mere contact with the mother (and not licking or mother's milk) could provide a label. Therefore, alien kids were included that had no contact with their mother but were bottle-fed on mother's milk ("milk isolate" alien kids). Unlabeled aliens had no contact with other goats and were fed skim milk.

The results, depicted in Fig. 3, indicate that both licking and mother's milk alone are sufficient to provide a label and that mere contact between mother and young does not ("milk isolate" results). The low acceptance of labeled alien kids (those both licked and nursed by their mother) compared with lick-only and milk-

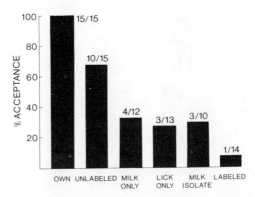

Figure 3. Percentage of mothers accepting labeled and unlabeled kids (Gubernick, unpublished data).

only alien kids suggests that the combination of licking and milk intake provides a more potent label. Some of our other data suggest that a label develops within 4 hr of contact between a mother and her kid.

Interestingly, feral goats tend to hide their kids for the first 2–3 days postpartum, periodically returning to them during that time (Rudge, 1970). This would probably assure among other things that the mother learns to recognize her young possibly through labeling them. Further, after kids are 4 days old, mothers may recognize their own kids' vocalization, since prior to this time kids' vocalizations are acoustically similar (Lenhardt, 1977). Kids at this age may recognize their mother at a distance, probably on the basis of the mother's vocalization (Gubernick, unpublished data; Rudge, 1970). Thus by the time feral goat kids join and follow their mother in the herd they already recognize each other. With increasing age of the young, other cues probably become important for the mother in recognizing her kids, especially from a distance. It has not yet been established at what age or on what basis kids first form an attachment to their mother.

In contrast to other ungulates, Suidae produce large litters of altricial young and females may occasionally share the same nest and nurse one another's offspring (Eisenberg and Lockhart, 1972; Frädrich, 1965, cited in Lent, 1974). Domestic pigs, *Sus domesticus,* seem to recognize their own young by 2 days postpartum (Signoret *et al.,* 1975). Recognition appears to be based on olfaction, since olfactory bulbectomized sows accepted in the presence of their own litters a single alien piglet from 1 through 6 weeks postpartum, whereas intact sows did not (Meese and Baldwin, 1975). However, at 6 weeks postpartum some normal sows rejected a litter of alien piglets, while other sows nursed the strange litter. Nursing of alien litters seems to be a common occurrence in domestic pigs (Signoret *et al.,* 1975). These results are difficult to interpret. Perhaps recognition of young differs from the stimulus conditions which induce a sow to nurse (Meese and Baldwin,

1975). Piglets, however, do not appear to recognize their own mother. The intensive artifical selection on domestic pigs may have affected mother and infant
attachment. More information is sorely needed of free-ranging domestic and feral
pigs.

6.3.1. Ungulate Summary

Thus maternal attachment develops rapidly and on the basis of olfaction in
both sheep and goats. Lambs and kids develop an attachment by 1 week of age
which appears to be mediated primarily by vision and audition (see Table 2).
Although pigs seem to recognize their own young after several days postpartum
and on the basis of olfaction, they will also nurse indiscriminately. There is no
evidence for filial attachment in domestic pigs.

6.4. Mechanisms of Parent and Infant Attachment: Summary

Attachment was demonstrated to be associated with mobility of young and
the social structure of a species (see Section 5). Although species of different social
structure and mobility of the young may be similar in that they exhibit parent and
infant attachment (see Table 1), they may very well differ in the mechanisms and
developmental timing of attachment. The selected species of rodents, primates, and
ungulates discussed above represent different social structures and mobility of
young. They illustrate a diversity of strategies of parent and infant attachment.
They also exhibit differences in the sensory basis and developmental timing of
attachment, which are briefly summarized next (see also Table 2).

Although there is some doubt as to whether attachment is found in guinea
pigs, the evidence suggests that parent and infant attachment is absent in the other
rodents. These latter species appear to have developed another strategy for helping
maintain proximity to mother and nest site, that of attraction to a maternal
pheromone.

In contrast to these rodents, primates carry their semi-altricial young with
them for extended periods of time. Infants can develop an attachment to their
mothers at least by 1–3 months postpartum on the basis of olfaction and vision.
That olfaction may mediate filial attachment is not surprising given the early and
close contact infant primates have with their mother. How early such infant
attachment occurs has not been determined. We know even less about the sensory
bases of maternal attachment and when such attachments develop. We would
expect that olfaction and vision underlie maternal attachment in those primates
where mothers and infants maintain a face-to-face orientation. It would be worthwhile to further investigate attachment in primates that leave their young in a nest
as do rodents [e.g., galagos, mouse lemurs, dwarf lemurs (cf. Klopfer and Boskoff,
1979)].

In comparison to the primates, the ungulates discussed above form rapid maternal and filial attachments as expected given the mobility of the young (see Section 5). Maternal attachment may develop within a few hours postpartum (except in pigs), while filial attachment may take several days. Maternal attachment is mediated initially by olfaction and subsequently also by vision and audition. Filial attachment develops primarily on the basis of vision and audition, which would be expected given that the precocial young are not carried but rather follow their mother.

7. Concluding Remarks

Several questions regarding parent and infant attachment have been considered in this chapter: what is attachment, why form attachments, and where is attachment found?

Discussion of the first question dealt with attachment as a construct (i.e., "bond"). The difficulties encountered in using such a construct can be avoided by viewing attachment as a descriptive label applied to the preferential responding between parents and infants as demonstrated by various operational critera. One such operation, separation from the caregiver, highlights this argument. An infant's reaction to separation has often been attributed to its attachment to the caregiver. An alternative explanation, however, is that an infant's response to separation is a function of the withdrawal from the infant of stimulation provided by the caregiver. The alleviation of separation distress by substitute caregivers may then be the result of the substitute's providing the same stimulation normally provided by the mother.

The question was then raised of why form an attachment if others could provide the same stimulation. This question must be viewed from an evolutionary perspective which distinguishes between ultimate and proximate functions. The ultimate function of attachment (measured in terms of fitness) is probably not protection from predation as some have argued but rather insurance that care is provided for one's own young and not those of another. Since the same ultimate and proximate consequences are, in some species, achieved without forming attachment, it is necessary to assess the conditions favoring the evolution of parent and infant attachment.

Based on the assumed function of attachment, predictions can be made regarding the likelihood of finding attachment under varying conditions of mobility of the young and social structure. A survey of the available evidence suggests that attachment does in fact occur in those situations where the chances of misdirecting parental care seems high.

Although parent and infant attachment is found in many species, the mechanisms and developmental timing of such attachments may differ, as illustrated in several species of rodents, primates, and ungulates. The sensory basis and developmental timing of infant and parent attachment have been studied in only a few species. Relatively little is known about possible hormonal correlates of attachment in any species. We know even less about the ecological constraints on parent and infant attachment.

Attention in the past has focused on the development of filial attachment. But questions on the evolution and functional significance of attachment have been relatively ignored. The combined use of convergent approaches and different levels of analysis (e.g., ecological, developmental, hormonal) is obviously needed to further our understanding of parent and infant attachment in mammals.

ACKNOWLEDGMENTS. This chapter has greatly benefited from the comments and criticisms of Rob Gendron and Tim Keith-Lucas, who for some reason occasionally agreed with me. My thanks also to Peter H. Klopfer, Nadav Nur, and Joe Travis for their comments on earlier drafts and to Sharon Grubb and David Maddox for sharing some beer and some of my late-night rambling. My appreciation to S. Grubb for typing the mansuscript and to Sally Anderson for the graphics. Writing of this chapter was supported in part through an NIMH postdoctoral fellowship (1 F32 MH07188).

References

Ainsworth, M. D. S., 1969, Object relations, dependency, and attachment: A theoretical review of the infant-mother relationship, *Child Dev.* **40**:969–1025.

Ainsworth, M. D. S., 1972 Attachment and dependency: A comparison, in: *Attachment and Dependency* (J. L. Gewirtz, ed.), pp. 97–137, Winston, Washington, D.C.

Ainsworth, M. D. S., 1973, The development of infant-mother attachment, in: *Review of Child Development Research,* Vol 3 (B. M. Caldwell and H. N. Ricciuti, eds.), pp. 1–94, University of Chicago Press, Chicago.

Ainsworth, M. D. S., 1977, Attachment theory and its utility in cross-cultural research, in: *Culture and Infancy* (P. H. Leiderman, S. R. Tulkin, and A. Rosenfeld, eds.), pp. 49–67, Academic Press, New York.

Ainsworth, M. D. S., Blehar, M. C., Waters, E., and Wall, S., 1978, *Patterns of Attachment: A Psychological Study of the Strange Situation,* Erlbaum, Hillsdale, N.J.

Alexander, G., 1977, Role of auditory and visual cues in maternal recognition between ewes and lambs in Merino sheep, *Appl. Anim. Ethol.* **3**:65–81.

Alloway, T., Pliner, P., and Krames, L. (eds.), 1977, *Attachment Behavior, Advances in the Study of Communication and Affect,* Vol. 3, Plenum Press, New York.

Altmann, J., 1978, Infant independence in yellow baboons, in: *The Development of Behavior: Comparative and Evolutionary Aspects* (G. M. Burghardt and M. Bekoff, eds.), pp. 253–277, Garland Press, New York.

Altmann, M., 1958, Socialization of the moose calf, *Anim. Behav.* **6***:155–159.*

Altmann, M., 1963, Naturalistic studies of maternal care in moose and elk, in: *Maternal Behavior in Mammals* (H. L. Rheingold, ed.), pp. 233–253, Wiley, New York.

Andrews, R. S., and Boggess, E. K., 1978, Ecology of Coyotes in Iowa, in: *Coyotes:*

Biology, Behavior, and Management (M. Bekoff, ed.), pp. 249–265, Academic Press, New York.

Arnold, G. W., Boundy, C. A. P., Morgan, P. D., and Bartle, G., 1975, The roles of sight and hearing in the lamb in the location and discrimination between ewes, *Appl. Anim, Ethol.* **1**:167–176.

Baldwin, B. A., and Shillito, E. E., 1974, The effects of ablation of the olfactory bulbs on parturition and maternal behavior in Soay sheep, *Anim. Behav.* **22**:220–223.

Baldwin, J. D., 1971, The social organization of a semi-free ranging troop of squirrel monkeys, *Folia Primatol.* **14**:23–50.

Barnett, S. A., 1963, *The Rat: A Study in Behaviour,* Aldine, Chicago.

Barrette, C., 1977*a*, Some aspects of the behaviour of muntjacs in Wilpattu National Park, *Mammalia* **41**:1–34.

Barrette, C., 1977*b*, The social behaviour of captive muntjacs *Muntiacus reevesi* (Ogilby 1939), *Z. Tierpsychol.* **43**:188–213.

Bartholomew, G. A., 1959, Mother-young relations and the maturation of pup behaviour in the Alaska fur seal, *Anim. Behav.* **7**:163–171.

Bartholomew, G. A., and Hoel, P. G., 1953, Reproductive behavior of the Alaska fur seal, *Callorhinus ursinus, J. Mammal.* **34**:417–436.

Beach, F. A., 1939, Maternal behavior of the pouchless marsupial *Marmosa cinerea, J. Mammal.* **20**:315–322.

Beach, F. A., and Jaynes, J., 1954, The effects of early experience upon the behavior of animals, *Psychol. Bull.* **51**:239–263.

Beach, F. A., and Jaynes, J., 1956, Studies of maternal retrieving in rats, I: Recognition of young, *J. Mammal.* **37**:177–188.

Beauchamp, G. K., and Hess, E. H., 1971, The effects of cross-species rearing on the social and sexual preferences of guinea pigs, *Z. Tierpsychol.* **28**:69–76.

Bernal, J., 1974, Attachment: Some problems and possibilities, in: *The Integration of a Child into a Social World* (M. P. M. Richards, ed.), pp. 153–166, Cambridge University Press, London.

Bertram, B. C. R., 1976, Kin selection in lions and in evolution, in: *Growing Points in Ethology* (P. P. G. Bateson and R. A. Hinde, eds.), pp. 281–301, Cambridge University Press, London.

Bischof, N., 1975, A systems approach toward the functional connections of attachment and fear, *Child Dev.* **46**:801–817.

Blurton-Jones, N., 1972, Comparative aspects of mother–child contact, in: *Ethological Studies of Child Behaviour* (N. Blurton-Jones, ed.), pp. 305–328, Cambridge University Press, Cambridge.

Bouissou, M. F., 1968, Effet de l'ablation des bulbes olfactifs sur la reconnaissance du jeune par sa mère, *Rev. Comp. Anim.* **3**:77–83.

Bowlby, J., 1969, *Attachment, Attachment and Loss,* Vol. 1, Basic Books, New York.

Bowlby, J., 1973, *Separation, Attachment and Loss,* Vol. 2, Basic Books, New York.

Bradbury, J. W., 1977, Social organization and communication, in: *Biology of Bats,* Vol. 3 (W. A. Wimsatt, ed.), pp. 1–72, Academic Press, New York.

Breen, M. F., and Leshner, A. I., 1977, Maternal pheromone: A demonstration of its existence in the mouse *(Mus musculus)*, *Physiol. Behav.* **18**:527–529.

Brown, P., 1976, Vocal communciation in the pallid bat, *Antrozous pallidus, Z. Tierpsychol.* **41**:34–54.

Cairns, R. B., 1966, Attachment behavior of mammals, *Psychol. Rev.* **73**:409–426.

Cairns, R. B., 1972, Attachment and dependency: A psychobiological and social learning synthesis, in: *Attachment and Dependency* (J. L. Gewirtz, ed.), pp. 29–80, Winston, Washington, D.C.

Camenzind, F. J., 1978, Behavioral ecology of coyotes on the National Elk Refuge, Jackson, Wyoming, in: *Coyotes: Biology, Behavior, and Management* (M. Bekoff, ed.), pp. 267–294, Academic Press, New York.

Carter, C. S., and Marr, J. N., 1970, Olfactory imprinting and age variables in the guinea-pig, *Cavia porcellus, Anim. Behav.* **18**:238–244.

Caughley, G., 1966, Mortality patterns in mammals, *Ecology* **47**:906–918.

Clark, A. B., 1978, Sex ratio and local resource competition in a prosimian primate, *Science* **201**:163–165.

Coe, C. L., Mendoza, S. P., Smotherman, W. P., and Levine, S., 1978, Mother-infant attachment in the squirrel monkey: Adrenal response to separation, *Behav. Biol.* **22**:256–263.

Cohen, L. J., 1974, The operational definition of human attachment, *Psychol. Bull.* **81**:207–217.

Collias, N. E., 1956, The anaylsis of socialization in sheep and goats, *Ecology* **37**:228–239.

Conely, L., and Bell, R. W., 1978, Neonatal ultrasounds elicited by odor cues, *Dev. Psychobiol.* **11**:193–197.

Crook, J. H., 1970, The socio-ecology of primates in: *Social Behaviour in Birds and Mammals* (J. H. Crook, ed.), pp. 103–168, Academic Press, New York.

Crook, J. H., 1977, On the integration of gender strategies in mammalian social systems, in: *Reproductive Behavior and Evolution* (J. S. Rosenblatt and B. R. Komisaruk, eds.), pp. 17–39, Plenum Press, New York.

Crook, J. H., Ellis, J., and Goss-Custard, J., 1976, Mammalian social systems: Structure and function, *Anim. Behav.* **24**:261–274.

Crowcroft, P., and Rowe, F. P., 1963, Social organization and territorial behavior in the wild house mouse *(Mus musculus* L.), *Proc. Zool. Soc. London* **140**:517–531.

Darling, F. F., 1937, *A Herd of Red Deer,* Oxford University Press, London.

David, J. H. M., 1975, Observations on mating behaviour, parturition, suckling and the mother-young bond in the Bontebok (*Damaliscus dorcas dorcas*), *J. Zool.* **177**:203–223.

Davis, R. B. Herreid, C. F., II, and Short, H. L., 1962, Mexican free-tailed bats in Texas, *Ecol. Monogr.* **32**:311–346.

Davis, W. H., Barbour, R. W., and Hassell, M. D., 1968, Colonial behavior of *Eptesicus fuscus, J. Mammal.* **49**:44–50.

Dawson, G. A., 1977, Composition and stability of social groups of the tamarin, *Sanguines oedipus geoffroyi,* in Panama: Ecological and behavioral implications, in: *The Biology and Conservation of the Callitrichidae* (D. G. Kleiman, ed.), pp. 23–37, Smithsonian Institution Press, Washington, D.C.

Denenberg, V. H., and Whimby, A. E., 1963, Behavior of adult rats modified by experiences their mothers had as infants, *Science* **142**:1192–1193.

DeVore, I., 1963, Mother-infant relations in free-ranging baboons, in: *Maternal Behavior in Mammals* (H. L. Rheingold, ed.), pp. 305–335, Wiley, New York.

DeVore, I., and Konner, M. J., 1974, Infancy in hunter-gatherer life: An ethological perspective, in: *Ethology and Psychiatry* (N. F., White, ed.), pp. 113–141, University of Toronto Press, Toronto.

DeVos, A., Brokx, P., and Geist, V., 1967, A review of social behavior of the North American cervids during the reproductive period, *Am. Midl. Nat.* **77**:390–417.

Doane, H. M., and Porter, R. H., 1978, The role of diet in mother-infant reciprocity in the spiny mouse, *Dev. Psychobiol.* **11**:271–277.

Dolhinow, P., 1978, Commentary, *Behav. Brain Sci.* **2**:443–444.

Douglas-Hamilton, I., 1973, On the ecology and behaviour of the Lake Manyara elephants, *E. Afr. Wild. J.* **11**:401–403.

Doyle, G. A., and Bearden, S. K., 1977, The galagines of South Africa, in: *Primate Conservation* (His Serene Highness Prince Rainier III of Monaco and G. H. Bourne, eds.), pp. 1–35, Academic Press, New York.

Eaton, R. L., 1970, Group interactions, spacing and territoriality in Cheetahs, *Z. Tierpsychol.* **27**:481–491.

Eisenberg, J. F., 1962, Studies on the behavior of *Peromyscus maniculatus gambelii* and *Peromyscus californicus parasiticus, Behaviour* **19**:177–207.

Eisenberg, J. F., 1963, The intraspecific social behavior of some cricetine rodents of the genus *Peromyscus, Am. Midl. Nat.* **69**:240–246.

Eisenberg, J. F., 1966, The social organization of mammals, *Hand. Zool.* **10**:1–92.

Eisenberg, J. F., 1968, Behavior patterns, in: *Biology of Peromyscus (Rodentia)* (J. A. King, ed.), pp. 451–495, *Spec. Publ. Am. Soc. Mammal.* No. 2.

Eisenberg, J. F., 1977, The evolution of the reproductive unit in the Class Mammalia, in: *Reproductive Behavior and Evolution* (J. S. Rosenblatt and B. R. Komisaruk, eds.), pp. 39–71, Plenum Press, New York.

Eisenberg, J. F., and Maliniak, E., 1967, The breeding of *Marmosa* in captivity, *Int. Zoo Yearb.* **7**:78–79.

Eisenberg, J. F., and Lockhart, M., 1972, An ecological reconnaissance of Wilpattu National Park, Ceylon, *Smithson. Contrib. Zool.* No. 101, pp. 1–118.

Eisenberg, J. F., Muckenhirn, N., and Rudran, R., 1972, The relationship between ecology and social structure in primates, *Science* **176**:863–874.

Epple, G., 1975, The behavior of marmoset monkeys (Callithricidae), in: *Primate Behavior* (L. A. Rosenblum, ed.), pp. 195–239, Academic Press, New York.

Espmark, Y., 1971, Individual recognition by voice in reindeer mother-young relationship. Field observation and playback experiments, *Behaviour* **40**:295–301.

Estes, R. D., 1974, Social organization of the African Bovidae, in: *The Behaviour of Ungulates and Its Relation to Management*, Vol. 1 (V. Geist and F. Walther, eds.), pp. 166–205, IUCN, Morges, Switzerland.

Estes, R. D., 1976, The significance of breeding synchrony in the wildebeest, *E. Afr. Wildl. J.* **14**:135–152.

Estes, R. D., and Estes, R. K., 1979, The birth and survival of wildebeest calves, *Z. Tierpsychol.* **50**:45–95.

Ewer, R. F., 1961, Further observations on suckling behaviour in kittens, together with some general considerations of the interrelations of innate and acquired responses, *Behaviour* **17**:247–260.

Ewer, R. F., 1968, *Ethology of Mammals,* Elek Science, London.

Ewer, R. F., 1973, *The Carnivores,* Cornell University Press, Ithaca, N.Y.

Fogden, S. C. L., 1968, Suckling behaviour in the Grey seal *(Halichoerus grypus)* and the Northern elephant seal *(Mirounga angustirostris)*, *J. Zool.* **154**:415–420.

Fogden, S. C. L., 1971, Mother–young behaviour at Grey seal breeding beaches, *J. Zool. (London)* **164**:61–92.

Frädrich, H., 1974, A comparison of behaviour in the Suidae, in: *The Behaviour of Ungulates and Its Relation to Management,* Vol. 1, (V. Geist and F. Walther, eds.), pp. 133–143, IUCN, Morges, Switzerland.

French, N. R., Stoddart, D. M., and Bobek, B., 1975, Patterns of demography in small mammal population in: *Small Mammals: Their Productivity and Population Dynamics* (F. B. Golley, K. Petrusewicz, and L. Ryszkowski, eds.), pp. 73–102, Cambridge University Press, London.

Fuller, J. L., 1967, Experiential deprivation and later behavior, *Science* **158**:1645–1652.

Fullerton, C., Berryman, J. C., and Porter, R. H., 1974, On the nature of mother-infant interactions in the guinea-pig *(Cavia porcellus)*, *Behaviour* **48**:145–156.

Geist, V., 1963, On the behaviour of the North American moose (*Alces alces andersoni* Peterson 1950) in British Columbia, *Behaviour* **20**:377–416.

Geist, V., 1971, *Mountain Sheep: A Study in Behavior and Evolution,* University of Chicago Press, Chicago.

Geist, V., and Walther, F., eds., 1974, *The Behaviour of Ungulates and Its Relation to Management,* IUCN, Morges, Switzerland.

Gewirtz, J. L., ed., 1972a, *Attachment and Dependency,* Winston, Washington, D.C.

Gewirtz, J. L., 1972b, Attachment, dependence, and a distinction in terms of stimulus control, in: *Attachment and Dependency* (J. L. Gewirtz, ed.), pp. 139–177, Winston, Washington, D.C.

Gewirtz, J. L., 1972c, On the selection and use of attachment and dependence indices, in: *Attachment and Dependency* (J. L. Gewirtz, ed.), pp. 179–215, Winston, Washington, D.C.

Gier, H. T., 1975, Ecology and behavior of the coyote *(Canis latrans)*, in: *The Wild Canids* (M. W. Fox, ed.), pp. 247–262, Van Nostrand, New York.

Gould E., 1971, Studies of maternal-infant communication and development of vocalizations in the bats *Myotis* and *Eptesicus, Commun. Bhav. Biol.* **5**:263–313.

Gubernick, D. J., 1980, Maternal "imprinting" or maternal "labelling" in goats? *Anim. Behav.* **28**:124–129.

Gubernick, D. J., Jones, K. C., and Klopfer, P. H., 1979, Maternal "imprinting" in goats? *Anim. Behav.* **27**:314–315.

Gurski, J. C., 1977, Multiple mothering in mice: Effects on maternal and offspring behaviors, Paper presented at the Eighty-Fifth Annual Meeting of the American Psychological Association, San Francisco.

Gurski, J. C., and Scott, J. P., Individual vs. multiple mothering in mammals, in: *Maternal Influences and Early Behavior* (R. W. Bell and W. P. Smotherman, eds.), Spectrum Press, Holliswood, N.Y. (in press).

Hamilton, W. D., 1964a, The genetical theory of social behaviour, I., *J. Theor. Biol.* **7**:1–16.

Hamilton, W. D., 1964b, The genetical theory of social behaviour. II., *J. Theor. Biol.* **7**:17–52.

Harlow, H. F., 1958, The nature of love, *Am. Psychol.* **13**:673–685.

Harlow. H. F., 1962, The heterosexual affectional system in monkeys, *Am. Psychol.*
    17:1–9.
Harlow, H. F., and Harlow, M. K., 1965, The affectional systems, in: *Behavior of Non-
    human Primates,* Vol. 2 (A. M. Schrier, H. F. Harlow, and F. S. Stollnitz, eds.),
    pp. 287–334, Academic Press, New York.
Harlow,.H. F., and Harlow, M. K., 1969, Effects of various mother-infant relationships
    on rhesus monkey behavior, in: *Determinants of Infant Behaviour,* Vol. 4 (B. M.
    Foss, ed.), pp. 219–256, Wiley, New York.
Harlow, H. F., and Suomi, S. J., 1970, Nature of love—simplified, *Am. Psychol.* **25**:161–
    168.
Harlow, H. F.,and Suomi, S. J., 1974, Induced depression in monkeys, *Behav. Biol.*
    12:273–296.
Harlow, H. F., and Zimmermann, R. R., 1959, Affectional responses in the infant mon-
    key, *Science* **130**:421–432.
Harper, L. V., 1970, Role of contact and sound in eliciting filial responses and develop-
    ment of social attachments in domestic guinea pigs, *J. Comp. Physiol. Psychol.*
    73:427–435.
Hendrichs, H., and Hendrichs, U., 1971, Freilanduntersuchungen zur Ökologie und
    Ethologie der Zwerg-Antilope *Modoqua (Rhynchotragus) kirki* (Gunther, 1880),
    in: *Dikdik und Elefanten* (W. Wickler, ed.), pp. 9–75, Piper Verlag, Müchen.
Hennessy, M. B., Kaplan, J. N., Mendoza, S. P., Lowe, E. L., and Levine, S., 1979,
    Separation distress and attachment in surrogate-reared squirrel monkeys, *Physiol.
    Behav.* **23**:1019–1023.
Hersher, L., Moore, A. U., and Richmond, J. B., 1958, Effect of postpartum separation
    of mother and kid on maternal care in the domestic goat, *Science* **128**:1342–1343.
Hersher, L., Richmond, J. B., and Moore, A. U., 1963, Maternal behavior in sheep and
    goats, in: *Maternal Behavior in Mammals* (H. L. Rheingold, ed.), pp. 203–232,
    Wiley, New York.
Hess, E. H., 1973, *Imprinting: Early Experience and the Developmental Psychobiology
    of Attachment,* Van Nostrand, New York.
Hill, J. L., 1977, Space utilization of *Peromyscus:* Social and spatial factors, *Anim.
    Behav.* **25**:373–389.
Hill, S. D., McCormack, S. A., and Mason, W. A., 1973, Effects of artificial mothers
    and visual experience on adrenal responsiveness of infant monkeys, *Dev. Psychobiol.*
    6:421–429.
Hinde, R. A., 1975, The concept of function in: *Function and Evolution of Behavior* (G.
    Baerends, C. Beer, and A. Manning, eds.), pp. 3–15, Clarendon Press, Oxford.
Hinde, R. A., 1977, On assessing the bases of partner preferences, *Behaviour* **62**:1–9.
Hinde, R. A., and Davies, L. M., 1972, Changes in mother–infant relationship after
    separation in rhesus monkeys, *Nature (London)* **239**:41–42.
Hinde, R. A., and McGinnis, L. M., 1977, Some factors influencing the effects of tem-
    porary mother-infant separation—Some experiments with rhesus monkeys, *Psychol.
    Med.* **7**:197–212.
Hinde, R. A., and Spencer-Booth, Y., 1967, Behavior of socially living rhesus monkeys
    in their first two-and a-half years, *Anim. Behav.* **15**:169–196.
Hinde, R. A., and Spencer-Booth, Y., 1971, Effects of brief separation from mother on
    rhesus monkeys, *Science* **773**:111–118.
Hofer, M. A., 1978, Hidden regulatory processes in early social relationships, in: *Social*

Behavior Perspectives in Ethology, Vol. 3 (P. P. G. Bateson and P. H. Klopfer, eds.), pp. 135–166, Plenum Press, New York.

Houston, D. B., 1974, Aspects of the social organization of moose, in: *The Behaviour of Ungulates and Its Relation to Management*, Vol. 2, (V. Geist and F. Walther, eds.), pp. 690–696, IUCN, Morges, Switzerland.

Hrdy, S. B., 1976, Care and exploitation of nonhuman primate infants by conspecifics other than the mother, *Adv. Study Behav.* **6**:101–158.

Hunsaker, D., II, and Shupe, D., 1977, Behavior of New World marsupials, in: *The Biology of Marsupials* (D. Hunsaker, II, ed.), pp. 279–347, Academic Press, New York.

Ingram, J. C., 1977, Interactions between parents and infants, and the development of independence in the common marmoset *(Callithrix jacchus)*, *Anim. Behav.* **25**:811–827.

Itani, J., 1959, Paternal care in the wild Japanese monkey, *Macaca fuscata fuscata*, *Primates* **2**:61–93.

Jarman, J. V., 1974, The social organization of antelope in relation to their ecology, *Behaviour* **48**:215–267.

Jarman, M. V., 1976, Impala social behaviour: Birth behaviour, *E. Afr. Wildl. J.* **14**:153–167.

Jay, P., 1963, Mother-infant relations in langurs, in: *Maternal Behavior in Mammals* (H. L. Rheingold, ed.), pp. 282–304, Wiley, New York.

Jensen, G. D., 1965, Mother-infant relationship in the monkey *Macaca nemestrina:* Development of specificity of maternal response to own infant, *J. Comp. Physiol. Psychol.* **59**:305–308.

Jensen, G. D., and Tolman, C. W., 1962, Mother-infant relationship in the monkey, *Macaca nemestrina:* The effect of brief separation and mother-infant specificity, *J. Comp. Physiol. Psychol.* **55**:131–136.

Jolly, A., 1966, *Lemur Behavior: A Madagascar Field Study*, University of Chicago Press, Chicago.

Jones, C., 1967, Growth, development and wing loading in the evening bat, *Nycticeius humeralis. J. Mammal* **48**:1–19.

Kaplan, J., 1977, Perceptual properties of attachment in surrogate-reared and mother-reared squirrel monkeys, in: *Primate Bio-Social Development: Biological, Social and Ecological Determinants* (S. Chevalier-Skolnikoff and F. E. Poirier, eds.), pp. 225–234, Garland Press, New York.

Kaplan, J., and Schusterman, R. J., 1972, Social preferences of mother and infant squirrel monkeys following different rearing experiences, *Dev. Psychobiol.* **5**:53–59.

Kaplan, J., and Russell, M., 1974, Olfactory recognition in the infant squirrel monkey, *Dev. Psychobiol.* **7**:15–19.

Kaplan, J. N., Cubicciotti, D., III, and Redican, W. K., 1977, Olfactory discrimination of squirrel monkey mothers by their infants, *Dev. Psychobiol.* **10**:447–453.

Kaplan, J. N., Winship-Ball, A., and Sim, L., 1978, Maternal discrimination of infant vocalizations in squirrel monkeys, *Primates* **19**:187–193.

Kaufman, I. C., 1973, Mother-infant separation in monkeys: An experimental model, in: *Separation and Distress* (J. P. Scott and E. C. Senay, eds.), pp. 33–52, *Publ. Am. Assoc. Adv. Sci.* No. 94.

Kaufman, I. C., and Rosenblum, L. A., 1967a, Depression in infant monkeys separated from their mothers, *Science* **155**:1030–1031.

Kaufman, I. C., and Rosenblum, L. A., 1967*b,* The reaction to separation in infant monkeys: Anaclitic depression and conservation-withdrawal, *Psychosom. Med.* **29**:648–675.

Kaufman, I. C., and Rosemblum, L. A., 1969, Effects of separation from mother on the emotional behavior of infant monkeys, *Ann. N. Y. Acad. Sci.* **159**:681–695.

Kaufmann, J. H., 1962, Ecology and social behavior of the coati, *Nasua narica* on Barro Colorado Island Panama, *Univ. Calif. Berkley Pub. Zool.* **60**:95–222.

Kaufmann, J. H., 1966, Behavior of infant rhesus monkeys and their mothers in a free-ranging band, *Zoologıca N.Y.* **51**:17–27.

Kaufmann, J. H., 1974*a,* Social ethology of the whiptail wallaby, *Macropus parryi,* in northeastern New South Wales, Australia, *Anim. Behav.* **22**:281–369.

Kaufmann, J. H., 1974*b,* The ecology and evolution of social organization in the kangaroo family (Macropodidae), *Am. Zool.* **14**:51–62.

Kawamura, S., 1958, The matriarchal social order in the Minoo-B troop: A study on the rank system of Japanese monkeys, *Primates* **1**:148–156.

King, D. L., 1966, A review and interpretation of some aspects of the infant-mother relationship in mammals and birds, *Psychol. Bull.* **65**:143–155.

King, J. A., 1955, Social behavior, social organization and population dynamics in a black-tailed prairie dog town in the Black Hills of South Dakota, *Contrib. Lab. Vertebr. Biol. Univ. Mich.* **67**:1–123.

King, J. A., 1956, Social relations of the domestic guinea pig living under semi-natural conditions, *Ecology* **37**:221–228.

King, J. A., 1963, Maternal behavior in *Peromyscus,* in: *Maternal Behavior in Mammals* (H. L. Rheingold, ed.), pp. 58–93, Wiley, New York.

Klaus, M. H., and Kennell, J. H., eds., 1976, *Maternal-Infant Bonding,* Mosby, St. Louis, Mo.

Kleiman, D. G., 1969, Maternal care, growth rate, and development in the noctule *(Nyctalus noctula),* pipistrelle *(Pipistrellus pipistrellus),* and serotine *(Eptesicus serotinus)* bats, *J. Zool.* **157**:187–211.

Kleiman, D. G., 1972, Maternal behaviour of the green acouchi *(Myoprocta pratti* Pocock), a South American caviomorph rodent, *Behaviour* **43**:48–84.

Kleiman, D. G., 1977*a,* Monogamy in mammals, *Q. Rev. Biol.* **52**:39–69.

Kleiman, D. G., ed., 1977*b, The Biology and Conservation of the Callithricidae,* Smithsonian Institution Press, Washington, D.C.

Kleiman, D. G., and Eisenberg, J. F., 1973, Comparisons of canid and felid social systems from an evolutionary perspective, *Anim. Behav.* **21**:637–659.

Klingel, H., 1974, A comparison of the social behaviour of the Equidae, in: *The Behaviour of Ungulates and Its Relation to Management,* Vol. 1 (V. Geist and F. Walther, eds.), pp. 124–132, IUCN, Morges, Switzerland.

Klopfer, P. H., 1970, Discrimination of young in Galagos, *Folia Primatol* **13**:137–143.

Klopfer, P. H., 1971, Mother love: What turns it on? *Am. Sci.* **59**:404–407.

Klopfer, P. H., and Boskoff, K. J., 1979, Maternal behavior in prosimians, in: *The Study of Prosimian Behavior* (G. A. Doyle, and R. D. Martin, eds.), pp. 123–156, Academic Press, New York.

Klopfer, P. H.,and Gamble, L., 1966, Maternal "imprinting" in goats: Role of chemical senses, *Z. Tierpsychol.* **23**:588–592.

Klopfer, P. H., Adams, D. K., and Klopfer, M. S., 1964, Maternal "imprinting" in goats, *Proc. Natl. Acad. Sci. U.S.A.* **52**:911–914.

Koford, C. B., 1963*a*, Group relations on an island colony of rhesus monkeys, in: *Primate Social Behavior* (C. H. Southwick, ed.), pp. 136–152, Van Nostrand, Princeton, N.J.

Koford, C. B., 1963*b*, Rank of mothers and sons in bands of rhesus monkeys, *Science* **141**:356–357.

Kruuk, H., 1972, *The Spotted Hyena,* University of Chicago Press, Chicago.

Kummer, H., 1968, *Social Organization of Hamadryas Baboons,* University of Chicago Press, Chicago.

Lamb, M. E., 1974, A defense of the concept of attachment, *Hum. Dev.* **17**:376–385.

Lamb, M. E., 1976, Proximity seeking attachment behaviors: A critical review of the literature, *Genet. Psychol. Monogr.* **93**:63–89.

Lancaster, J., 1971, Play-mothering: The relations between juvenile females and young infants among free-ranging vervet monkeys *(Cercopithecus aethiops), Folia Primatol.* **15**:161–182.

Langman, V. A., 1977, Cow-calf relationships in giraffe *(Giraffa camelopardalis giraffa),* *Z. Tierpsychol.* **43**:264–286.

Laws, R. M., 1974, Behaviour, dynamics and management of elephant populations, in: *The Behaviour of Ungulates and Its Relation to Management,* Vol. 2, (V. Geist and F. Walther, eds.), pp. 513–529, IUCN, Morges, Switzerland.

Laws, R. M., and Parker, I. S. C., 1968, Recent studies on elephant populations in East Africa, *Symp. Zool. Soc. London* **21**:319–359.

LeBoeuf, B. J., and Briggs, K. T., 1977, The cost of living in a seal harem, *Mammalia* **41**:167–195.

LeBoeuf, B. J., Whiting, R.J., and Gantt, R. F., 1972, Perinatal behavior of Northern elephant seal females and their young, *Behaviour* **43**:121–156.

Lenhardt, M. L., 1977, Vocal contour cues in maternal recognition of goat kids, *Appl. Anim. Ethol.* **3**:211–219.

Lent, P. C., 1966, Calving and related social behavior in the barren-ground caribou, *Z. Tierpsychol.* **23**:702–756.

Lent, P. C., 1974, Mother-infant relationships in ungulates, in: *The Behaviour of Ungulates and Its Relation to Management,* Vol. 1 (V. Geist and F. Walther, eds.), pp. 14–53, IUCN, Morges, Switzerland.

Leon, M., 1974, Maternal pheromone, *Physiol. Behav.* **13**:441–453.

Leon, M., 1975, Dietary control of maternal pheromone in the lactating rat, *Physiol. Behav.* **14**:311–319.

Leon, M., 1978, Filial responsiveness to olfactory cues in the laboratory rat, *Adv. Study Behav.* **8**:117–153.

Leon, M., and Moltz, H., 1971, Maternal pheromone: Discrimination by pre-weaning albino rats, *Phsyiol. Behav.* **7**:265–267.

Leon, M., and Moltz, H., 1972, The development of the pheromonal bond in the albino rat, *Phsyiol. Behav.* **8**:683–686.

Leuthold, W., 1977, *African Ungulates,* Springer-Verlag, Berlin.

Levine, S., 1967, Maternal and environmental influences on the adrenocortical response to stress in weanling rats, *Science* **157**:258–260.

Levine, S., Coe, C. L., Smotherman, W. P., and Kaplan, J. N., 1978, Prolonged cortisol elevation in the infant squirrel monkey after reunion with mother, *Physiol. Behav.* **20**:7–10.

Lindsay, D. R. and Fletcher, I. C., 1968, Sensory involvement in the recognition of lambs by their dams, *Anim. Behav.* **16**:415–417.

Lorenz, K., 1935, Der Kumpan in der Umwelt des Vogels [Translated from the original, in: Lorenz, K., 1970, *Studies in Animal and Human Behavior,* Vol. 1, pp. 101–258, Harvard University Press, Cambridge, Mass.].

Marlow, B. J., 1975, The comparative behaviour of the Australasian sea lions *Neophoca cinerea* and *Phocartos hookeri* (Pinnipedia: Otariidae) *Mammalia* **39**:159–230.

Martin, R. D., 1966, Tree shrews: Unique reproductive mechanism of systematic importance, *Science* **152**:1402–1404.

Martin, R. D., 1973, A review of the behaviour and ecology of the lesser mouse lemur (*Microcebus murinus,* J. F. Miller, 1977), in: *Comparative Ecology and Behaviour of Primates* (R. P. Michael and J. H. Crook, eds.), pp 1–68, Academic Press, London.

Mason, W. A., 1970, Motivational factors in psychosocial development, *Nebr. Symp. Motiv.* **18**:35–67.

Mason, W. A., 1978, Social experience and primate cognitive development, in: *The Development of Behavior: Comparative and Evolutionary Aspects* (G. M. Burghardt and M. Bekoff, eds.), pp. 233–251, Garland Press, New York.

Mason, W. A., and Berkson, G., 1975, Effects of maternal mobility on the development of rocking and other behaviors in rhesus monkeys: A study with artificial mothers, *Dev. Psychobiol.* **8**:197–211.

Mason, W. A., Hill. S. D., and Thomsen, C. E., 1974, Perceptual aspects of filial attachment in monkeys in: *Ethology and Psychiatry* (N. F. White, ed.), pp. 84–93, University of Toronto Press, Toronto.

Matthews, L. H., 1971, *The Life of Mammals,* Vol. 2, Weidenfeld and Nicolson, London.

McCabe, T. T., and Blanchard, B. D., 1950, *Three Species of Peromyscus,* Rood Associates, Santa Barbara, Calif.

McCarley, H., 1966, Annual cycle, population dynamics and adaptive behavior of *Citellus tridecemlineatus, J. Mammal.* **47**:294–316.

McHugh, T., 1958, Social behavior of the American buffalo *(Bison bison),* Zoologica **43**:1–40.

McHugh, T., 1972, *The Time of the Buffalo,* Knopf, New York.

McKay, G. M., 1973, Behavior and ecology of the Asiatic elephant in southeastern Ceylon, *Smithson. Contrib. Zool.* No. 125, pp. 1–113.

McKenna, J., 1979, Aspects of infant socialization, attachment, and maternal caretaking patterns among primates: A cross-disciplinary review, *Yearb. Phys. Anthropol.* **22**:250–286.

McManus, J. J., 1970, Behavior of captive opossums, *Didelphis marsupialis virginiana, Am. Midl. Nat.* **84**:144–169.

Mech, L. D., 1970, *The Wolf,* Natural History Press, New York.

Meese, G. B., and Baldwin, B. A., 1975, Effects of olfactory bulb ablation on maternal behaviour in sows, *Appl. Anim. Ethol.* **1**:379–386.

Mendoza, S. P., Smotherman, W. P., Miner, M. T., Kaplan, J., and Levine, S., 1978,

Pituitary-adrenal response to separation in mother and infant squirrel monkeys, *Dev. Psychobiol.* **11**:169–175.

Merchant, J. C., and Sharman, G. B., 1966, Observations on the attachment of marsupial pouch young to the teats and on the rearing of pouch young by foster-mothers of the same or different species, *Aust. J. Zool.* **14**:593–609.

Meyer, J. S., Novak, M. A., Bowman, R. E., and Harlow, H. F., 1975, Behavioral and hormonal effects of attachment object separation in surrogate-peer-reared and mother-reared infant rhesus monkeys, *Dev. Psychobiol.* **8**:425–435.

Michener, G. R., 1971, Maternal behaviour in Richardson's ground squirrel, *Spermophilus richardsonii richardsonii*: Retrieval of young by lactating females, *Anim. Behav.* **19**:653–656.

Michener, G. R., 1972, Social relationships between adult and young in Richardson's ground squirrel, *Spermophilus richardsonii richardsonii,* unpublished Ph.D. thesis, University of Saskatchewan.

Michener, G. R., 1973, Field obsrevations on the social relationships between adult female and juvenile Richardson's ground squirrels, *Can. J. Zool.* **51**:33–38.

Michener, G. R., 1974, Development of adult-young identification in Richardson's ground squirrel, *Dev. Psychobiol.* **7**:375–384.

Michener, G. R., and Sheppard, D. H., 1972, Social behavior between adult female Richardson's ground squirrels *(Spermophilus richardsonii)* and their own and alien young, *Can. J. Zool.* **50**:1343–1349.

Mineka, S., and Suomi, S. J., 1978, Social separation in monkeys, *Psychol. Bull.* **85**:1376–1400.

Misanin, J. R., Zawacki, D. M., and Krieger, W. G., 1977, Differential maternal behavior of the rat dam toward natural and foster pups: Implication for nutrition research, *Bull. Psychon. Soc.* **10**:313–316.

Mitchell, G., and Brandt, E. M., 1972, Paternal behavior in primates, in: *Primate Socialization* (F. E. Poirier, ed.), pp. 173–206, Random House, New York.

Moltz, H., 1971, The onotgeny of maternal behavior in some selected mammalian species, in: *Ontogeny of Vertebrate Behavior* (H. Moltz, ed.), pp. 265–313, Academic Press, New York.

Morgan, P. D., Boundy, C. A. P., Arnold, G. W., and Lindsay, D. R., 1975, The roles played by the senses of the ewe in the location and recognition of lambs, *Appl. Anim. Ethol.* **1**:139–150.

Napier, J. R., and Napier, P. H., 1967, *A Handbook of Living Primates,* Academic Press, New York.

Neal, E., 1970, The banded mongoose, *Mungos mungo* Gmelia, *E. Afr. Wildl. J.* **8**:63–71.

Neyman, P. F., 1977, Aspects of the ecology and social organization of free-ranging cotton-top tamarins *(Saguinus oedipus)* and the conservation status of the species, in: *The Biology and Conservation of the Callithrichidae* (D. G. Kleiman, ed.), pp. 39–71, Smithsonian Institution Press, Washington, D.C.

Neimitz, C., 1979, Outline of the behavior of *Tarsius bancanus,* in: *The Study of Prosimian Behavior* (G. A. Doyle and R. D. Martin, eds.), pp. 631–660, Academic Press, New York.

Owens, D. D., and Owens, M. J., 1979a, Communal denning and clan associations in brown hyenas (*Hyaena brunner,* Thunberg) of the Central Kalahari Desert, *Afr. J. Ecol.* **17**:35–44.

Owens, D., and Owens, M., 1979*b*, Notes on social organization and behavior in brown hyenas *(Hyaena brunnea), J. Mammal.* **60**:405–408.

Owens, M. J., and Owens, D. D., 1978, Feeding ecology and its influence on social organization in brown hyenas (Hyaena brunner, Thunberg) of the central Kalahari Desert, *E. Afr. Wildl. J.* **16**:113–135.

Peterson, R. S., 1968, Social behavior in pinnipeds, in: *The Behavior and Physiology of Pinnepeds* (R. J. Harrison, R. C. Hubbard, R. S. Peterson, C. E. Rice, and R J. Schusterman, eds.), pp. 3–53, Appleton-Century-Crofts, New York.

Peterson, R. S., and Bartholomew, G. A., 1967, The natural history and behaviour of the California sea lion, *Spec. Publ. Am. Soc. Mammal.* No. 1.

Petrinovich, L., 1974, Individual recognition of pup vocalization by northern elephant seal mothers, *Z. Tierpsychol.* **34**:308–312.

Pettijohn, T. F., 1977, Reaction of parents to recorded infant guinea pig distress vocalization, *Behav. Biol.* **21**:438–442.

Pettijohn, T. F., 1978, Development of social behavior in young guinea pigs *(Cavia porcellus), J. Gen. Psychol.* **99**:81–86.

Pettijohn, T. F., 1979*a*, Attachment and separation distress in the infant guinea pig, *Dev. Psychobiol.* **12**:73–81.

Pettijohn, T. F., 1979*b*, Social attachment of the infant guinea pig to its parents in a two-choice situation, *Anim. Learn. Behav.* **7**:263–266.

Pettijohn, T. F., Wont, T. W., Ebert, P. D., and Scott, J. P., 1977, Alleviation of separation distress in 3 breeds of young dogs, *Dev. Psychobiol.* **10**:373–381.

Pianka, E. R., 1970, On r-and K-selection, *Am. Nat.* **104**:592–597.

Poindron, P., 1976, Mother-young relationships in intact or anosmic ewes at the time of sucking, *Biol. Behav.* **2**:161–177.

Poindron, P., and Carrick, M. J., 1976, Hearing recognition of the lamb by its mother, *Anim. Behav.* **24**:600–602.

Poindron, P., Martin, G. B., and Hooley, R. D., 1979, Effects of lambing induction on the sensitive period for the establishment of maternal behavior in sheep, *Physiol. Behav.* **23**:1081–1087.

Poirier, F. E., 1968, The Nilgiri langur *(Presbytis johnii)* mother-infant dyad, *Primates* **9**:45–68.

Porter, R. H., and Doane, H. M., 1976, Maternal pheromone in the spiny mouse *(Acomys cahirinus), Physiol. Behav.* **16**:75–78.

Porter, R. H., and Doane, H. M., 1977, Dietary-dependent cross-species similarities in maternal chemical cues, *Physiol. Behav.* **19**:129–131.

Porter, R. H., and Doane, H. M., 1978, Studies of maternal behavior in spiny mice *(Acomys cahirinus), Z. Tierpsychol.* **47**:225–235.

Porter, R. H., and Etscorn, F., 1974, Olfactory imprinting resulting from brief exposure in *Acomys cahirinus, Nature (London)* **250**:732–733.

Porter, R. H., and Ruttle, K., 1975, The responses of one-day old *Acomys cahirinus* pups to naturally occurring chemical stimuli, *Z. Tierpsychol.* **38**:154–162.

Porter, R. H., Berryman, J. C., and Fullerton, C., 1973*a*, Exploration and attachment behaviour in infant guinea pigs, *Behaviour* **45**:312–322.

Porter, R. H., Fullerton, C., and Berryman, J. C., 1973*b*, Guinea-pig maternal-young attachment behaviour, *Z. Tierpsychol.* **32**:489–495.

Porter, R. H., Doane, H. M., and Cavallaro, S. A., 1978, Temporal parameters of responsiveness to maternal pheromone in *Acomys cahirinus, Physiol. Behav.* **21**:563– 566.

Rabb, G. B., Woolpy, J. H., and Ginsburg, B. E., 1967, Social relationships in a group of captive wolves, *Am. Zool.* **7**:305–311.

Rajecki, D. W., Lamb, M. E., and Obmascher, P., 1978, Toward a general theory of infantile attachment: A comparative review of aspects of the social bond, *Behav. Brain Sci.* **3**:417–464.

Rand, R. W., 1955, Reproduction in the female Cape fur seal *Arctocephalus pusillus* (Schreber), *Proc. Zool. Soc. London* **124**:717–740.

Rasa, O. A. E., 1977, The ethology and sociology of the dwarf mongoose *(Helogale undulata rufula), Z. Tierpsychol.* **43**:337–406.

Redican, W. K., and Kaplan, J. N., 1978, Effects of synthetic odors on filial attachment in infant squirrel monkeys, *Physiol. Behav.* **20**:79–85.

Reite, M. Kaufman, I. C., Pauley, J. D., and Stynes, A. J., 1974, Depression in infant monkeys: Physiological correlates, *Psychosom. Med.* **36**:363–370.

Reite, M., Short, R., Kaufman, I. C., Stuynes, A. J., and Pauley, J. D., 1978*a*, Heart rate and body temperature in separated monkey infants, *Biol. Psychiatry* **13**:91–105.

Reite, M., Short, R., and Seiler, C., 1978*b*, Physiological correlates of maternal separation in surrogate-reared infants: A study in altered attachment bonds, *Dev. Psychobiol.* **11**:427–435.

Richards, M. P. M., 1966, Maternal behaviour in the golden hamster: Responsiveness to young in virgin, pregnant, and lactating females, *Anim. Behav.* **14**:310–313.

Rood, J. P., 1972, Ecological and behavioural comparisons of three genera of Argentine Cavies, *Anim. Behav. Monogr.* **5**:1–83.

Rood, J. P., 1974, Banded mongoose males guard young, *Nature (London)* **248**:176.

Rood, J. P., 1978, Dwarf mongoose helpers at the den, *Z. Tierpsychol.* **48**:277–287.

Rosenblatt, J. S., Turkenwitz, G., and Schneirla, T. C., 1962, Development of sucking and related behavior in neonate kittens, in: *Roots of Behavior* (E. L. Bliss, ed.), pp. 198–210, Harper, New York.

Rosenblum, L. A., 1968, Mother-infant relations and early behavioral development in the squirrel monkey, in: *The Squirrel Monkey* (L. A. Rosenblum and R. W. Cooper, eds.), pp. 207–234, Academic Press, New York.

Rosenblum, L. A., 1971, The ontogeny of mother-infant relations in macaques, in: *The Ontogeny of Vertebrate Behavior* (H. Moltz, ed.), pp. 315–367, Academic Press, New York.

Rosenblum, L. A., and Alpert, S., 1974, Fear of strangers and specificity of attachment in monkeys, in: *The Origins of Fear* (M. Lewis and L. A. Rosenblum, eds.), pp. 165–193, Wiley, New York.

Rosenblum, L. A., and Alpert, S., 1977, Response to mother and stranger: A first step in socialization in: *Primate Bio-Social Development* (S. Chevalier-Skolnikoff and F. E. Poirier, eds.), pp. 463–477, Garland Press, New York.

Rosenblum, L. A., and Kaufman, I. C., 1967, Laboratory observations of early mother-infant relations in pigtail and bonnet macaques, in: *Social Communication among Primates* (S. Altmann, ed.), pp. 33–42, University of Chicago Press, Chicago.

Rosenthal, M. K., 1973, Attachment and mother-infant interaction: Some research impasse and a suggested change in orientation, *J. Child Psychol. Psychiatry* **14**:201– 207.

Rowell, T. E., 1961, The family group in golden hamsters: Its formation and break-up, *Behaviour* **17**:81–94.

Rowell, T. E., Hinde, R. A., and Spencer-Booth, Y., 1964, "Aunt"–infant interaction in captive rhesus monkeys, *Anim. Behav.* **12**:219–226.

Roy, M. A. (ed.), 1980, *Species Identity and Attachment,* Garland Press, New York.

Rudge, M. R., 1970, Mother and kid behaviour in feral goats (*Capra hircus* L.), *Z. Tierpsychol.* **27**:687–692.

Rudnai, J. A., 1973, *The Social Life of the Lion,* Washington Square East, Wallingford, Pa.

Russell, E. M., 1970, Observations on the behaviour of the red kangaroo *(Megaleia rufa)* in captivity, *Z. Tierpsychol.* **27**:385–404.

Sackett, G. P., Griffin, G. A. Pratt, C. L., Joslyn, W. D., and Ruppenthal, G. C., 1967, Mother-infant and adult female choice behavior in rhesus monkeys after various rearing experiences, *J. Comp. Physiol. Psychol.* **63**:376–381.

Sade, D. S., 1965, Some aspects of parent-offspring and sibling relations in a group of rhesus monkeys, with a discussion of grooming, *Am. J. Phys. Anthropol.* **23**:1–17.

Sadleir, R. M. F. S., 1969, *The Ecology of Reproduction in Wild and Domestic Mammals,* Methuen, London.

Sauer, G. F., 1967, Mother-infant relationship in galagos and the oral child transport among primates, *Folia Primatol.* **7**:127–149.

Sayler, A., and Salmon, M., 1971, An ethological analysis of communal nursing by the house mouse *(Mus musculus),* *Behaviour* **40**:60–85.

Schaffer, H. R. (ed.), 1977, *Studies in Mother–Infant Interactions,* Academic Press, New York.

Schaller, G. B., 1963, *The Mountain Gorilla,* University of Chicago Press, Chicago.

Schaller, G. B., 1972, *The Serengeti Lion,* University of Chicago Press, Chicago.

Schaller, G. B., 1977, *Mountain Monarchs: Wild Sheep and Goats of the Himalaya,* University of Chicago Press, Chicago.

Schilling, A., 1979, Olfactory communication in prosimians, in: *The Study of Prosimian Behavior* (G. A. Doyle and R. D. Martin, ed.), pp. 461–542, Academic Press, New York.

Schneirla, T. C., Rosenblatt, J. S., and Tobach, E., 1963, Maternal behavior in the cat, in: *Maternal Behavior in Mammals* (H. L. Rheingold, ed.), pp. 122–168, Wiley, New York.

Scott, J. P., 1963, The process of primary socialization in canine and human infants, *Monogr. Soc. Res. Child Dev.* **28**:1–47.

Scott, J. P., 1971, Attachment and separation in dog and man: Theoretical propositions, in: *The Origins of Human Social Relations* (H. R. Schaffer, ed.), pp.227–246, Academic Press, New York.

Scott, J. P., and Marston, M. V., 1950, Critical periods affecting the development of normal and maladjustive social behavior in puppies, *J. Genet. Psychol.* **77**:25–60.

Scott, J. P., Stewart, J. M., and DeGhett, V. J., 1974, Critical periods in the organization of systems, *Dev. Psychobiol.* **7**:489–513.

Seay, B., Alexander, B. K., and Harlow, H. F., 1964, Maternal behavior of socially deprived rhesus monkeys, *J. Abnorm. Soc. Psychol.* **69**:345–354.

Seay, B., Hansen, E., and Harlow, H. F., 1962, Mother-infant separation in monkeys, *J. Child Psychol. Psychiatry* **3**:123–132.

Seyfarth, R. M., 1976, Social relationships among adult female baboons, *Anim. Behav.* **24**:917–938.

Sharman, G. B., and Calaby, J. H., 1964, Reproductive behaviour in the red kangaroo, *Megaleia rufa*, in captivity, *CSIRO Wildl. Res.* **9**:58–85.

Shillito, E. E., 1975, A comparison of the role of vision and hearing in lambs finding their own dams, *Appl. Anim. Ethol.* **1**:369–397.

Shillito, E., and Alexander, G., 1975, Maternal recognition amongst ewes and lambs of four breeds of sheep *(Ovis aries)*, *Appl Anim. Ethol.* **1**:151–165.

Signoret, J. P., Baldwin, B. A., Frase, D., and Hafez, E. S. E., 1975, The behaviour of swine, in: *The Behaviour of Domestic Animals,* 3rd. ed. (E. S. E. Hafez, ed.), pp. 295–329, Williams and Wilkins, Baltimore.

Simonds, P. E., 1965, The bonnet macaque in South India, in: *Primate Behavior* (I. DeVore, ed.), pp. 175–196, Holt, Rinehart and Winston, New York.

Sinclair, A. R. E., 1977, *The African Buffalo,* University of Chicago Press, Chicago.

Sluckin, W., 1968, Imprinting in guinea-pigs, *Nature (London)* **220**:1148.

Sluckin, W., and Fullerton, C., 1969, Attachment of infant guinea pigs, *Psychon. Sci.* **17**:179–180.

Smith, E. A., 1968, Adoptive sucking in the greay seal, *Nature (London)* **217**:762–763.

Smith, F. V., 1965, Instinct and learning in the attachment of lamb and ewe, *Anim. Behav.* **13**:84–86.

Smith, F. V., Van-Tollen, C., and Boyes, T., 1966, The 'critical period' in the attachment of lambs and ewes, *Anim. Behav.* **14**:120–125.

Smotherman, W. P., Mendoza, S. P., and Levine, S., 1977a, Ontogenetic changes in pup-elicited maternal pituitary-adrenal activity: Pup age and stage of lactation effects, *Dev. Psychobiol.* **10**:365–371.

Smotherman, W. P., Wiener, S. G., Mendoza, S. P., and Levine, S., 1977b, Maternal pituitary-adrenal responsiveness as a function of differential treatment of rat pups, *Dev. Psychobiol.* **10**:113–122.

Smotherman, W. P., Hunt, L. E., McGinnis, L. M., and Levine, S., 1979, Mother-infant separation in group-living rhesus macaques: A hormonal analysis, *Dev. Psychobiol.* **12**:211–218.

Smythe, N., 1970, The adaptive value of the social organization of the coati *(Nasua narica)*, *J. Mammal.* **51**:840–849.

Smythe, N., 1978, The natural history of the central American agouti *(Dasyprocta punctata)*, *Smithson. Contrib. Zool.* No. 257, pp. 1–52.

Spencer-Booth, Y., 1970, The relationship between mammalian young and conspecifics other than mothers and peers: A review, *Adv. Study Behav.* **3**:119–194.

Sroufe, L. A., and Waters, E., 1977, Attachment as an organizational construct, *Child Dev.* **48**:1184–1199.

Struhsaker, T. T., 1975, *The Red Colobus Monkey,* University of Chicago Press, Chicago.

Suomi, S. J., Harlow, H. F., and Domek, C. J., 1970, Effects of repetitive infant-infant separation of young monkeys, *J. Abnorm. Psychol.* **76**:161–172.

Suomi, S. J., Eisele, C. D., Grady, S. A., and Tripp, R. L., 1973, Social preferences of monkeys reared in an enriched laboratory social environment, *Child Dev.* **44**:451–460.

Tedman, R. A., and Bryden, R. A., 1979, Cow-pup behaviour of the Weddell seal, *Lep-*

tonychotes weddelli (Pinnipedia), in McMurdo Sound, Antarctica, *Aust. Wildl. Res.* 6:19-37.

Thomas, J. A., and Birney, E. C., 1979, Parental care and mating systems of the prairie vole, *Microtus ochrogaster, Behav. Ecol. Sociobiol.* 5:171-186.

Thrasher, J. D., Barenfus, M., Rich, S. T., and Shupe, D. V., 1971, The colony management of *Marmosa mitis,* the pouchless opossum, *Lab Anim. Sci.* 21:526-536.

Tileston, J. V., and Lechleitner, R. R., 1966, Some comparisons of the black-tailed and white-tailed prairie dogs in north-central Colorado, *Am. Midl. Nat.* 75:292-316.

Tinbergen, N., 1963, On aims and methods in ethology, *Z. Tierpsychol.* 20:410-433.

Trivers, R. L., 1971, The evolution of reciprocal altruism, *Q. Rev. Biol.* 46:35-57.

Turner, D. A., Shaughnessy, A., and Gould, E., 1972, Individual recognition between mother and infant bats *(Myotis),* in: *Animal Orientation and Navigation* (S. R. Galler, R. Sidney, K. Schmidt-Koenig, G. J. Jacobs, and R. E. Bellville, eds.), pp. 365-371, *NASA Spec.Publ.* No. 262, Washington, D.C.

Tyler, S. J., 1972, The behaviour and social organization of the New Forest ponies, *Anim. Behav. Monogr.* 5:85-196.

van Lawick, H., and VanLawick-Goodall, J., 1971, *Innocent Killers,* Houghton Mifflin, Boston.

van Lawick-Goodall, J., 1967, Mother-offspring relationships in free-ranging chimpanzees, in: *Primate Ethology* (D. Morris, ed.), pp. 287-346, Aldine, Chicago.

van Lawick-Goodall, J., 1968, The behavior of free-living chimpanzees in the Gombe Stream Reserve, *Anim. Behav. Monogr.* 1:165-311.

van Lawick-Goodall, J., 1971, *In the Shadow of Man,* Houghton Mifflin, Boston.

Verts, B. J., 1967, *The Biology of the Striped Skunk,* University of Illinois Press, Urbana, Ill.

Wade, T. D., 1979, Inbreeding, kin selection, and primate social evolution, *Primates* 20:355-370.

Waring, G. H., Wierzbowski, S., and Hafez, E. S. E., 1975, The behaviour of horses, in: *The Behaviour of Domestic Animals,* 3rd ed. (E. S. E. Hafez, ed.), pp. 330-369, Williams and Wilkins, Baltimore.

Weinraub, M., Brooks, J., and Lewis, M., 1977, The social network: A reconsideration of the concept of attachment, *Human Dev.* 20:31-47.

West Eberhard, M. J., 1975, The evolution of social behavior by kin selection, *Q. Rev. Biol.* 50:1-34.

Wickler, W., 1976, The ethological analysis of attachment, *Z. Tierpsychol.* 42:12-28.

Wilson, E. O., 1975, *Sociobiology: The New Synthesis,* Harvard University Press, Belnap Press, Cambridge, Mass.

Wolski, T. R., Houpt, K. A., and Aronson, R., 1980, The role of the senses in mare-foal recognition, *Appl. Anim. Ethol.* 6:121-138.

Wooldridge, F.L., 1971, *Colobus guereza:* Birth and infant development in captivity, *Anim. Behav.* 19:481-485.

Yahner, R. H., 1978, Some features of mother-young relationships in Reeve's muntjac *(Muntiacus reevesi), Appl. Anim. Ethol.* 4:379-388.

Yamada, M., 1963, A study of blood-relationship in the natural society of the Japanese macaque, *Primates* 4:43-66.

Yeaton, R. I., 1972, Social behavior and social organization in Richardson's ground squirrel *(Spermophilus richardsonii)* in Saskatchewan, *J. Mammal.* 53:139-147.

Zajonc, R. B., 1971, Attraction, affiliation, and attachment, in: *Man and Beast: Comparative Social Behavior* (J. F. Eisenberg and W. S. Dillon, eds.), pp. 141–179, Smithsonian Institution Press, Washington, D.C.

Zarrow, M. X., Denenberg, V. H., and Anderson, C. O., 1965, Rabbit: Frequency of suckling in the pup, *Science* **150**:1835–1836.

Zimmerman, G. D., 1974, Cooperative nursing behavior observed in *Spermophilus tridecemlineatus* (Mitchill), *J. Mammal.* **55**:680–681.

8

Mammalian Sibling Interactions

Genes, Facilitative Environments, and the Coefficient of Familiarity

Marc Bekoff

Cruelty and Compassion come with the chromosomes/All men are merciful & all are murderers. (Huxley, 1948, p. 55)

. . . it is of the essence of behavior that it is forever attempting to transcend itself and that it thus supplies evolution with its principal motor. (Piaget, 1978, p. 139)

All correct reasoning is a grand system of tautologies, but only God can make direct use of that fact. (Simon, 1969, p. 15)

Today's ad hoc hypothesis is tomorrow's law of nature. (Hull, 1978*a*, p. 53)

1. Introduction

When I first began to review the literature for this chapter I naively believed that I was embarking on a rather smooth journey into a field in which there were a plethora of data. Simply, I was wrong. While there are fairly substantial amounts of data relating to the social interaction patterns of young and old "family members" alike, I was struck by the excessive use of the terms "probably," "maybe," "we shall assume," "it is thought," etc., to refer to genealogical relationships. In an area of behavioral biology in which rigorous assessments of genetic relatedness are absolutely critical, tenuous, forced *ex post facto* reconstructions of genealogies are (too) frequently being used not only to summarize the results of particular

---

MARC BEKOFF • Department of Environmental, Population, and Organismic Biology, Behavioral Biology Group, University of Colorado, Boulder, Colorado 80309.

studies, but furthermore, and perhaps in a more injurious way, also to rewrite and reformulate evolutionary theory. It also is the case, I believe, particularly for mammals, that theoretical arguments in many areas of inquiry of social biology have severely outstripped empirical studies. Although theories about the way things "ought to be" have provided a solid axis around which basic research can revolve, especially in areas of social biology, only very few significant questions have been investigated and in only very few mammalian species. This is most unfortunate. One more notable trend surfaced as I reviewed papers covering rather disparate species and superficially dissimilar problems searching for some thoughts, or better, data on sibling interactions. I became struck by the large number of eponymous theories and the relative absence of "minority" views, although many available data could have easily, and in some cases, more expeditiously, been used to generate new and interesting hypotheses (e.g., Kaplan, 1978; McCracken and Bradbury, 1977; Rood, 1978), rather than simply providing weak support for extant ideas. Has intellectual censorship reached behavioral biology as it might have once affected ecology (van Valen and Pitelka, 1974)? I strongly doubt this. However, trying to fit square blocks into round holes may be the result of the almost sermonlike dialogues espousing that *this* is right and *that* is wrong, discourses that might have a rather intimidating effect on readers or listeners. Thus, I was pleased to read in the concluding paragraph in a recent paper (Horn, 1978, p. 429): "I began with questions; I have ended with questions. It is a measure of progress that the two lists of questions are different."

In this chapter I shall consider certain aspects of mammalian sibling interactions. The minimum criterion for calling two (or more) individuals genetic siblings will be the known fact that they at least have the same mother. Full siblings, for whom the coefficient of genetic relatedness, r, averages ½, must have the same father as well. It will become clear that my argument is going to deemphasize the *determinative* effects of r and stress the importance of *facilitative* social environments having general and specific effects (Bateson, 1976, 1978a) that favor, in a *probabilistic* way, sibling interactions at different stages of development throughout life (but see Section 3.1.1). The ideas below stress the importance of the immediate and cumulative effects of early social experience but not to the exclusion of the genetic *consequences* of sibling interactions (Hamilton, 1964; Hines and Maynard Smith, 1979). Although some of the thoughts are not new, the time seems appropriate for rehashing and reanalyzing alternative views to "r determinism" (see also Kaplan, 1978; and Rood, 1978, p. 284). I shall introduce the term *coefficient of familiarity, f,* which in many natural situations is closely related to r but may also vary independently as well. Consequently, the characteristics differentiating sibling from nonsibling interactions that would be predicted by r are not necessarily always applicable, though the genetic consequences of social interactions do revolve around the relatedness of the individuals concerned. I shall also try to show that it is necessary to augment studies of adult social organization with analyses of the antecedent infant and juvenile social organizations from which they emerge. Infant and juvenile social relationships can be rather difficult to untangle, yet much light can be shed on adult behavior by comparatively studying species and individual differences in behavioral development (Bekoff, 1977a; Happold, 1976; Hinde, 1971; Kruuk, 1972; Schaller, 1972; Stamps, 1978). Such analyses are particularly helpful when comparing closely related species that demonstrate contrasting life styles (Bekoff, 1977a; Happold, 1976). In addition, a life-span perspective stresses the point that infants are not simply inefficient or little adults, and that there are immediate as well as cumulative (and delayed) costs and benefits associated with their behavioral patterns (see also Galef, this volume).

2. Conditions Favoring Sibling Interaction

Mammalian young are born into a wide variety of social situations, ranging from being isolated from all other individuals except their mother (and possibly other siblings) to being born into large social groups. Although siblings do interact in a wide variety of species having different life histories, there are certain conditions, almost all of which have to do with the developmental environment, that will favor a biased occurrence of interactions between littermates and/or different-aged siblings. It will be argued later that it is these, and perhaps other, conditions that *predispose* (in a probabilistic way) siblings to interact with one another. However, if two (or more) very young unrelated individuals (assume conspecifics for simplicity) are exposed to these conditions, they too will behave like siblings. That is, although r and f are tightly linked in many mammals, it is f that can override r, rather than the reverse.

The conditions that appear to be most favorable for promoting sibling interactions include the following: (1) There is more than one individual per litter, and early mortality spares at least two littermates. In situations in which only one individual is born, other mechanisms (see below) may also be instrumental in favoring transgenerational sibling interactions. (2) Individuals are born relatively immature (altricial), and there is a corresponding protracted period of dependency on caregivers such that the infants are relieved, so to speak, from having to care for themselves for some time after birth. (3) Litters remain intact during den or group movements, dens are not shared with other litters, and there is no communal nursing. (4) The social groups remains closed to outsiders, and emigration by young is rare. Young (at least twins) may even be born in isolation from all individuals other than the mother, such as occurs in many ungulates (Eisenberg, 1966; Langman, 1977; Lent, 1974).

2.1. Litter Size and Mortality

In the best of all possible worlds, the possibility for sibling interactions would be enhanced by having a litter size of at least two. This is the case in many mammals, excluding pinnipeds, most ungulates, as well as most primates [with the exception of the South American Callitrichidae (Kleiman, 1977a) and some Lemuridae], in which twinning is rare. Also, there is a general trend in mammalian evolution such that large species tend to have small litters (Asdell, 1964; Ballinger, 1978; Sadleir, 1973), but since most of these species also produce altricial young that tend to have long lifespans, the opportunity to interact with older siblings exists under certain conditions (see below). Litter size also varies with birth season, parity of the mother, size of the mother's own litter (Asdell, 1964), food supply, and a host of proximate factors, though the only highly predictable correlation seems to be between body size and the number of young produced.

Mortality rate is also important to consider, especially since mortality among young, prereproductive mammals tends to be high (Caughley, 1966; Emlen, 1970; Smith, 1978) and may reduce the possibility for littermates to interact even in multi-individual litters. Although it might be advantageous to siblings and parents if defective young die early (Emlen, 1970), if sibling interactions are important in behavioral ontogeny, one would expect to see at least two individuals being spared from early death (unless there are overlapping generations). Also, if it is the case that the sex with the most rapid growth rate has the highest mortality (Case, 1978a), one would expect sex ratios at birth to be biased in the direction of the fastest growing sex. This would help to ensure that siblings would have intrasexual as well as intersexual social interactions. Kleiman (1979) summarizes interesting data for African wild dogs *(Lycaon pictus)* for which the sex ratio at birth favors males. The sex ratio of adult wild-dog packs is even more skewed toward males, suggesting even heavier mortality in young females. Clearly, factors other than growth rate are important to consider, such as the nature of intrasexual versus intersexual aggression, the latter of which appears to be more severe in wild dogs (Kleiman, 1979).

2.2. Altricial versus Precocial Young

Theoretical aspects and evolutionary consequences of species differences in the degree of maturity at birth have been considered by Case (1978a,b), Daly (1976), and Gould (1977). With respect to sibling interactions, altricial species born in multi-individual litters are very likely to interact first with siblings (besides parents or other caregivers, who might be siblings as well) as soon as they are able. These early interactions may simply involve mouthing or olfactory and auditory exchange due to huddling for thermoregulatory purposes (Rosenblatt, 1976), but may be extremely important in promoting sibling preferences in later social interactions (see below). Rood (1972), studying three genera of Argentine caviomorph rodents, observed that altricial species had more opportunity than precocial species to interact with siblings simply because immature young could not move away.

Usually associated with relative immaturity at birth is a protracted period of dependency of altricial young on their mother and/or other caregivers. The young are thus relieved of providing for themselves for some period of time and may invest their time and energy into socially interacting and learning and/or refining a wide variety of skills (Ewer, 1960; Mason, 1977). If alloparental care, or care provided by individuals other than the parents, is provided by older siblings, then additional sibling interactions will occur as well (see McKenna, this volume).

2.3. Intact Litters

Sibling interactions will be facilitated if litters remain intact at least until the age after which individuals are able to recognize other individuals (i.e., littermates) with whom they have had contact. However, litters do not always remain intact, and may be split during den movements [e.g., red foxes, *Vulpes vulpes* (Storm, *et al.,* 1976)]. This situation would provide a beautiful natural experiment for assessing how much early experience is necessary to result in sibling preferences in later social interactions during which an individual could choose between a sibling and a stranger (see Section 3.1).

Communal nursing (e.g., Sayler and Salmon, 1971) and den sharing (Kruuk, 1972; van Lawick and van Lawick-Goodall, 1971) may also dilute sibling relationships, although litters may remain intact. The position taken here would lead to the prediction that unrelated (or distantly related) individuals that are reared in sibling-like social environments would behave like siblings at later ages.

2.4. "Closed" Social Groups

If young are born away from all other individuals excluding a parent(s) (and possibly other caregivers), they will, by necessity, interact among themselves. Such is the case for many ungulates [Geist (1971) points out that twin bighorn sheep (*Ovis* spp.) lambs may be at a disadvantage when compared with singletons during severe weather]. Moreover, in many groups of social carnivores there is a high probability that group members are closely related and that the group is an "extended family" (Bekoff, 1978a; Bertram, 1976, 1978; van Lawick and van Lawick-Goodall, 1971; Mech, 1970). Therefore, in relatively closed social groups into which immigration by young individuals is rare and from which emigration during early ontogeny is unlikely, interactions among siblings would be favored. This situation would provide for the necessary social experience and exposure that would maximize f and result in social preferences for group members, with the important exception of pair bonding for mating, during which siblings, at least, may be avoided (Bateson, 1978b). Thus, as will be discussed below, there is a potential conflict in that the individuals to whom one becomes bonded early in life are those who are later avoided during mating.

3. Sibling Recognition and the Coefficient of Familiarity: Does $\overline{X}r = \frac{1}{2}\,\overline{X}$ Anything?

> A very elaborate social organization is required before the survival of an individual's genotype becomes so dependent upon survival of the group that natural selection will encourage individual sacrifice for the sake of the community. In most cases we may expect this to occur only within the family group which is one of the reasons for the strong emphasis on individual information in communication signals. (Marler, 1961, p. 313)

> Statistical forecasts of behaviour are possible, but there is no sense in saying that organisms are in principle determinant. (Young, 1978, p. 24).

> I propose that organisms do not wait, passively, for repetitions of an event (or two) to impress or impose upon their memory the existence of a regularity. . . . Rather, organisms actively try to impose guessed regularities (and, with them, similarities) upon the world. (Popper, 1977, in Popper and Eccles, 1977, p. 137)

> A difference which makes no difference, is no difference. (Bernstein, 1974, p. 291)

In general, the same social conditions that favor sibling interactions are also closely associated with the development of the ability of an individual to discriminate between siblings and other individuals. As Sherman (1980) points out, the likelihood that a relative can be recognized is one of at least three limiting factors on differential nepotism, the other two being population demography and resource competition. Although there are sufficient data that show that siblings (or kin) do preferentially interact with one another during social play, grooming, and feeding (for example), the most extensive data base being for nonhuman primates (e.g., Altmann, 1968; Berman, 1978; Cheney, 1978a,b; Fady, 1969; Fedigan, 1972; Nash, 1978; Norikoshi, 1974; Owens, 1975; Sade, 1965; Seyfarth, *et al.*, 1978; Stevenson, 1978; Voland, 1977; Yamada, 1963; see also Waldman and Adler, 1979 for a discussion of sibling interactions in the American toad *Bufo americanus*), most studies can be criticized because they fail to provide *expected* (chance) values of such interactions for each pair of individuals within the group (Altmann and Walters, 1978a; but see Wu *et al.*, 1980). This information is necessary to show that individuals do distribute their behavior selectively among potential recipients based on kinship. Also, in most studies spatial relationships may be biased such that close kin happen to be closer to one another and more accessible than more distantly related individuals (Altmann and Walters, 1978b). (This proximity might also indicate partner preference, in turn). Lastly, the problem of assigning paternity arises (Kurland, 1977), and in some cases, maternity must also be inferred. Precise genealogies must be established, especially for species in which only different-aged siblings have the opportunity to interact (e.g., most primates).

The recognition of one's siblings (and other kin) is a *probabilistic* problem that may be facilitated in a number of nonmutually exclusive ways. These include (1) the use of "recognition alleles" [for which there seems to be no evidence (Sherman, 1980)] that predispose an individual to respond differentially to siblings; (2) phenotypic resemblance that *may* be linked to the amount of shared genes (Barash *et al.*, 1978; but see Dawkins, 1978, p. 65); (3) closed social groups or spatial isolation; (4) delayed emigration of young at least until after they have had the opportunity to interact socially among themselves; (5) the presence of a common reference individual (most likely the mother) who serves as a recognition cue; Maynard Smith (1978a, p. 140) suggests that individuals raised by the same parents would be treated as siblings whether or not they were genetic siblings; (6) various types of early social experience during which there is an exchange of information that may be important in learning individual identities.

The importance of visual phenotypic resemblance in sibling (kin) recognition has yet to be studied in any systematic way. Besides the fact that an individual knows almost nothing about what it looks like, in many mammals social relationships develop between littermates before eye opening. Also, in many mammals, social rituals requiring close inspection of various parts of the body seem to be important in assessing familiarity even between individuals who are not strangers (Bekoff, 1979; Dunbar, 1977; Lockwood, 1976; Marler, 1976; Rasa, 1973; Wemmer and Scow, 1977). It seems unlikely that visual phenotypic resemblance plays any role in the ontogeny of sibling recognition. However, the possibility that phenotypes such as specific scents (see below) and vocalizations may be used in kin recognition (see Treisman, 1978, for a discussion of birds) by mammals in which there are clear-cut individual differences (e.g., Caldwell *et al.*, 1973; Harrington and Mech, 1978; Marler and Hobbitt, 1975) and no significant changes in vocalizations, at least after the individuals part, is likely and requires further investigation. The possibility that phenotypic resemblance might reflect genotypic similarity and affect social interaction patterns (Barash *et al.*, 1978) is interesting but unlikely (Maynard Smith, 1978a).

Although there has been a tendency to deemphasize the importance of learning in the perceptual development of nonhuman animals (Marler, 1978; see also Peterson, 1978), and indeed, the ability to recognize individuals is a perceptual problem (Humphrey, 1974), there are sufficient data that support the notion that learning is important in the development of individual and sibling recognition, and that much of it takes place during very early life (even before eye opening) during suckling, during huddling for homeostatic purposes (Rosenblatt, 1976), and during "passive" nonlocomotor social play. Early bonding may have subsequent effects on patterns of social play (Bekoff, 1977a, 1978a; Rosenblatt, 1976) and bonds may be strengthened during social play as well (Bekoff, 1972, 1977a; Bekoff and Byers, 1981). Indeed, Allison (1971) found that just before red fox pups began to disperse, there was a brief resurgence of social play. It is interesting to speculate on whether such play is important in reinforcing bonds prior to separation that might have some influence on subsequent individual (in this case, sibling) recognition. Olfactory cues also seem to be very important in the development of sibling relationships in rodents (Kalkowski, 1972; Porter *et al.*, 1978; Sheppard and Yoshida, 1971; Wilson and Kleiman, 1974; see also Gilder and Slater, 1978) and in individual recognition in African dwarf mongooses *(Helogale undulata rufula)* (Rasa, 1973) and small Indian mongooses *(Herpestes auropunctatus)* (Gorman, 1976). Even in a primitive eusocial bee *(Lasioglossum zephyrum)*, learning odors early in life appears to play a role in kin group (nest) recognition (Kukuk *et al.*, 1977; see also Wilson, 1971).

3.1. The *f* Factor and Sibling Recognition

The problems involved in sibling recognition are grossly similar to those involved in any type of individual recognition. Assuming the ability to discriminate between individuals is not hard-wired genetically, an individual must recall prior experiences and discriminate familiar from unfamiliar individuals when given a choice. There is evidence that such discriminatory abilities arise early in life and that individuals may also become better discriminators over time (Humphrey, 1974). In pronghorn *(Antilocapra americana)*, even when fawn groups of unrelated individuals are formed, Autenrieth and Fichter (1975) noted six behavioral features that could be used to differentiate sibling from nonsibling encounters. Prior to fawn group formation, siblings interact almost exclusively with one another (besides their mother). Apparently as a result of this experience, siblings tend to stay together in fawn groups, bed together, and approach nonsiblings differently than they approach one another (see also Bromley, 1967).

The view taken herein is that there is nothing "special" about mammalian sibling identification, although the consequences of sibling interactions *are* significant, since full siblings, at least, may serve as the major vehicles of shared gene propogation to future generations. In other words, a sibling is regarded as a sibling by virtue of early social experience and exposure during which learned associations are formed (for reviews of the social effects of exposure see Collias, 1962; Hill, 1978; Zajonc, 1971), and differential survival of genes (or replicators, Dawkins, 1978 or whatever) is accomplished for the most part through developmental pathways. Thus, if genetically unrelated, or distantly related, individuals grow up under the same environmental conditions that characterize a species-typical sibling environment, it is predicted that they will behave as siblings because of a shared past history of repeated exposure and consequent familiarity with one another. The key proximate mechanism appears to be degree of familiarity that usually is strongly related to r, especially in species in which there are multi-individual litters and protracted periods of dependency on a parent(s) and possible other caregivers, and by which fitness is enhanced. A major question, of course, is how preferences for interacting with familiar individuals are behaviorally translated so as to benefit fitness? Reduced agonism, coordinated hunting, group defense of specific resources, and alloparental care to siblings may be some of the ways by which fitness is enhanced.

Although it is not easy to estimate quantitatively the coefficient of familiarity, f, the field is wide open for research endeavors that attempt to measure precisely how much prior exposure is necessary to produce observed preferential responses for familiar individuals. For example, f might be estimated as the ratio of t_p/t_t, where t_p is the amount of time that individual 1 has spent with individual 2 and t_t represents the total amount of time that 1 has spent with all other individuals, including 2 (f may vary in value from 0 to 1). In order to apply this formula in a rigorous fashion, it would be necessary to determine which senses are important in the development of individual and/or group recognition (in order to define precisely a critical distance within which certain cues might be effective) and at what age young individuals become sensitive to (and emit) these stimuli. In some cases, $t_p \simeq t_t \simeq$ the age of the individual, and $f \simeq 1$ at least for individuals reared together from birth. But, the relationship between f and r could vary independently, especially in situations where littermates are highly developed at birth (precocial) and have differential exposure to one another because of increased mobility (when compared to more altricial species) and in species in which only different-aged siblings have the possibility of interacting due to the production of single individual litters. Experimentally, t_p and t_t could be manipulated (with r held constant) and a critical value of f could be determined beyond which there are no changes in the probability distribution of responses in a situation in which an individual would have to choose either to associate with another individual with whom it has had previous contact or one with whom it has had less or no prior association (Fig. 1). It is also interesting to speculate on whether there would be a critical value of f above which a particular individual or phenotype (Bateson, 1978b; see Section 3.1.1) would be avoided as a potential mate.

The above ideas about familiarity and sibling recognition are supported by available data. In spotted hyenas *(Crocuta crocuta)* all cubs of a large clan may share a den and play together regardless of family ties (Kruuk, 1972). Autenrieth and Fichter (1975) reported that two nonsibling pronghorn that were artificially brought together at an early age behaved as siblings. In spiny mice *(Acomys cahirinus)*, Porter et al. (1978) found that familiar siblings preferred one another more than a strange conspecific agemate. However, if two unrelated individuals were housed together before they were subsequently exposed to a sibling as well as to another stranger, discriminative responding to siblings was no longer evident. Porter et al. (1978) also found that olfactory cues were critical in littermate recognition. Anosmic animals did not discriminate between siblings and nonsiblings (see also Alberts, 1976, and Cheal, 1975). They concluded, at least for captive spiny mice, that "for sibling recognition to have any impact on evolution in *A.*

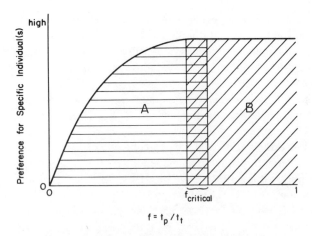

Figure 1. A hypothetical relationship between the coefficient of familiarity, $f(= t_p/t_t,$ where t_p = the amount of time that individual 1 has spent with individual 2 and t_t = the total amount of time that 1 has spent with all other individuals, including 2; f varies from 0 to 1) and the social preference shown for a specific individual(s) (with the possible exception of mate selection; see Section 3.1.1). For example, in region A we observe an increase in social preference as f increases. After f reaches some critical value (cross-hatched area) we see no further increase in social preferences. The critical region may also refer to values of f greater than which we see avoidance of particular individuals during mate selection (see Bateson, 1978b, and Section 3.1.1). Experimentally, t_p and t_t could be manipulated with r (the coefficient of genetic relatedness) held constant (see Section 3.1).

cahirinus, it may be necessary for animals to remain in proximity to one another over an extended period of their lives" (p. 67). Subsequently, Porter and Wyrick (1979) reported that sibling recognition in spiny mice is most likely the result of *exposure learning* during early development.

Familiarity not only is important in sibling interactions, but also appears to be a key variable in the reduction of agonistic behavior among conspecifics of certain species (Calhoun, 1948; Davis, 1975; Greenberg, 1979; King, 1954; Marler, 1976; Maynard Smith, 1972; Sherman, 1980; Thompson, 1978; Yeaton, 1972) and in interactions between neighboring individuals of different species as well (Barash, 1974; Vestal and Hellack, 1978). Familiarity may also affect the frequency distribution (and plausability) of deceptive communication, with less misinformation being passed between familiar individuals (Otte, 1974; Smith, 1977).

3.1.1. Familiarity and Mate Selection

As mentioned above (Section 2.4), familiarity may lead to both the development of partner preferences in social interactions as well as the avoidance of otherwise preferred members of the opposite sex during mate selection. In *Peromys-*

cus, prepubertal familiarity may interfere with the later establishment of a sexual relationship (Hill, 1974; see also Hasler and Nalbandov, 1974). Along these lines, Bateson (1978*b*) has proposed an interesting model supported by data on Japanese quail *(Coturnix coturnix japonica)* that predicts that outbreeding in birds (and mammals?) may result from an assessment of familiarity based on early experience with siblings, after which a male chooses a female slightly different (phenotypically) from the females with whom he was reared (see also Gilder and Slater, 1978). Young (1978, p. 221) proposes that there is a "maturational device that provides at adolescence for avoidance for those with whom the childhood has been spent, whether parents or sibs or unrelated individuals" and there is no reason to explain incest avoidance as being produced by any special capacity to recognize blood relatives (see also Maynard Smith, 1978*a*, p. 140). In lions *(Panthera leo)* and other mammals, it appears that incest avoidance comes about as a by-product of the social system (Bertram, 1978). Inbreeding in natural populations appears to be a rare event and strongly selected against in most species (Bengtsson, 1978; Bertram, 1978; Hill, 1974; Maynard Smith, 1978*a*; Wade, 1979) presumably because of the high probability of producing offspring suffering from inbreeding depression (Bengtsson, 1978; Packer, 1977*b*, cited in Bertram, 1978). However, there may be some exceptions to "inbreeding penalties" [e.g., wolves, *Canis lupus* (Woolpy and Eckstrand, 1979); Wasps, *Euodynerus foraminatus* (Cowan, 1979); it should also be noted that all hamsters used in research in this country until at least the early 1970s originated from three individuals imported in 1931 (Murphy, 1971)]. Eisenberg (1977, p. 220) suggests that "the structure of natural populations often is a mosaic of subpopulations which are highly inbred." Inbreeding might also play an important role in chromosomal evolution (Wilson, *et al.,* 1975). What with the increasing concern to identify individuals and to construct genealogies in field studies, data on inbreeding versus outbreeding will certainly be forthcoming, as will assessments of asymmetries in relatedness in different types of mating systems (Flesness, 1978). Whether or not siblings provide a template against which future possible mates are compared is a very interesting idea that requires corroborative data from mammalian species.

3.2. Probabilistic Sibling Recognition and Some Predicted Species Differences

Briefly, I shall try to tie together some of the above material concerning conditions favoring sibling interactions and sibling recognition. Obviously, both are closely linked with one another and intraspecific and interspecific differences in life histories can either favor, or select against, sibling interactions and subsequent kin-selected behaviors such as cooperation and "altruism." In many social groups of animals, kin selection could provide a strong selective pressure favoring the evolution and elaboration of different forms of cooperation (Bertram, 1978; West

	(1) m, a, pc + sh, or pc	(2) m, p, pcm	(3) m, p, little pc	(4) s, a, pc + sh, or pc	(5) s, p, pc + sh, or pc	(6) s, p, pcm	(7) s, p, little pc
(1)	−	≥	>	≥	≥	>	>
(2)	−	−	≥	≤	≤	>	>
(3)	−	−	−	≤	≤	>	>
(4)	−	−	−	−	=	>	>
(5)	−	−	−	−	−	>	>
(6)	−	−	−	−	−	−	=

Figure 2. Predicted relationships among seven groups (see Section 3.2) showing differences in litter size, degree of maturity at birth, parental care, and alloparental care by older siblings, with respect to the probability that siblings recognize and show interaction preferences for one another. Similar predictions might be made for the relative effects of kin selection as well. Key: > means that the group on the *left* is more likely to show sibling recognition and preferences than the group listed across the top of the table with which it is matched, < means that the group on the left is less likely to show such a relationship than the group with which it is matched, and ≤ and ≥ mean less than or equally likely and greater than or equally likely, respectively. m, multi-individual litter; s, single-individual litter; p, precocial, a, altricial; pc, parental care in addition to nursing; pcm, parental care by mother; sh, help provided by older siblings.

Eberhard, 1975). However, it is important to keep in mind that proximate factors are able to produce differences that might not be predicted by solely considering ultimate selective pressures (e.g., Kaplan, 1978; Rood, 1978).

For convenience, and perhaps with a bit of oversimplification, a number of different major categories can be recognized in which sibling interaction, sibling recognition, and kin (sibling) or family selection (Wade 1979; Williams and Williams, 1957) would be more or less likely to occur. Using the four variables (1) litter size, (2) degree of maturity at birth, and the nature of (3) parental care and (4) care by siblings, 16 different categories may be constructed. However, some would be very unlikely to occur in other than extraordinary conditions (e.g., multi-individual litters, precocial young, no parental care, care by siblings; multi-individual litters, altricial young, no parental care, care by siblings), and some have not evolved at all (e.g., multi-individual litters, altricial young, no parental care, no care by siblings). Therefore, the 7 categories listed below and in Figure 2 represent the most likely combinations of the above four variables to have evolved, with full recognition of the fact that the systems could become noisy (variable) due to the influence of proximate factors. The seven groups about which I shall be concerned include the following:

1. Multi-individual litters, altricial young, care by a parent(s) and older siblings or just by a parent(s) (if one parent provides care it is invariably the mother) [e.g., many carnivores (for reviews, see Bertram, 1976, 1978, and Rood, 1978), rodents, and South American Callitrichidae].

2. Multi-individual litters, precocial young, care provided exclusively by the mother [e.g., pronghorn (Autenrieth and Fichter, 1975; Bromley, 1967; Ingold, 1969)].

3. Multi-individual litters, precocial young, little or no care provided by parents other than nursing, no care by sibling helpers [e.g., some Argentine caviomorph rodents (Rood, 1972); some hystricomorph rodents (Kleiman, 1974); domestic guinea pigs *(Cavia porcellus)* (Fullerton *et al.*, 1974)].

4. Single-individual litters, altricial young, care provided by a parent(s) and older siblings or just by a parent(s) (most primates; assigning paternity is often an extremely difficult problem).

5. Single-individual litters, precocial young, care provided by a parent(s) and older siblings or just by a parent(s) [e.g., African elephants (*Loxodonta africana* (I. Douglas-Hamilton and O. Douglas-Hamilton, 1975; Laursen and Bekoff, 1978); some equids (Klingel, 1974)].

6. Single-individual litters, precocial young, care provided (almost) exclusively by the mother [many ungulates (Altmann, 1958; Dubost, 1975; Jungius, 1971; Lent, 1974; Schenkel and Schenkel-Hulliger, 1969; Stringham, 1974); northern elephant seals *(Mirounga angustirostris)* (Reiter *et al.*, 1978; see also Le Boeuf and Briggs, 1977); milk stealing from other females may occur].

7. Single-individual litters, precocial young, little or no parental care besides nursing [e.g., Alaskan fur seals *(Callorhinus ursinus)* (Bartholomew, 1959)].

It should be emphasized that the assignment of a particular species to any one of the above categories is not necessarily hard and fast. Rather, I have relied on what available information I could find and have tried to do a "best fit" type of analysis. Figure 2 presents predicted relationships among the seven groups in terms of relative (unquantified) estimates of f among siblings and subsequent sibling preferences in social interactions. Also, similar predictions might be made for the relative effects of kin selection on these groups. A whole host of proximate variables (e.g., food availability, weather, and other factors that can influence litter size—see below and Asdell, 1964, and Sadleir, 1973) could make a shambles of the predicted relationships; however, they do present a series of testable ideas from a comparative–developmental perspective. Briefly, species falling into group 1 would be the most likely of all to show sibling recognition and preferences, while those in groups 6 and 7 would be unlikely to show such behavioral patterns or any considerable effects of kin selection, since there is no early contact between siblings. Those species in groups 2–5 might show differences as suggested in Figure 2; however, the present data base is too scanty and the influence of proximate factors rather too unpredictable to support any firm predictions at the moment— hence the large number of $\leq$ and $\geq$ signs. It should be mentioned that the present argument is very similar to that presented by Alexander (1975) in which he pictorially represented correlations between social relationships, patterns of reciprocity, and factors selecting for altruism in primitive human cultures (see also Barash, 1977, p. 316). As genetic relatedness decreases, one expects to see fewer instances of reciprocity and altruism, and this appears to be borne out by existing data. I have simply added ontogeny as the key factor promoting familiarity and the elaboration of kin-selected behavior patterns.

3.3. Does *Familiarity* Breed Contempt?

Since I am proposing that familiarity is a key ingredient affecting the probability distribution of siblings' responses to one another, it would be unwise to ignore totally this cliché. In most cases, the answer to the question is "no." Cooperation and kin-selected altruism seem to flourish between familiar individuals. Anthropomorphically, we might imagine that familiar nonhumans might "let down their guard" and not go through the usual niceties upon meeting, soliciting play, or courting. However, this does not seem to apply in most instances. For example, familiar individuals will engage in time-consuming courtship rituals and formalized greetings (Marler, 1976) and do not show more contempt towards one another than do unfamiliar individuals. We should be careful not to insult our nonhuman counterparts (Bekoff, 1977c).

3.3.1. Does Familiarity *Breed* Contempt?

Now, we can also analyze this cliché from another perspective. Namely, does familiarity *breed* contempt? That is, are offspring of sibling matings (or other types of inbreeding) necessarily contemptuous individuals? While there appear to be no available data that speak to this question, it seems unlikely that inbreeding should result in the biased production of contemptuous individuals. Alexander (1974, p. 354) suggests that members of inbred species will display more tolerance and cooperativeness than those of outbred species, with inbreeding being an "'acceptable' detriment deriving incidentally from group living. . . ." Scott (1977) similarly argues that genotypic similarity is an advantage in situations demanding coordinated behavior. Of course, in relatively closed social groups, it could prove most difficult to separate out the variables, specifically, familiarity based on early experience and genotypic similarity. Studies in which strange infants are introduced to social groups would provide a good test of the hypothesis that it is familiarity, and not genotypic similarity, that is the proximate factor in generating tolerance and cooperative behavior.

4. What Do Siblings Do?

Although siblings may be thought of as "gene machines" (Dawkins, 1976) for one another, overall there appear to be no unique behavioral patterns that exclusively differentiate sibling interactions from those between nonsiblings. The frequency and rates of occurrence of sibling interactions might be higher due to social preferences (see above), and these preferences may have cumulative effects that bias siblings to behave differently towards one another than they do toward unfamiliar and/or unrelated individuals. Furthermore, there are no data to suggest that siblings can learn anything from one another that could not at least theoretically be learned from other individuals. However, they are more likely to learn certain skills from one another simply due to proximity early in life. Briefly, siblings may huddle together to fulfill homeostatic needs such as thermoregulation (Rosenblatt, 1976), nurse together, play socially, groom one another, fight and establish dominance relationships [in which younger siblings frequently dominate older ones (Moore, 1978; Rasa, 1972)], share food, guard and babysit (van Law- ick-Goodall, 1968; Rasa, 1977, Rood, 1978; Rowell *et al.*, 1968), kidnap one another (van Lawick-Goodall, 1968), engage in cannibalism (Fox, 1975), retrieve younger stray siblings [even from water (Hediger, 1970)], provide alloparental care (see Section 5), and breed. Siblings may even suppress one another's growth and reproduction (Batzli *et al.*, 1977; Hill, 1974). In humans, older siblings appear to serve as foci of attention for younger siblings and may be important in providing experiences for these younger individuals that are unlikely to occur in the course of parent–infant interactions (Lamb, 1978).*

Sibling interactions can also be subjected to the same sources of variability that affect nonsibling relationships, as well as others that are unique to the early environments in which siblings usually are reared. Factors that can affect sibling (and other) relationships include food supply [which may also be affected by numerous variables (Ballinger, 1978; Crook, 1970; Kleiman and Eisenberg, 1973; Łomnicki, 1978; Schaller, 1972; White, 1978)], population density (Jannett, 1978), weather (Jonkel and McTaggart Cowan, 1971; Rose and Gaines, 1978), soil quality (Hill, 1972a,b), age specificity of mortality (Emlen, 1970; Smith, 1978), parity of the mother (Fleming and Rauscher, 1978), maternal style [restrictive or laissez-faire (Altmann, 1978)], litter size (Teicher and Kenny, 1978)

---

*The literature on human birth-order effects may contain certain information that is relevant to this section. However, I found this material to be unwieldy and relatively indigestible because of its enormous scope. It proved impossible and seemed unwise to attempt to condense all of the findings and counterfindings into a short space. For reviews, I suggest Neisser's (1951) and Sutton-Smith and Rosenberg's (1970) books, and the following articles: Bates (1975), Belmont *et al.* (1975, 1978), Davis *et al.* (1977), Schooler (1972), and Zajonc and Markus (1975), and Grotevant *et al.* (1977) for a critique of Zajonc and Markus (1975). Grotevant *et al.* (1977) point out the necessity for concentrating on individual families and not on entire populations. There is also a phenomenon known as deprivation dwarfism (Gardner, 1972) that is of interest here. It seems that some children raised in emotionally deprived environments suffer from abnormal sleep patterns and dwarfism, possibly due to a reduction in growth hormone, which is normally produced during sleep. How this effects sibling interactions is unknown. However, there is no evidence the siblings selfishly or spitefully keep one another awake. Finally, with respect to the issue of the applicability of biological principles to human behavior, Hull (1978b, p. 695) raises an interesting point: "Of course, the human species is unique; all species are. The issue is whether *Homo sapiens,* in contrast to all other species, is unique in just those ways to preclude the entension of the relevant biological principals to them."

(which could be affected by some of these and other factors), the nature of the environment [e.g., arboreal or terrestrial (Eisenberg, 1975; Ferron, 1977; Sussman, 1977)], and when during the whelping seasons litters are born (Morton *et al.,* 1974; Schneider *et al.,* 1971). Sex ratios and litter size may also vary depending on the amount of external population control (e.g., predator management) to which a group is subjected (Kleiman and Brady, 1978; Knowlton, 1972). The number of young that disperse from their natal site and the distance that they travel, both of which can be affected by food supply and population density, can also affect sibling relationships. For example, Mazurkiewicz and Rajaska (1975) found that dispersal distance in bank voles *(Clethrionomys glareolus)* was inversely related to population density. Under conditions of high density 46% of the young remained in their place of birth, while during times of low density only 18% stayed where they were born (see also Gadgil, 1971a; Lidicker, 1962, 1975; Łomnicki, 1978). It would be interesting to know whether high density leads to increased inbreeding or additional reproductive suppression (see Section 5.2).

Internal factors such as hormones can also affect behavior in which siblings engage early in life, such as social play (Loy *et al.,* 1978; Quadagno *et al.,* 1977). Also, the nature of parental care can influence sibling interactions (Altmann, 1978) especially with respect to food allocation (this is best documented in birds; see O'Connor, 1978, and references therein).

I would now like to consider briefly and selectively some of the types of interactions in which siblings may partake, recognizing full well that (1) not all mammalian young do engage in such interactions. (2) nonsiblings under certain conditions also partake of such interactions, and (3) the reason that certain types of interactions are thought to be sibling-specific is because of the social environment in which the young are reared and not because of any determinative effects of shared or insightful (inciteful?) genes.

4.1. Nursing Orders

Excluding possible *in utero* "competition" between siblings, most mammalian young born in multi-individual litters first interact socially, and in some instances, competitively, during suckling. In some species, young form nursing orders via serious fights (Fraser, 1975; Hemsworth *et al.,* 1976; Hoeck, 1977; McBride, 1963; McVittie, 1978), after which there is a reduction in aggression. In large litters, teat orders tend to be less stable than in small litters (Hemsworth *et al.,* 1976). Although nursing orders can have long-lasting effects with respect to the ability of adult animals to adjust to new and changing environments and in problem solving (Lien and Klopfer, 1978), it is not well known how nursing order can affect later patterns of social interaction among littermates (Pfeifer, 1979). When young animals suckle, there is active exchange of information, albeit limited in variety perhaps, via olfactory, auditory, and possible visual and tactile channels. Such information may be important in later sibling recognition and may also affect

on-going social interactions and development. Pfeifer (1979) studied nursing relationships in three captive mountain lion *(Felis concolor)* kittens. She found that
the location of the preferred nipple appeared to influence social interactions
between siblings. The kitten that nursed in the middle of its two siblings initiated
the greatest number of social interactions with the other two kittens. She suggests
that this trend might have been due to a variation in the degree of familiarity
between siblings established during nursing. Studies such as this should be conducted on a wider range of species in which nursing orders are established during
early life to test further this interesting idea.

4.2. Social Play, Cooperation, and Play Fighting

The importance of social play in behavioral evolution and ontogeny has been
expressed by a number of people, including Wilson (1977), who lists it as one of
five major areas of importance in research in social biology. Recent reviews of
mammalian social play may be found in Bekoff and Byers (1981) and Symons
(1978b). Except for the not-so-surprising (Altmann and Walters, 1978b) play
partner preferences between siblings that may result from "choosing" to play with
an evenly matched individual or because of accessibility due to spatial proximity
[Klopfer and Dugard, 1976, reported that in captive lemurs (Lemur variegatus),
siblings were nearest neighbors during 80% of all active periods; see also Berger,
1978], there has been little more than lip service given to possible differences
between sibling and nonsibling social play (Bekoff, 1978a; Fagen, 1978; Symons,
1978a,b). Social play appears to be very important in the development and maintenance of social bonds (Bekoff and Byers, 1981; Hamer and Missakian, 1978;
Wilson, 1973) that may be important in sibling recognition (Bekoff, 1978a) and
in the cooperation necessary for the procurement and defense of food (Bowen,
1978; Kleiman and Eisenberg, 1973; Lamprecht, 1978) by family groups (Wells
and Bekoff, unpublished data). Along these lines, it has been suggested that social
play has evolved primarily as a cooperative venture (Bekoff, 1978a). Fagen (1978)
has modeled social play using Maynard Smith's (1974, 1976) various strategies
[e.g., hawk, dove, etc.; such categories may not be entirely appropriate for social
play, and there are numerous other applicable strategies as well (Bekoff *et al.*,
unpublished data)] and found that play should be stable against cheating and will
evolve in family living animals in which there are protracted periods of immaturity. [The temptation to cheat (e.g., injure a partner during play) would be greatly
increased in situations in which a single encounter would have major effects on
future reproductive success.] In addition, there are no data (Bekoff, 1978a;
Symons, 1978a) to support the contention that play relations among unrelated
individuals have evolved to afford practice of agonistic behaviors (Konner, 1975).
Nonetheless, Geist (1978, p. 4) has provided an interesting perspective on play
that deserves further attention. He suggests that play may be less "innocent" than
it seems to be and that body growth might be reduced by the stresses (social and

energetic) that may result from play. (Play may also be a risky activity.) Geist's suggestion contains the notion that there may be more subtle forms of cheating than have heretofore been considered (see Rose, 1978, for a discussion of the ways in which infinitesimal cheating can be evolutionarily stable against rule-obeying strategies). A review of available data strongly suggests that arguments about cheating in play are far more tenuous than those that favor the cooperative nature of social play (Bekoff, 1978a; see also Smith, 1977, p. 238).

In line with the argument developed herein, I would suggest that familiarity and not genetic relatedness have the greatest influence on play patterns (i.e., what is done during play and who plays with whom). This thought is consistent with Symon's (1978a, p. 213) contention that "fear is the foe of play. . . ."

4.3. Sibling Competition and "Selfishness"

Siblings are often the closest relatives within a population, and they are often one another's most direct competitors as well. Thus the extremes of cooperation and competition may both be represented in their interactions. (Alexander, 1974, p. 340)

Though it has become somewhat in vogue in some camps to concentrate (myopically, I daresay) on sibling rivalry, and to invoke all-inclusive (tautological) theories to explain its significance, there are *very few substantive data* concerning selfish sibling rivalry in nonhuman mammals. Siblings, like any other individuals, will compete with one another and appear to act selfishly in a variety of social situations, the intensity of which, at least theoretically, might vary depending on whether the siblings are born simultaneously or sequentially (Metcalf *et al.,* 1979). Under usual conditions, especially those under which natural selection would have primarily operated, it would be predicted, and it has been found, that siblings or other close relatives and unrelated individuals who are familiar with one another fight less and are less likely to injure one another than are unfamiliar individuals (Barash, 1974; Dunford, 1977; Marler, 1976; Maynard Smith, 1972; Sheppard and Yoshida, 1971; Sherman, 1980; Triesman, 1977; Vestal and Hellack, 1978: Yamada, 1963; Yeaton, 1972). However, there are noted exceptions to the lack of serious competitive fighting between siblings in some young canids (Bekoff, 1972, 1974, 1977a, 1978b; Fox, 1969; Wandrey, 1975), pigs (McBride, 1963), some primates (Kleiman, 1979), and some birds (O'Connor, 1978; Watts and Stokes, 1971). Both the lack of injurious agonism and sibling cooperation as well as intense sibling fighting that may often result in death have been explained by the theory of kin selection (Alexander, 1974; O'Connor, 1978; Watts and Stokes, 1971), according to which the overall fitness of the kin (family) group might be enhanced by the adoption of either behavioral strategy. Conditions might be such that an excess of individuals, related or not, could reduce family group fitness [e.g., low food supply and a surplus of helpers (Brown, 1978)].

I shall now consider briefly two case studies in sibling competition, the adaptive significance of which is still unclear. The immediate as well as future consequences of sibling rivalry will be discussed.

4.3.1. Coyotes

Coyote siblings usually engage in serious, unritualized, rank-related agonistic encounters between 3 and 6 weeks of age (Bekoff, 1972, 1974, 1978b; Fox, 1969; Simpson, 1975) *regardless* of rearing conditions (i.e., hand-reared without the mother, mother-reared with no human intervention, reared apart and allowed daily controlled social interaction, reared socially either by hand or mother, etc.). Fights between siblings and nonsiblings are indistinguishable (they last the same amount of time, there are approximately the same mean number of exchanges, and the likelihood of an individual's emerging dominant is the same) and follow the same course of development (Bekoff, unpublished data). Social play typically does not appear until after fighting has begun, and play occurs in its highest frequency after rank relations have been formed (and declines in frequency after the coyotes are about 2–3 months of age). Furthermore, individuals of different social rank play in different amounts [the same is true for captive wolves (Lockwood, 1976)] and are not equally successful in soliciting play from other littermates (Bekoff, 1977a). High-ranking individuals tend to be less successful in soliciting play, and subordinate individuals, frequently scapegoat targets of aggression from higher-ranking siblings, engage in social play only rarely. High-ranking individuals tend to be avoided by littermates, while very subordinate littermates show a strong tendency to avoid interacting with siblings. The immediate effects of attaining high intralitter social rank are unclear. It appears that high-ranking individuals are more free to walk about [the same is true for domestic dogs, *Canis familiaris* (Scott, 1962)] and are the focus of attention of more subordinate siblings (Bekoff, unpublished data), but the high-ranking individuals do not have greater access to food and simply seem to be avoided more than other littermates.

The tendency for certain individuals to be avoided and others to avoid interaction may have cumulative effects, and these possibilities invite interesting and testable hypotheses, many of which are being analyzed in free-ranging, individually identified coyotes in the Grand Teton National Park, Jackson, Wyoming (Bekoff and Wells, 1980). Briefly, it is hypothesized that social play is important in the development and maintenance of social bonds, and that individuals that play together will be more likely to form longer-lasting bonds. Species profiles of select canids (Bekoff, 1977a), various rodents (Happold, 1976; Wilson and Kleiman, 1974), and ungulates (Berger, 1978, 1979) strongly suggest that early social experience can have profound effects on later adult social organization. Specifically, the more social species (or populations) show more amicable behavior and social play early in life than do the more solitary species. Now, if play and other affiliative behaviors are important in the formation and maintenance of social bonds, one would predict that the individuals that are less tightly bonded to their social group (litter) would be the most likely to emigrate (disperse) first (see, also Gubernick, this volume). With respect to coyotes and individual selection, one could argue that the individuals that disperse first would maximize their own fitness, since by and large, yearling coyotes are able to breed successfully, depending on local conditions (Bekoff, 1977b; Gier, 1968; Kennelly, 1978). Of course, these individuals might also suffer from higher mortality. However, conclusive data are not available and may be less likely to apply to coyotes than perhaps to rodents, for which there are data that indicate that early dispersers do suffer from higher mortality (Kozakiewicz, 1976; Smith, 1974a,b; Windberg and Kieth, 1976) due to predation or stress, and are less well physically developed than individuals who disperse at a later age. [It is known that subadult lions ousted from their pride often have trouble procuring food (Elliott and McTaggart Cowan, 1978; Schaller, 1972).] Individuals that do not disperse during their first potential breeding season may serve as helpers for their younger siblings (see below). It is possible that these nondispersers are sexually mature and actually forego mating; however, there are no data from wild populations of coyotes (and most other mammals) that can be used either to support or to refute this idea. For the very subordinate scapegoat individuals, it remains possible that they may show a stress-related delay in sexual maturity or are socially suppressed (see below) and are unable to breed during their first season if they remain with their littermates. Hence, it would not be surprising to find out that these individuals, who have been avoiding interaction with siblings (and perhaps also their more dominant littermates who had been avoided), were the first to leave their natal site. If the subordinate individuals are socially suppressed, this suppression might be removed simply by leaving the individuals responsible for the suppression. A similar mechanism may be operating in wolves (Medjo and Mech, 1976). Thus, early aggression and the development of rank may be proximate causes for differential dispersal (Bekoff, 1977a).

4.3.2. South American Callitrichidae

Marmosets and tamarins, unlike most other primates, live in extended family groups in which there frequently are same-aged siblings (Brown and Mack, 1978; Ingram, 1977a,b; Kleiman, 1979; Rothe, 1975; Stevenson, 1978; for reviews of various aspects of Callitrichidae biology see Kleiman, 1977a). Kleiman (1979) studied family conflict in four families of captive lion tamarins *(Leontopithecus rosalia)*. In these animals, as in common marmosets *(Callithrix jacchus)*, sibling conflict is not uncommon. Kleiman's observations are particularly important because they suggest that intrasexual conflict in female tamarins (and marmosets) is more injurious than in males and that as a result of this aggression, females are more likely to emigrate than are males. She suggests that the reproduction of subordinate females may be more damaging to the family group than reproduction by males, since rearing more than one litter per family group could have negative effects on the fitness of all related individuals. Furthermore, Kleiman suggests that greater intrasexual aggression among females and forced emigration of females by siblings and/or parents may be two of three "universal" traits of monogamous mammals that develop extended families and in which related individuals provide care to infants (the third trait being a skewed sex ratio favoring males). Predictions such as these might provide a tangible base from which future research can depart, it is hoped that comparative data from varied mammalian groups will be forthcoming.

5. Alloparental Care

Since alloparental care is covered in Chapter 6 and is the subject of two excellent recent reviews (Brown, 1978; Emlen, 1978; these reviews specifically concern birds; however, they contain much food for thought for mammalian studies), I shall only briefly comment on some aspects of helper social biology that are related to sibling interactions. Alloparental care by siblings would not be unexpected from a genetical point of view (Hamilton, 1964; but see Rood, 1978), and there are data that show that siblings born even two litters apart provide help, and do it differently depending on their age (Ingram, 1977*a,b*). For example, in common marmosets, subadults are larger than adolescents, and the former group of animals carry and take care of their younger siblings more often than do adolescents, although adolescents do provide care. But, factors other than degree of relatedness that are peculiar to a given species may be stronger determinants for helping. For example, in dwarf mongooses *(Helogale parvula)*, Rood (1978, p. 284) found that "[my] results indicate that age and sex may be stronger determinants for helping at the den than degree of relationship and that even unrelated mongooses help to raise pack young." Even in cases where helpers are closely related to the young to whom they provide care, no substantive data for mammals show that helping actually benefits the helper (increases its fitness) [data for a few species of birds suggest that helpers do benefit (J. L. Brown and E. R. Brown, 1979; Emlen, 1978)]. Although the rentention of sibling helpers could possibly have negative effects by increasing local food consumption and attracting predators, for example (Brown, 1978, lists six potentially harmful effects related to the presence of helpers), there are no data to suggest that siblings are behaving selfishly (or spitefully) by remaining with their parents through a subsequent breeding season(s).

Since a basic prerequisite for alloparental care by older siblings is that there are overlapping generations, for mammals (and birds) in which young typically disperse before breeding, one might expect to see a behavioral polymorphism (Bekoff, 1977*a*) for dispersal, as a result of which some young do not disperse at the age at which they are at least potentially reproductively active, while others do. Two points that I would like to pursue in a bit more detail involve differential dispersal and possible suppression of reproductive activity. It has been suggested that reproductive suppression of young in the presence of their parents might be a trend specific to mammals exhibiting obligate monogamy (Kleiman, 1977*b*; a lone female cannot rear a litter without aid from conspecifics, but only one female can breed, because the carrying capacity of the habitat cannot support more than one breeding female) and alloparental care (Kleiman, 1979).

5.1. Differential Dispersal, Individual Decisions, and Food

As mentioned above, for coyotes at least, there may be ontogenetic reasons for differential dispersal by juveniles that may be rank related (see Emlen, 1978, pp. 259–260, for various dispersal "options" in birds). Environmental factors are also important to consider (Brown, 1978; Emlen, 1978; Hamilton and May, 1976; Horn, 1978; Lidicker, 1975; Łomnicki, 1978; Stacey, 1978). In red foxes, an abundance or preferred food items may delay dispersal (Ables, 1975). If a species does display dispersal strategies such that only a certain number of littermates disperse and attempt to breed on their own as soon as they are potentially reproductively active, it is possible that some of the other siblings that do not disperse will provide some type of care to younger siblings. It would be interesting to speculate on whether within a litter, for example, the "best" representatives for each behavioral phenotype (i.e., early disperser or helper) are somehow sorted out during early life and used in each capacity for the benefit of each individual in the entire kin group. From the point of view of the individual, helping to rear younger siblings would be beneficial if breeding was a risky affair during the first reproductive season. In coyotes, most yearlings can breed; however, they are much more vulnerable (especially females) to environmental perturbations than are older individuals (Gier, 1968). In coyote populations, the greatest annual variation in the number of breeding females is due to the number of juveniles that become sexually mature (Kennelly, 1978; Knowlton, 1972). It is also known that adult coyotes typically reach breeding condition 3–4 weeks before juveniles (Kennelly, 1978), and consequently juveniles may have more time during which they could better assess their own chances of reproducing.

Differential dispersal by juveniles is not at all a simple matter. Dispersal mechanisms and environmental factors such as food supply must also be considered. If it is the case that some juveniles disperse while others do not, but remain instead and provide care to younger siblings, one might now inquire as to the nature of the dispersal process. For example, which individuals stay and which leave, are sex (Bertram, 1978; L. H. Frame and G. W. Frame, 1976; Greenwood and Harvey, 1976; Harcourt, 1978; Kleiman, 1979) and/or social rank (Bekoff, 1977*a*; Gauthreaux, 1978) important variables in dispersal, are dispersers forced to leave either by siblings and/or by parents (Kleiman, 1979) or do they leave of their own accord? Should parents and offspring agree (Horn, 1978)? And, to add complexity to the situation, food availability will undoubtedly be a proximate factor that will make predictions more difficult. Does the number of young that the parents "try" to keep around as helpers depend on the parents' nutritional state? Does a juvenile in poor nutritional condition at the beginning of the breeding season "decide" that breeding will be too risky and attempt to remain with its parents? It is possible that some individuals make a "decision" not to breed based on an assessment of the type of year that it is and is going to be (Low, 1978). The list of questions seems endless. However, analyses of the relationships between behavioral ontogeny, parent–offspring interactions, differential dispersal, and alloparental care will provide exciting insights into the evolution of behavior and the way in which proximate and ultimate variables may influence behavioral decisions (Fig. 3).

I would like to make a brief digression here to discuss a point that seems to need clarification. The above argument about dispersal suggests that early experience among siblings and other social factors (see also Fairbairn, 1978) play a very large role in determining who disperses and at what age dispersal occurs. Genetic contributions to the dispersal process are not necessarily totally ignored. Rather, their impact is lessened. It has been suggested that certain individuals within a population have "instincts" to disperse (Howard, 1960) for the good of the species (Howard, 1965, p. 472). Although there are genetic correlates of dispersal in some rodent species (Krebs *et al.,* 1973; Myers and Krebs, 1974), direct causative genetic contributions to dispersal are extremely hard to assess (Fairbairn, 1978). Hilborn's (1975) data showing that there are similar dispersal tendencies among siblings in vole *(Microtus)* species are very interesting but do not implicate genes in any causative way. Yet, for some reason, it appears that some authors force their data into the instinct mold. For example, in an outstanding study of dispersal of four species of desert rodents (French *et al.,* 1968, p. 272), we read in the abstract that "the results support the hypothesis that some members of the population have an instinct to disperse. . . ." Yet, later on the following statement is made (p. 280): "Whether there is a genetic basis for this behavior . . . or whether all individuals are subject to this urge at a certain stage of their lives or season of the year, is uncertain."

5.2. Suppressed Maturation and Reproduction

In many mammals that show obligate monogamy and alloparental care, it has been observed that some individuals show suppressed maturation and reproduction in the presence of their parents (Kleiman, 1977*b*, 1979). There is good evidence that this occurs in captive marmosets (Brown and Mack, 1978; Hearn, 1977; Rothe, 1975), and the possibility that it also occurs in wolves has been considered as well (Medjo and Mech, 1976). Even in species (particularly rodents) in which these two behavioral patterns are not found, social factors have been implicated in maturational suppression (Batzli *et al.,* 1977; Boonstra, 1978;

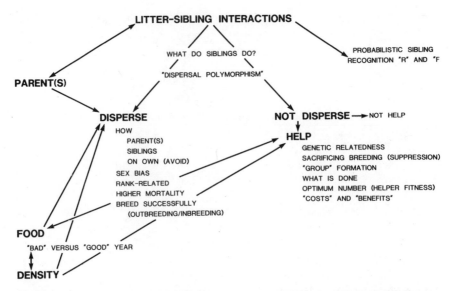

Figure 3. A summary diagram of some possible relationships between behavioral development, dispersal, ecology, and helping. As discussed in this chapter, sibling interactions can have effects on (1) future recognition or siblings, (2) patterns of dispersal, and (3) behavior of parents. Parents may also influence sibling interactions as well as dispersal of their offspring. Food availability and population density, among other ecological variables, can affect dispersal patterns and perhaps helper social biology. Under dispersal, one must consider how it comes about (i.e., the roles played by parents, siblings, and the dispersing individual itself), if there are any biases as to which sex disperses, is dispersal related to social (dominance) rank, if early dispersers suffer higher mortality than individuals who disperse at a later age, and if dispersers breed successfully after they leave their natal site and with whom (a relative or unrelated individual; that is, does dispersal favor outbreeding?). With respect to helping, one must consider the genetic relationship between the helper(s) and the young individuals to whom care is provided, whether helpers are actually sacrificing reproducing on their own, how helpers are incorporated in the "group" (I hesitate to use the word "pack" too loosely because of the implications of the word from descriptions of wolf packs, in which there is coordinated hunting, divisions of labor, a high degree of social organization, etc.), what helpers do (e.g., nurse, provide other types of food, guard, carry, play), if there is an optimum number of helpers after which no additional young are successfully reared and beyond which resources may become overexploited by the presence of too many individuals, and the "costs" and "benefits" to the *helper's* reproductive fitness. Needless to say, the alternatives presented here are rather complex but not exhaustive. For mammals, there is very little information currently available that can be used to tie together the areas represented in this diagram, though such data would be extremely useful and important for providing insights into the ways that proximate and ultimate factors may influence behavioral patterns and "decisions."

Drickamer, 1979; Jannett, 1978; King, 1973; Krebs *et al.*, 1976; Viitala, 1977). The importance of this finding to species in which alloparental care is found is that *if* sexual activity is suppressed and an individual is not actually foregoing mating by providing help, then it becomes clear why the individual might help to rear younger siblings. Even if the suppression of maturation is short-lived such that it renders an individual only temporarily out of phase so that it cannot pair with a conspecific of the opposite sex, it would be in the best interests of an individual to help to rear siblings. Also, the genetic interests of its parents and siblings would be served.

To the best of my knowledge no data are available for wild-mammal populations that bear on the question of whether helpers are actually suppressed reproductively and thereby not sacrificing reproducing on their own. We are currently trying to study this problem for wild coyotes by analyzing urine specimens of nonbreeding helpers. If all individuals in a litter, for example, are reproductively active as juveniles, and if helping does result in increasing the survivorship of younger siblings, then we are back to the problems mentioned above concerning the mechanisms responsible for some individuals dispersing and trying to breed on their own with others remaining with their parents to provide alloparental care.

6. Parents and Siblings

It is clear that parent–offspring interactions are two-way affairs (Alexander, 1974; Bell and Harper, 1977; Trivers, 1974) and that young individuals can affect parents (and other caregivers) as well as vice versa (e.g., Altmann *et al.*, 1978; Berman, 1978; Harper, this volume; Ingram, 1977*a,b*; Priestnall, 1972; Savidge, 1974; Spencer-Booth, 1968). With the recent surge of interest in parent–offspring interaction, especially conflict, it is unfortunate that many models are not directly applicable to mammals in which sibling interactions are important, since they deal with sequentially produced offspring that do not interact (Parker and MacNair, 1978; Stamps *et al.*, 1978) and with parents that die after producing one set of offspring (Stamps *et al.*, 1978; but see Metcalf *et al.*, 1979). And, besides the fact that there appear to be no general solutions to parent–offspring conflict (Stamps *et al.*, 1978), Maynard Smith (1978*b*) points out that yet another difficulty in analyzing parental behavior arises because the optimal behavior for one parent depends on what the other parent is doing. Also, it is very important to consider the age of a parent(s) and its present and future reproductive potential (Dawkins and Carlisle, 1976; Hirshfield and Tinkle, 1975; Smith and Fretwell, 1974) as well as how much total energy is available for reproduction (Ballinger, 1978).

6.1. How Can Parents Influence Sibling Interactions?

Very few detailed data for mammals are available concerning this interesting question. Obviously, parents can and do provide care during early life and theoretically might delegate out resources to prevent overcrowding and starvation (Brockelman, 1975), to limit sibling selfishness (Metcalf, *et al.*, 1979), and/or to promote independence of their young (Davies, 1978). As mentioned above, parents could possibly serve as reference individuals fostering sibling recognition. While it is theoretically exciting to speculate about whether parents can manipulate siblings to behave selfishly, competitively, or cooperatively, few data are available. It has been suggested that Phillippe, the Duke of Orleans, was raised as a female so as not to be a rival to his brother, the future Louis XIV, and that Phillippe led his soldiers into battle wearing high heels, a perfumed wig, and jewelry, but no hat so as not to mess up his (?) hair (Wallechinsky and Wallace, 1978, p. 908).

A parent(s) might also interfere in social interactions between its own offspring and unrelated individuals (Bekoff, 1978a) and be more permissive toward an infant's older siblings than toward group members of the same age (Owens, 1975). However, it is interesting to note that although parents should oppose sibling aggression (Blurton-Jones, 1975), at least in coyotes this does not occur even during intense sibling fights (Bekoff, 1978b). In crocodiles *(Crocodylus moreleti)* in which young individuals are frequently attacked by older individuals (Hunt, 1977), mothers will interfere in fights between differentially matched siblings (R. H. Hunt, personal communication). The interesting speculations provided by Metcalf *et al.,* (1979) concerning parental manipulation of sibling selfishness by producing individuals of different sizes need to be quantitatively documented.

7. Conclusions

> Explanations of lifestyles are like potato chips. People insist on eating them until the whole bag is gone. (Harris, 1974, p. vi)

> Nothing in nature is quite so separate as two mounds of expertise. (Harris, 1974, p. vii)

> What we know . . . is that something unknown is doing what we don't know what. (Sir Arthur Eddington, quoted by Watts, 1977, p. 41)

I began by criticizing excessive theorizing, and in some ways I am guilty of adding to the pile. But, detailed data on mammalian sibling interactions are scarce. Theories and models are necessary and useful. However, I share Williams's (1975, p. 7) belief that "for answering questions on function in biology, comparative evidence is more reliable than mathematical reasoning." Therefore, the consumption of ethology by insatiable predators such as sociobiology and behavioral ecology (Wilson, 1975, p. 6) might be a dangerous tactic. The hand that is feeding social biologists should not be bitten off unless the intention is to prevent the accumulation of raw data from which grand schemes can be developed (Bekoff, 1980). As Eibl-Eibesfeldt (1979, p. 52) has written: "By concentrating on the question of how social structures and behavior contribute to genetic survival, in the sense of contributing to inclusive fitness, the new field of sociobiology has certainly justified its own existence as a part of ethology."

In addition to the need for comparative analyses of a wide variety of behavioral phenotypes, there is also a great need for positive identification of individuals comprising sibships for which techniques such as radioisotope labeling (Wolff and Holleman, 1978) may be very useful. We must be "pigheaded" about demanding precise genealogical analyses. Long-term field studies of identified individuals of known parentage and known genetic relationships are essential to test theories about the adaptive significance of social behavior (McKinney, 1978).

Finally, I have proposed that familiarity is the major axis around which sibling (and other) interactions revolve. Although univariate explanations of anything tend to oversimplify reality, familiarity has emerged as a key proximate factor influencing diverse types of social interactions in a wide range of species. I have concluded that there is nothing "special" about the way in which sibling relationships develop. However, sibling relationships are very special with respect to the genetic consequences of such interactions.

ACKNOWLEDGMENTS.   Some of my research has been supported by grants, from the Public Health Service and the National Science Foundation. The following people provided information on pertinent references and/or unpublished data, and to all of them I extend a hearty thanks: B. Beck, D. Cheney, A. Collins, E. Gould, R. H. Hunt, J. Johnson, D. Kleiman, R. Plomin, D. Sade, C. Southwick, P. Stacey, and J. Stamps. I would also like to thank A. Bekoff, J. Berger, M. Breed, J. Byers, V. DeGhett, D. Duvall, V. Lipetz, S. Pfeifer, P. Stacey, D. Symons, and M. Tyrrell for reading an earlier draft of this chapter and Dion McMain for typing and retyping portions of the manuscript. Some of the ideas developed herein were stimulated by discussions with various people at the symposium on behavioral development held at the University of Bielefeld, West Germany (1977–1978). I am grateful to Professor Klaus Immelmann for supporting me during my stay in Bielefeld during March 1978. Needless to say, I have trespassed into many fields and I apologize for any oversights, omissions of pertinent references, and misinterpretations.

References

Ables, E. D., 1975, Ecology of the red fox in North America, in: *The Wild Canids* (M. W. Fox, ed.), pp. 216–236, Van Nostrand Reinhold, New York.

Alberts, J. R., 1976, Olfactory contributions to behavioral development in rodents, in: *Mammalian Olfaction, Reproductive Processes, and Behavior* (R. L. Doty, ed.), pp. 67–94, Academic Press, New York.

Alexander, R. D., 1974, The evolution of social behavior, *Ann. Rev. Ecol. Syst.* 4:325–383.

Alexander, R. D., 1975, The search for a general theory of behavior, *Behav. Sci.* 20:77–100.

Allison, L. M., 1971, Activity and behaviour of red foxes in central Alaska, M.Sc. thesis, University of Toronto.

Altmann, J., 1978, Infant independence in yellow baboons, in: *The Development of Behavior: Comparative and Evolutionary Aspects* (G. Burghardt and M. Bekoff, eds.), pp. 253–277, Garland, New York.

Altmann, J., Altmann, S. A., and Hausfater, G., 1978, Primate infant's effects on mother's future reproduction, *Science* **201**:1028–1030.

Altmann, M., 1958, Social Integration of the moose calf, *Anim. Behav.* **6**:155–159.

Altmann, S. A., 1968, Sociobiology of rhesus monkeys. IV. Testing Mason's hypothesis of sex differences in affective behavior, *Behaviour* **32**:49–69.

Altmann, S. A., and Walters, J., 1978a, Critique of Kurland's "Kin selection in the Japanese monkey" (unpublished manuscript).

Altmann, S. A., and Walters, J., 1978b, Review of Kurland, 1977, *Man* **13**:324–325.

Asdell, S. A., 1964, *Patterns of Mammalian Reproduction,* Cornell University Press, Ithaca, N.Y.

Autenrieth, R. E., and Fichter, E., 1975, On the behavior and socialization of pronghorn fawns, *Wildl. Monogr.* **42**:1–111.

Ballinger, R. E., 1978, Variation in and evolution of clutch and litter size, *in: The Vertebrate Ovary: Comparative Biology and Evolution* (R. E. Jones, ed.), pp. 789–825, Plenum Press, New York.

Barash, D. P., 1974, Neighbor recognition in two "solitary" carnivores: The raccoon *(Procyon lotor)* and the red fox *(Vulpes fulva)*, *Science* **185**:794–796.

Barash, D. P., 1977, *Sociobiology and Behavior,* Elsevier, New York.

Barash, D. P., Holmes, W. G., and Green, P. J., 1978, Exact versus probabilistic coefficients of relationships: Some implications for sociobiology, *Am. Nat.* **112**:355–363.

Bartholomew, G. A., 1959, Mother-young and the maturation of pup behaviour in the Alaska fur seal, *Anim. Behav.* **7**:163–171.

Bates, E., 1975, Peer relations and the acquisition of language, in: *Friendship and Peer Relations* (M. Lewis and L. A. Rosenblum, eds.), pp. 259–292, Wiley, New York.

Bateson, P. P. G., 1976, Specificity and the origins of behavior, *Adv. Study Behav.* **6**:1–20.

Bateson, P. P. G., 1978a, How does behavior develop? in: *Perspectives in Ethology,* Vol. 3 (P. P. G. Bateson and P. H. Klopfer, eds.), pp. 55–66, Plenum Press, New York.

Bateson, P. P. G., 1978b, Sexual imprinting and optimal outbreeding, *Nature (London)* **273**:659–660.

Batzli, G. O., Getz, L. L., and Hurley, S. S., 1977, Suppression of growth and reproduction of microtine rodents by social factors, *J. Mammal.* **58**:583–591.

Bekoff, M., 1972, The development of social interactions, play, and metacommunication in mammals: An ethological perspective, *Rev. Biol.* **47**:412–434.

Bekoff, M., 1974, Social play and play-soliciting by infant canids, *Am. Zool.* **14**:323–340.

Bekoff, M., 1977a, Mammalian dispersal and the ontogeny of individual behavioral phenotypes, *Am. Nat.* **111**:715–732.

Bekoff, M., 1977b, The coyote, *Canis latrans* Say., *Mammal. Spec.* **79**:1–9.

Bekoff, M., 1977c, "Man" and "animal": A sociobiological dichotomy? *Biologist* **59**:1–10.

Bekoff, M., 1978a, Social play: Structure, function, and the evolution of a cooperative social behavior, in: *The Development of Behavior: Comparative and Evolutionary Aspects* (G. Burghardt and M. Bekoff, eds.), pp. 367–383, Garland, New York.

Bekoff, M., 1978b, Behavioral development in coyotes and eastern coyotes, in: *Coyotes:*

Biology, Behavior, and Management (M. Bekoff, ed.), pp. 97–126, Academic Press, New York.

Bekoff, M., 1979, Scent marking by free ranging domestic dogs: Olfactory and visual components, *Biol. of Behav.* **4**:123–139.

Bekoff, M., 1980, Human ethology, biological determinism, directive genes, and trees, *Behav. Brain Sci.* **2** (in press).

Bekoff, M., and Byers, J. A., 1981, A critical reanalysis of the ontogeny and phylogeny of mammalian social play: An ethological hornet's nest, in: *Behavioral Development in Animals and Man: The Bielefeld Conference* (K. Immelmann, G. Barlow, M. Main, and L. Petrinovich, eds.), Cambridge University Press, New York (in press).

Bekoff, M., and Wells, M. C., 1980, Social ecology of coyotes, *Sci. Am.* **242**:130–148.

Bell, R. Q., and Harper, L. V., 1977, *Child effects on Adults,* L. Erlbaum, Hillsdale, N. J.

Belmont, L., Stein, Z. A., and Susser, M. W., 1975, Comparison of associations of birth order with intelligence test score and height, *Nature (London)* **255**:54–56.

Belmont, L., Stein, Z., and Zybert, P., 1978, Child spacing and birth order: Effect on intellectual ability in two-child families, *Science* **202**:995–996.

Bengtsson, B. O., 1978, Avoiding inbreeding: At what cost? *J. Theor. Biol.* **73**:439–444.

Berger, J., 1978, Social ontogeny and reproductive strategies in bighorn sheep. Ph.D. dissertation, University of Colorado.

Berger, J., 1979. Social ontogeny and behavioural diversity: Consequences for bighorn sheep inhabiting desert and mountain environments, *J. Zool.* **188**:251–266.

Berman, C. M., 1978, The analysis of mother-infant interaction in groups: Possible influence of yearling siblings, *Rec. Adv. Primatol.* **1**: 111–113.

Bernstein, I. S., 1974, Principles of primate group organization, in: *Perspectives in Primate Biology* (A. B. Chiarelli, ed.), pp. 283–298, Plenum Press, New York.

Bertram, B. C. R., 1976, Kin selection in lions and in evolution, in: *Growing Points in Ethology* (P. P. G. Bateson and R. A. Hinde, eds.), pp. 281–301, Cambridge University Press, New York.

Bertram, B. C. R., 1978, Living in groups: Predators and Prey, in: *Behavioural Ecology: An Evolutionary Approach* (J. R. Krebs and N. B. Davies, eds.), pp. 64–96, Sinauer, Sunderland, Mass.

Blurton-Jones, N., 1975, Ethology, anthropology, and childhood, in: *Biosocial Anthropology* (R. Fox, ed.), pp. 69–92, Halsted Press, New York.

Boonstra, R., 1978, Effect of adult townsend voles *(Microtus townsendii)* on survival of young, *Ecology* **59**:242–248.

Bowen, D., 1978, Prey size and coyote social organization, PH.D. dissertation, University of British Columbia.

Brockelman, W. Y., 1975, Competition, the fitness of offspring, and optimal clutch size, *Am. Nat.* **109**:677–699.

Bromley, P. T., 1967, Pregnancy, birth, behavioral development of the fawn and territoriality in the pronghorn (*Antilocapra americana* Ord) on the national bison range, Moiese, Montana, M.A. thesis, University of Montana.

Brown, J. L., 1978, Avian communal breeding system, *Ann. Rev. Ecol. Syst.* **9**:123–155.

Brown, J. L., and Brown, E. R., 1980, Kin selection and individual selection in babblers, *Misc. Publ. Mus. Zool. Univ. Mich.* (in press).

Brown, K., and Mack, D.S., 1978, Food sharing among captive *Leontopithecus rosalia,* *Folia Primatol.* **29**:268–290.

Caldwell, M.C., Caldwell, D.K., and Miller, J. F., 1973, Statistical evidence for individual signature whistles in the spotted dolphin, *Stenella plagiodon, Cetology* **16**:1–21.

Calhoun, J.B., 1948, Mortality and movement of brown rats *(Rattus norvegicus)* in artificially super-saturated populations, *J. Wildl. Manage.* **12**:167–171.

Case. T. J., 1978a, On the evolution and adaptive significance of postnatal growth rates in the terrestrial vertebrates, *Quart. Rev. Biol.* **53**:243–282.

Case, T. J., 1978b, Endothermy and parental care in the terrestrial vertebrates, *Am. Nat.* **112**:*861–874.*

Caughley, G., 1966, Mortality patterns in mammals, *Ecology* **47**:906–918.

Cheal, M., 1975, Social olfaction: A review of the ontogeny of olfactory influences on vertebrate behavior, *Behav. Biol.* **14**:1–25.

Cheney, D. L., 1978a. Interactions of immature male and female baboons with adult females, *Anim. Behav.* **26**:389–408.

Cheney, D. L., 1978b, The play partners of immature baboons, *Anim. Behav.* **26**:1038–1050.

Collias, N., 1962, Social development in birds and mammals, in: *Roots of Behavior* (E. L. Bliss, ed.), pp. 264–273, Hafner, New York.

Cowan, D. P., 1979, Sibling matings in a hunting wasp: Adaptive inbreeding? *Science* **205**:1403–1405.

Crook, J. H., 1970, The socio-ecology of primates, in: *Social Behaviour in Birds and Mammals* (J. H. Crook, ed.), pp. 103–166, Academic Press, New York.

Daly, M., 1976, Behavioral development in three hamster species, *Dev. Psychobiol.* **9**:315–323.

Davies, N. B., 1978, Parental meanness and offspring independence: An experiment with hand-reared great tits *Parus major, Ibis* **120**:509–514.

Davis, D. J., Cahan, S., and Bashi, J., 1977, Birth order and intellectual development: The confluence model in light of cross-cultural evidence, *Science* **196**:1470–1472.

Davis, T. M., 1975, Effects of familiarity on agonistic encounter behavior in male degus *(Octodon degus)*, *Behav. Biol.* **14**:511–517.

Dawkins, R., 1976, *The Selfish Gene,* Oxford University Press, New York.

Dawkins, R., 1978, Replicator selection and the extended phenotype, *Z. Tierpsychol.* **47**:61–76.

Dawkins, R., and Carlisle, T. R., 1976, Parental investment, mate desertion and a fallacy, *Nature (London)* **262**:131–133.

Douglas-Hamilton, I., and Douglas-Hamilton, O., 1975, *Among the Elephants* Viking, New York.

Drickamer, L., 1979, Acceleration and delay of first estrus in wild *Mus musculus, J. Mammal.* **60**:215–216.

Dubost, G., 1975, Du comportement du chevrotain africain, *Hymoschus aquaticus* Ogilby (Artiodactyla, Ruminantia), *Z. Tierpsychol.* **37**:403–501.

Dunbar, I., 1977, Olfactory preferences in dogs: The response of male and female beagles to conspecific odors, *Behav. Biol.* **20**:471–481.

Dunford, C., 1977, Social system of round-tailed ground squirrels *Anim. Behav.* **25**:885–906.

Eibl-Eibesfeldt, I., 1979, Human ethology—Concepts and implications for the study of man, *Behav. Brain Sci.* **2**:1–57.

Eisenberg, J. F., 1966, The social organization of mammals, *Handbuch der Zoologie,* Vol VIII (10/7), pp. 1–92, De Gruyter, Berlin.

Eisenberg, J. F., 1975, Phylogeny, behavior, and ecology in the mammalia, in:*Phylogeny of the Primates* (W. P. Luckett and F. S. Szalay, eds,), pp. 47–68, Plenum Press, New York.

Eisenberg, J. F., 1977, Comments, in: *The Biology and Conservation of the Callitrichidae* (D. G. Kleiman, ed.), p. 220, Smithsonian Institution Press, Washington, D.C.

Elliott, J. P., and Cowan, I. McTaggart, 1978, Territoriality, density, and prey of the lion in Ngorongoro Crater, Tanzania, *Can. J. Zool.* 56:1726–1734.

Emlen, J. M., 1970, Age specificity and ecological theory, *Ecology* 51:588–601.

Emlen, S. T., 1978, The evolution of cooperative breeding in birds, in: *Behavioural Ecology: An Evolutionary Approach* (J. R. Krebs and N. B. Davies, eds.), pp. 245–281, Sinauer Sunderland, Mass.

Ewer, R. F., 1960, Natural selection and neoteny, *Acta Biotheor.* **13***:161–184.*

Fady, J. C., 1969, Les jeux sociant; le compagnon de leux chez les jeunes. Observation chex *Macaca irus, Folia Primatol.* **11***:134–143.*

Fagen, R. M., 1978, Evolutionary biological models of animal play behavior, in: *The Development of Behavior: Comparative and Evolutionary Aspects* (G. Burghardt and M. Bekoff, eds.), pp. 385–404, Garland, New York.

Fairbairn, D. J., 1978, Behaviour of dispersing deer mice *(Peromyscus maniculatus),* *Behav. Ecol. Sociobiol.* 3:265–282.

Fedigan, L., 1972, Social and solitary play in a colony of vervet monkeys *(Cercopithecus aethiops),* *Primates* 13:347–364.

Ferron, J., 1977, Comparative ontogeny of behaviour in some ground and tree squirrel species (Sciuridae): An evolutionary approach, Paper presented at the XVth International Ethological Conference, Bielefeld, West Germany.

Fleming, T. H., and Rauscher, R. J., 1978, On the evolution of litter size in *Peromyscus leucopus, Evolution* 32:45–55.

Flesness, N. R., 1978, Kinship asymmetry in diploids, *Nature (London)* 276:495–496.

Fox, L. R., 1975, Cannibalism in natural populations, *Ann. Rev. Ecol. Syst.* 5:87–106.

Fox, M. W., 1969, The anatomy of aggression and its ritualization in Canidae: A developmental and comparative study, *Behaviour* 35:242–258.

Frame, L. H., and Frame, G. W., 1976, Female African wild dogs emigrate, *Nature (London)* 263:227–229.

Fraser, D., 1975, The "teat order" of suckling pigs. II. Fighting during suckling and the effects of clipping the eye teeth, *J. Agric. Sci.* 84:393–399.

French, N. R., Tagami, T. Y., and Hayden, P., 1968, Dispersal in a population of desert rodents, *J. Mammal.* 49:272–280.

Fullerton, C., Berryman ,J. C., and Porter, R. H., 1974, On the nature of mother-infant interactions in the guinea pig *(Cavia porcellus),* *Behaviour* 48:145–156.

Gadgil, M., 1971, Dispersal: Population consequences and evolution, *Ecology* 52:253–261.

Gardner, L. I., 1972, Deprivation dwarfism, *Sci. Am.* 227:76–82.

Gauthreaux, S. A., 1978, The ecological significance of behavioral dominance, in: *Per-*

spectives in Ethology, Vol. 3 (P. P. G. Bateson and P. H. Klopfer, eds.), pp. 17–54, Plenum Press, New York.

Geist, V., 1971, *Mountain Sheep,* University of Chicago Press, Chicago.

Geist, V., 1978, On weapons, combat, and ecology, in: *Aggression, Dominance, and Individual Spacing* (L. Krames, P. Pliner, and T. Alloway, eds.), pp. 1–30, Plenum Press, New York.

Gier, H. T., 1968, Coyotes in Kansas, *Kansas State Coll. Agric. Exp. Stn. Appl. Sci. Bull.* 393.

Gilder, P. M., and Slater, P. J. B., 1978, Interest of mice in conspecific male odours is influenced by degree of kinship, *Nature (London)* **274**:364–365.

Gorman, M. L., 1976, A mechanism for individual recognition by odour in *Herpestes auropunctatus* (Carnivora: Viverridae), *Anim. Behav.* **24**:141–145.

Gould, S. J., 1977, *Ontogeny and Phylogeny,* Harvard University Press, Cambridge, Mass.

Greenberg, L., 1979, Genetic Component of bee odor in kin recognition, *Science* **206**:1095–1097.

Greenwood, P. J., and Harvey, P. H., 1976. The adaptive significance of variation in breeding area fidelity of the blackbird *(Turdus merula L.)*, *J. Anim. Ecol.* **45**:887–898.

Grotevant, H. D., Scarr, S., and Weinberg, R. A., 1977, Intellectual development in family constellations with adopted and natural children: A test of the Zajonc and Markus model, *Child Dev.* **48**:1699–1703.

Hamer, K. H., and Missakian, E., 1978, A longitudinal study of social play in Synanon/peer-reared children, in: *Social Play in Primates* (E. O. Smith, ed.), pp. 297–319, Academic Press, New York.

Hamilton, W. D., 1964, The genetical evolution of social behaviour, *J. Theor. Biol.* **7**:1–52.

Hamilton, W. D., and May, R. M., 1976, Dispersal in stable habitats, *Nature (London)* **269**:578–581.

Happold, M., 1976, The ontogeny of social behaviour in four conilurine rodents (Muridae) of Australia, *Z. Tierpsychol.* **40**:265–278.

Harcourt, A. H., 1978, Strategies of emigration and transfer by primates, with particular reference to gorillas, *Z. Tierpsychol.* **48**:401–420.

Harrington, F. H., and Mech, L. D., 1978, Wolf vocalization, in: *Wolf and Man: Evolution in Parallel* (R. L. Hall and H. S. Sharp, eds.), pp. 109–132, Academic Press, New York.

Harris, M., 1974. *Cows, Pigs, Wars, and Witches: The Riddles of Culture,* Random House, New York.

Hasler, M. J., and Nalbandov, A. V., 1974, The effect of weanling and adult males on sexual maturation in female voles *(M. ochrogaster)*, *Gen. Comp. Endocrinol.* **23**:277–238.

Hearn, J. P., 1977, The endocrinology of reproduction in the common marmoset *Callithrix jacchus,* in: *The Biology and Conservation of the Callitrichidae* (D. G. Kleiman, ed.), pp. 163–171, Smithsonian Institution Press, Washington, D.C.

Hediger, J., 1970, Zum Fortpflanzungsverhalten des Kanadischen Bibers *(Castor fiber canadensis),* *Forma Functio* **2**:336–351.

Hemsworth, P. H., Winfield, C. G., and Mullaney, P. D., 1976. A study of the development of the teat order of piglets, *Appl. Anim. Ethol.* **2**:225–233.

Hilborn, R., 1975, Similarities in dispersal tendency among siblings in four species of voles *(Microtus)*, *Ecology* **56**:1221-1225.

Hill, E. P., 1972*a*, Litter size in Alabama cottontails as influenced by soil fertility, *J. Wildl. Manage.* **36**:1199-1209.

Hill, E. P., 1972*b*, The cottontail in Alabama, *Auburn Univ. Agr. Exp. Stn. Bull.* No. 440.

Hill, J. L., 1974. Peromyscus: Effect of early pariring on reproduction, *Science* **196**:1042-1044.

Hill, W. F., 1978, Effects of mere exposure on preferences in nonhuman mammals, *Pschol. Bull.* **85**:1177-1198.

Hinde, R. A., 1971, Development of social behavior, in: *Behavior of Nonhuman Primates* (A. M. Schrier and F. Stollnitz, eds.), pp. 1-68, Academic Press, New York.

Hines, W., and Maynard Smith, J., 1974, Games between relatives, *J. Theoret. Biol.* **79**:19-30.

Hoeck, H., 1977, "Teat order" in Hyrox *(Procavia johnston* and *Heterohyrox brucei)*, *Z. Saeugetierkd.* **42**:42-115.

Hirshfield, M. F., and Tinkle, D. W., 1975, Natural selection and the evolution of reproductive effort, *Proc. Natl. Acad. Sci. U. S. A.* **72**:2227-2231.

Horn, H. S., 1978, Optimal tactics of reproduction and life-history, in: *Behavioural Ecology: An Evolutionary Approach* (J. R. Krebs and N. B. Davies, eds.), pp. 441-428, Sinauer, Sunderland, Mass.

Howard, W. E., 1960, Innate and environmental dispersal of individual vertebrates, *Am. Midl. Nat.* **63**:152-161.

Howard, W. E., 1965, Interaction of behavior, ecology, and genetics of introduced mammals, in: *The Genetics of Colonizing Species* (H. G. Baker and G. L. Stebbins, eds.), pp. 461-480, Academic Press, New York.

Hull, D. L., 1978*a*, Sociobiology: Scientific bandwagon or traveling medicine show? *Transaction* September/October, pp. 50-59.

Hull, D. L., 1978*b*, Altruism in science: A sociobiological model of cooperative behaviour among scientists, *Anim. Behav.* **26**:685-697.

Humphrey, N. K., 1974, Species and individuals in the perceptual world of monkeys, *Perception* **3**:105-114.

Hunt, R. H., 1977, Aggressive behavior by adult morelet's crocodiles *Crocodylus crocodylus* toward young, *Herpetologica* **33**:195-201.

Huxley, A., 1948, *Apes and Essence,* Harper and Row, New York.

Ingold, D. A., 1969, Social organization and behavior in a Wyoming pronghorn population, Ph.D. dissertation, University of Wyoming.

Ingram, J. C., 1977*a*, Parent–infant interactions in the common marmoset *(Callithrix jacchus)*, in: *The Biology and Conservation of the Callitichidae* (D. G. Kleiman, ed.), pp. 281-291, Smithsonian Institution Press, Washington, D.C.

Ingram, J. C., 1977*b*, Interactions between parents and infants, and the development of independence in the common marmoset *(Callithrix jacchus)*, *Anim. Behav.* **25**:811-827.

Jannett, F. J., 1978, The density-dependent formation of extended maternal families of the montane vole, *Microtus montanus nanus, Behav. Ecol. Sociobiol.* **3**:245-263.

Jonkel, C. J., and McTaggart Cowan, I., 1971, The black bear in the spruce-fir forest, *Wildl. Monogr.* **27**:1-57.

Jungius, H., 1971, The biology and behaviour of the reedbuck (*Redunca arundinum* Boddaert 1785) in the Kruger National Park, *Mamm. Depicta Beih. Z. Saevgetierkd.* pp. 1–106.

Kalkowski, W., 1972, Social ties among immature white mouse siblings, *Folia Biol.* **20**:245–261.

Kaplan, J. R., 1978, Fight interference and altruism in rhesus monkeys, *Am. J. Phys. Anthropol.* **49**:241–250.

Kennelly, J. J., 1978, Coyote reproduction, in: *Coyotes: Biology, Behavior, and Management* (M. Bekoff, ed.), pp. 73–93, Academic Press, New York.

King, J. A., 1954, Closed social groups among domestic dogs, *Proc. Am. Philos. Soc.* **98**:327–336.

King, J. A., 1973, The ecology of aggressive behavior, *Ann. Rev. Ecol. Syst.* **4**:117–138.

Kleiman, D. G., 1974, Patterns of behaviour in hystricomorph rodents, *Symp. Zool.* **34**:171–209,

Kleiman, D. G. (ed.), 1977*a*, *The Biology and Conservation of the Callitrichidae*, Smithsonian Institution Press, Washington, D.C.

Kleiman, D. G., 1977*b*, Monogamy in mammals, *Quart. Rev. Biol.* **52**:39–69.

Kleiman, D. G., 1979, Parent-offspring conflict and sibling competition in a monogamous primate, *Am. Nat.* **114**:753–759.

Kleiman, D.G., and Brady, C. A., 1978, Coyote behavior in the context of recent canid research: Problems and perspectives, in: *Coyotes: Biology, Behavior and Management* (M. Bekoff, ed.), pp. 163–188, Academic Press, New York.

Kleiman, D. G., and Eisenberg, J. F., 1973, Comparisons of canid and felid social systems from an evolutionary perspective, *Anim. Behav.* **21**:637–659.

Klingel, H., 1974, A comparison of the social behaviour of the Equidae, in: *The Behaviour of Ungulates and Its Relation to Management* (V. Geist and F. Walther, eds.), pp. 124–132, IUCN, Morges, Switzerland.

Klopfer, P. H., and Dugard, J., 1976, Patterns of maternal care in lemurs. III. *Lemur variegatus, Z. Tierpsychol.* **40**:210–220.

Knowlton, F. F., 1972, Preliminary interpretation of coyote population mechanics with some management implications, *J. Wildl. Manage.* **36**:369–382.

Konner, M., 1975, Relations among infants and juveniles in comparative perspective, in: *Friendship and Peer Relations* (M. Lewis and L. A. Rosenblum, eds.), pp. 99–129, Wiley, New York.

Kozakiewicz, M., 1976, Migratory tendencies in populations of bank voles and description of migrants, *Acta Theriol.* **21**:321–338.

Krebs, C. J., Gaines, M. S., Keller, B. L., Myers, J. H., and Tamarin, R. H., 1973, Population cycles of small rodents, *Science* **179**:35–41.

Krebs, C. J., Wingate, I., LeDuc, J., Redfield, J. A., Taitt, M., and Hilborn, R., 1976, *Microtus* population biology: Dispersal in fluctuating populations of *M. townsendii. Can. J. Zool.* **54**:79–95.

Kruuk, H., 1972, *The Spotted Hyena,* University of Chicago Press, Chicago.

Kukuk, P. F., Breed, M. C., Sobti, A., and Bell, W. J., 1977, The contributions of kinship and conditioning to nest recognition and colony member recognition in a primitively eusocial bee, *Lasioglossum zephyrum* (Hymenoptera: Halictidae), *Behav. Ecol. Sociobiol.* **2**:319–327.

Kurland, J. A., (1977), Kin selection in the Japanese monkey, *Contrib. Primatol.* **12**:1–145.

Lamb, M. E., 1978, Interactions between eighteen-month-olds and their preschool-aged siblings, *Child Dev.* **49**:51–59.

Lamprecht, J., 1978, The relationship between food competition and foraging size group in some larger carnivores, *Z. Tierpsychol.* **46**:337–343.

Langman, V. A., 1977, Cow-calf relationships in giraffe *(Giraffa camelopardalis giraffa)*, *Z. Tierpsychol.* **43**:264–286.

Laursen, L., and Bekoff, M., 1978, The African elephant, *Loxodonta africana* Blumenbach 1797, *Mamm. Species* **92**:1–8.

Le Boeuf, B. J., and Briggs, K. T., 1977, The cost of living in a seal harem, *Mammalia* **41**:167–195.

Lent, P. C., 1974, Mother-infant relationships in ungulates, in: *The Behaviour of Ungulates and its Relation to Management* (V. Geist and F. Walther, eds.), pp. 14–55, IUCN, Morges, Switzerland.

Lidicker, W. Z., 1962, Emigration as a possible mechanism permitting the regulation of a population density below carrying capacity, *Am. Nat.* **96**:29–34.

Lidicker, W. Z., 1975, The role of dispersal in the demography of small animals, in: *Small Mammals: Their Productivity and Population Dynamica* (F. Golley, K. Petrusewicz, and L. Rysziwski, eds.), pp. 103–128, Cambridge University Press, New York.

Lien, J., and Klopfer, F. D., 1978, Some relations between stereotyped suckling in piglets and exploratory behaviour and discrimination reversal learning in adult swine. *Appl. Anim. Ethol.* **4**:223–233.

Lockwood, R., 1976, An ethological analysis of social structure and affiliation in captive wolves, PH. D. dissertation, Washington University.

Łomnicki, A., 1978, Individual differences between animals and the natural regulation of their numbers, *J. Anim. Ecol.* **47**:461–475.

Low, B., 1978, Environmental uncertainty and the parental strategies of marsupials and placentals, *Am. Nat.* **112**:197–213.

Loy, J., Loy, K., Patterson, D., Keifer, G., and Conaway, C., 1978, The behavior of gonadectomized rhesus monkeys. I. Play, in: *Social Play in Primates* (E. O. Smith, ed.), pp. 49–78, Academic Press, New York.

Marler, P., 1961, The logical analysis of animal communication, *J. Theor. Biol.* **1**:295–317.

Marler, P., 1976, On animal aggression: The roles of strangeness and familiarity, *Am. Psychol.* **31**:239–246.

Marler, P., 1978, Perception and innate knowledge, in: *The Nature of Life* (W. H. Heidcamp, ed.), pp. 111–139, University Park Press, Baltimore.

Marler, P., and Hobbett, L., 1975, Individuality in a long-range vocalization of wild chimpanzees, *Z. Tierpsychol.* **38**:97–109.

Mason, W., 1977, The evolution of immaturity, Paper presented at the Conference on Behavioral Development in Man and Animals, University of Bielefeld, Bielefeld, West Germany.

Maynard Smith, J., 1972, Game theory and the evolution of fighting, in: *On Evolution* (J. Maynard Smith, ed.), pp. 8–28, Edinbrugh University Press, Edinburgh.

Maynard Smith, J., 1974, The theory of games and the evolution of animal conflicts, *J. Theor. Biol.* **47**:209–221.

Maynard Smith, J., 1976, Evolution and the theory of games, *Am. Sci.* **64**:41–45.

Maynard Smith, J., 1978*a, The Evolution of Sex,* Cambridge University Press, New York.

Maynard Smith, J., 1978*b*, The ecology of sex, in: *Behavioural Ecology: An Evolutionary Approach* (J. R. Krebs and N. B. Davies, eds.), pp. 159–179, Sinauer, Sunderland, Mass.

Mazurkiewicz, M., and Rajska, E., 1975, Dispersion of young bank voles from their place of birth, *Acta Theriol.* **20**:71–81 .

McCracken, G., and Bradbury, J., 1977, Paternity and genetic heterogeneity in the polygonous bat, *Phyllostomus hastatus, Science* **198**:303–306.

McBride, G., 1963, The "teat order" and communication in young pigs, *Anim. Behav.* **11**:53–56.

McKinney, F., 1978, Comparative approaches to social behavior in closely related species of birds, *Adv. Study Behav.* **8**:1–38.

McVittie, R., 1978, Nursing behavior of snow leopard cubs, *Appl. Anim. Ethol.* **4**:159–168.

Mech, L. D., 1970, *The Wolf,* Natural History Press, New York.

Medjo, D. C., and Mech, L. D., 1976. Reproductive activity in nine- and ten-month-old wolves, *J. Mammal.* **57**:406–408.

Metcalf, R. A., Stamps, J. A., and Krishnan, V. V., 1979, Parent-offspring conflict that is not limited by degree of kinship, *J. Theor. Biol.* **76**:99–107.

Moore, J., 1978, Dominance relations among free-ranging female baboons in Gombe National Park, Tanzania, *Rec. Adv. Primatol.* **1**:67–70.

Morton, M. L., Maxwell, C. S., and Wade, W. E., 1974, Body size, body composition, and behavior of juvenile belding ground squirrels, *Great Basin Nat.* **34**:121–134.

Murphy, M., 1971, Natural history of the Syrian golden hamster— A reconnaissance expedition, *Am. Zool.* **11**:632.

Myers, J. H., and Krebs, C. J., 1974, Population cycles in rodents, *Sci. Am.* **230**:38–46.

Nash, L. T., 1978, Kin preference in the behavior of young baboons, *Rec. Adv. Primatol.* **1**:71–73.

Neisser, E. G., 1951, *Brothers and Sisters*, Harper and Row, New York.

Norikoshi, K., 1974, The development of peer-mate relationships in Japanese macaque infants, *Primates* **15**:39–46.

O'Connor, R. J., 1978, Brood reduction in birds: Selection for fratricide, infanticide and suicide? *Anim. Behav.* **26**:79–96.

Otte, D., 1974, Effects and function in the evolution of signaling systems, *Ann. Rev. Ecol. Syst.* **5**:385–417.

Owens, N.W., 1975, Social play in free living baboons, *Papio anubis, Anim. Behav.* **23**:387–408.

Parker, G. A., and MacNair, M. R., 1978, Models of parent-offspring conflict. I. Monogamy, *Anim. Behav.* **26**:97–110.

Peterson, E., 1978, Discussion, in *Behavior and Neurology of Lizards* (N. Greenberg and P. D. MacLean, eds.), p. 225, National Institute of Mental Health, Rockville, Md.

Pfeifer, S., 1980, Role of the nursing order in social development of mountain lion kittens *(Felis concolor)*, *Dev. Psychobiol.* **13**:47–54.

Piaget, J., 1978, *Behavior and Evolution,* Pantheon, New York.

Popper, K. R., and Eccles, J. C., 1977, *The Self and Its Brain,* Springer International, New York.

Porter, R. H., and Wyrick, M., 1979, Sibling recognition in spiny mice *(Acomys cahirinus):* Influence of age and isolation, *Anim. Behav.* **27**:761–766.

Porter, R. H., Wyrick, M., and Pankey, J., 1978, Sibling recognition in spiny mice *(Acomys cahirinus), Behav. Ecol. Sociobiol.* **3**:61–68.

Priestnall, R., 1972, Effects of litter size on the behaviour of lactating female mice *(Mus musculus), Anim. Behav.* **20**:304.

Quadagno, D. M., Briscoe, R., Quadagno, J. S., 1977, Effect of perinatal gonadal hormones on selected nonsexual behavior patterns: A critical assessment of the nonhuman and human literature, *Psychol. Bull.* **84**:62–80.

Rasa, O. A. E., 1972, Aspects of social organization in captive dwarf mongooses, *J. Mammal.* **53**:181–185.

Rasa, O. A. E., 1973, Marking behaviour and its social significance in the African dwarf mongoose, *Helogale undulata rufula, Z. Tierpsychol.* **32**:293–318.

Rasa, O. A. E., 1977, The ethology and sociology of the dwarf mongoose *(Helogale undulata rufula), Z. Tierpsychol.* **43**:337–406.

Reiter, J., Stonson, N. L., and Le Boeuf, B., 1978, Northern elephant seal development: The transition from weaning to nutritional independence, *Behav. Ecol. Sociobiol.* **3**:337–367.

Rood, J. P., 1972, Ecological and behavioural comparisons of three genera of Argentine cavies, *Anim. Behav. Mongr.* **5**:1–83.

Rood, J. P., 1978, Dwarf mongoose helpers at the den, *Z. Tierpsychol.* **48**:277–287.

Rose, M. R., 1978, Cheating in evolutionary games, *J. Theor. Biol.* **75**:21–34.

Rose, R. K., and Gaines, M. S., 1978, The reproductive cycle of *Microtus ochrogaster* in eastern Kansas, *Ecol. Mongr.* **48**:21–42.

Rosenblatt, J. S., 1976, Stages in the early behavioural development of altricial young of selected species of non-primate mammals, in: *Growing Points in Ethology* (P. P. G. Bateson and R. A. Hinde, eds.), pp. 345–383, Cambridge University Press, New York.

Rothe, H., 1975, Some aspects of sexuality and reproduction in groups of captive marmosets *(Callithrix jacchus), Z. Tierpsychol.* **37**:255–273.

Rowell, T. E., Din, N. A., and Omar, A., 1968, The social development of baboons in their first three months, *J. Zool.* **155**:461–483.

Sade, D. S., 1965, Some aspects of parent-offspring and sibling relations in a group of rhesus monkeys, with a discussion of grooming, *Am. J. Phys. Anthropol.* **23**:1–18.

Sadleir, R. M. S. S., 1973, *The Reproduction of Vertebrates,* Academic Press, New York.

Savidge, I. R., 1974, Social factors in dispersal of deer mice *(Peromyscus maniculatus)* from their natal site, *Am. Midl. Nat.* **91**:395–405.

Sayler, A., and Salmon, M., 1971, An ethological analysis of communal nursing by the house mouse *(Mus musculus), Behaviour* **40**:62–85.

Schaller, G. B., 1972, *The Serengeti Lion,* University of Chicago Press, Chicago.

Schenkel, R., and Schenkel-Hulliger, L., 1969, Ecology and behaviour of the black rhinoceros *(Diceros bicornis L.). Mamm. Depicta Beih. Z. Saeugetiericd.,* pp. 1–101.

Schneider, D. G., Mech, L.D., and Tester, J. R., 1971, Movements of female raccoons and their young as determined by radio-tracking, *Anim. Behav. Mongr.* **4**:1–43.

Schooler, C., 1972, Birth order effects: Not here, not now, *Psychol. Bull.* **78**:161–175.

Scott, J. P., 1962, Hostility and aggression in animals, in: *Roots of Behavior* (E. L. Bliss, ed.), pp. 167–178, Hafner, New York.

Scott, J. P., 1977, Social genetics, *Behav. Genet.* **7**:327–346.

Seyfarth, R. M., Cheney, D. L., and Hinde, R. A., 1978. Some principles relating social interactions and social structure among primates, *Rec. Adv. Primatol.* **1**:39–51.

Sheppard, D. H., and Yoshida, S. M., 1971, Social behavior in captive Richardson's ground squirrels, *J. Mammal.* **52**:793–799.

Sherman, P., 1980, The limits of nepotism, in: *Sociobiology: Beyond Nature/Nurture* (G. Barlow and J. Silverberg, eds.), pp. 505–544, Westview Press, Boulder, Colo.

Simon, H. A., 1969, *The Sciences of the Artificial,* The M.I.T. Press, Cambridge, Mass.

Simpson, T. R., 1975, The ontogeny of social behavior in coyotes, M.Sc. thesis, Texas A & M University.

Smith, A. T., 1974*a*, The distribution and dispersal of pikas: Consequences of insular population structure, *Ecology* **55**:1112–1119.

Smith, A. T., 1974*b*, The distribution and dispersal of pikas: Influences of behavior and climate, *Ecology* **55**:1368–1376.

Smith, A. T., 1978, Comparative demography of pikas *(Ochonta)*: Effect of spatial and temporal age-specific mortality, *Ecology* **59**:133–139.

Smith, C. C., and Fretwell, S. D., 1974, The optimal balance between size and number of offspring, *Am. Nat.* **108**:499–506.

Smith, W. J., 1977, *The Behavior of Communcating: An Ethological Approach,* Harvard University Press, Cambridge, Mass.

Spencer-Booth, Y., 1968, The behaviour of twin rhesus monkeys and comparisons with the behaviour of single infants, *Primates* **9**:75–84.

Stacey, P., 1978, The ecology and evolution of communal breeding in the acorn woodpecker, Ph.D. dissertation, University of Colorado.

Stamps, J. A., 1978, A field study of the ontogeny of social behavior in the lizard *Anolis aeneus, Behaviour* **66**:1–31.

Stamps, J. A., Metcalf, R. A., and Krishman, V. V., 1978, A genetic analysis of parent-offspring conflict, *Behav. Ecol. Sociobiol.* **3**:369–392.

Stevenson, M. F., 1978, The ontogeny of playful behaviour in family groups of the common marmoset, *Rec. Adv. Primatol.* **1**:139–143.

Storm, G. L., Andrews, R. D., Phillips. R. L., Bishop, R. A., Siniff, D. B., and Tester, J. R., 1976, Morphology, reproduction, dispersal, and mortality of midwestern red fox populations, *Wildl. Monogr.* **49**:1–82.

Stringham, S. F., 1974, Mother-infant relations in moose, *Nat. Can. (Ottawa)* **101**:325–369.

Sussman, R. W., 1977, Socialization, social structure, and ecology of two sympatric species of Lemur, in: *Primate Bio-Social Development* (S. Chevalier-Skolnikoff and F. E. Poirier, eds.), pp. 515–528, Garland, New York.

Sutton-Smith, B., and Rosenberg, B. G., 1970, *The Sibling,* Holt, Rinehart and Winston, New York.

Symons, D., 1978*a,* The question of function: Dominance and play, in: *Social Play in Primates* (E. O. Smith, ed.), pp. 193–230, Academic Press, New York.

Symons, D., 1978b, *Play and Aggression: A Study of Rhesus Monekeys*, Columbia University Press, New York.

Teicher, M. H., and Kenny, J. T., 1978, Effect of reduced litter size on the suckling behaviour of developing rats, *Nature (London)* **275**:644–646.

Thompson, D. C., 1978, The social system of the grey squirrel, *Behaviour* **64**:305–328.

Treisman, M., 1977, The evolutionary restriction of aggression within a species: A game theory analysis, *J. Math. Psychol.* **16**:167–203.

Treisman, M., 1978, Bird song dialects, repertoire size, and kin association, *Anim. Behav.* **26**:814–817.

Trivers, R., 1974, Parent–offspring conflict. *Am. Zool.* **14**:249–264.

van Lawick, H., and van Lawick-Goodall, J., 1971, *Innocent Killers*, Houghton Mifflin, Boston.

van Lawick-Goodall, J., 1968, The behaviour of free-living chimpanzees in the Gombe Stream Reserve, *Anim. Behav. Monogr.* **1**:161–311.

van Valen, L., and Pitelka, F. A., 1974, Commentary—Intellectual censorship in ecology, *Ecology* **55**:925–926.

Vestal, B. M., and Hellack, J. J., 1978, Comparison of neighbor recognition in two species of deer mice *(Peromyscus)*, *J. Mammal.* **59**:339–346.

Viitala, J., 1977, Social organization in cyclic subarctic populations of the voles (*Clethrionomys rufocanus* Sund.) and *Microtus agrestis* (L.), *Acta Zool. Fenn.* 14:53–93.

Voland, E., 1977, Social play behavior of the common marmoset (*Callithrix jacchus* Erxl., 1977) in captivity, *Primates* **18**:883–901.

Wade, T. D., 1979, Inbreeding, kin selection, and primate social evolution, *Primates,* **20**:355–378.

Waldman, B., and Adler, K., 1979, Toad tadpoles associate preferentially with siblings, *Nature* **282**:611–613.

Wallenchinsky, D., and Wallace, I., 1978, *The People's Almanac,* Vol. 2, Morrow, New York.

Wandrey, R., 1975, Contribution to the study of social behaviour of golden jackals (*Canis aureus* L.), *Z. Tierpsychol.* **39**:365–402.

Watts, A., 1977, *Nonsense,* Dutton, New York.

Watts, C. R., and Stokes, A. W., 1971, The social order of turkeys, *Sci. Am.* **224**:112–118.

Wemmer, C., and Scow, K., 1977, Communication in the felidae with emphasis on scent marking and contact patterns, in: *How Animals Communicate* (T. A. Sebeok, ed.), pp. 749–766, Indiana University Press, Bloomington, Ind.

West Eberhard, M. J., 1975, The evolution of social behavior by kin selection, *Quart. Rev. Biol.* **50**:1–33.

White, T. C. R., 1978, The importance of relative shortage of food in animal ecology, *Oecologia (Berlin)* **33**:71–86

Williams, G. C., 1975, *Sex and Evolution,* Princeton University Press, Princeton, N. J.

Williams, G. C., and Williams, D. C., 1957, Natural selection of individually harmful social adaptations among sibs with special reference to social insects, *Evolution* **11**:32–39.

Wilson, A. C., Bush, G. L., Case, S. M., and King, M. C., 1975, Social structuring of mammalian populations and rate of chromosomal evolution, *Proc. Natl. Acad. Sci. U.S.A.* **72**:5061–5065.

Wilson, E. O., 1971, *The Insect Societies,* Harvard University Press, Cambridge, Mass.

Wilson, E. O., 1975, *Sociobiology: The New Synthesis,* Harvard University Press, Cambridge, Mass.

Wilson, E. O., 1977, Animal and human sociobiology, in: *The Changing Scenes in the Natural Sciences, 1776–1976* (C. E. Goulden, ed.), pp. 273–281, Academy of Natural Sciences, Philadelphia.

Wilson, S., 1973, The development of social behaviour in the vole *(Microtus agrestis),* *Zool. J. Linn. Soc.* **52**:45–62.

Wilson, S., and Kleiman, D. G., 1974, Eliciting and soliciting play, *Am. Zool.* **14**:341–370.

Windberg, L. A., and Keith, L. B., 1976, Experimental analyses of dispersal in snowshoe hare populations, *Can. J. Zool.* **54**:2061–2081.

Wolff, J. O. and Holleman, D. F., 1978, Use of radioisotope labels to establish genetic relationships in free-ranging small mammals, *J. Mammal.* **59**:859–860.

Woolpy, J. H., and Eckstrand, I., 1979, Wolf pack genetics, a computer simulation with theory, in: *The Behavior and Ecology of Wolves* (E. Klinghammer, ed.), pp. 206–224, Garland, New York.

Wu, H. M. H., Holmes, W. G., Medina, S. R., and Sackett, G. P., 1980, Kin preference in infant *Macaca nemestrina, Nature* **285**:225–227.

Yamada, M., 1963, A study of the blood-relationship in natural society of the Japanese macaque, *Primates* **4**:43–65.

Yeaton, R. I., 1972, Social behavior and social organization in Richardson's ground squirrels *(Spermophilus richardsonii)* in Saskatchewan, *J. Mammal.* **53**:139–147.

Young, J. Z., 1978, *Programs of the Brain,* Oxford University Press, New York.

Zajonc, R. B., 1971, Attraction, affiliation and attachment, in: *Man and Beast: Comparative Social Behavior* (J. F. Eisenberg, W. S. Dillon, and S. D. Ripley, eds.), pp. 141–179, Smithsonian Institution Press, Washington, D.C.

Zajonc, R. B., and Markus, G. B., 1975, Birth order and intellectual development, *Psychol. Rev.* **82**:74–88.

9

The Evolution of Male Parental Investment in Mammals

Devra G. Kleiman and James R. Malcolm

1. Introduction

A variety of approaches have been used to understand the evolution of male parental care. General frameworks are provided by Trivers' theory of sexual selection (1972), the theory of life history strategies (see Horn, 1978; Stearns, 1976) and game theory (Grafen and Sibly, 1978; Maynard Smith, 1977). The factors invoked to explain male parental investment have varied with the level of analysis; intrinsic biological factors such as internal versus external fertilization (Dawkins and Carlisle, 1976; Ridley, 1978) or the capacity to invest (Orians, 1969) have been used to illuminate differences between large taxonomic units such as the vertebrate classes; ecological factors such as harshness (Wilson, 1975), richness (Jenni, 1974), and unpredictability (Pitelka et al., 1974) have all been invoked to explain the presence of unusual levels of male investment in smaller taxonomic units.

In this paper, we will review the types of male parental investment seen in mammals and their taxonomic distribution. We will then consider the evolution of male parental investment and assess its role in molding the social and breeding systems of different species. We will try to determine if there are certain conditions when male care may place constraints on the breeding system, or particular ecological pressures which promote male parental investment. Unlike Spencer-Booth (1970), we will not be concerned with the motivation underlying male parental care or the stimuli eliciting this behavior.

DEVRA G. KLEIMAN AND JAMES R. MALCOLM • Department of Zoological Research, National Zoological Park, Smithsonian Institution, Washington, DC. 20008.

1.1.   The Definition of Male Parental Investment

We conceive of male parental investment as *any increase in a prereproductive mammal's fitness attributable to the presence or action of a male.* This can include behavior directed at a pregnant female between conception and birth. It also includes behavior of the parental type by males other than the known (or presumed) father. Redican (1976) has used the term "paternalistic" in this context since in many species the actual father is unknown. We will not try to differentiate between parental behavior exhibited by fathers versus other males, since the same behavior patterns are usually shown with only quantitative differences in the behavior of fathers and nonfathers.

Our concept of parental investment is somewhat different than Trivers's (1972), i.e., "any investment by the parent in an individual that increases the offspring's chance of surviving (and hence reproductive success) at the cost of the parent's ability to invest in other offspring" (p. 139). First, there are many types of parental investment, especially protection and babysitting, where guarding of one individual does not necessarily reduce the fitness of another member of the same clutch or litter (or even other relatives). Second, in iteroparous species, and especially seasonal breeders, the extent to which investing in one set of offspring precludes future investment may be minimal. Whereas Trivers appears to view investment as a fixed quantity which is apportioned during an individual's lifetime, we tend to consider investment at different times as being at least partially independent. Third, as pointed out by Trivers and Hare (1976) and J. H. Ligon and S. H. Ligon (1978), some investment in young may increase an individual's reproductive success by raising future helpers. Fourth, Trivers excludes investment by individuals other than the parents.

1.2.   Categories of Male Parental Investment

We will later try to enumerate some of the myriad ways in which male mammals can and do help young of their species. First, however, we will define two major axes along which male parental investment can vary which aid both in organizing the data and in understanding how male parental investment can affect the social organizations of the species in which they occur.

The first distinction is between *direct* and *indirect* forms of male parental investment. *Direct* investment by a male includes those acts which a male performs towards young that have an immediate physical influence on them which increases survivorship. Feeding or carrying infants falls into this category as does sleeping with young, grooming young, or playing with young. References to male parental investment in the literature usually involve *direct* investment (we will occasionally use the term male parental care when referring to direct male investment).

Indirect male parental investment includes those acts a male may perform in the absence of the young which increase the latter's survivorship. These acts may have delayed effects on survivorship of young and include such behaviors as the acquisition, maintenance, and defense of critical resources within a home range or territory by the elimination of competitors, the construction of shelters, and actions which improve the condition of pregnant or lactating females. Many forms of male parental investment that are indirect are also *incidental* to the species' breeding system, ecology, or social organization. These are activities which males would perform regardless of the presence of young. For example, breeding male zebras *(Equus burchelli)* or gorillas *(Gorilla gorilla)* defend harem females and offspring when the latter are threatened. Presumably, a male would display such protective behavior both in the presence and absence of young. Similarly, behaviors such as scent marking and long-distance vocalizations which aid in the spacing of individuals or groups, and thus may maintain critical resources for eventual use by young, should also be considered as indirect forms of male parental investment which appear to be an incidental consequence of social organization and ecology. There are suggestions in the literature that the frequencies of these behaviors alter in response to the existence of young, even when they are not performed in the direct presence of young. Thus, male behavior may be modified when young are born, even if direct male care is not seen.

Indirect and incidental forms of investment are often ignored, but may be important. Ewer (1973) suggested that in carnivores such as otters, where males hold territories that overlap the ranges of several females, the young may benefit considerably by the territorial male's exclusion of other males from the preferred feeding areas. In birds, Wittenberger (1978) has related the various breeding systems seen in grouse to differences in the advantages which a female can gain from male territoriality. He suggests that monogamy will evolve in those species where a female can increase her foraging efficiency by feeding on a male's territory.

The second axis which will be used to differentiate between patterns of male parental investment* relates to the degree to which investment in one individual or litter precludes investment in others, a parameter that has been recognized as important by other authors [e.g., Maynard Smith (1977) and Ridley (1978)]. Altmann *et al.* (1977) have coined the terms "depreciable" and "non-depreciable." They define depreciable as follows: "A depreciable contribution is like a nonrenewable or slowly renewable resource: it is reduced in availability to one individual to the extent that it is expended on or used by another" (p. 409). In this paper, *individual* will refer to a single female or her current litter. Although the distinction between depreciable and nondepreciable is clear in extreme cases, most patterns of male investment have both depreciable and nondepreciable components.

Crouching with or huddling over multiple offspring provides the commonest example of nondepreciable direct investment, while regurgitation or carrying of young is usually depreciable. However, if resources being devoted to the young,

---

*It should be obvious that these axes can also be used to differentiate types of maternal care.

such as meat carried to a den, are in such abundance or renew so fast that assistance to one individual or litter has little or no effect on the contribution to others, then even acts of regurgitation may sometimes be considered as nondepreciable. It will be argued below that when depreciable investment is a sufficiently valuable resource for females to compete over it, the social system may be profoundly affected.

It should be emphasized that the same behavior pattern may be considered as depreciable under some conditions (or in some species) and nondepreciable in others. The categorization depends on the species' social organization and reproductive characteristics. For example, in a polygynous, but nongregarious species where females raise young separately, huddling by a male with young may be a depreciable investment since only a single young or litter can be nested with at a time, while in a polygynous social species where females rear young communally, the male may be able to huddle simultaneously with a number of litters. The behavior may be nondepreciable under the latter circumstances.

Most forms of indirect male parental investment are nondepreciable in that they benefit several young without differentiating among them. One exception is the activity of a male feeding a pregnant or lactating female. Only one litter can benefit from this behavior at a time.

1.3. Measuring Male Parental Investment

Real problems exist in measuring the costs to a male, in terms of energy and risks of injury, in parental behavior, and the benefits in terms of increased survivorship accruing to the young. A simple list of the presence or absence of each pattern of investment is clearly insufficient. Measuring the frequencies or durations of patterns of male parental investment is a better solution but still imperfect, since we cannot measure the real energetic costs and risks. For indirect and particularly incidental forms of investment, there is a problem in deciding if the male really incurs costs by his contribution.

One method would be to consider each behavior pattern separately and attempt to gauge the relative difference between the sexes in performance of the pattern. For example, one can measure the time in the nest with young for a mated male and female and determine which sex spends relatively more time with the young. This approach results in a different male : female ratio for each behavior pattern performed by a species. Although such ratios may be meaningful in comparisons among closely related species where the behavior patterns of parental care are similar, they will be less useful when comparing broad taxonomic groups. However, data on six muroid rodents presented by Hartung and Dewsbury (1979) suggest that in most cases the amount of parental care exhibited by the male and female of a species will be correlated, i.e., there will be species differences in overall parental investment which will be reflected in both male and female parental care. Thus, the use of ratios to compare even closely related species may still be inappropriate.

Also, some highly significant male behaviors, which may be crucial to development, can occur over a very limited time span. For example, male canids may exhibit little direct care to young prior to weaning but then contribute significantly to the weaning process by bringing solid foods to the young.

Measuring indirect investment, especially the role of males in securing and maintaining resources, is also difficult. When males and females hold partially, or totally, overlapping, long-term territories, the male may aid the female and young by excluding competitors and detecting predators. However, many forms of land tenure in mammals do not conform to this simple pattern. In the analysis of male parental investment in carnivores presented below, males were only recorded as aiding females by securing resources if females also defended the same resources from conspecifics. This criterion excludes many "lek-type" territories in males.

Even in those cases where females and males appear to defend a common area or resource, a decision has to be made on whether a resident male (or males) limits or exacerbates competition for resources. In some species, such as lions *(Panthera leo)* (Schaller, 1972), resident males may appropriate food acquired by females. In other species, resident males may not completely exclude other males (Rood and Waser, 1978) or may defend areas only temporarily (Leyhausen, 1965). Carnivore species where active intersexual competition for food or resources has been reported were not recorded as showing indirect male investment in the analysis that follows.

1.4.   The Available Data on Male Parental Investment

The broad definition of male parental investment used here makes a comprehensive survey of the mammals impossible. A very large number of species probably show some form of incidental, indirect investment, such as antipredator and sentinel behavior or resource maintenance and defense.

Our vast ignorance of the social life of most species in the rodents and bats reduced the problem of attempting a comprehensive survey. In this review, we sought evidence for both direct and indirect forms of male investment in the carnivores and summarized the literature on direct male care in other mammals. For the summary of male parental investment in primates, we have relied extensively on reviews by Hrdy (1976), Mitchell and Brandt (1972), Mitchell (1969), and Redican (1976), and have not, for the most part, cited original sources. The summary of primate male parental investment is therefore not exhaustive, but the reader can find more recent references in the bibliography prepared by Williams (1978).

Many of the descriptions of male parental care are based on laboratory or zoo observations. Although the captive setting usually provides the opportunity for more quantitative data collection, the conditions of confinement may seriously distort natural behavior, especially when the social milieu is abnormal. The successful induction of parental care, including grooming, retrieving, and nest building in male rats *(Rattus norvegicus)* after a 6- to 7-day exposure to infants (Rosenblatt, 1967), exemplifies the manner in which caregiving behavior towards young can be elicited by manipulation of the social and physical setting.

A recent debate on male parental care in domestic gerbils *(Meriones unguiculatus)* highlights other problems with captive observations. While several authors (e.g., Elwood, 1975; Elwood and Broom, 1978; Gerling and Yahr, 1979) have shown that male parental care is common in gerbils and does not negatively affect pup survivorship, Ahroon and Fidura (1976) found that the presence of a male significantly increased pup mortality. More recently, several explanations have been proposed for these contradictory findings [Klippel (a.k.a. Ahroon), 1979], most of which relate to the transport and housing of the experimental animals in the Ahroon and Fidura (1976) study.

Descriptions of male parental care is captivity should be treated with caution, unless the persistence of male parental care in a variety of individuals under a variety of conditions indicates that such behavior could be occurring in nature or there is some corroborative evidence from the field.

Available data on male parental investment in mammals fall into three categories. Some sources refer to the existence of male parental investment without detailing the specific behavior patterns. A second category names the behavior pattern observed but provides no quantitative data to indicate how common the act is. Third, and most rarely, authors have presented the behavior patterns involved in male parental investment as well as quantitative supporting data. The information available from different descriptions is rarely comparable. Also, authors infrequently describe changes in or refer to indirect forms of male parental investment.

Another problem involves the evaluation of negative evidence. In some studies, male parental investment could not be seen or expressed due to housing or observation conditions. In some, it could have occurred, but was simply not mentioned. The most reliable negative sources are obviously those in which the conditions were adequate for the behavior to be performed and seen, and the author specifically indicates that no male parental care was observed. Such references are as rare as those in which an author not only describes but quantifies both direct and indirect male parental investment. Thus, the available literature is, for the most part, inadequate.

2. Results

2.1. What Can Male Mammals Do?

Male mammals can neither gestate young nor lactate; thus their intrinsic ability to aid offspring is more limited than the female's. However, if these sex differences are disregarded, it appears as though males have the potential to display the same parental care patterns as females and that their ability to aid offspring is constrained by the same factors limiting female parental care.

For example, herbivorous mammals of either sex could rarely increase their offspring's survival or fitness by attempting to carry or hoard grass and leaves. However, where food comes in large, energy-rich packets which can be either stored for later use or carried, the potential for parental investment by both male and female increases. Thus, beaver *(Castor fiber)* can create a food hoard for postweaning feeding of young, and some larger carnivores are able to kill and transport large prey to developing offspring.

In any species in which environmental manipulation (i.e., nest building, trail formation) may be a critical factor in increasing both individual and offspring fitness, the male's role could approximate the female's. Interestingly, this is one behavior category where the males of different taxonomic groups may vary most in their investment relative to the females. For example, few carnivore males are reported to take a major role in burrow and nest construction. However, in a recent comparison of male parental behavior in six muroid rodents, Hartung and Dewsbury (1979) provide data suggesting that males of some muroids exhibit as much manipulation of nesting materials as females or in some conditions more.

Similarly, the maintenance of the trail system by clearing debris off pathways in rufous elephant shrews *(Elephantulus rufescens)* is more common in the male than in the female (Rathbun, 1979). Those behaviors which are usually exhibited more by males than by females are typically indirect and also contribute to the fitness of the male.

Two other factors may increase an offspring's fitness, but are not typically considered as evidence for male parental investment. First, for a large segment of artiodactyls whose young are hiders, the pattern of "avoiding offspring" may increase offspring survivorship. Indeed, among hiders, females similarly avoid young as an antipredator device. Unfortunately, males which actively avoid young and males which ignore young are difficult to distinguish, even though the former may be an evolved characteristic.

Second, in some cases tolerance towards juveniles may be an evolved attribute which contributes to offspring fitness. The males of highly territorial species who attack intruder conspecifics, but who do not attack maturing young within the territory, may have evolved a differential response to conspecific young. Male aggression towards young may also be differentially inhibited or suppressed, depending on the relationship of the male to the young (Hrdy, 1976). That aggression towards young is usually inhibited in male mammals when it is known that the young could provide a nutritional food source for the males of some species through cannibalism (Sherman, 1979) further suggests an evolved response. Of course, genes promoting male aggression towards, and cannibalism of, young would not be easily spread in most species, since the young most immediately available to a male are more likely to be related than unrelated. Thus a male would be reducing his own fitness by regular cannibalism.

The tendency of males to harm young may be inhibited by counterstrategies evolved by the young. For example, there are several descriptions of male mammals treating conspecific young of both sexes as though they were females. Geist

(1971) provides quantitative evidence for this phenomenon in Stone sheep *(Ovis dalli)*, and Rood (1972) describes it for the caviid genera *Cavia* and *Microcavia*. Smythe (1978) and Kleiman (1971) similarly report that individual males of the dasyproctid genera *Dasyprocta* and *Myoprocta* court and urinate over infants as though they were females. In some species, young may bear odors which resemble the adult female (they may even be transferred from the mother during early ontogeny) or, at least, smell unlike a mature male. Clearly, it is to the advantage of the young to develop and maintain mechanisms for not provoking aggression from adult males.

2.2. What Do Male Mammals Do?

Since male and female mammals have the potential to be similar in parental investment patterns (disregarding gestation and lactation), a summary of male parental behavior will resemble a summary of female parental behavior. Figures 1–4 present categories of indirect and direct investment which may be observed. Resource acquisition, maintenance, and defense (Fig. 1) refers to those behaviors (scent marking, patrolling, vocalizing, food hoarding, expulsion of intruders) which contribute indirectly to ensuring that necessary resources are available for use by young. Figure 2 presents additional indirect forms of male parental investment, including provisioning the female, shelter construction, and antipredator behaviors. Of all indirect forms of male parental investment, only investment in the female does not contribute to male as well as juvenile survivorship.

Figures 3 and 4 detail direct male parental care in mammals. Huddling encompasses all behaviors associated with resting or sleeping in contact with young; by increasing body temperatures of young such behavior increases growth rates of deer mice *Peromyscus californicus* (Dudley, 1974a,b). Retrieval refers to either carrying or leading young back to a shelter or secure site. The transport of young includes carrying young on a regular or irregular basis during ordinary movements through the home range. Groom and clean young includes licking, nibbling, and other cleaning movements as well as ingesting excreta of young. Providing food for young consists of regurgitation of food, carrying food to young, and permitting young to take food in the male's possession (through food sharing or food stealing). It may also include leading young to a rich food source. Babysitting refers to remaining with young during the absence of the mother. Playing and socializing with young is a broad category encompassing all social interactions with young that contribute to the social development and social integration of young. Behaviors such as mutual sniffing, greeting, wrestling, and scent marking young are included in this category. Active defense of young includes only those aggressive behaviors performed when young are being harassed and threatened by conspecific or nonconspecific intruders.

Table 1 is a summary of direct male parental care in mammals, exluding the

Resource acquisition

Resource maintenance

Resource defense

Figure 1. Categories of male parental behavior: Indirect male investment. I. Behavior associated with securing resources for female and young.

carnivores. Indirect male investment was included in the remarks section only if it was conspicuously mentioned in a reference. Table 2 presents all available data on both direct and indirect male parental investment in carnivores, as well as references which specifically indicate the absence of male investment.

Shelter construction and maintenance

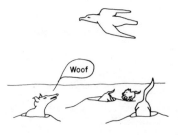

Sentinel and antipredator behavior

Care of female

Figure 2. Categories of male parental behavior: Indirect male investment. II. Other forms of indirect male parental investment.

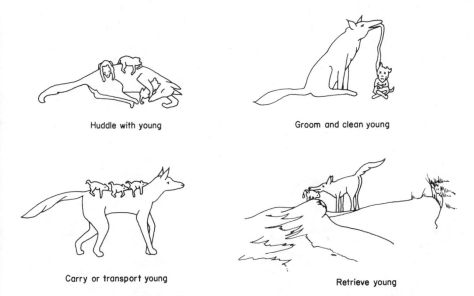

Huddle with young

Groom and clean young

Carry or transport young

Retrieve young

Figure 3. Categories of male parental behavior: Direct male investment. I. Behavior often shown to young before they are weaned.

Figure 5 summarizes the data from Table 1 (and part of Table 2). The percent of genera within each mammalian order for which direct male parental care has been described is presented. Figure 6 presents the percent of genera in each order for which direct male parental care has been described relative to the total number of mammalian genera (from Walker, 1975). A comparison of the two figures reveals some interesting patterns. In several large orders direct male care has been recorded at low frequencies, regardless of how the data are presented; these include the marsupials, chiropterans, cetaceans, and artiodactyls. Several orders with few genera have a disproportionately large number of genera exhibiting direct male parental investment, especially the Perissodactyla, but the percent of genera are small relative to all mammalian genera. By contrast, although only a small percentage of rodent genera (6.4%) have been described as exhibiting direct male investment, the percentages for the rodents and primates are the same (2.2%), when all mammalian general are considered.

Nearly 40% of primate genera have been reported as exhibiting direct male parental care, the highest for any individual order. Yet, the carnivores show the greatest percentage of genera (Fig. 6), when all genera of mammals are considered.

Provide food to young

Active defense of young

Babysitting

Play or socialization with young

Figure 4. Categories of male parental behavior: Direct male investment. II. Behavior often shown to young between weaning and independence.

Although the above figures are probably biased because our knowledge of different mammalian groups varies and the carnivores were researched more thoroughly than other orders, they may represent real trends. Thus, based on our current knowledge of the life histories and social systems of the marsupials, chiropterans, artiodactyls, and pinnipeds, it is unlikely that further research would reveal a much higher frequency of direct male parental care. A large percentage of the species within these orders tend to exhibit either or both of two characteristics; they are typically polygynous in their mating systems, and maternal care patterns are not very complex. By contrast the insectivores, once they are better known, might exhibit higher frequencies of direct male parental investment since maternal care patterns are complex and polygyny may be found to be less common than appears to be the case currently. Undoubtedly, once the rodents are better known, the genera of rodents in which direct male parental investment has been described will exceed the primate and carnivore genera, since the carnivores and primates are very well studied relative to the rodents.

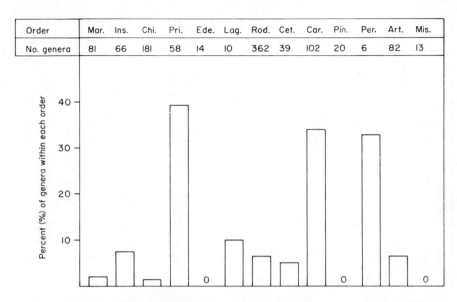

Figure 5. The proportion of genera within each mammalian order in which direct male parental care has been recorded. Abbreviations: Mar., Marsupialia; Ins., Insectivora; Chi., Chiroptera; Pri., Primates; Ede., Edentata; Lag., Lagomorpha; Rod., Rodentia; Cet., Cetacea; Car., Carnivora; Pin., Pinnipedia; Per., Perissodactyla; Art., Artiodactyla; Mis. (miscellaneous) Monotremata, Dermoptera, Pholidota, Proboscoidea, Hyracoidea, and Sirenia.

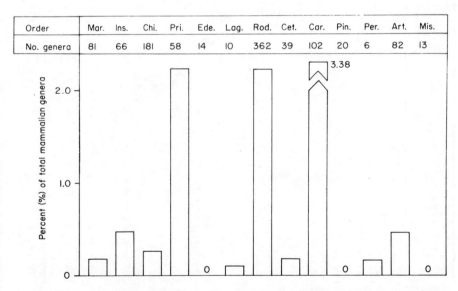

Figure 6. The genera of mammals in each order recorded as showing direct male parental care as a percentage of the total number of mammalian genera. Abbreviations as in Fig. 5.

Table 1. Direct Male Parental Investment in Mammals (Excluding Carnivora)[a,b]

Mammal	Depreciable	Nondepreciable	Source of observations	Modal mating system	Remarks	References
MARSUPIALIA						
Phalangeridae						
Petaurus breviceps	RT	DF, HD, BS	C	PGN		Schultze-Westrum (1965)
Sugar glider						
Dasyuridae						
Sarcophilus harrisii	CL		C	PGN		Turner (1970)
Tasmanian devil						
EUTHERIA						
Soricidae						
Suncus etruscus	RT	HD	C		SC	Fons (1974)
Etruscan shrew						
Cryptotis parva	RT		C		SC	Conaway (1958)
Least shrew						
Tenrecidae						
Hemicentetes semispinosus		HD	C		SC	Gould and Eisenberg (1966)
Streaked tenrec						
Tenrec ecaudatus	PF		C		SC	Louwman (1973)
Tenrec						
Macroscelididae						
Elephantulus rufescens	CL	HD, DF, PL	C, F	M	AP	Rathbun (1979) (personal communication)
Rufous elephant shrew						
Phyllostomatidae						
Phyllostomus discolor	TR		F	PGN		Bradbury (1977)
Spear-nosed bat						
Vampyrum spectrum	PF	HD	C, F	M		Greenhall (1968); Vehrencamp et al. (1977)
False vampire						
Pteropodidae						
Rousettus sp.	CL	HD	C	PBGM		Kulzer (1958)
Rousette bats						

Taxon					Reference
Indridae					
Propithecus verreauxi Sifaka	CL, TR, HD	HD	F	PGN	Mitchell (1969)
Galagidae					
Galago senegalensis Senegal bush baby	PL, HD, DF CL, RT	HD	C		Mitchell and Brandt (1972)
Tarsiidae					
Tarsius syrichta Philippine tarsier	TR, HD		C	M	Schreiber (1968)
Callitrichidae					
Callithrix jacchus Common marmoset	RT, TR, PF, HD		C	M	Mitchell (1969); Redican (1976)
Cebuella pygmaea Pygmy marmoset	RT, TR, HD		C	M	Mitchell (1969); Redican (1976)
Saguinus spp. Tamarins	RT, TR, HD		C	M	Mitchell (1969); Redican (1976)
Leontopithecus rosalia Lion tamarin	RT, TR, HD		C	M	Mitchell (1969); Redican (1976)
Cebidae					
Cebus albifrons Capuchin	TR, PL	PL	C	PGN	Mitchell (1969)
Saimiri sciureus Squirrel monkey	TR, PL, RT	PL	C	PBGM	Mitchell (1969); Hrdy (1976)
Alouatta palliata Howler monkey	TR, PL, RT	PL	F	PGN	Mitchell (1969)
Callicebus moloch Titi monkey	TR, RT, HD		F, C	M	Mitchell (1969)
Aotus trivirgatus Night monkey	TR, RT, HD		F, C	M	Mitchell (1969)

aIndirect male investment is indicated in "Remarks" only if it is very conspicuous. Definitions of mating systems are from Selander (1972).

bKey: Observation source: C, captive observations; F, field observations. Mating system: M, monogamy; PGN, polygyny; PBGM, polybrachygamy. Direct care: HD, huddle with young; CL, groom and clean; RT, retrieve; TR, carry and transport; PF, provide food to young; PL, play and socialize; DF, active defense; BS, babysitting. Indirect care: RA, resource acquisition; RM, resource maintenance; RD, resource defense; AP, antipredator, sentinel behavior; SC, shelter construction and maintenance; CF, care of female.

Table 1. *(Continued)*

Mammal	Depreciable	Nondepreciable	Source of observations	Modal mating system	Remarks	References
Cercopithecidae						
Macaca mulatta Rhesus monkey	PL, CL, HD, TR DF	PL, HD, DF	F, C	PBGM	Infrequent male care	Mitchell (1969); Redican (1976); Hrdy (1976)
Macaca fascicularis Crab-eating macaque	PL, DF	PL, HD, DF	C	PBGM		Mitchell (1969); Mitchell and Brandt (1972)
Macaca sylvana Barbary macaque	TR, RT, HD, CL BS, PL	BS, PL, HD	C, F	PBGM	Agonistic buffering	Mitchell (1969); Hrdy (1976); Redican (1976)
Macaca radiata Bonnet macaque	TR, RT, PL, DF	PL, DF	C	PBGM		Mitchell (1969); Mitchell and Brandt (1972)
Macaca fuscata Japanese macaque	TR, RT, PL, DF, CL, HD	PL, DF, HD	F	PBGM	Agonistic buffering, mainly with juveniles	Mitchell (1969); Hrdy (1976); Redican (1976)
Macaca nemestrina Pig-tailed macaque	DF	DF	C	PBGM		Mitchell (1969); Redican (1976)
Macaca arctoides Stump-tailed macaque	TR, RT, PL, CL, HD	PL, HD	C	PBGM		Estrada and Sandoval (1977); Redican (1976)
Theropithecus gelada Gelada baboon	TR, PL	PL	F	PGN		Redican (1976)
Papio hamadryas Hamadryas baboon	TR, RT, DF, HD, BS	DF, HD, BS	F	PGN	Agonistic buffering adoption	Mitchell (1969); Hrdy (1976); Redican (1976); Mitchell and Brandt, (1972)
Papio anubis Anubis baboon	TR, RT, DF, BS	DF, BS	F	PBGM	Agonistic buffering adoption	Mitchell (1969); Hrdy (1976); Redican (1976)
Papio cynocephalus Yellow baboon	CL		F	PBGM		Mitchell and Brandt (1972)
Erythrocebus patas Patas monkey		DF	F	PGN	AP	Redican (1976)

Species						Reference
Presbytis entellus Grey langur	RT		F	PGN		Hrdy (1976)
Presbytis johnii Nilgiri langur	PL	PL	F	PGN		Hrdy (1976)
Pongidae						
Hylobates lar White-handed gibbon	TR, CL, PL		F	M		Mitchell (1969); Redican (1976)
Symphalangus syndactylus Siamang	TR, HD, BS, CL		F	M		Redican (1976)
Pan troglodytes Chimpanzee	RT, DF		F	PBGM	Adoption	Mitchell (1969); Hrdy (1976)
Gorilla gorilla Gorilla	TR, HD, CL	HD	F, C	PGN	Adoption	Mitchell (1969); Redican (1976); Tilford and Nadler (1978)
Pongo pygmaeus Orangutan	PL		C	PBGM		Zucker et al. (1978)
Leporidae						
Oryctolagus cuniculus Rabbit	CL		C	PGN	Scent mark young	Mykytowycz (1959, 1965)
Sciuridae						
Marmota caligata Hoary marmot	PL	HD	F	PGN	AP	Barash (1975)
Hystricidae						
Hystrix sp. Crested porcupine	CL, RT	HD	C	PGN	AP	Mohr (1965)
Caviidae						
Dolichotis patagonum Mara		HD, DF	C	M	AP	Dubost and Genest (1974)
Dolichotis (Pediolagus) salincola Salt desert cavy		HD	C	M	AP	Kleiman (1974)
Dasyproctidae						
Cuniculus (Agouti) paca Paca	CL, HD		C	M	Scent mark young	Kleiman (1974)
Dasyprocta punctata Agouti			F, C	M	Scent mark and mount young	Smythe (1978)
Myoprocta pratti Acouchi	CL	HD, BS	C	M	SC, scent mark young	Kleiman (1969, 1972)

Table 1. (*Continued*)

Mammal	Depreciable	Nondepreciable	Source of observations	Modal mating system	Remarks	References
Chinchillidae						
Chinchilla lanigera						
Chinchilla		HD	C	PGN		Weir (personal communication)
Octodontidae						
Octodon degus						
Degu	CL	HD, BS	C	PGN	Scent mark young, AP	Wilson and Kleiman (1974); Wilson (personal communication)
Octodontomys gliroides						
Choz choz	CL	PL, HD	C		Scent mark young	Wilson and Kleiman (1974); Wilson (personal communication)
Capromyidae						
Capromys melanurus						
Hutia	CL	HD, DF	C	M		Bucher (1937)
Bathyergidae						
Heterocephalus glaber						
Naked mole rat		HD	F, C	M	RA, RM, RD, SC	Jarvis (1978)
Castoridae						
Castor fiber						
Beaver	CL, PF, TR	HD, BS	F, C	M	RA, RM, RD, SC, AP	Hodgdon and Larson (1973); Wilsson (1971)
Cricetidae						
Peromyscus melanocarpus	RT, CL	HD	F, C	M	SC, RD	Rickart (1977)
Peromyscus mexicanus						
Mexican deer mouse	RT, CL	HD	F, C	M	SC, RD	Rickart (1977)
Peromyscus californicus						
California deer mouse	RT, CL	HD, BS	C	M		Dudley (1974a,b)
Peromyscus maniculatus						
Deer mouse	RT, CL	HD, BS	C, F		SC	Howard (1949); Horner (1947); Hartung and Dewsbury (1979)
Peromyscus leucopus						
White-footed deer mouse	RT, CL	HD	C			Horner (1947); McCarty and Southwick (1977); Hartung and Dewsbury (1979); Smith (1966)
Peromyscus polionotus						
Old-field mice	RT, CL	HD, BS, DF	C, F	M		
Baiomys taylori	RT, CL	BS	C, F	M		Blair (1941)

Species						References
Reithrodontomys humulus Eastern harvest mouse	RT, CL	DF	C		SC	Layne (1959)
Onychomys leucogaster Northern grasshopper mouse	PF		C	M		Ruffer (1966)
Onychomys torridus Southern grasshopper mouse	RT, CL	HD	C	M	SC	Horner and Taylor (1968); Horner (1961); McCarty and Southwick (1977)
Meriones unguiculatus Clawed jird (gerbil)	RT, CL	HD	C		SC	Elwood (1975); Elwood and Broom (1978)
Meriones crassus Fat jird		DF	C		Prevent juvenile fighting	Fiedler (1973)
Meriones tamariscinus Tamarisk gerbil	RT, CL	DF, HD			Young lick and ingest female saliva; prevent juvenile fighting	Fiedler (1973)
Microtus ochrogaster Prairie vole	RT, CL	HD, BS	C	M	SC, RA	Thomas and Birney (1979); Wilson (in preparation); Hartung and Dewsbury (1979)
Microtus pennsylvanicus Meadow vole	RT, CL	HD, BS	C		SC	Hartung and Dewsbury (1979); Wilson (in preparation); Yardeni-Yaron (1952)
Microtus guentheri Levante vole		HD	C			
Microtus californicus California vole	RT, CL	HD	C		SC	Hartung and Dewsbury (1979)
Microtus montanus Montane vole	RT, CL	HD	C		SC	Hartung and Dewsbury (1979)
Muridae						
Mus musculus House mouse	RT, CL	HD	C	PGN	SC	Beniest-Noirot (1958)
Rattus fuscipes Rat	CL	HD	C	PBGM	SC	Horner and Taylor (1969)
Rattus (Myomys) daltoni Dalton's rat	RT	HD	C		SC	Anadu (1979)
Notomys alexis Hopping mice	RT, CL	HD	C	M		Stanley (1971); Happold (1976)

Table 1. (*Continued*)

Mammal	Depreciable	Nondepreciable	Source of observations	Modal mating system	Remarks	References
Pseudomys albocinereus Native mice	RT, CL	HD	C	PGN		Happold (1976)
Delphinidae						
Tursiops truncatus Bottle-nosed dolphin	BS		C	PBGM		M. C. Caldwell and D. K. Caldwell (1966)
Balaenopteridae						
Megaptera novaeangliae Humpback whale		DF	F			M. C. Caldwell and D. K. Caldwell (1966)
Equidae						
Equus caballus Horse	RT		F	PGN	AP	Feist and McCullough (1975)
Equus hemionus Kulan		DF	F	PGN	AP	Formozov (1966)
Equus burchelli Common Zebra		DF	C, F	PGN	AP	Wackernagel (1965); Klingel (1972)
Tapiridae						
Tapirus indicus Malay tapir	PL		C			Seitz (1970)
Rhinocerotidae						
Diceros bicornis Black rhinoceros	PL		C	PBGM		Dittrich (1967)
Suidae						
Phacochoerus aethiopicus Wart hog	HD		F	PGN		Geigy (1955)
Potamochoerus porcus Bush pig		BS	F			Skinner et al. (1976)
Bovidae						
Madoqua phillipsi Phillip's dik-dik	CL, DF		C, F	M	RA, RM	Simonetta (1966)
Oreotragus oreotragus Klipspringer	DF		C			Cuneo (1965)
Ovibos moschatus Musk-ox	RT	DF	F	PGN	AP	Tener (1965)

Table 2. The Presence and Absence of Male Parental Investment in the Order Carnivora[a,b]

Mammal	Males other than "father"	Any direct care	Huddle and sleep with	Groom and clean	Retrieve	Carry	Provide food	Active defense	Babysit	Play and socialize	Secure resources	Shelter construction	Antipredator sentinel	Care to female	References
Canidae															
Canis lupus	+						+	+	+	+	+		+	+	Murie (1944); Haber (1977)
Canis latrans	+						+	+	+	+	+		+	+	Ryden (1974, 1975)
Canis aureus	+			+			+	+	+	+	+		+	+	van Lawick (1970); Moehlman (personal communication)
Canis mesomelas	+				+	+	+	+	+	+	+	—	+	+	Moehlman (1979) (personal communication)
Lycaon pictus	+		—				+	+	+		+		+	+	van Lawick (1974); Malcolm (personal observation)
Cuon alpinus	+						+	+		C					Davidar (1974)
Alopex lagopus							+							+	MacPherson (1969); Kleiman (1968)
Vulpes vulpes				C			+				+			+	Macdonald (personal communication); Tembrock (1957)
Vulpes corsac								C							Dathe (1966)
Vulpes velox					C		+								Egoscue (1962)
Fennecus zerda					C		+						C	C	Weiher (1976); Roberts, (personal communication); Koenig (1970)
Nyctereutes procyonoides					C		+							C	Stroganov (1962); Roberts (personal communication)
Otocyon megalotis							+		+	+			+	+	Lamprecht (1979)
Dusicyon culpaeus							+								Housse (1949)
Dusicyon griseus							+	C							Housse (1949)
Cerdocyon thous				C			C	C	C						Brady (1978)
Chrysocyon brachyurus				C			C	C	C	C					Brady (personal communication)
Speothos venaticus			C	C			C	C	C	C				C	Porton (personal communication); Drüwa (1977)
(17 spp.— no data)															

[a]Care in captive animals is not recorded if it has also been seen in the field. Absence of care in captivity was not recorded, as many forms of care, such as resource defense, cannot be displayed in captive conditions. The distribution of species in genera follows Walker (1975).

[b]Key: +, recorded in field; —, recorded absent in field; ?, possible field record; C, recorded as present in captivity.

(*continued*)

Table 2 (Continued)

Mammal	Males other than "father"	Direct									Indirect				References
		Any direct care	Huddle and sleep with	Groom and clean	Retrieve	Carry	Provide food	Active defense	Babysit	Play and socialize	Secure resources	Shelter construction	Antipredator sentinel	Care to female	
Ursidae															
Ursus arctos		—									—				Stonorov and Stokes (1972)
Euarctos americanus		—									—				Rogers (1977)
Thalarctos maritimus		—									—				Perry (1966)
Melursus ursinus		—									—				Laurie and Seidensticker (1977)
(3 spp.—no data)															
Procyonidae															
Bassariscus astutus		—													Grinnell *et al.* (1937)
Potos flavus		—					C			C	—				Poglayen-Neuwall (1976)
Nasua narica		—									—				Kaufmann (1962)
Procyon lotor		?		C											Schneider *et al.* (1971)
Ailurus fulgens		?													Roberts (1975, personal communication)
(12 spp.—no data)															
Mustelidae															
Mustela erminea		?									+				Erlinge (1977); Hamilton (1933); Powell (1978)
Mustela nivalis											+				Lockie (1966); King (1975)
Mustela rixosa															Hamilton (1933)
Mustela frenata															Quick (1944)
Mustela sibirica															Stroganov (1962)
Mustela putorius															Stroganov (1962)
Mustela lutreola															Gerell (1970); Novikov (1962)
Martes martes							+		+	+	+				Balharry (1978); Krott (1973)
Martes americana		—									+				Hawley and Newby (1957); Grinnell *et al.* (1937)
Martes flavigula		?									+				Stroganov (1962); Lekagul and McNeely (1977)
Gulo gulo		?									+				Kott (1959); Gipson (personal communication)

Species	References
Tayra barbara	Poglayen-Neuwall (1978)
Ictonyx striatus	Rowe Rowe (1978)
Meles meles	Middleton et al. (1974)
Arctonyx collaris	Parker (1979)
Mephitis mephitis	Verts (1967)
Lutra lutra	Erlinge (1968)
Lutra canadensis	Grinnell et al. (1937); Liers (1951)
Lutra maculicollis	Kingdon (1977)
Lutrogale perspicillata	Wayre (1978); Desai (1974)
Amblonyx cinerea	Wayre (1978); Leslie (1970)
Pteronura brasiliensis	Duplaix (1980)
Enhydra lutris	Kenyon (1969)
(45 spp.—no data)	
Viverridae	
Viverra zibetha	Lekagul and McNeely (1977)
Civettictis civetta	Rahm (1966); Kingdon (1977); Mallinson (1969)
Genetta genetta	Malcolm (personal observation)
Genetta tigrina	Carpenter (1970); Roberts (personal communication)
Nandinia binotata	Charles-Dominique (1978)
Arctictis binturong	Gensch (1962); Roberts (personal communication)
Fossa fossa	Albignac (1973)
Eupleres goudoti	Albignac (1973, 1974)
Galidia elegans	Albignac (1973)
Galidia fasciata	Albignac (1973)
Mungotictis decemlineata	Albignac (1973)
Salanoia concolor	Albignac (1973)
Herpestes pulverulentes	Rowe Rowe (1978)
Herpestes sanguineus	Rood and Waser (1978)
Mungos mungos	Rood (1974, personal communication)
Helogale parvula	Rood (1978); Kingdon (1977); Rasa (1972)
Ichneumia albicauda	Waser (personal communication)
Suricata suricatta	Ewer (1963, 1973); Roberts (personal communication)
Cryptoprocta ferox	Albignac (1973)
(53 spp.—no data)	

(continued)

Table 2 (Continued)

Mammal	Males other than "father"	Direct — Any direct care	Huddle and sleep with	Groom and clean	Retrieve	Carry	Provide food	Active defense	Babysit	Play and socialize	Indirect — Secure resources	Shelter construction	Antipredator sentinel	Care to female	References
Hyaenidae															
Hyaena hyaena							+			+	+				Mills (1978); Novikov (1962); Kruuk (1976)
Hyaena brunnea							+			+	+				Mills (1978)
Crocuta crocuta										+	+				Kruuk (1972)
Proteles cristatus		−									+				Kruuk and Sands (1972); Ketelhodt (1966)
Felidae															
Felis silvestris			C							C	+			C	Leuw (1957); Condé and Schauenberg (1969)
Felis libyca				C						C					Hemmer (1978)
Felis nigripes				C			C			C					Schürer (1978)
Felis serval		−													Geertsema (1976, personal communication)
Felis viverrina										C	+				Ulmer (1968)
Felis bengalensis							C			C					Dathe (1966); Roberts (personal communication)
Felis aurata			C							C					Tonkin and Kohler (1978)
Felis temmincki										C					Louwman and Van Oyen (1968)
Felis concolor										C	+				Seidensticker et al. (1973)
Felis geoffroyi										C					Scheffel and Hemmer (1975)
Felis yagouaroundi										C					Hulley (1976)
Lynx lynx										+					Berrie (1973); Dathe (1966)
Panthera leo			−	−	−	−	+	−	−		+				Schaller (1972); Bertram (1978)
Panthera tigris							+				+				Schaller (1972)
Panthera pardus		−									−				Bertram (1978)
Uncia uncia				C			C	C		C	+	−	−	−	Schaller (1972); Freeman and Hutchins (1980)
Neofelis nebulosa											+				Geidel and Gensch (1976)
Acinonyx jubatus		−								C	−				Schaller (1972); Bertram (1978)

In total, about 9-10% of mammalian genera have been described as exhibiting direct male parental investment. This percentage is quite high, considering that male mammals cannot lactate or aid during the early stages of development. The percentage is higher than the estimated 7.7% of genera in which monogamy occurs (Kleiman, 1977), suggesting that the potential for direct male parental investment in mammals is considerable and is not necessarily tied to a monogamous mating system. Indeed, many of the primates in which direct male parental care occurs neither are monogamous nor live in closed harems, the two social systems in which one would predict the greatest amount of male parental investment, based on certainty of paternity. Of course, the descriptions of direct male parental care we have used do not differentiate between species in which direct male parental care is commonly observed and those for which the behaviors are rarely seen. In many of the primates in which direct male parental care has been described, especially those species living in multimale groups, the behavior occurs erratically and idiosyncratically. Moreover, interactions with infants are often used to decrease aggression from more dominant males, a behavior which has been termed "agonistic buffering" (Deag and Crook, 1971) (see Table 1). Nevertheless, the capacity for demonstrating direct male parental investment appears to be common to a high percentage of primate genera.

Table 2 presents and Fig. 7 summarizes the distribution of both direct and indirect male parental investment in carnivores. References that stated that no male investment was seen in a field study were also included. Some data were found for 91 of the 232 species listed by Ewer (1973). Information is most sparse for the mustelids, viverrids, and procyonids. The data are biased towards large, diurnal species living in open habitats. Indirect investment, particularly male defense of a home range containing several females, will probably be recorded more frequently as more field studies are undertaken.

Male investment of some form has been reported, at least occasionally, in all four species of hyaenid, although it is not well developed in *Crocuta* and *Proteles*. Male investment has not been reported as absent from any canid, although there is considerable variation in the extent of male parental behavior within the family ranging from the almost solitary maned wolf *(Chrysocyon brachyurus)* to the highly social African wild dog *(Lycaon pictus)*. High levels of direct male care occur in tropical otters (Duplaix, 1980; Wayre, 1978) and in group foraging mongooses (Ewer, 1963; Rood, 1974, 1978). Direct male care in the form of playing with young and letting them take food items has been recorded in a number of felids in captivity. However, even in the species involved, not all males tolerate females and young. It is unclear whether these records are an artifact of captivity or whether some felids will show male parental care in nature under certain conditions.

Direct male care in mammals does not appear to be positively correlated with any particular diet type, although herbivores in general are underrepresented. It is not restricted to any mode of life, being seen in terrestrial, aquatic, fossorial, and arboreal species. Male parental care has been recorded in more temperate than tropical rodents, but this probably reflects a sampling bias. Male parental investment may be more common in tropical than temperate ungulates. Male parental investment is not restricted to any particular mating system, although monogamous species are overrepresented. In particular, rodents reported as showing male parental behavior have usually been considered monogamous. This may reflect only the housing conditions of captivity.

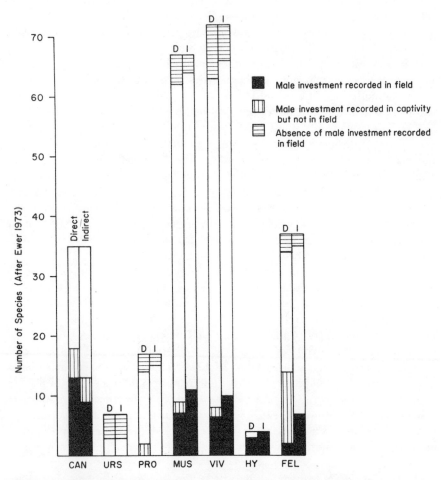

Figure 7. The presence and absence of recorded cases of male parental investment in all species of carnivores. Abbreviations: CAN., Canidae; URS., Ursidae; PRO., Procyonidae; MUS., Mustelidae; VIV., Viverridae; HY., Hyaenidae; FEL., Felidae.

In summary, high levels of direct male care appear to have evolved several times independently among the mammals. It is common in carnivores, perissodactyls, and primates and may be found to be more common among rodents and insectivores.

3. Discussion

3.1. Phylogenetic Considerations

Whether the mammals are considered to show a high or low incidence of species with male parental investment is largely subjective depending on how male parental investment is defined and with which other taxonomic groups the mammals are compared. Amongst the vertebrate classes, direct male investment in mammals is clearly less frequent than in the birds (Lack, 1968), probably more frequent than in the reptiles and amphibians (McDiarmid, 1978; Ridley, 1978), and comparable to or a little lower than in fishes (Breder and Rosen, 1966; Ridley, 1978).

In attempting to explain this pattern, Dawkins and Carlisle (1976) suggested that the evolution of internal fertilization and consequent susceptibility of the female to desertion might account for the low levels of male parental investment seen in mammals. However, this explanation does not help to elucidate the high levels of male parental care in birds.

A similar problem exists in the argument that internal fertilization lowers paternity certainty. Some of the most extreme forms of male parental care occur in birds such as the ratites and the American Jacana *(Jacana spinosa)* (Jenni, 1974) where the paternity certainty is lower than that in most monogamous species. However, differences in the internal development time between birds and mammals may affect certainty of paternity and thus have affected levels of male investment.

However, as Maynard Smith (1977) and Orians (1969) have noted, the evolution of lactation in addition to extended internal development does seem to have restricted the role that many male mammals can play. [Why male mammals do not lactate has been discussed by Daly (1979) and will not be considered here.] Male mammals cannot guard eggs, the most common form of male parental care in animals, and in many mammals the young are nearly independent at the time of weaning.

3.2. Factors That May Predispose Male Mammals to Care for Young

3.2.1. Intrinsic Ability to Aid Offspring

As has already been discussed, probably the most important factor explaining the distribution of male parental behavior relates to the male's ability to exhibit parental care. Instances in which it appears males could provide care but do not have been mentioned above, e.g., in shelter construction, and will be further discussed below

3.2.2. Sociality

Permanent group living often seems to have fostered the evolution of male parental behavior. Incidental or indirect male investment, particularly alarm calls and the warding off of predators, is common in species living in large groups. More direct male care characterizes small "closed" groups often composed of genetically related individuals. Male primates constitute an exception, since male parental care is described for several species in which male immigration at puberty decreases the likelihood of males being related to all infants and juveniles.

The effects of sociality are very clear among the mongooses (Gorman, 1979). Species are divided into two distinct types; the diurnal, group foraging species such as *Mungos mungos, Helogale parvula,* and *Suricata suricatta* which show extensive direct male care including food provisioning, protection, and babysitting by males (Rasa, 1977; Rood, 1974, 1978), and a larger number of solitary, usually nocturnal forms where male investment seldom extends beyond mere tolerance of the young (e.g., Rood and Waser, 1978). Cooperative group foraging also seems important in the extreme case of communal breeding reported in naked mole rats *(Heterocephalus glaber)* (Jarvis, 1978), and in the wolf *(Canis lupus)* and other pack hunting canids. In the social species, males other than the father often provide care. In these cases the advantages of living in a group presumably outweigh the benefits of independent reproduction.

3.2.3. High Costs to Polygyny

As Maynard Smith (1977) and Trivers (1972) have stressed, the benefits to an offspring from male parental investment have to be counted by a male against his chances of mating again. Although in some cases a female would prefer a male who invests over one who deserts, there is also a point at which a female would prefer to mate with a successful polygynist who will give her effectively polygynous sons. ("Successful" and "effective" probably relate ultimately to the heritability of the traits leading to polygyny.) Except in cases of exceptional male investment (Grafen and Sibley, 1978), a male who mates with a number of females will outreproduce a monogamist.

Certain cases of direct male care in mammals may have evolved in situations in which a male could not consistently mate with more than one female, even if he deserted, because females are so widely dispersed and so irregularly in estrus. A number of small, dispersed, tropical forms such as dik-dik (*Madoqua* spp.) and other small ungulates, elephant shrews (Macroscelididae), and dasyproctids (*Agouti, Dasyprocta,* and *Myoprocta*) may fall into this category. In these cases it does not appear that the habitat is "harsh" if measured in terms of the rate of adult mortality, but individuals would probably risk very high costs if they strayed beyond the limits of a territory which they know with great accuracy. In the case of the rufous elephant shrew, *E. rufescens,* the male devotes considerable time to maintaining a network of trails through the territory. In these species the intrinsic ability of a male to make a direct contribution to raising young seems limited. Indeed, maternal care is restricted by the precocial nature of the neonates and the absentee parental care system. However, the young may reap considerable indirect rewards by the use both of resources in the territory and of the system of escape trails (Rathbun, 1979).

3.2.4. Paternity Certainty

Few cases have been reported to date in which males invest in offspring to which they are distantly related or unrelated genetically. However, males other than the father certainly care for young both in those species with cooperative breeding ("helpers at the nest") and in some social species with a multimale group structure, e.g., baboons (*Papio* spp.) and capuchins *(Cebus nigrivittatus)* (Robinson, personal communication). In addition, there are species in which a male appears to be able to identify his offspring, e.g., species with male infanticide (Hrdy, 1977), but in which the male provides little or no care. Paternity certainty may represent a necessary but not sufficient condition for the evolution of male investment, and it is unsurprising that male parental care is commoner in species which are monogamous or live in one-male "closed"groups.

3.3. The Role of Male Parental Investment in Molding the Social Organization of Species

There are obvious correlations between a species social organization defined in terms of the mating and land tenure systems and the relative contribution of the sexes to raising young. Trivers (1972) suggested that the relative parental investment of the sexes was the factor "governing" or directing the operation of sexual selection. This may be useful when considering the operation of sexual selection at a single point in time. However, when the evolution of patterns of parental investment is considered, it is seldom clear if the relative parental investment of the sexes is a cause or a consequence of some antecedent pattern of sexual selection. In many cases the relative parental investment of the sexes seems to be constrained by some other aspect of the natural history of the species. Therefore, it is probably more useful to consider the correlations between mating system, variance in reproductive success, and differential parental investment as the result of co-evolution between the factors rather than attempt to erect one factor as directing the evolution of the other two.

For many indirect forms of male parental investment, it is unlikely that the behavior or distribution of females would be significantly altered if the males provided no investment. Klingel (1972) provides evidence that groups of female zebras *(E. burchelli)*, like many primate groups, retain their integrity even in the absence of males despite the fact that zebra males may actively defend their harems from predators. Similarly, the dispersion of females in many "solitary" species such as the mustelids and cats appears to be independent of male dispersion, although the males may provide important indirect benefits by excluding other males.

Even in some cases where a male direct its care to the offspring of a single female, it appears that the male's investment need not play an important role in the dispersion, rearing strategy, or intrasexual selection in the female. Rathbun (1979) reports that for a species of elephant shrew *(E. rufescens)* the male and the female rarely interact and most territorial defense is directed towards members of the same sex. It seems unlikely that the behavior of the female would differ if the males' ranges overlapped those of several females as is more usual in mammals generally.

For the species mentioned above, it appears that the mating systems have evolved in response to factors in the ecology or natural history of the species independent of the capacity of the males to invest in young. Even in species with monogamy, it is usually not the male parental investment which appears to restrict a male to mating with a single female. Male parental investment such as sentinel and antipredator behavior, and occasionally playing with young, which are seen in a number of group-living species especially, are probably displayed by males only to the extent that they do not interfere with a polygynous mating system. It is unlikely that a male in such species would ever give up an opportunity to consort or mate with a female in order to care for young.

The situation is different in species with high levels of depreciable male investment. Important components of the social organization and mating systems of these species can only be understood as consequences of the high levels of male parental care. These species, which include golden lion tamarins *(Leontopithecus rosalia)*, African wild dogs, and wolves *(Canis lupus)*, share a complex of behavioral characters which include more intense competition between females than males (Kleiman, 1979; Frame *et al.*, 1979), sex ratios tending towards males, disproportionate female emigration, and often care provided by adults in addition to the father. It appears that in these species females compete for access to male investment, which in tamarins involves carrying the young and also providing food (Hoage, 1977, 1978) and in canids involves providing food and protection for an extended period of dependence. The species involved are all classed as "obligate" monogamists by Kleiman (1977). The extent to which other obligate monogamous species show the same traits may depend on the extent to which the contribution of males to the young is depreciable.

4. Conclusions

Virtually no data exist for mammals which relate male parental investment to the fitness of offspring in a quantitative way. Some results (Malcolm, 1979; Moehlman, 1979), show that behavior of the parental type provided by animals in addition to the father increases the fitness of young born into the group, from which one presumes that parental behavior by the father also increases the fitness of the young. Mugford and Nowell (1972) have shown that male mice raised with the father are more aggressive during encounters (after a period of isolation) than mice raised alone with the mother. Such behavior might improve a male's fitness, although it is not known what behavior on the part of the father could cause this increased aggression.

Captive studies have produced confusing results with respect to the effects of male parental care on fitness. As already mentioned, in gerbils the male has been claimed to increase, decrease, or have no effect on the fitness of offspring (Ahroon and Fidura, 1976; Elwood and Broom, 1978; Gerling and Yahr, 1979; Klippel, 1979), but the different experimental conditions apparently greatly affected the results. In other experiments, huddling and retrieving by virgin mice housed with mothers (Sayler and Salmon, 1971) decreased the mean weaning weight of the young, suggesting that apparently caregiving behaviors may have deleterious effects.

For many species the reason why they do or do not show male parental care remains unclear. However, it is probably not useful to look for global explanations in terms such as richness or harshness of the habitat. In many cases there are probably two or more predisposing factors that act in concert. For instance, the advantages of cooperative hunting combined with the canid ability to regurgitate may have led to the high levels of male parental investment seen in the social canids. The absence of male parental behavior is often as surprising as its occurrence. Few male primates or bats share the prolonged burden of carrying the young, and large numbers of carnivores do not share food with their offspring.

To understand why certain species show male parental investment, it is necessary to know the alternative ways in which a male's reproductive effort could be channeled. Recent studies by Owen-Smith (1977) and Popp (1978) have related different male reproductive strategies in ungulates and baboons, respectively, to ecological factors. In particular, Owen-Smith argues that territoriality in ungulates represents a low-cost/low-benefit form of effort. These results suggest two ways in which the evolution of male parental investment in mammals could be investigated.

First, some modeling of the distribution of a male's reproductive effort into various channels, some of which include direct or indirect parental investment, would be useful. The usual models typically assume some unitary mode of reproductive effort which varies monotonically with fitness. However, as argued above, some forms of male parental investment may not interfere with polygyny. The models would also have to include the role of heritability in the evolution of traits leading to polygyny (see Weatherhead and Robertson, 1979).

Second, field data could be collected on groups which show interesting patterns of variation in male parental behavior. Coyotes *(Canis latrans)* may show male parental care in parts of their range [e.g., Wyoming (Camenzind, 1978)] but not in others [e.g., Minnesota, (Berg and Chesness, 1978)]. The otters might also be a suitable group, as temperate forms are polygynous while most tropical species seem to live in monogamous groups, with one species *(Lutra maculicollis)* perhaps living in larger social groups (Kingdon, 1977; Proctor, 1962). The high levels of male tolerance and food sharing in captive small felids also suggest that they would be interesting to study, especially to compare a species at low and high population density where the ability to find mates might differ.

ACKNOWLEDGMENTS. We would like to thank the following for helpful discussions and access to unpublished data: R. Brownell, N. Duplaix, J. F. Eisenberg, D. Macdonald, P. Moehlman, K. Ralls, M. Roberts, J. Robinson, and J. Seidensticker. We are also greatful to G. Hill for typing several drafts of the manuscript, and to V. Garber for typing the final draft. D. G. Kleiman is, in part, supported by NIMH 27241. J. R. Malcolm was, in part, supported by the Friends of the National Zoo. The figures were prepared by S. James.

References

Ahroon, J. K., and Fidura, F. G., 1976, The influence of the male on maternal behaviour in the gerbil *(Meriones unguiculatus)*, *Anim. Behav.* **24**:372–375.

Albignac, R., 1973, *Mammiferes Carnivores*, Faune de Madagascar, CNRS, Paris.

Albignac, R., 1974, Observations éco-éthologiques sur le genre *Eurpleres,* Viverride de Madagascar, *Terre Vie* **28**:321–351.

Altmann, S. A., Wagner, S. S., and Lenington, S., 1977, Two models for the evolution of polygyny, *Behav. Ecol. Sociobiol.* **2**:397–410.

Anadu, P. A., 1979, Gestation period and early development in *Myomys daltoni* (Rodentia: Muridae), *Terre Vie* **33**:59–70.

Balharry, R., 1978, The private life of the pine marten, *Wildlife* **20** (2):74–77.

Barash, D. A., 1975, Ecology of paternal behavior in the hoary marmot *(Marmota caligata)*: An evolutionary interpretation, *J. Mammal.* **56**:613–617.

Beniest-Noirot, E., 1958, Analyse du comportement dit 'maternal' chez la souris, *Monogr. Fr. Psychol.* No. 1., CNRS, Paris.

Berg, W. E., and Chesness, R. A., 1978, Ecology of coyotes in Northern Minnesota, in: *Coyotes* (M. Bekoff, ed.), pp. 299–247, Academic Press, New York.

Berrie, P. M., 1973, Ecology and status of the lynx in interior Alaska, in: *The World's Cats,* Vol. I (R. L. Eaton, ed.), pp. 4–41, World Wildlife Safari, Winston, Oregon.

Bertram, B. C. L., 1978, *Pride of Lions,* Scribner's, New York.

Blair, W. F., 1941, Observations on the life history of *Baiomys taylori subater, J. Mammal* **22**:378–383.

Bradbury, J. W., 1977, Social organization and communication, in: *Biology of Bats,* Vol. III (W. A. Wimsatt, ed.), pp. 1–72, Academic Press, New York.

Brady, C. A., 1978, Reproduction, growth and parental care in crab-eating foxes *Cerdocyon thous* at the National Zoological Park, Washington, *Int. Zoo Yearb.* **18**:130–134.

Breder, C. M., Jr., and Rosen, D. E., 1966, *Modes of Reproduction in Fishes,* Natural History Press, New York.

Bucher, G. C., 1937, Notes on life-history and habits of *Capromys, Mem. Soc. Cubana Hist. Nat.* **11**:93–107.

Caldwell, M. C., and Caldwell, D. K., 1966, Epimeletic (care-giving) behavior in Cetacea, in: *Whales, Dolphins, and Porpoises* (K. S. Norris, ed.), University of California Press, Berkeley and Los Angeles.

Camenzind, F. J., 1978 , Behavioral ecology of coyotes on the National Elk Refuge, Jackson, Wyoming, in: *Coyotes* (M. Bekoff, ed.), pp. 267–294, Academic Press, New York.

Carpenter, G. C., 1970, Observations on the rusty spotted genet, *Lammergeyer* **11**:60–63.

Charles-Dominique, P., 1978, Ecologie et vie sociale de *Nandinia binotata* (Carnivores. Viverrides): comparison avec les prosimiens sympatriques de Gabon, *Terre Vie* **32**:477–528.

Conaway, D. H., 1958, Maintenance, reproduction, and growth of the least shrew in captivity, *J. Mammal.* **39**:507–512.

Condé, B., and Schauenberg, P., 1969, Reproduction du Chat forestier d'Europe (*Felis silvestris* Schreber) en captivité, *Rev. Suisse Zool.* **76**:183–212.

Cuneo, F., 1965, Observations on the breeding of the klipspringer antelope, *Oreotragus oreotragus,* and the behaviour of their young at the Naples Zoo, *Int. Zoo Yearb.* **5**:45–48.

Daly, M., 1979, Why don't male mammals lactate, *J. Theor. Biol.* **78**:325–346.

Dathe, H., 1966, Breeding the Corsac fox *(Vulpes corsac)* at East Berlin Zoo, *Int. Zoo Yearb.* **6**:166–169.

Dathe, H., 1968, Breeding the Indian leopard cat *(Felis bengalensis)* at East Berlin Zoo, *Int. Zoo Yearb.* **8**:42–44.

Davidar, E. R. C., 1974, Observations at the dens of the dhole or Indian wild dog *(Cuon alpinus), J. Bombay Nat. Hist. Soc.* **71**:183–187.

Dawkins, R., and Carlisle, T. R., 1978, Parental investment: A fallacy, *Nature (London)* **262**:131–133.

Deag, J. M., and Crook, J. H., 1971, Social behaviour and agonistic buffering in the wild barbary macaque *(Macaca sylvana), Folia Primatol.* **15**:183–201.

Desai, J. H., 1974, Observations on the breeding habits of the Indian smooth otter in captivity, *Int. Zoo Yearb.* **14**:123–124.

Dittrich, L., 1967, Breeding the black rhinoceros at Hanover Zoo, *Int. Zoo Yearb.* **7**:161–162.

Drüwa, P., 1977, Beobachtungen zur Geburt and natürlichen Aufzucht von Waldhunden *(Speothos venaticus)* in des Gefangenschaft, *Zool. Gart. N.F.* **47**:109–137.

Dubost, G., and Genest, H., 1974, Le comportement social d'une colonie de Maras *Dolichotis patagonum* Z. dans le Parc de Branféré, *Z. Tierpsychol.* **35**:225–302.

Dudley, D., 1974*a*, Contributions of paternal care to the growth and development of the young in *Peromyscus californicus, Behav. Biol.* **11**:155–166.

Dudley, D., 1974*b*, Paternal behavior in the California mouse, *Peromyscus californicus, Behav. Biol.* **11**:247–252.

Duplaix, N., 1980, The ecology and behaviour of the giant Brazilian otter in Suriname, A preliminary study, *Terre Vie* (in press).

Egoscue, H. J., 1962, Ecology and life history of the kit fox in Tooele County, Utah, *Ecology* **43**:481–497.

Elwood, R. W., 1975, Paternal and maternal behavior in the Mongolian gerbil, *Anim. Behav.* **23**:766–773.

Elwood, R. W., and Broom, D. M., 1978, The influence of litter size and parental behavior on the development of Mongolian gerbil pups, *Anim. Behav.* **26**:438–454.

Erlinge, S., 1968, Territoriality of the otter *(Lutra lutra), Oikos* **19**:81–98.

Erlinge, S., 1977, Spacing strategy in the stoat, *Mustela erminea, Oikos* **28**(1):32–42.

Estrada, A., and Sandoval, J. M., 1977, Social relations in a free-ranging troop of stumptail macaques *(Macaca arctoides):* Male care behaviour, 1, *Primates* **18**(4):793–813.

Ewer, R. F., 1963, The behaviour of the meerkat, *Suricata suricatta* (Schreber), *Z. Tierpsychol.* **20**:570–607.

Ewer, R. F., 1973, *The Carnivores,* Cornell University Press, Ithaca, N.Y.

Feist, J. D., and McCullough, D. R., 1975, Reproduction in feral horses, *J. Reprod. Fertil.* **23**:13–18.

Fiedler, U., 1973, Beobachtungen zur Biologie einiger Gerbillinen, insbesondere *Gerbillus (Dipodillus) dasyurus* (Myomorpha, Rodentia) in Gefangenschaft I. Verhalten, *Z. Säugetierkd.* **38**:321–340.

Fons, R., 1974, The behavior patterns of the Etruscan shrew, *Suncus etruscus* (Savi 1822), *Terre Vie* **28**(1):131–157.

Formozov, A. N., 1966, Adaptive modifications of behavior in mammals of the Eurasian steppes, *J. Mammal.* **47**:208–223.

Frame, L. H., Malcolm, J. R., Frame, G. W., and van Lawick, H. 1979, Social organization of African wild dogs *(Lycaon pictus)* on the Serengeti Plains, Tanzania, 1967–1978. *Z. Tierpsychol.* **50**:225–249.

Freeman, H., and Hutchins, M., 1980, Captive management of snow leopard cubs: An overview, *Zool. Gart. N.F.* (in press).

Geertsema, A., 1976, Impressions and observations on serval behavior in Tanzania, East Africa, *Mammalia* **40**:13–19.

Geidel, B., and Gensch, W., 1976, The rearing of clouded leopards *(Neofelis nebulosa)* in the presence of the male, *Int. Zoo Yearb.* **16**:124–126.

Geigy, R., 1955, Observation sur les Phacochères du Tanganyika, *Rev. Suisse Zool.* **62**:139–163.

Geist, V., 1971, *Mountain Sheep: A Study in Behavior and Evolution,* University of Chicago Press, Chicago.

Gensch, W., 1962, Successful rearing of the binturong, *Int. Zoo Yearb.* **4**:79–80.

Gerell, R., 1970, Home ranges and movements of the mink, *Mustela vison,* in southern Sweden, *Oikos* **21**:160–173.

Gerling, S., and Yahr, P., 1979, Effect of the male parent on pup survival in Mongolian gerbils, *Anim. Behav.* **27**:310–311.

Gorman, M., 1979, Dispersion and foraging of the Small Indian mongoose, *Herpestes auropunctatus* (Carnivora: Viverridae) relative to the evolution of social viverrids *J. Zool.* **187**:65–73.

Gould, E., and Eisenberg, J. F., 1966, Notes on the biology of the Tenrecidae, *J. Mammal.* **47**:660–686.

Grafen, A., and Sibly, R., 1978, A model of mate desertion, *Anim. Behav.* **26**:645–652.

Greenhall, A. M., 1968, Notes on the behavior of the false vampire bat, *J. Mammal* **49**:337–340.

Grinnell, J., Dixon, J. S., and Linsdale, J. M., 1937, *Fur-bearing Mammals of California,* Vol. 1, University of California Press, Berkeley and Los Angeles.

Haber, G. C., 1977, Socio-ecological dynamics of wolves and prey in a subarctic ecosystem, Ph.D. dissertation, University of British Columbia.

Hamilton, W. J., 1933, The weasels of New York, *Am. Midl. Nat.* **14**:289–344.

Happold, M., 1976, Social behavior of the conilurine rodents (Muridae) of Australia, *Z. Tierpsychol.* **40**:113–182.

Hartung, T. G., and Dewsbury, D. A., 1979, Paternal behavior in six species of muroid rodents, *Behav. Neural Biol.* **26**:466–478.

Hawley, V. D., and Newby, F. E., 1957, Marten home ranges and population fluctuations, *J. Mammal.* **38**:174–184.

Hemmer, H., 1978, Were the leopard cat and the sand cat among the ancestry of domestic cat races? *Carnivore* **1**:106–108.

Hoage, R. J., 1977, Parental care in *Leontopithecus rosalia rosalia:* Sex and age differences in carrying behavior and the role of prior experience, in: *The Biology and Conservation of the Callitrichidae* (D. G. Kleiman, ed.), pp. 293–305, Smithsonian Institution Press, Washington, D.C.

Hoage, R. J., 1978, Biosocial development in the golden lion tamarin, *Leontopithecus rosalia rosalia,* (Primates: Callitrichidae), Ph.D. dissertation, University of Pittsburgh.

Hodgdon, H. E., and Larson, J. S., 1973, Some sexual differences in behaviour within a colony of marked beavers *(Castor canadensis),* *Anim. Behav.* **21**:147–152.

Horn, H. S., 1978, Optimal tactics of reproduction and life history, in: *Behavioural Ecology* (J. R. Krebs and N. B. Davies, eds.), pp. 411–429, Sinauer, Sunderland, Mass.

Horner, B. E., 1947, Paternal care of young mice of the genus *Peromyscus, J. Mammal.* **28**:31–36.

Horner, B. E., 1961, Paternal care of the young and convulsive seizures in the grasshopper mouse, *Am. Zool.* **1**:360.

Horner, B. E., and Taylor, J. M., 1968, Growth and reproductive behavior in the Southern grasshopper mouse, *J. Mammal.* **49**:644–660.

Horner, B. E., and Taylor, J. M., 1969, Paternal behavior in *Rattus fuscipes, J. Mammal.* **50**:803–805.

Housse, R. P. R., 1949, Los Zorros de Chile o Chacales Americanos, *Rev. Univ. Chile* **34**(1):33–56 *(An. Acad. Chil. Cienc. Nat.* No. 14).

Howard, W. E., 1949, Dispersal, amount of inbreeding and longevity in a local population of prairie deermice on the George Reserve, Southern Michigan, *Contrib. Lab. Vertebr. Biol. Univ. Mich.* **43**:1–50.

Hrdy, S. B., 1976, Care and exploitation of nonhuman primate infants by conspecifics other than the mother, *Adv. Study Behav.* **6**:101–158.

Hrdy, S. B., 1977, *The Langurs of Abu,* Harvard University Press, Cambridge, Mass.

Hulley, J. T., 1976, Maintenance and breeding of captive jaguarundis at Chester Zoo and Toronto, *Int. Zoo Yearb.* **16**:120–122.

Jarvis, J. U. M., 1978, Energetics of survival in *Heterocephalus glaber* (Rüppell), the naked mole-rat (Rodentia:Bathyergidae), *Bull. Carnegie Mus.* **6**:81–87.

Jenni, D. A., 1974, Evolution of polyandry in birds, *Am. Zool.* **14**:129–144.

Kaufmann, J. H., 1962, Ecology and social behavior of the coati *(Nasua narica)* on Barro Colorado Island, Panama, *Univ. Calif. Berkeley, Publ. Zool.* **60**(3):95–222.

Kenyon, K. W., 1969, *The Sea Otter in the Eastern Pacific Ocean, U.S. Dep. Inter, Bur. Sport Fish. Wildl. North Am. Fauna* **68**:1–352.

Ketelhodt, H. F. von, 1966, Der Erdwolf (Sparrman, 1783), *Z. Säugetierkd.* **31**:300–308.

King, C. M., 1975, The home range of the weasel *(Mustela nivalis)* in an English woodland, *J. Anim. Ecol.* **44**:639–668.

Kingdon, J., 1977, *East African Mammals,* Vol. IIIA, Academic Press, New York.

Kleiman, D. G., 1968, Reproduction in the Canidae, *Int. Zoo Yearb.* **8**:3–8.

Kleiman, D. G., 1969, *The Reproductive Behaviour of the Green Acouchi,* Ph.D. thesis, University of London.

Kleiman, D. G., 1971, The courtship and copulatory behavior of the green acouchi,*Myoprocta pratti, Z. Tierpsychol.* **29**:259–278.

Kleiman, D. G., 1972, Maternal behaviour of the green acouchi, *Myoprocta pratti,* a South American caviomorph rodent, *Behaviour,* **43**:48–84.

Kleiman, D. G., 1974, Patterns of behaviour in hystricomorph rodents, *Symp. Zool. Soc. London* **34**:171–209.

Kleiman, D. G., 1977, Monogamy in mammals,*Quart. Rev. Biol.* **52**:39–69.

Kleiman, D. G., 1979, Parent-offspring conflict and sibling competition in a monogamous primate, *Am. Nat.* **114**(5):753–760.

Klingel, H., 1972, Social behaviour of African Equidae, *Zool. Afr.* **7**:175–185.

Klippel, J. A., 1979, Does the male gerbil parent *(Meriones unguiculatus)* contribute to pup mortality? A reply, *Anim. Behav.* **27**:311–312.

Koenig, L., 1970, Zur Fortpflanzung and Jugendentwicklung des Wüsten fusches *(Fennecus zerda,* Zimm. 1780), *Z. Tierpsychol.* **27**:205–246.

Krott, P., 1959, Die Vielfrass *(Gulo gulo L.* 1758), *Monogr. Wildsäuget.* **13**:1–159.

Krott, P., 1973, Die Fortpflanzung des Edelmarders *(Martes martes L.)* in freier Wildbahn, *Z. Jagdwiss.* **19**: 113–117.

Kruuk, H., 1972, *The Spotted Hyena,* University of Chicago Press, Chicago.

Kruuk, H., 1976, Feeding and social behaviour of the striped hyaena *(Hyaena vulgans* Desmarest), *E. Afr. Wildl. J.* **14**:91–111.

Kruuk, H., and Sands, W. A., 1972, The aardwolf *(Proteles cristatus* Sparrman, 1783) as a predator of termites, *E. Afr. Wildl. J.* **10**:211–227.

Kulzer, E., 1958, Untersuchungen über die Biologie von Flughunden der Gattung *Rousettus* Gray, *Z. Morphol. Oekol. Tiere* **47**:374–402.

Lack, D., 1968, *Ecological Adaptations for Breeding in Birds,* Methuen, London..

Lamprecht, J., 1979, Field observations on the behaviour and social system of the bateared fox, *Otocyon megalotis* Desmarest, *Z. Tierpsychol.* **49**:260–284.

Laurie, A., and Seidensticker, J., 1977, Behavioural ecology of the sloth bear *(Melursus ursinus)*, *J. Zool.* **182**:187–204.

Layne, J. N., 1959, Growth and development of the eastern harvest mouse, *Reithrodontomys humulus, Bull. Fl. State Mus. Biol. Ser.* **4**(2):61–82.

Lekagul, B., and McNeely, J., 1977, *Mammals of Thailand,* Karusapa Press, Bangkok.

Leslie, G., 1970, Observations on the oriental short-clawed otter *(Amblonyx cinerea)* at Aberdeen Zoo, *Int. Zoo Yearb.* **10**:79–81 .

Leuw, A., 1975, Die Wildkatzen, *Merkbl. Niederwildauss Dtse. Jagdschutzverb.* **16**.

Leyhausen, P., 1965, The communal organization of solitary mammals, *Symp. Zool. Soc. London* **14**:249–263.

Liers, E. E., 1951, My friends the otters,*Nat Hist.* **60**:320–326.

Ligon, J. D., and Ligon, S. H., 1978, Communal breeding in green wood-hoopoes as a case for reciprocity, *Nature (London)* **276**:496–498.

Lockie, J. D., 1966, Territory in small carnivores, *Symp. Zool. Soc. London* **18**:143–165.

Louwman, J. W. W., 1973, Breeding the tailess tenrec *(Tenrec ecaudatus)* at Wassenaar Zoo, *Int. Zoo Yearb.* **13**:125–126.

Louwman, J. W. W., and Van Oyen, W. G., 1968, A note on breeding Temminck's golden cat *(Profelis temmincki)* at Wassenaar Zoo, *Int. Zoo Yearb.* **8**:47–49.

MacPherson, A. H., 1969, The dynamics of Canadian arctic fox populations, *Can. Wildl. Rep. Ser.* No. 8.

Malcolm, J. R., 1979, Social organization and communal rearing in African wild dogs *(Lycaon pictus)*, Ph.D. dissertation, Harvard University.

Mallinson, J. J. C., 1969, Notes on breeding the African civet *(Viverra civetta)* at Jersey Zoo, *Int. Zoo Yearb.* **9**:92–93.

Maynard Smith, J., 1977, Parental investment: A prospective analysis, *Anim. Behav.* **25**:1–9.

McCarty, R. and Southwick, C. H., 1977, Patterns of parental care in two cricetid rodents, *Onychomys torridus and Peromyscus leucopus, Anim. Behav.* **25**:945–948.

McDiarmid, R. W., 1978, Evolution of parental care in frogs, in: *Development of Behavior* (G. M. Burghardt and M. Bekoff, eds.), pp. 127–147, Garland Press, New York.

Middleton, A. L. V., and Paget, R. J., 1974, *Badgers of Yorkshire and Humbleside,* Morley and Sons, York, England,.

Mills, M. G. L., 1978, The comparative socio-ecology of the Hyaenidae, *Carnivore* **1**(1):1–6.

Mitchell, G., 1969, Paternalistic behavior in primates, *Psych. Bull.* **71**:399–417.

Mitchell, G., and Brandt, E. M., 1972, Paternal behavior in primates, *in: Primate Socialization* (F. E. Poirier, ed.), pp. 173–206, Random House, New York.

Moehlman, P. D., 1979, Jackal helpers and pup survival, *Nature (London)* **277**:382–383.

Mohr, E., 1965, *Altweltliche Stachelschweine,* Die Neue Brehm-Bücherei, A. Ziemsen Verlag, Wittenberg Lutherstadt.

Mugford, R. A., and Nowell, N. W., 1972, Paternal stimulation during infancy: Effects upon agression and open-field performance of mice, *J. Comp. Physiol. Psych.* **79**:30–36.

Murie, A., 1944, The wolves of Mt. McKinley, *Fauna Nat. Parks U.S. Fauna Ser.* **5**:1–238.

Mykytowycz, R., 1959, Social behaviour of an experimental colony of wild rabbits, *Oryctolagus cuniculus* (L.), II. First breeding season, *CSIRO Wildl. Res.* **4**:1–13.

Mykytowycz, R., 1965, Further observations on the territorial function and histology of the submandibular cutaneous (chin) glands in the rabbit, *Oryctolagus cuniculus* (L.), *Anim. Behav.* **13**(4):400–412.

Novikov, G. A., 1962, *Carnivorous Mammals of the USSR,* Fauna of the USSR No. 62, Zoological Institute of the Academy of Sciences of the USSR, Israeli Program for Scientific Translation.

Orians, G. H., 1969, On the evolution of mating systems in mammals and birds, *Am. Nat.* **103**:589–603.

Owen-Smith, N., 1977, On territoriality in ungulates and an evolutionary model, *Q. Rev. Biol.* **52**:1–38.

Parker, C., 1979, Birth, care and development of Chinese hog badgers *(Arctonyx collaris)* at Metro Toronto Zoo, *Int. Zoo Yearb.* **19**:182–185.

Perry, R., 1966, *The World of the Polar Bear,* University of Washington Press, Seattle.

Pitelka, F. A., Holmes, R. T., and MacLean, S. A., Jr., 1974, Ecology and evolution of social organization in arctic sandpipers, *Amer. Zool.* **14**:185–204.

Poglayen-Neuwall, I., 1976, Zur Fortpflanzungsbiologie und Jugendentwicklung von *Potos flavus* (Schreber, 1774), *Zool. Gart. N.F.* **46**:237–283.

Poglayen-Neuwall, I., 1978, Breeding, rearing, and notes on the behaviour of tayras *(Eira barbara),* *Int. Zoo Yearb.* **18**:134–140.

Popp, J. L., 1978, Male baboons and evolutionary principles, Ph.D. dissertation, Harvard University.

Powell, R. A., 1978, Zig-zag zap, *Anim. Kingdom* **81**(6):20–25.

Proctor, J., 1962, A contribution to the natural history of the spotted-necked otter (*Lutra maculicollis* Lichtenstein), in Tanganyika *E. Afr. Wildl. J.* **1**:92–102.

Quick, H. F., 1944, Habits and economics of the New York weasel in Michigan, *J. Wildl. Manage.* **8**:71–78.

Rahm, U., 1966, Les mammifères de la fôret équatoriale, de l'est du Congo, *Ann. Mus. R. Afr. Centr. Tervuren Sci. Zool.* **149**.

Rasa, O. A. E., 1972, Aspects of the social organization in captive dwarf mongooses, *J. Mammal.* **53**:181–185.

Rasa, O. A. E., 1977, The ethology and sociology of the dwarf mongoose, *Helogale undulata rufula, Z. Tierpsychol.* **43**:337–406.

Rathbun, G. 1979, The social structure and ecology of elephant shrews, *Z. Tierpsychol. Suppl.* **20**:1–76.

Redican, W. K., 1976, Adult male-infant interactions in nonhuman primates, in: *The Role of the Father in Child Development* (M. E. Lamb, ed.), pp. 345–385, Wiley, New York.

Rickart, E. A., 1977, Reproduction, growth and development in two species of cloud forest *Peromyscus* from southern Mexico, *Occas. Pap. Mus. Nat. Hist. Univ. Kans.* **67**:1–22.

Ridley, M., 1978, Paternal care, *Anim. Behav.* **26**:904–932.

Roberts, M. S., 1975, Growth and development of mother reared red pandas *(Ailurus fulgens),* *Int. Zoo Yearb.* **15**:57–63.

Rogers, L. L., 1977, Social relationships, movements, and population dynamics of black bears in northeastern Minnesota, Ph.D. dissertation, University of Minnesota.

Rood, J. P., 1972, Ecological and behavioural comparisons of three genera of Argentine cavies, *Anim. Behav. Monogr.* **5**:1–83.

Rood, J. P., 1974, Banded mongoose males guard young, *Nature (London)* **248**:176.

Rood, J. P., 1978, Dwarf mongoose helpers at the den, *Z. Tierpsychol.* **48**:277–287.

Rood, J. P., and Waser, P. M., 1978, The slender mongoose in the Serengeti, *Carnivore* **1**(3):54–58.

Rosenblatt, J. S., 1967, Nonhormonal basis of maternal behavior in the rat, *Science* **156**:1512–1514.

Rowe Rowe, D. T., 1978, The small carnivores of Natal, *Lammergeyer* **25**:1–50.

Ruffer, D. G., 1966, Sexual behaviour of the northern grasshopper mouse *(Onychomys leucogaster)*, *Anim. Behav.* **13**:447–452.

Ryden, H., 1974, The 'lone' coyote likes family life, *Nat. Geogr.* **146**:278–294.

Ryden, H., 1975, *God's Dog*, Coward, McCann and Geoghegan, New York.

Saylor, A., and Salmon, M., 1971, An ethological analysis of communal nursing by the house mouse *(Mus musculus)*, *Behaviour* **40**: 62–85.

Schaller, G. B., 1972, *The Serengeti Lion*, University of Chicago Press, Chicago.

Scheffel, W., and Hemmer, H., 1975, Breeding Geoffroy's cat *(Leopardus geoffroyi)* in captivity, *Int. Zoo. Yearb.* **15**: 152–154.

Schneider, D. G., Mech, L. D., and Tester, J. R., 1971, Movements of female raccoons and their young as determined by radio-tracking, *Anim. Behav. Monogr.* **4**(1):1–43.

Schreiber, G. R., 1968, A note on keeping and breeding the Philippine tarsier at Brookfield Zoo, Chicago, *Int. Zoo Yearb.* **8**:114–115.

Schultze-Westrum, T., 1965, Innerartliche verständigung durch düfte beim gleitbeutler *Petaurus breviceps papuanus* Thomas (Marsupialia, Phalangeridae), *Z. Vergl. Physiol.* **50**:151–220.

Schürer, U., 1978, Breeding the black-footed cat in captivity. *Carnivore* **1**:109–111.

Seidensticker, J. C., IV, Hornocker, M. G., Wiles, W. V., and Messick, J. P., 1973, Mountain lion social organization in the Idaho Primitive Area, *Wildl. Mongr.* **35**:1–60.

Seitz, A., 1970, Beitrag zur Haltung des Schabrackentapirs (*Tapirus indicus* Desmarest 1819), *Zool. Gart. N.F.* **39**:271–283.

Selander, R. K., 1972, Sexual selection and dimorphism in birds, in: *Sexual Selection and the Descent of Man, 1871–1971* (B. Campbell, ed.), pp. 180–230, Aldine Press, Chicago.

Sherman, P. W., 1980, Infanticide in Belding's ground squirrels, in: *Natural Selection and Social Behavior: Recent Research and New Theory* (R. D. Alexander and D. W. Tinkle, eds.) Chiron Press, New York (in press).

Simonetta, A. M., 1966, Osservazioni etologiche ed ecologiche sui dik-dik (gen. *Madoqua;* Mammalia: Bovidae) in Somalia, *Monit. Zool. Ital.* **74**:1–33.

Skinner, J. D., Breytenbach, G. J., and Maberly, C. T. A., 1976, Observations on the ecology and biology of the bush pig *Potamochoerus porcus,* *S. Afr. J. Wildl. Res.* **6**:123–128.

Smith, M. H., 1966, The evolutionary significance of certain behavioral, physiological, and morphological adaptations of the old-field mouse, *Peromyscus polionotus*, Ph.D. dissertation, University of Florida.

Smythe, N., 1978, The natural history of the Central American agouti *(Dasyprocta punctata)*, *Smithson. Contrib. Zool.* **257**:1–52.

Spencer-Booth, Y., 1970, The relationship between mammalian young and conspecifics other than the mothers and peers: A review, in: *Advances in the Study of Behaviour,* Vol. 3 (D. S. Lehrman, R. A. Hinde, and E. Shaw, eds.), pp. 120–194, Academic Press, New York.

Stanley, M., 1971, An ethogram of the hopping mouse, *Notomys alexis, Z. Tierpsychol.* **29**:225–258.

Stearns, S. C., 1976, Life-history tactics: A review of the ideas, *Quart. Rev. Biol.* **51**:3–47.

Stonorov, D., and Stokes, A. W., 1972, Social behavior of the Alaskan brown bear, in: *Bears—Their Biology and Management* (S. M. Herrero, ed.), pp. 232–242, *IUCN Publ.*

Stroganov, S. U., 1962, *Carnivorous Mammals of Siberia,* Published for the Smithsonian Institution and the National Science Foundation by the Israeli Program for Scientific Translation, Tel Aviv.

Tembrock, G., 1957, Zur Ethologie des Rotfuchses (*Vulpes vulpes* (L.)), unter besonderer Berücksichtigung des Fortpflanzung, *Zool. Gart.* **23**:289–532.

Tener, J. S., 1965, *Muskoxen in Canada,* Canadian Wildlife Service, Queen's Printer, Ottawa, Canada.

Thomas, J. A., and Birney, E. C., 1979, Parental care and mating system of the prairie vole, *Microtus ochrogaster, Behav. Ecol. Sociobiol.* **5**:171–186.

Tilford, B. L., and Nadler, R. D., 1978, Male parental behavior in a captive group of lowland gorillas (*Gorilla gorilla gorilla*), *Folia Primatol.* **29**:218–228.

Tonkin, B. A., and Kohler, E., 1978, Breeding the African golden cat in captivity, *Int. Zoo Yearb.* **18**:147–150.

Trivers, R. L., 1972, Parental investment and sexual selection, in: *Sexual Selection and the Descent of Man 1871–1971* (B. Campbell, ed.), pp. 136–179, Aldine Press, Chicago.

Trivers, R. L., and Hare, H., 1976, Haplodiploidy and the evolution of the social insects, *Science* **191**:249–261.

Turner, K., 1970, Breeding Tasmanian devils (*Sarcophilus harrisii*) at Westbury Zoo, *Int. Zoo Yearb.* **10**:65.

Ulmer, F. A., 1968, Breeding fishing cats (*Felis viverrina*) at Philadelphia Zoo, *Int. Zoo Yearb.* **8**:49–55.

van Lawick, H., 1970, Golden jackals in: *Innocent Killers* (H. van Lawick and J. van Lawick-Goodall), pp. 105–145, Collins, London.

van Lawick, H., 1974, *Solo: The Story of an African Wild Dog,* Houghton Mifflin, Boston.

Vehrencamp, S. L., Stiles, F. G., and Bradbury, J. W., 1977, Observations on the foraging behavior and avian prey of the neotropical carnivorous bat, *Vampyrum spectrum, J. Mammal.* **58**:469–478.

Verts, B. J., 1967, *The Biology of the Striped Skunk,* University of Illinois Press, Urbana, Ill.

Wackernagel, H., 1965, Grant's zebra, *Equus burchelli boehmi,* at Basle Zoo—a contribution to breeding biology, *Int. Zoo Yearb.* **5**:38–41.

Walker, E. P., 1975, *Mammals of the World,* 3rd ed., Johns Hopkins University Press, Baltimore.

Wayre, P., 1978, Status of otters in Malaysia, Sri Lanka and Italy, in: *Otters* (N. Duplaix, ed.), pp. 152–155, *IUCN Publ. New Ser.*

Weatherhead, P. J., and Robertson, R. J., 1979, Offspring quality and the polygyny threshold: "The sexy son hypothesis," *Am. Nat.* **113**:201–208.

Weiher, E., 1976, Hand rearing fennec foxes, *Int. Zoo Yearb.* **16**:200–202.

Williams, J. B., 1978, *Adult Male–Infant Interaction in Nonhuman Primates: A Bibliography with Primate Index,* 2nd ed., Primate Information Center, Regional Primate Research Center, University of Washington, Seattle.

Wilson, E. O., 1975, *Sociobiology: The New Synthesis,* Harvard University Press, Cambridge, Mass.

Wilson, S. C., and Kleiman, D. G., 1974, Eliciting play: A comparative study *(Octodon, Octodontomys, Pediolagus, Phoca, Choeropsis, Ailuropoda)*, *Am. Zool.* **14**:341–370.

Wilsson, L., 1971, Observations and experiments on the ethology of the European beaver *(Castor fiber* L.), *Viltrevy (Stockholm)* **8**:117–266.

Wittenberger, J. F., 1978, The evolution of mating systems in grouse, *Condor* **80**:126–137.

Yardeni-Yaron, E., 1952, Reproductive behaviour of the levante vole *(Microtus guentheri* D. A.), *Bull. Res. Counc. Isr.* **1**:96–98.

Zucker, E. L., Mitchell, G., and Maple, T., 1978, Adult male-offspring play interactions within a captive group of orang-utans *(Pongo pygmaeus)*, *Primates* **19**(2):379–384.

10

Primate Infant Caregiving Behavior

Origins, Consequences, and Variability with Emphasis on the Common Indian Langur Monkey

James J. McKenna

1. Introduction

Field and laboratory data collected upon the Colobinae subfamily of Old World monkeys and, specifically, the Indian langurs *(Presbytis entellus)* will be used here to raise important questions about how the study of primate parenting might be approached. While the data on the Indian langur monkey describe only one of several parenting strategies which could be examined, the behavior and socioecology of this species have been studied by a great many investigators in diverse environmental settings. Consequently, a broad comparative perspective permits us to generate hypotheses about the origins of langur maternal care with some degree of confidence. It is possible that important variables elucidated by langur researchers may prove relevant to interpretations of infant care in other species. This proposition is justified because it is clear that langur maternal behavior emerged from a basic design common to all primates; however, certain features of that design such as degrees of maternal restrictiveness, rates of infant maturity, degrees of female–female competition, and so forth have been altered in specific ways primarily due to unique evolutionary circumstances. The challenge facing us is to elucidate how the processes of natural (including kin and sexual) selection produced and sculpted particular species differences, presumably to increase individual fitness. Moreover, we need to determine the underlying morphologic, physi-

JAMES J. McKENNA • Department of Anthropology and Sociology, Pomona College, Claremont, California 91711.

ologic, and anatomic adaptations influencing the form and nature of social structure within which maternal care develops (see Swartz and Rosenblum, this volume).

The questions mentioned above first will be approached by dealing with them on a general level. For example, what are the pertinent primate adaptations within which systems of infant care should be analyzed? Secondly, and on a more specific and comparative level, how might langur maternal care, which includes extensive allomothering, contrast with other species of Old World monkeys and what do the differences reveal about the evolution of these particular taxa? Finally, after having examined both proximate and ultimate explanations, it will be possible to propose several theoretical guidelines which appear to be useful in interpreting the evolutionary origins of parental care.

2. The Context of Primate Parenting Behavior: What Makes the Primates Different?

Even a quick survey of the behavior and ecology of the primates will reveal that relative to the other mammals they are an exceedingly diverse order. Of course, taxonomic distinctions are only relative insofar as some primate traits are exhibited to varying degrees by other forms. But when one considers rates of brain maturity (neurogenesis), fecundity, life span, reproductive span, relationships between social and biological maturity, the importance of learning, and durations of offspring dependency, it is obvious that mammals vary enormously, especially as one moves from the study of laboratory rats, to wolves, to chimpanzees, and finally to humans. Consider that a lamb exhibits sexual mounting and batting behavior at 10 days, while a young wolf, jackal, or hyena participates in cooperative group hunts in less than 6 weeks; but a chimpanzee female cannot reproduce or begin to exhibit fully adult social roles until 12–14 years of age. Such basic differences must constantly be borne in mind when the kind, quality, and duration of infant care are compared across different mammalian species, and most especially when primates are examined. It is fair to say that langur monkeys are first and foremost mammals, but it also is necessary to point out that they are primate mammals with stronger morphological, biochemical, and behavioral commonalities with prosimians, apes, and other monkeys rather than with, for example, carnivores or rodents.

Whenever he discussed the behavior and morphology of the primates, Clark (1959) articulated the theme of diversity among unity. It has only been in the last decade, however, that the extent of this diversity has been realized. Primates are known to fill either nocturnal (see Charles-Dominique, 1977) or diurnal (see Clutton-Brock, 1977; Hamburg and McCown, 1979) niches; feeding habits range from frugivorous, heavily insectivorous species such as the tarsiers and lorises to herbivorous–omnivorous species such as baboons, macaques, and humans. Divergent but flexible locomotor patterns are exhibited and range from the saltating or vertical clinging and leaping of some prosimian, strepsirhini primates to human bipedalism.

Despite the wide range of adaptations, particular evolutionary trends having significant implications for behavior unite the primates and are important in fully understanding parental behavior. For example, early in primate evolution, perhaps 70 million years ago, and due to requirements of a nocturnal, insectivorous niche (after Cartmill, 1974) a grasping hand evolved that included a replacement of claws with fingernails and the emergence of autonomous, opposable digits with sensitive tactile pads orginarily found on the terminal phalange. These hand modifications among primates promoted manipulative skills leading to an increasingly elaborate cerebellum. Migration of the eyes to the front of the face permitted binocular, three-dimensional vision, with accompanying changes in the lateral geniculate nucleus and visual striate regions in the brain. All of these adaptations accommodated both color perception and improved hand–eye coordination.

Most significantly, as one ascends the primate order leading to humans, the neocortex or neomammalian part of the brain, as McClean (1977) chooses to describe it, becomes more convoluted and voluminous. But increase in brain size relative to body size is not by itself diagnostic of superior intelligence and/or behavioral flexibility (see Jerison, 1973). Most important is the contribution that newer, more neurologically sophisticated substrates of cortex make to behavior. Among primates, for example, it is clear that evolutionary changes in the cortex provided an increasingly important place for learning, rather than for the operation of genetically fixed responses. In fact, behavior is expressed among primates not so much due to the action of the older part of the brain, i.e., the limbic or paleomammalian (after McClean, 1977) system operating under genetic control, but rather due to cerebral functioning which makes continuous reference to the animals' lifetime learning experiences (see Gibson, 1977). This last statement is not meant to negate the influence of hormones and particular neurotransmitters on behavior (see Cooper *et al.*, 1978). Yet it must be realized that to interpret primate behavior accurately, learned behaviors and related social factors can mediate against the expression of a particular species-typical behavior (see Slob *et al.*, 1978). In other words, there is a difference between the potential for a behavior to be developed and whether or not it actually does.

While it is important, then, to stress the role of learning in the development and expression of primate behavior, it is altogether a more difficult task to ask why learning evolved in the first place. Would not innate behaviors have been easier (see also Alcock, 1975)? Symons (1979) points out that learning can be favored among species when it is adaptive for an animal to exhibit different behavior patterns, or to make different discriminations in environments which can change quickly and frequently. Further, he suggests that innate behaviors will be favored when it is adaptive for an animal to exhibit the same behavior pattern to a particular stimulus in all the environments it will confront. Symons' discussion makes sense insofar as both the paleontological evidence for primate evolution and the current ethological data indicate that throughout their evolution primates have adapted to exceedingly diverse environments within which socioecological dynamics changed over relatively short periods of time. Thus, flexible social behavior made possible by an emphasis upon learning would, in the long run, prove adaptive.

Important changes in reproductive patterns and rates of infant maturity have evolved among primates which also differentiate them relatively from other mammals. For example, excluding several of the lemur species and members of the Callithricidae family of New World monkeys (the marmosets), we find that the majority of primate females give birth to one infant at a time. The reduction in litter size constitutes one reproductive strategy, so it seems, for maximizing the survival chances of the neonate, presumably because the mother can deliver maximum care and attention to her offspring without competition from litter mates (see also Klopfer, this volume). The amount and quality of care a primate mother delivers to her young is especially critical, since primate neonates are born much less precocious than other K-selected mammals (see Pianka, 1978). In fact, primate neonates are relatively helpless at birth, particularly insofar as locomotor behavior is concerned. Ordinarily, the mother must provide transportation for the infants for weeks, but more usually for months, following birth.

It was learned years ago that monkeys, apes, and humans mature exceedingly slowly, both socially and biologically. Even among the less phylogenetically advanced species such as the lorisids and galagos this is true. Although these species generally mature within 2 years to become reproductively active, this duration assumes significance when it is considered that very likely prosimians have a lifespan much shorter than the higher primates such as the monkeys. Charles-Dominique (1977) estimates that within specific demes galagos replace themselves approximately every 6 years. Among other primates it is possible for individuals to spend 20 or 30% of their lives in nonreproducing stages (Beach, 1978; Schultz, 1978. It is ordinarily assumed that this delayed social and biological maturity (neoteny) is due in part to the learning demands faced by the individuals. But regardless of why neoteny evolved, it is clear that it has greatly influenced the nature and form of social structures and organizations and, as will be discussed later, may limit and constrain the directions in which primate species can change.

Inasmuch as primate females invest a great deal of time and energy in offspring due to long gestations and even longer periods of postnatal care, fecundity is relatively low. It has been demonstrated, for example, that the long periods of infant dependency affect how soon a mother will resume cycling and become pregnant again (Altmann et al., 1978). With this in mind it is not surprising that among primates natural selection has produced a relatively close fit between periods of infant helplessness and highly motivated caregivers who ordinarily respond quickly and effectively to their needs (see Hrdy, 1976, for an excellent discussion).

It is against this general background of related biosocial adaptations that primate parenting systems must be evaluated. The majority of primate young are born into intensely social environments within which they must learn and refine basic lessons of survival (see McKenna, 1979a, for complete discussion). They mature slowly and, in order to survive, require immediate social, emotional, transportational, and nutritional support. It is to a specific, detailed example of how one species cares for infants that we now turn.

3. Indian Langurs: A Case Study in Maternal Care

3.1. Classification

The common Indian langur monkey *(P. entellus entellus)* is a haplorhine primate (McKenna, 1975), classified among the Old World superorder, Cercopithecoidea, and is one of six genera in the herbivorous, leaf-eating family Colobidae (J. R. Napier and P. H. Napier, 1967). Most frequently the subfamily designation (Colobinae) is ued, which can be contrasted with the other half of the Old World monkeys, the Cercopithecines, which includes the macaques, mangabeys, guenons, and baboons.

As members of the Colobine subfamily, langurs evolved a diverticulate four-chambered stomach capable of digesting mature, senescent leaves and extracting cellulose, primarily through a fermentation process involving intestinal symbionts (Bauchop and Martucci, 1968; Kay *et al.,* 1976; Moir, 1968). Unlike the Cercopithecines, who do not exhibit these digestive characteristics, the Colobines are able to go without water for considerably longer periods of time (Ripley, 1967) and are able to ingest some known secondary (toxic) compounds (Hladik, 1977; Oates, 1977a,b.) These dietary adaptations are thought to produce feeding advantages at least in arid regions closed to the Cercopithecine macaques and possibly when preferred (primary) food resources are unavailable (McKenna, 1979a; Moreno-Black and Maples, 1977). Although traditionally the Colobines have been described as the leaf-eating specialists, recent field studies have revealed that flexible, if not rather catholic, feeding habits exist, especially among the Indian, Nepalese, and Ceylon langurs (Bishop, 1979; Hrdy, 1977a; Curtin, 1975; Hladik, 1975, 1977; Sugiyama, 1976).

Except for the western Indian desert, the common langur monkey inhabits varying regions of the Indian subcontinent. Areas of dry scrub as well as wet or dry deciduous forests and coniferous forests are some of the ecological zones to which langurs can adapt, with elevations ranging between sea level (in Sri Lanka) to regions over 11,000 feet in the Himalayas (Bishop, 1979, 1979; Boggess, 1976; Curtin, 1975; Ripley, 1965; Sugiyama, 1976). The *P. entellus* species are, perhaps, the most ground-dwelling of the Asian Colobines–a group containing between 23 and 30 different species (Jay, 1965).

3.2. Social Organization

Relative to other species of this subfamily, field studies conducted on the common langur are numerous and indicate that langur society is exceedingly flexible. Studies by Sugiyama (1964, 1965, 1967), Yoshiba (1968), Rahaman (1973), Curtin (1975), Boggess (1976), Jay (1963, 1965), Bishop (1979), and Hrdy (1977a) clearly reveal the extent to which local demographic and ecological conditions, including human interference, have influenced langur social organization and behavior. Home ranges can vary dramatically in their size as can langur population densities. For example, Curtin (1977) reports that langurs living in North India at the sites of Kaukori, Utar Pradesh; Orcha, Maydhya Pradesh; and Junbesi, and in Solu Khumdu, Nepal, inhabit home ranges of 770 ha, 380 ha, and 1200 ha, respectively. Population densities range from 1/km² in Junbesi to 7/km² in Orcha and Kaukori (see Curtin, 1977, for complete discussion). But langurs living primarily in southern India and Sri Lanka exhibit striking differences with respect to these figures. In Dharwar, Mysore, India, Sugiyama (1965) and Yoshiba (1968) report a mean home range size of 16.8 ha with a population density of 134/km². Hrdy (1977a), studying the langurs of Abuv, had less dramatic findings, with home ranges averaging 38 ha with an estimated population density of 50/km². Ripley (1965, 1967) reports that in Sri Lanka home ranges fell between 32 and 128 ha, with a density estimate of 58/km².

The composition of Indian langur troops may be bisexual or unisexual (all male). In North India the modal pattern of social organization appears to be multimale–at least in regions of relatively low langur densities. Societies are organized around fluid dominance hierarchies of males, females, and their young. In the south, however, it is more common for langur troops to be less stable. All-male bachelor groups can invade troops which are typically organized around one dominant adult male who controls the reproductive activities of his associated female troop members. Resident males are thought to be replaced approximately every 5 years (Sugiyama, 1965) by a new adult male who, along with a raiding band of bachelor males, expels the old male (Hrdy, 1977a; Mohnot, 1971; Sugiyama, 1965).

Such male replacements precipitate a great deal of intragroup aggression. Raids by bachelor bands have been reported to occur among other Colobine species as well (see Hrdy, 1977c and 1979, for review). Of major interest is the phenomenon of infanticide, the systematic killing of dependent offspring by the new resident adult male which can accompany the replacement process in some, but not all langur troops. But presently the mechanisms governing this phenomenon are not entirely understood and the issue of langur infanticide continues to be a subject of major controversy. Hrdy (1977a,b, 1979) argues that infanticide is one of several reproductive strategies exhibited by males to maximize their reproductive fitness. The death of a nursing infant hastens the mother's return to estrus, thereby increasing the chances that the new male can impregnate her and see his offspring grown before a new male enters to repeat the cycle. Adopting a different explanation, Curtin and Dolhinow (1978) and Curtin (1977) argue that langur infanticide is an aberrant response to rapidly deteriorating environmental conditions such as, for example, the encroachment by humans on natural langur home ranges, thereby causing overpopulation of langurs. They point out that very few infant killings have actually been observed directly by researchers. Chapman and Hausfater (1979) provide a mathematical model of infanticide showing how this behavior can increase male reproductive fitness over time in certain circumstances. However, their model analyzes infanticide as a single strategy evolving with all other behavioral strategies held constant. Until more data are available on this issue it seems sensible to assume, as did Jay (1965), that "social behavior of all langurs is the result not so much of truly diverse forms of behavior as degrees of emphasis in patterns common to all langurs" (p. 247). It should be borne in mind that this species is an exceedingly flexible one and that rigid phylogenetic restraints on behavior patterns, including social organization and structure, are not apparently characteristic.

3.3. Langur Infant-Caretaking Patterns: Who Does It?

The social environmental setting in which the young langur infant matures is a most interesting one, primarily because so many different individuals other than the mother can deliver care to the infant and establish early social relationships with it. Data presented in Tables 1, 2, and 3 are based upon approximately 101 hr of focal animal sampling drawn from over 1500 hr of observation collected on a colony of langurs housed in a research facility at the San Diego Zoo, and away from public view. Focal animal samples were taken on two male infants during the first 3 months of life. The results show dramatically that infants are not the exclusive social property of their biological mothers and that before the infants are even 3 months old, they have had extensive contact with female caretakers that are young and old and experienced and inexperienced, and that include an extraordinary array of personality types.

Data collected on this specific group, which contained one adult male, eight adult females, four immature males, four immature females, and of course, the two infants, reveal that juvenile and subadult females in particular (ranging in age from 3 to 4.5 years old) were highly motivated to obtain custody, to carry, to groom, and to fondle 1- to 3- month-old brown neonates (Table 2).

It must be mentioned that such infant transfer data can very enormously not only among females of a single group (some females will be more restrictive than others, e.g., Table 2) but also between females of different groups. Dolhinow's (1977, 1978a,b) study of mothering behavior among colony living langurs at the U.C. Berkeley Animal Research Station shows that mothers may vary with respect to their permissiveness or restrictiveness from one infant to the next. The data Dolhinow has collected also show that among the Berkeley group of langurs, adult females rather than subadult females, as I report here, were much more successful in obtaining neonates, and the decline in general interest in the infants occurred much earlier.

Similar to my data (Table 3) but not altogether identical, either, were Hrdy's findings (1977a) that for females living in the Toad Rock troop at Abu, India, both nulliparous and pregnant females were significantly overrepresented among the classes of animals taking, and attempting to take, neonates. She reports, that nulliparous females over 15 months of age held infants for an average of 9.97 min (based upon 108 instances), while nulliparous females under 15 months of age held infants for an average of 2 min (based upon 7 instances). Pregnant females near term held infants for approximately an average of 5 min (based upon 102 instances).

While the average length of time multiparous females held the infants was almost identical to data on captive multiparous females (Table 3), nulliparous

Table 1. Summary of Infant Transfers for Two Male Langur Infants from Week 1 to Week 12 of Life[a]

Infant	Week 1	Week 2	Week 3	Week 4	Week 5	Week 10	Week 12
A							
Mean frequency of transfers	0.8 (0.5)[b]	1.2 (0.3)	0.9 (0.5)	0.7 (0.4)	0.3 (0.2)	0.1 (0.04)	0.1 (0.06)
Total number of transfers	43	63	45	41	17	6	5
Number of 15-min samples	35	30	30	35	35	32	33 = total 3450 min
B							
Mean frequency of transfers	n.d.[c] —	0.9 (0.8)	0.4 (0.3)	0.1 (0.02)	0.1 (0.01)	0.2 (0.04)	0.2 (0.07)
Total number of transfers	—	39	33	22	21	28	25
Number of 15-min samples	n.d.	32	30	34	33	40	15 = total 2610 min

[a]Observation time is 6060 min or 101 hr; 15 min, per observation session.
[b]Standard deviation.
[c]n.d., no data.

Table 2. Percent of Total Time Observed per Week of Life That Langur Infants Spent in the Custody of Caretakers Other Than the Mother

Number of individuals	Week 1	Week 2	Week 3	Week 4	Week 5	Week 10	Week 12
Infant A (male)							
8 Adult females	21	18	8	10	18	11	8
2 Subadult females	26	34	33	12	4	6	14
2 Juvenile females	9	5	1	12	4	6	3
4 Juvenile males	-	-	-	-	-	-	1
Percent of time away from mother	56	57	42	46	37	28	26
Infant B (male)							
8 Adult females	n.d.[a]	11	13	6	10	13	10
2 Subadult females	n.d.	29	25	16	8	3	4
2 Juvenile females	n.d.	2	5	3	-	3	0.4
4 Juvenile males	n.d.	-	-	-	-	-	-
Percent of time away from mother	n.d.	42	43	35	18	19	14.4

[a]n.d., No data.

wild langur females held the infants for periods averaging longer than what occurred in captivity. This may be related to the fact that wild langurs can, of course, move further away from other potential allomothers, diminishing the chances that the infant will be quickly taken by another. Interestingly, Hrdy determined that 49% of the observed time the infant was in association with females other than the mother. This figure is roughly comparable to that computed for captive langurs over the first 4-week period of the infant's life (Table 2).

Table 3. Mean Duration in Minutes of Infant Holding per Animal per Reproductive Class Based on 15-min Focal Animal Samples of Two Male Infants, during the First 12 Weeks of Life

N	Subject females and reproductive state	Week 1	Week 2	Week 3	Week 4	Week 5	Week 10	Week 12
4	Nulliparous	1.1	1.2	4.3	3.6	2.8	5.2	2.3
6	Multiparous	3.3	2.7	2.2	1.9	0.4	0.7	1.1
2	Primiparous	0.9	2.1	1.3	1.4	1.6	1.2	0.6

More specifically, my own findings reveal that the langur neonate is transferred most frequently during the 2nd and 3rd weeks of life, exclusively by females. Subadult nulliparous females (between the ages of 3 and 4.5 years who have not apparently cycled) were involved in the greatest amount of allomothering behavior, followed by multiparous and primiparous adult females, respectively. While subadult nulliparous females exhibited a sustained interest in the neonates over the 12-week period, there was a decline in the frequencies and percentage contribution of allomothering after the 4th week (see Tables 1 and 2).

While subadult nulliparous females hold the infant more frequently than females of other reproductive classes, the mean duration of infant holding per animal per reproductive class is consistently less than what it is computed for the multiparous adult females (Table 3). In fact, in the first week of life when one could expect that the biological mother is more closely monitoring the extramaternal caretaking of its infant, multiparous and presumably "experienced" females held the neonates three times longer on the average per allomothering episode than the nulliparous ("inexperienced") females.

Most significant to consider are success rates. After I compared the total instances in which multiparous and primiparous adults and nulliparous females successfully took infants relative to their attempts, my data revealed that nulliparous females were unsuccessful in a higher proportion of the time, though not at rates which were statistically significant. In other words, while adult females engaged in allomothering less often, their attempts to do so were almost always successful, but this was not the case for nulliparous females.

Both duration and frequency data show interesting differences when the responses to infants by allomothers at week 1 is compared with the general responses shown at week 12. Both the amount of time the infant is held by individual allomothers and the frequency with which they are transferred (per unit of observation) decrease gradually. To interpret these findings it is important to consider that the infant is playing an increasingly significant role in monitoring its own care, and in determining who has access to it as the weeks progress. Moreover, the mother's restrictiveness can change dramatically as she observes her own infant's maturation, even though for the first 3 months of life the infants of this group, but not necessarily other groups, were almost always in ventral–ventral contact with a caretaker. Furthermore, all of the real and potential allomothers play an important role in determining the allomothering scores from any one parity subgroup. For example, the multiparous adults of this particular study group were most interested in obtaining the infant in the first few weeks of life. Because they generally retain relaatively, though not absolutely, higher statuses than juvenile and subadult nulliparous females, and because they generally exhibit greater maternal competence, as presumably noted by the infant's mother, they gain the infant more frequently during this period. This fact has significance in making interpretations of the scores of the primiparous and nulliparous females who may be more interested in obtaining the infant than their scores indicate.

Another important dimension to allomothering behavior is the amount of time spent by the infant away from its own mother. As Table 2 reveals, the week-old neonate can spend as much as 57% of the time observed in the care of a female other than the mother. Even at week 12, when allomothering levels have declined, the infants spend between 26 and 14% of their time in ventral–ventral association with females other than the mother. As Dolhinow (1977, 1978a,b) has carefully documented, the activity of the biological mother throughout a 3-month period contrasts with the behavior of the allomothers, at least inasmuch as she visually monitors the whereabouts of her infant, protectively retrieves the infant, exhibits social and nonsocial restraining, and intercedes against third animals should the mother "fear" for her infant's safety. These kinds of "maternal" caretaking behaviors that reflect the strength of maternal attachment are not as frequently initiated by the substitute caretakers. Dolhinow's (1977, 1978a) data show a tremendous range of maternal styles and differences exhibited not only from one mother to the next, but even by the same mother across successive infants.

3.4. Forms of Infant Transfer

The form by which the infant transfer takes place also varies quite extensively from one transfer to another and, of course, depends to a large extent on the competence of the caregiver and the social context in which the transfer has occurred. It is also possible that the composite personality types in a group affect the form of transfer, i.e., is it relaxed or tense? Interestingly, in both the field and the laboratory setting the transfer experience for the neonate is not always very pleasant, nor is it always particularly safe for the neonate, especially in the first 2 or 3 weeks. Ordinarily, the transfer sequence begins with the approach by another female to the mother–infant or allomother–infant pair, wherein sometimes, but not always, either the infant or the infant's caretaker will be groomed. The potential allomother may lean forward toward the infant as if smelling it, and while lowering its head, the female taking the infant grasps the axillary or ventral region of the neonate and pulls it toward her own ventrum (see Fig. 1). Immediately after gaining physical control of the infant the allomother may turn away from, or carry the infant away from, the previous caregiver; or the female may immediately lick, groom, inspect, or simply cradle the neonate. During the transfer the infant may passively accept the transfer, or more often it will briefly struggle and/or vocalize as if in distress until it is firmly placed and possibly feels secure against the caregiver's ventrum. Infants are particularly distressed when juvenile females, often inept caregivers, take the infants and, without placing the infant centrally on their ventrums for transportation, gallop away while the infant screams, clinging tenaciously though not always successfully to the female's lower abdomen or higher thigh.

During my observations at the San Diego Zoo Research Colony, often infants were simultaneously tugged in opposite directions by two juvenile allomothers. In several instances, the infant's nose became bloody and tail or bottom was scraped, as an infant was unceremoniously dragged by an ambitious young female caregiver whose own physical size precluded a safe ride. Ordinarily the younger juveniles were not always able to placate the neonates, irrespective of what kind of care the infants appeared to be receiving. After approximately 15 sec in which clusters of adult females visually monitored a distressed infant, one of them might retrieve the infant. Retrieval of distressed infants either by mother or nonmother usually occurred without aggression on the part of the retriever. Most mothers simply followed the distressed infant and its caregiver until an opportunity arose allowing the infant to be gently but firmly taken.

A review of one rather eventful 15-min sample taken during the 1st week of life ought to demonstrate what a langur neonate can actually experience, and perhaps illustrate why risks are high in this system. One infant experienced ten transfers involving seven different individual caretakers from three different age groups that varied with respect to reproductive status and apparent caretaking competence. During this 15-min sample, the infant gave 32 distress vocalizations, struggled and/or attempted to resist seven of the ten transfers, and was forcefully withdrawn from its mother's nipples by different allomothers on three occasions. The infant was briefly groomed 6 times (less than a minute total), licked 13 times, and was turned upside down by allomothers 5 times, presumably to inspect its genitals. While the laboratory setting potentially can exaggerate the frequency of transfer

Figure 1.   A nulliparous (juvenile) allomother takes a 2-day-old neonate from its mother. These Indian langurs are, perhaps, the most permissive caregivers belonging to the Colobinae subfamily.

and these accompanying behaviors, this is not necessarily the case. Data collected in the field by both Jay (1962) and Hrdy (1977a,b,c) are consonant with the above laboratory observations.

4. Why Allomothering?

At least judging from the amount of time invested in these activities, allomothering appears to be an important aspect of Colobine and specifically langur socialication. But these data also show that there are certainly some serious disadvantages with this caregiving system, particularly for the neonate. Yet, in the long run, but not necessarily if the behavior is recent, we may assume that the net behavioral consequences of allomothering are beneficial; phrased another way, if the behavior is phylogenetically old, and is defined in fitness terms for the participants, the benefits of this behavior must exceed the costs or risks. If this were not the case, allomaternal behavior could not have evolved among this species or any other.

It is important, then, to determine what the benefits of this behavior may be, first to the infant and then to other participants. Consequences of the infant–transfer experience are especially significant for the infant because they should reveal a basic compatibility between specific socialization processes and the general social structure within which they operate and into which the infant must eventually be integrated. For example, the constant and sometimes vigorous activity of allomothering exposes the Colobine langur neonate to an exceedingly wide variety of social contacts who differ in age, parity, temperament, and caretaking competence (Dolhinow, 1977). The breadth of this social exposure is quite different from what the majority of Cercopithecine monkey neonates experience, that is, an exclusive, steady, and more predictable relationship with only one caregiver, the mother. It may be that the langur infant transfer pattern prepares the infant for a social life which, characteristically, is not based upon exclusive and long-lasting social ties with one's mother, or with one's lineage; instead McKenna (1979a) suggests that at least after the juvenile age grade the social relationships of the individual langur develop quite independently from those of mother. Both field and laboratory data suggest that langur social bonds (and possibly this is true of other Colobines) are neither built around, nor depend upon, specific kinship units *to the degree* characteristic of the Cercopithecine baboons, guenons, magabeys, and macaques. While the mother may serve as the central social focus of the langur infant and juvenile, the subadult and adult male and female langurs develop a social nexus quite separate from that established by the mother. This is especially different from the macaques, wherein matrilineages play a fundamental role in regulating female adult life (Eaton, 1976; Hanby, 1972; Kurland, 1977; Ransom and Rowell, 1972; Rosenblum, 1971; Simonds, 1974). While the Colobine langur may learn (at least in part) to predict certain kinds of behavioral interactions from its experiences during infant transfer, the macaque infant gradually learns to predict the effect of its actions based upon its defined membership within a particular kin unit (McKenna, 1979a). Clearly, these are two different kinds of systems, but each prepares the infant presumably for the appropriate behavioral strategies which will enable it to survive to reproduce (with a little luck) within the limitations of its own ecological niche.

There are other potential benefits to the infant which need mention. For example, after witnessing an adoption in the field, Jay (1965) proposed that allomothering increases the chances of infant adoption and, therefore, survival, should an infant's own mother die. Boggess (1976) observed a similar adoption among the Himalayan langurs, and Kaufman's (1977) research confirms the fact that, indeed, allomothering can be linked with the likelihood of an infant's surviving the loss of its mother.

A number of investigators propose that nulliparous females participating in allomothering activities acquire valuable maternal skills which can aid them in delivering effective care to their own infants. This has been called the "learning-to-mother" hypothesis and is discussed by Hrdy (1976), Gartlan (1969), Jay (1965), Lancaster (1971), and Struhsaker (1967). Another benefit of allomothering occurs at the group level, or so it has been proposed. Simply stated, the common attraction to infants and the female–female bonds established and maintained through infant sharing contributes to the general cohesion of the group (Horwich and Manski, 1975; Jolly, 1966, Struhsakeer, 1975). Finally, Hrdy (1976) suggests that allomothering may hasten socialization.

5. Separating Functions, Benefits, and Origins in Evolutionary Explanations

5.1. Levels of Explanation

As both Quiatt (1979) and McKenna (1979a,b) demonstrate, difficulties arise when explanations of allomothering examine only postulated benefits and stress functional consequences without also considering the ancestral preconditions from which a behavior emerged. Obviously, it is critical if not inevitable that a behavior's apparent usefulness as defined in fitness terms be considered, even when it may be that the present environment of adaptedness (as Bowlby, 1969, termed it) is not the same as, nor relevant to, the environment in which the behavior evolved (see McKenna, 1979a). But presently there appears to be a great deal of confusion and misunderstanding as to how to apply the concept of function; how does it differ from benefits or beneficial effects of a behavior? Furthermore, it appears that there is a difference of opinion as to what constitutes an explanation of the origins of behavior.

In attempts to clarify the nature of methodological, conceptual, and theoretical misunderstandings Daly and Wilson (1978) stress that important differences exist in the ways in which "why" questions about behavior can be answered. They stress that questions which deal with proximate mechanisms necessarily lead to developmental or ontogentic explanation; a behavior is explained by way of the more immediate physiological and social/environmental factors that underlie it and which influence its form and nature. In contrast, ultimate levels of explanation are concerned with a behavior's long-term adaptive significance (how a behavior contributes to individual fitness) and also with its phylogeny. Daly and Wilson point out that investigators do not always recognize the critical differences between these two levels of explanation, and consequently, unnecessary conflict and debate emerg (see also Kurland, 1977 for good discussion).

5.2. Function, Benefit, Effects and Origins: What Are the Distinctions?

Symons (1978), 1979) makes the same points but, in addition, has carefully shown that the social science concept of function as traditionally employed to explain nonhuman primate behavior is not only misleading in several ways, but has impeded progress in the field (see also Hrdy, 1977a). In order to delineate a behavior's function, he maintains, rather than simply describe its benefits or its effects, one must show it to be "the bases [by design] of differential reproduction among its ancestors" (Symons 1978, p. 194). Citing and reviewing earlier works by Williams (1966), Hinde (1975), and Richerson (1977), he points out that group benefits which can result from a particular behavior may simply be "incidental effects—statistical summations of individual adaptations" (Symons 1978, p. 195). Further, Symons states that "if effects can be explained as the result of physical laws or as fortuitous byproducts of an adaptation, it should not be called a function" (Symons 1978, p. 197). According to Williams (1966, p. 299, cited in Symons, 1978), "the demonstration of a benefit is neither necessary or sufficient in the demonstration of function , although it may sometimes prove an insight not otherwise obtainable." Finally, Symons maintains that beneficial effects cannot be equated with function and that "just because effect, but not function, can be demonstrated experimentally is not grounds for equating function and beneficial effect . . . and many beneficial effects are irrelevant to any significant biological problem" (1978, p. 198).

The perspective offered by Symons may be summarized in the following way: a behavior can be said to have a function if it can be shown that within the organism's natural environment it contributes to the individual's reproductive success and it has been designed to do so; any contribution that the behavior makes to the success of larger social units is secondary, if not incidental. To say that a behavior has a function is to say that it has been shaped and sculpted by natural selection. Beneficial effects and functions are not the same and must be analyzed differently.

In his discussion Symons moves from a critique of an antiquated social scientist's concept of function as it has been used to explain and interpret primate behavior to advocating individual selection models in order, if I read him correctly, to prevent the continued misinterpretation of a behavior's origins and adaptive significance. Interestingly, and arguing from the opposite direction, I found that individual and kin selection theories of primate behavior likewise confused beneficial effects with function and, most especially, inadvertently equated an understanding of a behavior's functional consequences with understanding how it arose. In other words, McKenna (1979a,b) reached similar, though not identical, conclusions about the misuse of terms in interpreting the evolution of behavior, but from two quite different positions. We both maintain that process is often sidestepped and that the terms "effects," "benefits," or "function" are not only used incorrectly, but often given analytic priority.

Perhaps the most important point to emerge here is that irrespective of the level of selection which is evoked (whether group or individual) the possibility of confusing the important differences between function (the adaptive qualities of a behavior as defined by contributions to individual fitness), beneficial effects (including side, or incidental, benefits—epiphenomena), and origins (the preexisting behavioral and physiologic–morphologic organizations out of which new behaviors emerge to become favorable selected) cannot be precluded.

5.3. The "Origins" of Allomothering: Ultimate Considerations

To illustrate my interpretation of the problem, we will examine some of Quiatt's (1979) work. He reevaluated the explanations of allomothering which emphasized possible benefits accruing to relatives and noted that while it may be useful to examine allomothering behavior within a framework of kin selection theory, allomaternal behavior cannot fully be explained in terms of kin selection (for examples see data in Kurland, 1977, and Hrdy, 1977a). He contends that among species in which a good deal of allomothering takes place, a potential for direct competition between mothers of varying skills, i.e., abilities to retrieve infants, is greater and, hence, natural selection operates along specific lines to eliminate offspring of deviant mothers. In addition, Quiatt suggests that "allomothering behavior may be no more than a fortuitous outcome for selection of a behavioral orientation toward conspecifics that is essential for maternal performance" (1969, p. 316).

It should also be emphasized that kin selection models used to explain allomothering actually tell us little about the context in which natural selection was working to promote this behavior. Thus, several relevant questions concerning the phylogeny of the behavior are left unanswered when the presumed functions of allomothering are inadvertently used as explanations. For example, from a theoretical perspective the functions and benefits of allomothering are as relevant to the Cercopithecine species as they are to the Colobines, but while female interest in infants among the Cercopithecines exists, as discussed earlier, except in special circumstances infant transfer does not take place for the majority of species. Why not? A discussion of either benefits or functions reveals little about the social dynamics involving ecological factors that allowed allomothering to be favorably selected among the Colobines but not among the Cercopithecines (see Hrdy, 1977). An answer to this question requires a different kind of analysis—one which moves beyond a mere description of what a behavior can accomplish, and how the benefits of the behavior can be translated into genetic calculus. An explanation of the "puzzle of infant sharing," as Hrdy (1977a) aptly phrases it, requires an analysis which considers more than the functional consequences of this behavior and more than inclusive fitness, or costs vs. benefits. The analysis must focus upon the process, rather than upon the results of the process, which is what we are attempting to explain. More specifically, to explain the *origins* of allomothering genetic benefits must be weighed against, and analyzed with reference to, the social system within which the animal behaves and its underlying morphologic, anatomic, and socioecologic bases. The final confrontation in obtaining an answer to the question of the origins of allomothering must deal with the interrelationships between such species characteristics as, for example, feeding habits, defense, mating, and dominance subsystems.

Elsewhere McKenna (1979a) argues that dominance among females and the existence of female–female competition in varying resource-specific contexts are the major obstacles preventing the emergence of infant–transfer patterns among the Cercopithecine. (see also Dolhinow *et al.*,1979). Simply stated, and as both Rowell *et al.* (1964) and Hrdy (1976) previously pointed out, safe infant transfer cannot take place in groups of animals in which the higher status of a female allomother can prevent a subordinate (the mother) from retrieving her infant. The infant, of course, would die of starvation. The clue to understanding the puzzle of infant sharing is not, then, obtained by considering what the behavior can do, but by trying to determine why dominance systems exist, or why they play a more powerful role in regulating Cercopithecine female behavior than they do among Colobine females.

One possibility is that early in Colobine evolution social structural differences which later permitted the gradual emergence of infant sharing developed, primarily due to feeding adaptations. Specifically, the Colobine digestive system and adaptations to efficient preparation of food made possible by Colobine dental morphology created, if not a wider spectrum of available foods, at least more efficient utilization of what limited variety of foods were available. In effect, reliance upon the mature leaf and a flexible browsing feeding pattern which included fruits, leaves, buds, and flowers reduced the amount of intragroup feeding competition. It is suggested that relative to the Cercopithecines, competition and the need to control access to food was reduced and, thus, female–female behaviors were influenced increasingly less by status differentials (dominance) but more by other integrating behaviors, such as, and most notably, infant care. The reduction of female–

female competition for food and the deemphasis of dominance as an organizing principle coincided with close–spatial feeding patterns and general cohesion, further facilitating the development of multiple female care of the young in the Colobines, but not in the Cercopithecines (McKenna, 1979a).

6. Langur Allomothering Behavior: Is It Unique?

Regardless of the bases of its origins, the amount and diversity of infant care administered by male or female primates other than the parents is impressive. Wilson (1975) has suggested that so great is the contribution made either by allomothers (female caregiving assistants) or by allofathers (male caregiving assistants) to infant development that surely the social evolution of the primates has been affected. Exactly in what way social evolution has been influenced by shared infant caretaking is not yet known, but it is clear that species other than the common langurs exhibit forms of alloparental behavior (this term includes both male and female caretaking assistants, after Wilson, 1975; see also Kleinman and Malcolm, this volume, regarding male parental investment). For example, while the prosimian primates tend not to exhibit alloparenting (see Erlich, 1977; and Klopfer and Boskoff, 1979, for exceptions), it can occur among certain New World monkeys, especially tamarin–marmoset species wherein males typically carry newborn, and among howler monkeys (Glander, 1975). It is also reported to take place occasionally and in certain contexts among the wooly, squirrel, and spider monkeys (for excellent reviews, see Hrdy, 1976; Redican, 1976; Roy, 1976, 1977).

But by far, allomothering behavior is expressed most frequently among the Old World monkeys, Inasmuch as female and sometimes male (see Fig. 2) interest in neonates is neither restricted nor socially limited to biological parents among many of these species, it is obvious that selection has placed a premium not only upon the mother–infant relationship *per se,* but also upon the entire social setting in which the Old World monkey matures. Among the macaques, for example, rhesus, bonnet, Japanese, and Barbary monkeys exhibit varying forms of alloparenting which sometimes involve males. Similarly, vervet monkeys *(Cercopithecus aethiops)* reportedly exhibit a great deal of allomothering behavior, while the Patas monkey *(Erythrocebus patas)* and gelada baboon exhibit some, but a great deal less than the other species.

It is essential, however, to evaluate these alloparenting behaviors carefully. Under close scrutiny inportant and significant differences can be exposed when the form, function, and social context of alloparenting among Cercopithecines and Colobines are compared.

For all the primates Hrdy (1976) carefully reviewed the range of infant-caretaking patterns. From her synthesis attention can be called to the fact that among Cercopithecine monkeys, but not necessarily among the Colobines, infant

Figure 2. While male participation in parenting is rare among the Indian langurs, this adolescent, 4-year-old male risks attacks from females as he competently cradles a 7-month-old infant.

sharing is often selective, occurs by way of physical force on the part of the allo-parent, involves more mobile infants, occurs less frequently, and, particularly when males are involved, reflects temporary changes in the status of the caretaker. For example, the "aunting" behavior originally described by Rowell *et al.* (1964) and by Hinde and Spencer-Booth (1967) for captive rhesus monkeys apparently takes place much less frequently in their natural habitat when interanimal spacing patterns are greater (Lindberg, 1971). Furthermore, the transfer of infants tends to occur only when the mother is dominant to the substitute caretake; subordinant females are extremely reluctant to release their infants to females whoe superior statuses may preclude their retrieval. Hinde and Spencer-Booth (1967) found that the presence or absence of "aunts" affected the degree to which a mother restricted her infant's movements. In the presence of other females, mothers retrieved their infants more frequently and limited the distances their infants could travel. Such behavior by the mother suggests that substitute infant caretaking is not always desired.

Among patas monkeys *(E. patas)* the "kidnapping" behaviors described by Hall and Mayer (1967) and by Chism (personal communication) represent infant-caretaking behaviors which differ substantially from that characteristic of the Colobines. During these interactions a dominant female may approach and, with great force and vigor, pull a neonate off the ventrum of a subordinate who, rather than risk attack, will give up her infant. As is true for many of the instances of female substitute caretaking among macaques, status differentials play a role in the transfer; the interaction is aggressive and tense, and death can result for the infant should its own mother be unable to retrieve it from a more dominant, non-lactating caretaker.

There are other differences as well. Unlike what is true for the majority of Colobine neonates, the "play mothering" behavior of juvenile female vervet monkeys *(C. aethiops)* described by Lancaster (1971) and previously observed by Struhsaker (1967) and Gartland (1969) typically does not occur in the first few days of life. While Struhsaker (1967) observed an exceptional case of a day-old infant transfer, the youngest infant Lancaster (1971) observed being held was 8 days old, and this occurred only once for a very brief period of time. The majority of caretaking by nonmothers occurs after 3 weeks of life, usually by juvenile females. Kriege and Lucas's (1974) study of vervets in Durban, South Africa, showed that infants between the ages of 1 and 6 weeks were most likely to be transferred and that "infant stealing" (the forced removal of an infant from its mother's ventrum) was a significant part of allomothering activities. Relative to the majority of transferred Colobine infants, the locomotor accomplishments of transferred vervet monkeys suggest that the vervets are developmentally more mature when transfer frequencies statistically peak. In many cases transferred vervet infants were able to wrestle away from their caretakers and return to their mothers.

Colobine infants are transferred most frequently during the first 3 weeks of life, and they are still quite immobile (Dolhinow, 1977; Horwich and Manski, 1975; Hrdy, 1976; McKenna, 1975). In contrast with vervet monkeys, at this point in their ontogeny the Colobine infants are unable to directly control their proximity to their mother or to necessarily control who has access to them or for how long.

7. Parental Investment: How Much is Too Much?

It should be clear that among a variety of primate species and not just langurs alloparenting takes place (though to limited degrees and in special circumstances) and provides a number of proximate benefits or effects, not necessarily designed by selection. For example, alloparenting may facilitate affiliative behaviors between animals, and as Kurland (1977, p. 118) notes infants can be used by females as well as by males to raise social statuses temporarily.

These data suggest that far from being an isolated behavioral subsystem, infant caretaking both influences and reflects many other aspects of social structure and organization. For any species we should expect the presence of a social structure which maximizes infant survivorship. This will come about by way of selection acting on individuals who live and breed within that structure (see Hrdy, 1979). But an overall assessment of evolved social and biological strategies especially among some hominoid species leads to an interesting observation: infant-caregiving behaviors and especially those exhibited primarily, though not exclusively, by mothers weigh prominently, if not disproportionately, in affecting the direction in which other aspects of primate sociality can change (see Swartz and Rosenblum, this volume). It may be that the special and long-term needs of monkey and ape offspring may act as a conservative, though not necessarily adaptive, line of resistance against changes in other life strategies such as intrasexual competition, birth intervals, defense, and even feeding strategies (see Crook *et al.*, 1976). The point being made here is that primate females invest an enormous amount of time and energy in child rearing (as noted earlier), perhaps far more time than what would be ideal or even, in the long run, adaptive. Trivers' (1972) concept of parental investment can be evoked here. The investment of the mother can be determined relatively by the length of time she spends increasing the fitness of her offspring at the expense of her own ability to invest in additional offspring. Consider, for example, that while chimpanzee birth intervals can vary somewhat from female to female and with the deaths of neonates (see Altmann *et al.*, 1978), this figure can easily exceed a 5 year period Goodall, 1971). Since female chimpanzees probably live to be only 30–40 years of age, 5 years between infants is considerable. With respect to maintaining respectable populations distributed over diverse (nontropical) ecological zones, generally speaking the pongids have not fared well. One could argue, therefore, that the exceedingly slow rate of biological and social maturity among chimpanzee infants is adaptive neither to the female whose fecundity remains low nor for the species, for which extinction becomes a strong possibility of populations decline statistically to unsafe levels.

Several examples can be offered which illustrate other ways in which infant needs can affect female behaviors and reproduction. Clark (1978) has shown that female galago offspring remain within their mother's home range and will compete with them for particularly nutritious resources when they are either pregnant or lactating. As a consequence of this mother–daughter competition, Clark suggests that galagos evolved sex ratios which favor male births, since sons, but not daughters, leave natal home ranges. Wrangham (1979) shows that female chimpanzees actually prove to be much less gregarious than males (see also Halperin,1979) and can exhibit territoriality while supporting older but dependent offspring. Special nutritional needs, particularly during different reproductive stages, may produce sex-differentiated foraging strategies. Interestingly, it has also been suggested that under certain conditions both chimpanzee (see Goodall, 1977, p. 227) and gorilla mothers (see Hartcourt, 1977) seek out the protection of individual males when their infants are very young. Such behaviors may function to increase protection from infanticide.

8. Opportunism and Adaptation: Some Theoretical Considerations

The examples discussed above and the data presented earlier on langur allomothering raise important questions about how the evolution of behavior, including parenting behavior, should be approached. The primary approach taken by ethologists and, now, sociobiologists initially is to consider and describe the possible adaptive merits of a given observed behavior; the presumption of its adaptiveness is made at least for more conspicuous behaviors (mating patterns, maternal care and defense, and so forth), and an analysis examining the ratio of costs to benefits to the actor and his or her relatives proceeds. This approach is, and can be, useful but only insofar as a number of other factors are considered. For example, in considering particular behavior strategies, it is common to project the genetic consequences of them as if all other behavioral strategies are held constant (e.g. Chapman and Hausfater, 1979; Kurland, 1977). But by doing so evolution is depicted, at best mechanistically, as a process which works to sculpt ideal or optimal behaviors in seemingly static environments. Yet, and as all would agree, natural selection works upon a spectrum of behaviors simultaneously, which means that the ideal projected consequences of viewing only one of them at a time may be irrelevant. Parental behaviors, defenses, affiliative behaviors, feeding, and foraging may certainly be viewed as separate subsystems for which the mathematically optimal strategies can be mapped, but when viewed as systems evolving together, and which together and not alone determine *survival,* including longevity (see Lancaster, 1979) and reproduction, it is known that the efficacy of any one behavioral subsystem is compromised (Crook *et al.,* 1976). It is my opinion that earlier papers, especially by sociobiologists, failed to appreciate the significance of these points. The real challenge is to determine for any species which behavioral subsystems are more important for survival, for reproduction, and for both.

In terms of reproductive success, nobody will argue that the net effects of behavior must be beneficial if the behavior in question is to be perpetuated or retained in the species geneotype. But this raises another point. Should it be presumed as often as it seems to be that behaviors emerge because they are adaptive, rather than because the opportunities were right (see Simpson, 1953; Williams, 1966)? A contemporary behavior may be relatively recent and may not necessarily be an adaptive strategy for the future. It may be selected against or be substituted for by a new, alternative behavior, or it could lead to extinction of the species.

The langur data can be used illustrate these points and to show that evolution is not always as clean, or as neat, as some evolutionary models emphasizing ideal or optimal strategies might suggest. Consider that allomothering can be exhibited ambitiously and at impressive rates among langurs. While debate may continue for a while as to what the functions may be and who actually benefits, it seems clear to me that at least in some and possibly many instances it is dangerous for the infant, at least if one considers immediate survival. Earlier I discussed how the allomothering experience may help prepare the Colobine infant for an adult social life which is independent of its geneological relations. But as Bowlby (1969) reminded us, it is critical to consider that no matter how effective early infant care may be with respect to guiding new individuals into adult social roles and activities, these processes are meaningless if the neonate dies during its period of relative helplessness. Though this must certainly fall within the realm of speculation, it could be that excessive allomothering will result in higher mortality rates for neonates whose mothers permit it. Therefore, and as Quiatt (1979) argues in part, natural selection may be working against extensive allomothering among langurs, or more specifically, against mothers who are not able to perform competently in controlling the quality of handling their infants receive from allomothers. It is possible that among the Common Indian langur experimentation is taking place as to the best possible pattern, but some of these patterns might be selected against.

For any given species, at any given point in time and space, the behavior patterns which can be ascribed to a species and even considered "species typical" *may* not be the long-term successful ones. It may be that langur allomothering behavior will prove to be beneficial as a long-term adaptation, but the counterpossibility exists and should receive as much attention as the approach which looks for and assumes its adaptiveness. By considering more than the functional attributes of a behavior (see earlier discussion), it is more likely that aspects of evolutionary opportunism will be revealed, as well as a proper ancestral context from which the emergence of a behavior from preexisting organizations can be placed. Simpson (1949) argued that a species morphology, its present behavioral capacities as well as its phylogenetic past ("phylogenetic inertia," after Wilson, 1975), sets the limits and provides the opportunities for change. He stressed that what can happen in evolution usually does happen, and that it does not necessarily have to be beneficial either for the individuals performing the behavior or for the species within which the behavior emerges. Utilizing this theoretical framework, and, indeed, perspective, an alternative exists for explaining the origins of a behavior and without emphasizing its possible, but as yet unproven, adaptiveness.

Much needs to be learned about the evolution of primate parenting behavior and, indeed, about how to use evolution to account for it. In fact, even though the common langur has been studied relatively extensively in both field and laboratory settings (Curtin, 1977), it is difficult to clearly delineate what constitutes their "species specific" behaviors. It has only been in the last few years that primate ethologists have begun to appreciate the amount of intragroup, not to mention interspecies, variability which characterizes primate societies. Like many other subsystems which constitute valid and important areas of investigation, parenting systems have been affected by local and complex ecological factors. With this in mind, it is my opinion that more and more the theme of evolutionary opportunism will be referred to in explanatory models of primate and other mammalian parental behavior.

References

Alcock, J., 1975, *Animal Behavior: An Evolutionary Approach,* Sinauer Associates, Sunderland, Mass.

Altmann, J., Altmann, S., and Hausfater, G. G., 1978, Primate infant's effects on mother's future reproduction, *Science* **201**:1028–1029.

Bauchop, T., and Martucci, R. W., 1968, Ruminant-like digestion of the Langur monkey, *Science* **161**:698–700.

Beach, F., 1978, Human sexuality and evolution, in: *Human Evolution* (S. L. Washburn and E. McCoun, eds.), pp. 123–153, Benjamin-Cummings, Menlo Park, Calif.

Bishop, N., 1979, Himalayan langurs: The temperate Colobines, *J. Hum. Evol.* **8**:251–281.

Boggess, J. E., 1976, Social behavior of the Himalayan langur *(Presbytis entellus)* in Eastern Nepal, Ph.D. dissertation, University of California.

Bowlby, J., 1969, *Attachment, Attachment and Loss,* Vol. I, Basic Books, New York.

Cartmill, M., 1974, Rethinking primate origins, *Science* **184**: 436–443.

Chapman, M., and Hausfater, G., 1979, The reproductive consequences of infanticide in langurs: A mathematical model, *Behav. Ecol. Sociobiol.* **5**:227–240.

Charles-Dominique, P., 1977, *Ecology and Behaviour of Nocturnal Primates,* Columbia University Press, New York.

Clark, A., 1978, Sex ratio and local resource competition in a prosimian primate, *Science* **201**:163–165.

Clark, L., 1959, *The Antecedents of Man,* University of Edinburgh Press, Edinburgh.

Clutton-Brock, T. (ed.), 1977, *Primate Ecology,* Academic Press, New York.

Cooper, J., Bloom, E., and Roth, R., 1977, *The Biochemical Basis of Neuropharmacology,* Oxford University Press, New York.

Crook, J. H., Ellis, J. E., and Goss-Custard, J. D., 1976, Mammalian social systems: Structure and function, *Anim. Behav.* **24**:261–273.

Curtin, R. A., 1975, The socioecology of the Common Langur *(Presbytis entellus)* in the Nepal Himalayas, Ph.D. dissertation, University of California.

Curtin, R. A., 1977, Langur social behavior and infant mortality, *Berkeley Pap. Phys. Anthropol.* No. 50, pp. 22–27.

Curtin, R., and Dolhinow, P., 1978, Primate behavior in a changing world, *Am. Sci.* **66**:4:468–475.

Daly, M., and Wilson, M., 1978, *Sex, Evolution and Behavior,* Duxbury Press, North Scituate, Mass.

Dolhinow, P., 1977, Caretaking patterns of the Indian Langur monkey, Paper presented at the annual meeting of the American Association of Physical Anthropologists, Seattle.

Dolhinow, P., 1978a, Mother-loss among Indian langur monkeys, Paper given at the annual meeting of the American Anthropological Association, Los Angeles.

Dolhinow, P., 1978*b*, Langur monkey mother loss and adoption, Commentary, *Behav. Brain Sci.* **3**:443–444.

Dolhinow, P., McKenna, J., and Laws, J. V., 1979, Rank and reproduction among female langur monkeys: Aging and improvement, *J. Aggress. Behav.* **5**:19–30.

Eaton, G., 1976, The social order of Japanese macaques, *Sci. Am.* **235**:96–106.

Gartlan, S., 1969, Sexual and maternal behavior of the Vervet monkey *(Cercopithecus aethrops)*, *J. Reprod. Fertil.* (Suppl.) **6**:137–150.

Gibson, K. R., 1977, Brain structure and intelligence in macaques and human infants from a Piagetian perspective, in: *Primate Bio-Social Development: Biological, Social, and Ecological Determinants,* (S. Chevalier-Skolnikoff and F. E. Poirier, eds.), pp. 113–157, Garland, New York.

Glander, K. E., 1975, Habitat and resource utilization: An ecological view of social organization in Mantled Howling monkeys, Ph.D. dissertation, University of Chicago.

Goodall, J., 1977, Infant killing and cannibalism in free-living chimpanzees, *Folia Primatol.* **28**:259–282.

Hall, K. R. L., and Mayer, B., 1967, Social interactions in a group of captive patas monkeys *(Erythrocebus patas)*, *Folia Primatol.* **5**:312–326.

Halperin, S. P., 1979, Temporary association patterns in free-ranging chimpanzees: An assessment of individual grouping preferences, in: *The Great Apes* (D. Hamburg and E. McCown, eds.), pp. 491–500, Benjamin-Cummings, Menlo Park, Calif.

Hamburg, D., and McCown, E. (eds.), 1979, *Perspectives on Human Evolution, The Great Apes,* Vol. 5, Benjamin-Cummings, Menlo Park, Calif.

Hamilton, W. D., 1964, The genetical evolution of social behavior. Parts I and II, *J. Theor. Biol.* **7**:1–16, 17–52.

Hanby, J., 1972, The sociosexual nature of mounting and related behaviors in a confined troop of Japanese macaques, Ph.D. dissertation, University of Oregon.

Hartcourt, A. H., 1977, Social relationships of wild mountain gorillas *(Gorilla gorilla beringei)*, Ph.D. thesis, Cambridge University.

Hinde, R., 1975, The concept of function, in: *Function and Evolution in Behaviour* (F. Baerends, C. Beer, and A. Manning, eds.), pp. 3–15, Oxford University Press, London.

Hinde, R. A., and Spencer-Booth, Y., 1967, The effect of social companions on mother-infant relations in rhesus monkeys, in: *Primate Ethology* (D. Morris, ed.), pp. 267–286, Aldine, Chicago.

Hladik, C. M., 1975, Ecology, diet, and social patterning in old and new world primates, in: *Socioecology and Psychology of Primates* (R. Tuttle, ed), pp. 3–35, Aldine, Chicago.

Hladik, C. M., 1977, A comparative study of the feeding strategies of two sympatric species of leaf monkeys: *Presbytis senex* and *Presbytis entellus,* in: *Primate Ecology* (T. H. Clutton-Brock, ed.), pp. 324–353, Academic Press, London.

Horwich, R. H., and Manski, D., 1975, Maternal care and infant transfer of two species of colobus monkeys, *Primates* **16**:49–73.

Hrdy, S. B., 1976, Care and exploitation of nonhuman primate infants by conspecifics other than the mother, in: *Adv. Study Behav.* **6**:101–158.

Hrdy, S. B., 1977*a, The Langurs of Abu,* Harvard University Press, Cambridge.

Hrdy, S. B., 1977*b*, Allomaternal choice of charges among Hanuman langurs, Paper presented at the annual meeting of the American Anthropological Association, Seattle,

Hrdy, S. B., 1977*c,* Infanticide as a primate reproductive strategy, *Am. Sci.* **65**(1):40–49.

Hrdy, S. B., 1979, Infanticide among animals: A review, classification, and examination of the implications for the reproductive strategies of females, *Ethnol. Sociobiol.* **1**:13–40.

Jay, P. 1962, Aspects of maternal behavior in langurs, *Ann. N.Y. Acad. Sci.* **102**:468–476.

Jay, P., 1963, The ecology and social behavior of the North Indian langur, Ph.D. dissertation, University of Chicago.

Jay, P., 1965, The common langur of North India, in: *Primate Behavior: Field Studies of Monkeys and Apes* (I. DeVore, ed.), pp. 197–249, Holt, Rinehart and Winston, New York.

Jerison, H. J., 1973, *Evolution of the Human Brain and Intelligence,* Academic Press, New York.

Jolly, A., 1966, *Lemur Behavior,* University of Chicago Press, Chicago.

Kaufman, I. C., 1977, Developmental considerations of anxiety and depression; psychobiological studies in monkeys, in: *Psychoanalysis and Contemporary Science* (T. Shapiro, ed.), pp. 317–363, International University Press, New York.

Kaufman, I. C., 1978, Evolution, interaction and object relationship. Commentary, *Behav. Brain Sci.* **3**:450–451.

Kay, R. W. B., Hoppe, P., and G. M. O. Maloiy, 1976, Fermentative digestion of food in the colobus monkey, *Colobus polykomous, Experientia* **32**:485–486.

Klopfer, P. H., and Boskoff, K. J., 1979, Maternal behavior in prosimians, in: *The Study of Prosimian Behavior* (G. A. Doyle and R. D. Martin, eds.), pp. 123–155, Academic Press, New York.

Kriege, P. D., and Lucas, J. W., 1974, Aunting behavior in an urban troop of *Cercopithecus aethiops, J. Behav. Sci.* **2**:55–61.

Kurland, J. A., 1977, Kin Selection in the Japanese Monkey, *Contrib. Primatol.* No. 12.

Lancaster, J., 1971, Play-mothering: The relations between juvenile females and young infants among free-ranging vervet monkeys, *Folia Primatol.* **15**:161–182.

Lancaster, J., 1979, Sex and gender in evolutionary perspective, in: *Human Sexuality: A Comparative and Developmental Perspective* (H. Kachadourian, ed.), pp. 43–90, Holt, Rinehart and Winston, New York.

Lindberg, D., The rhesus monkey of North India: An ecological and behavioral study, in: *Primate Behavior* (L. Rosenblum, ed.), pp. 103–140, Academic Press, New York.

McClean, P., 1977, The triune brain, in: *Human Evolution* (S. Washburn and E. McCown, eds.), pp. 3–58, Benjamin-Cummings, Menlo Park, Calif.

McKenna, J. J., 1975, An analysis of the social roles and behavior of seventeen captive Hanuman langurs *(Presbytis entellus)*, Ph.D. dissertation, University of Oregon.

McKenna, J. J., 1979*a,* The evolution of allomothering behavior among Colobine monkeys: Function and opportunism in evolution, *Am. Anthropol.* **81**:(4):4.

McKenna, J. J., 1979*b,* Aspects of infant socialization, attachment, and maternal caregiving patterns among primates: A cross-disciplinary review, in: *Yearb. Phys. Anthropol.* **22**:250–286.

Mohnot, I., 1971, Some aspects of social changes and infant-killings in the Hanuman langur, *Presbytis entellus* (Primates: Cercopithecidea) in Western India, *Mammalia* **35**:175–198.

Moir, R. J., 1968, Ruminant digestion and evolution, in: *Handbook of Physiology*, Section 6: "Alimentary Canal" (C. F. Code, ed.), pp. 2673–2694, American Physiological Society, Washington, D. C.

Moreno-Black, G., and Maples, W. R., 1977, Differential habitat utilization of four Cercopithecidae in a Kenyan forest, *Folia Primatol.* **27**:85–107.

Napier, J. R., and Napier, P. H., 1967, *A Handbook of Living Primates*, Academic Press, New York.

Oates, J., 1977a, The social life of a black-and-white Colobus monkey, *Z. Tierpsychol.* **45**:1–60.

Oates, J., 1977b, The guerza and its food, in: *Primate Ecology* (T. H. Clutton-Brock, ed.), pp. 276–319, Academic Press, London.

Pianka, E. R., 1978, On r- and K- selection, in: *Readings in Sociobiology* (T. H. Clutton-Brock and P. Harvey, eds.), pp. 45–51, W. H. Freeman, San Francisco.

Quiatt, D., 1979, Aunts and mothers: Adaptive implications of allomaternal behavior of nonhuman primates, *Am. Anthropol.* **81**(2):310–319.

Rahaman, H., 1973, The Langurs of the Gir Sanctuary, *J. Bombay Nat. Hist. Soc.* **70**:295–314.

Ransom, T., and Rowell, T. E., 1972, Early social development of feral baboons, in: *Primate Socialization* (F. Poirier, ed.), pp. 105–144, Random House, New York.

Redican, W. K., 1976, Adult male interactions in nonhuman primates, in: *The Role of the Father in Child Development* (M. Lamb, ed.), pp. 172–193, Wiley, New York.

Richerson, P. J., 1977, Ecology and human ecology: A comparison of theories in the biological and social sciences, *Am. Ethnol.* **4**:1–26.

Ripley, S., 1965, The ecology and social behavior of the Ceylon gray langur *(Presbytis entellus thersites)*, Ph.D. dissertation, University of California.

Ripley, S., 1967, Intertroop encounters among Ceylon gray langurs *(Presbytis entullus)*, in: *Social Communication among Primates* (S. A. Altmann and J. A. Altmann, eds.), pp. 237–253, University of Chicago Press, Chicago.

Rosenblum, L., 1971, Kinship interaction patterns in pigtail and bonnet macaques, in: *Proceedings of the Third International Congress of Primatology*, Vol. 3 (J. Biegert, ed.), pp. 79–84, Karger, Basel.

Rowell, T., Hinde, R. A., and Spencer-Booth, Y., 1964, "Aunt"-infant interaction in captive rhesus monkeys, *J. Anim. Behav.* **12**:219–226.

Roy, M., 1976, Early rearing of infrahuman primates with reduced conspecific contacts: A selected bibliography. Part I, *J. Biol. Psychol.* **18**(2):36–42.

Roy, M., 1977, Early rearing of infrahuman primates with reduced conspecific contacts: A selected bibliography. Part II, *J. Biol. Psychol.* **19**(1):21–30.

Schultz, A., 1978, Illustrations of the relation between primate ontogeny and phylogeny, in: *Human Evolution* (S. L. Washburn and E. McCown, eds.), pp. 255–283, Benjamin-Cummings, Menlo Park, Calif.

Simonds, P. E., 1974, *The Social Primates*, Harper and Row, New York.

Simpson, G. G., 1949, *The Meaning of Evolution*, Yale University Press, New Haven, Conn.

Simpson, G. G., 1953, *The Major Features of Evolution*, Simon and Schuster, New York.

Slob, A. K., Wiegand, S. J., Goy, R. W., and Robinson, J. A., 1978, Heterosexual interactions in laboratory-housed stumptail macaques *(Macaca arctoides):* Observations
during the menstrual cycles and after ovariectomy, *Horm. Behav.* **10**:193–211.

Struhsaker, T., 1967, Behavior of Vervet monkeys *(Cercopithecus aethrops), Publ. Zool.*
82:1–74.

Struhsaker, T., 1975, *The Red Colobus Monkey,* University of Chicago Press, Chicago.

Sugiyama, Y., 1964, Group composition, population density and some ecological observations of Hanuman langurs *(Presbytis entellus), Primates* **5**:7–37.

Sugiyama, Y., 1965, On the social change of Hanuman langurs *(Presbytis entellus), Primates* **6**:381–418.

Sugiyama, Y., 1967, Social organization of Hanuman langurs, in: *Social Communication
among Primates* (S. A. Altmann, ed.), pp. 221–236, University of Chicago Press,
Chicago.

Sugiyama, Y., 1976, Characteristics of the ecology of the Himalayan langurs, *J. Hum.
Evol.* **5**:249–277.

Symons, D., 1978, The question of function: Dominance and play, in: *Social Play in
Primates* (E. O. Smith, ed.), pp. 193–230, Academic Press, New York.

Symons, D., 1979, *The Evolution of Human Sexuality,* Oxford University Press, New
York.

Trivers, R., 1972, Parental investment and sexual selection, in: *Sexual Selection and the
Descent of Man 1871–1971* (B. Campbell, ed.), pp. 136–179, Aldine, Chicago.

Wade, T., 1976, Status and hierarchy in nonhuman primate societies, in: *Perspectives in
Ethology,* Vol. III (P. P. G. Bateson and P. H. Klopfer, eds.), pp. 109–134, Plenum
Press, New York.

Whitten, A., and Rumsey, T., 1973, 'Agonistic buffering' in the wild Barbary macaque,
Macaca sylvana, Primates **14**:421–426.

Williams, G. C., 1966, *Adaptation and Natural Selection,* Princeton University Press,
Princeton, N.J.

Wilson, E. O., 1975, *Sociobiology: The New Synthesis,* Harvard University Press, Cambridge, Mass.

Wrangham, R., 1979, Sex differences in chimpanzee dispersion, in: *The Great Apes* (D.
Hamburg and E. McCown, eds.), pp. 481–489, Benjamin-Cummings, Menlo Park,
Calif.

Yoshiba, K., 1968, Local and intertroop variability in ecology and social behavior of common Indian langurs, in: *Primates: Studies in Adaptation and Variability* (P. C. Jay,
ed.), pp. 217–242, Holt, Rinehart and Winston, New York.

11

The Social Context of Parental Behavior

A Perspective on Primate Socialization

Karyl B. Swartz and Leonard A. Rosenblum

1. Introduction

From birth, the nonhuman primate infant is an active participant in its social world. Although certain patterns of social behavior show ontogenetic changes indicative of the increasing social complexity of the infant, even the newborn is equipped with physical and behavioral capacities that initiate and influence social interactions. Generally, the first social partner the infant encounters is the mother, yet other members of the social group also contribute to the socialization of the infant. In the present chapter we will attempt to characterize and examine the social influences acting on the developing primate infant, starting with the infant itself and working outward to the mother and other immediate social companions. At each juncture we will attempt to specify some of the variables influencing the quantity and quality of social interactions provided to the infant and their influence on the infant's behavior. The focus will primarily be on parental behavior, broadly defined as infant-directed behavior, and its influence on the socialization of the nonhuman primate infant.

Socialization is usually defined as the process whereby the infant's broad range of social behavior is shaped to conform to behaviors that are in line with the norms of that infant's social group. More broadly, and in line with the comparative emphasis of the present chapter, the process of socialization can be perceived

KARYL B. SWARTZ • Department of Psychological Sciences, Purdue University, West Lafayette, Indiana 47907.   LEONARD A. ROSENBLUM • Primate Behavior Laboratory, Department of Psychiatry, Downstate Medical Center, Brooklyn, New York, 11203.

as the influence of a diversity of social, biological, and maturational factors over time that ultimately result in an appropriately functioning adult who is capable of interacting in a species-typical and predictable fashion with other members of its species. Such a definition is very broad and includes a number of separate behavioral systems, e.g., maternal behavior, sexual behavior, feeding, and dominance interactions (or social roles). A notion central to the following sections of the chapter is that each age/gender category of social companions provides some unique contribution to the infant's socialization by virtue of the unique interactive stimulus and response configurations presented to the infant by these social partners. An attempt will be made to specify, at least by inference, the role that each of these categories of social partners has in producing the ultimate product of their joint influences—an appropriately functioning adult organism.

Throughout the chapter we will be emphasizing the similarity of processes underlying social development across various primate species. In keeping with the availability of well-articulated data we will draw most heavily on material regarding several macaque species, but in addition, we will discuss species differences that serve to illustrate basic principles of primate attachment and social development. The main thrust of the chapter is an analysis of infant-directed behaviors that may contribute to the process of infant socialization. The assumption underlying such an approach is that there is a continuity of mechanisms underlying primate social development that is best exemplified through the interweaving of similarities and differences across species in such a way that the principles of social development emerge. Secondly, although we recognize the fact, illuminated by current field perspectives, that any partiular environment will impose its own constraints on the behaviors studied, it is our contention that the diverse variables potentially influencing the course and structure of mother–infant relations and infant socialization are best controlled within the confines of various laboratory settings. Therefore, in our attempts to disambiguate the role of these several factors as they influence development, laboratory data will provide the primary focus of the discussions that follow.

2. The Ontogeny of Specific Attachments*: Establishment and Maintenance of Maternal Behavior in the Context of the Mother–Infant Bond

The mother–infant attachment relationship is the first and probably the longest-lasting relationship that an infant primate forms; it is also one of the infant's most intense relationships. As such, the nature of the relationship certainly

---

*Attachment, as it is used in the present paper, refers to an underlying predisposition in both the mother and infant that guides their behavior, chiefly but not solely influencing their proximity to each other. This process can be conceptually divided into the mother's contribution and the infant's contribution, as they both serve in the establishment of the bond as well as its continued maintenance, once established. See Cairns (1972) for a further discussion of these issues.

affects social development and, consequently, all later relationships. The developing infant's ability to deal adequately with social encounters and the degree to which it engages in social interactions are greatly determined by the patterns of early social interactions between mother and infant. Individual differences in the mother and infant continuously affect and shape the nature of the relationship within a species; and across species, more broad-based, evolutionary variables such as species-specific behavior patterns may lead to rather striking differences in the form assumed by the relationship or in particulars of the infant's social development. In the following section, we will focus on maternal and infant factors that contribute to the *establishment* of the mother–infant bond; in later sections we will address individual and evolutionary differences that affect the *maintenance* of the relationship, as well as consider other social influences that affect the developing infant (see Gubernick, this volume, for further discussion of attachment).

2.1. Birth

The birth process plays a major part in the establishment of the mother–infant bond by setting up a situation in which the probability is high that the mother will interact with the newborn infant. During labor and delivery, the mother separates herself from the group and proceeds to deliver her own infant. Detailed descriptions of birth in several captive primate species (e.g., Bo. 1971; Bowden *et al.,* 1967; Brandt and Mitchell, 1971; Hartman, 1928; Rosenblum, 1971a; Tinklepaugh and Hartman, 1923) indicate that the following features are common to all births. The birth process can be roughly divided into three phases: (1) delivery of the infant with subsequent cleaning and attention to the infant; (2) delivery and, commonly but not always, ingestion of the placenta; and (3) further attention to the infant followed by sleep. The mother is usually very active during the delivery. When the infant's head begins to emerge she may grasp it and tug on it, thus directly aiding the delivery of the infant. As soon as the infant is born, the mother licks it and cleans the birth fluids from it. The mother's licking of the infant places the infant in a favorable position to make physical contact with the mother. When the placenta is delivered, the mother usually ignores the infant while she licks and eats the placenta. Although placental ingestion may be interrupted or only partially completed in some cases, no systematic data yet exist regarding the control of these behaviors in either primipara or multipara.

Following attention to the placenta, the mother may again attend to the infant. If the infant has not already established itself on the mother (ventrally for most species, dorsally for several New World forms, including the squirrel monkey), it will do so at this point. Chimpanzee and gorilla infants are unable to cling well at birth so they must rely on the mother's support during the first few days of life (Schaller, 1963; van Lawick-Goodall, 1968; Yerkes and Tomlin, 1935).

The role of hormones in mediating or stimulating maternal behavior at parturition is an interesting though as yet poorly studied question. Hartman (1928) suggested that ingestion of the placenta provides hormones which are important to lactation, and the hormonal importance of the placenta for caretaking behavior has been suggested by Cairns (1972); however, on numerous occasions we have observed quite adequate maternal behavior in pigtails, bonnets, and squirrel monkeys when the placenta has not been eaten. It should be noted here that failure to eat the placenta may occur when other factors are acting to debilitate the new mother (e.g., a long, difficult labor), and it would be an error to ascribe maternal failure in such instances to the lack of ingestion of the placenta (see Rosenblatt and Siegel, this volume, for further discussion of the hormonal bases of maternal behavior).

Two experimental studies have provided indirect evidence for hormonal enhancement of maternal interest in young infants. Cross and Harlow (1963) found that prior to parturition multiparous gravid rhesus females showed no preference for looking at a neonate over a 1-year-old; however, for the week following the birth of their infants the females spent more time viewing the neonate than they did the juvenile stimulus animal. These mothers were permanently separated from their infants on the infant's 3rd day of life, and the mothers' preference for viewing neonates declined soon after, suggesting that continuing maternal interest in infants is a function not only of changing hormonal state at parturition, but also of experience with an infant. Such an interaction between hormonal state and experience with infants has been well demonstrated in rats (Rosenblatt and Lehrman, 1963; Rosenblatt and Siegel, this volume). Further evidence for the combined contribution of hormonal and experiential factors in determining responsiveness to infants in nonhuman primates was provided by Sackett and Ruppenthal (1974). They found that adult female pigtails separated from their infants preferred a neonate over a juvenile and over an adult female on the day of separation, if the infants were very young at separation (under 2 months). This preference for neonates in the experimental social choice situation persisted for a week following separation only in those females who had lived with their infants for 2 weeks after birth. These data suggest that hormonal factors may cease to operate in adult females soon after birth, and that experience with an infant may contribute more to the female's interest in infants as the infant matures. Supportive of this idea is the finding, reported by Cross and Harlow (1963) and found in a second study by Sackett and Ruppenthal (1974), that, in contrast to nulliparous females, nonpregnant multiparous rhesus females consistently preferred infants in both choice situations over juvenile or adult animals. Thus, prior maternal experience with infants, occurring concurrently with hormonal influences, may contribute to later increased responsiveness to young infants, even in the absence of immediate hormonal influences.

Clearly, hormonal changes are not always necessary for the demonstration of interest in infants. In many species, subadult or nonpregnant adult females contribute significantly to the care of the infant (Jay, 1963; Lancaster, 1971; Poirier, 1968; Rowell *et al.*, 1964; Struthsaker, 1971; see also McKenna, this volume), and in other species, males are active caretakers (Burton, 1972; DeVore, 1963; T. W. Ransom and T. E. Ransom, 1971; see also Kleiman and Malcolm, this volume). Similarly, in squirrel monkeys, females of varying reproductive status ranging from adolescence to adulthood are responsive to young infants; moreover, adolescent males also seem quite attracted to infants in this species (Rosenblum, 1972). Hence, animals who lack the biochemical composition of the newly parturient female's blood can display adequate caretaking behavior, indicating that, although the hormonal factors may enhance interest in the newborn infant, they are not necessary to the initiation of those behaviors which may normally lead to the formation of the attachment bond. Thus, DeVore (1963) has cited the case of an adoption of a young female baboon by a dominant male at the death of the mother, and van Lawick-Goodall (1968) cited several cases in which orphaned chimpanzees were adopted by an older male or female sibling.

2.2. The Infant's Behavior at Birth

Monkey and ape infants are typically very active at birth. Observations of neonate macaques and baboons have shown that the eyes open during or just after birth; vocalizations often occur during birth; the rooting (head-turning and mouth-opening) response is active immediately although sucking may not begin until the second day; clinging is strong at birth in most species, although some mothers may have to provide support for the first few days (e.g., Bertrand, 1969; Hinde *et al.*, 1964; Ransom and Rowell, 1972). Newborn infants seem to be able to coordinate eye and head movements in relation to visual stimuli by the 2nd day of life (Mowbray and Cadell, 1962), and some infants have been observed to examine the mother's face during the first few days of life (Hines, 1942). By the 2nd week of life, the infant is actively exploring its environment visually (Hinde *et al.*, 1964; Rosenblum, 1968). Clinging, and the "searching" hand movements which accompany it, as well as rooting and sucking, provide tactile stimulation to the mother from the infant. In addition, the infant often jerks spasmodically, sometimes vocalizing simultaneously. Such behavior serves to immediately elicit the mother's attention (Bertrand, 1969; Hinde *et al.*, 1964). Young infants also have a tendency to climb upward, a behavior which serves to increase proximity to the mother's nipple, an important element in allowing the rooting reflex to function.

Infants of many species are born with a natal coat which is distinctly different from the coat of an adult [e.g., langurs (Jay, 1963; Poirier, 1968); vervets (Struthsaker, 1971); stumptail macaques (Bertrand, 1969); baboons (DeVore, 1963)]. In addition to coat color, skin color and the pink perineum of the infant distinguish it from other animals in the group. These characteristics do not disappear until the infant is at a developmental stage in which it can locomote well and can identify its mother. Infants of some species also have a distinctive cry which is not produced by any other member of the social group. This vocalization seems to occur spontaneously and is usually ignored by the mother and others in the group, unless the cry takes on a character which suggests the infant is in distress (Rowell and Hinde, 1962).

In addition to these reflexive behaviors, the infant macaque may contribute to the establishment of attachment by responding selectively to potential caretakers of its own species. Sackett (1970) has presented some highly provocative data which suggest that the very young infant macaque is predisposed to approach adult members of its own species and to ignore adults of other species. Rhesus macaque infants reared in partial isolation (visual and auditory contact, but no tactual access to peers) were observed in Sackett's "self-selection circus," a choice apparatus that allows the subjects to choose among live stimulus animals placed in compartments surrounding the subject's center compartment. Rhesus infants less than nine months of age oriented significantly more toward rhesus adults than toward either pigtail macaque or stumptail macaque adults. In addition, when given a choice between an adult male and an adult female rhesus, partial isolate-reared infants up to 44 months old preferred the adult female. These results suggest that the macaque infant can discriminate and responds differentially to certain cues provided by potential caretakers, even when the infant has had no social experience.

One major question, however, is: What are the effective cues that elicit this approach tendency in the neonate? To speculate in regard to the above question, any number of cues could be responsible for the infant's selective orientation and approach. The infant could have been approaching some visual or auditory (or other) characteristic found only in its own species, it could have been approaching a set of behavioral characteristics particular to the conspecific adult, or it could have been approaching the conspecific adult by default through its avoidance of the other stimulus animals. Crucial to this question is the fact that Sackett not only allowed the infants visual access to the stimulus animals, but also allowed the stimulus animals to see the infant subjects. Thus, the infant's approach may have been elicited by any of a complex of static visual and overt behavioral cues manifested by the conspecific adult, which may or may not have been related to the adult's reaction to the visual and behavioral cues presented by the infant. Although Sackett's procedure, as it stands, does not allow precise specification of the cues that his infant subjects were responding to, these data are, nonetheless, indicative of early selectivity in infant macaques and demonstrate the active role that the young infant may take in responding to social partners (see also Galef, Harper, this volume, on role of the young).

2.3. Behavior of the Mother toward the Neonate

With few exceptions [e.g., the squirrel monkey (Rosenblum, 1968)] nonhuman primate mothers are very active caretakers (Bertrand, 1969; DeVore, 1963; Hinde *et al.*, 1964; Jay, 1963; van Lawick-Goodall, 1967, 1968). The mother cradles her infant when they are sitting, and she is careful to provide support when she moves away. Eventually, the tendency to support the infant with one arm upon movement develops into a light touch which signals to the infant that the mother is leaving.

During the first few days of the infant's life, the mother spends much of her time grooming, licking, patting, and making eye contact with the infant. When she looks at the infant the mother macaque often lip-smacks, a common facial gesture which is friendly or placatory in nature. Observations of this early period suggest that the mother is very interested in the infant and that her behavior is directed toward maintaining maximal contact with the infant.

As the infant becomes active during the 1st or 2nd week of life, the mother attempts to restrain it when it tries to leave her. If the infant is successful in leaving, the mother follows the infant very closely, often with her hands hovering over the infant in a guarding fashion. She is very quick to retrieve the infant at the slightest sign of distress from the baby, or any sign of danger in the group (Bertrand, 1969; Hinde *et al.*, 1964; Jay, 1963; Struthsaker, 1971). When the infant is just beginning to walk during the 1st week of life, the mother may separate herself from the infant and, facing the infant, lip-smack and make chattering noises, which result in the infant's approaching her (Hinde *et al.*, 1964; Hines, 1942). This sequence has often been observed in macaques and has sometimes been interpreted as a communicative "training session" for the infant; however, Sackett's (1970) selection circus choice data would suggest that such approach training is not necessary in macaques.

2.4. The Infant as a Stimulus for Caretaking

That the infant is an effective stimulus for eliciting caretaking behavior should be evident from the fact that in most primate species, the newborn is very attractive to members of the social group (see Harper, this volume, for infant's effects on caretakers). "Aunts," nulliparous or nonpregnant females in the mother's social group, approach a newly parturient female and her infant, in attempts to touch, groom, or hold the infant (Jay, 1963; Poirier, 1968; Rowell *et al.*, 1964; Struthsaker, 1971; van Lawick-Goodall, 1968). In only a few species does the mother allow other females to hold or carry the newborn (Jay, 1963; Struthsaker, 1971; see McKenna, this volume); however, in most species the mother allows handling of a slightly older infant (Bertrand, 1969; Rosenblum, 1971a; Rowell *et al.*, 1964; van Lawick-Goodall, 1968). The adult males of several species, including savanna baboons (DeVore, 1963) and barbary apes (Burton, 1972; Lahiri and Southwick, 1966), also take an active interest in infants for long periods of time.

Laboratory observations of the effectiveness of the young infant in eliciting caretaking behavior were reported by Hansen (1966), who found that rhesus mothers in a playpen setting responded with positive behavior (cradling and grooming) to very young infants who entered the female's home cage. By the time the infants were 4 months old, however, the mothers responded punitively to strange infants.

In an experimental test of the effectiveness of young squirrel monkey infants in eliciting caretaking, Rosenblum (1972) presented social groups of squirrel monkeys with a young infant placed on the floor of their home pen. He found that although all members of the group directed exploratory behavior toward the infant, the highest levels of contact and active retrieval were shown by pregnant females near term and by immature females. Mature nonpregnant and early pregnant females showed some contact with, but little retrieval of, the infant, as did immature males. Adult males (both castrated and intact) and ovariectomized females showed little contact and no retrieval. These data suggest that even though the infant may be of interest to all members of the squirrel monkey social group, such biological factors as age or hormonal state can affect the stimulus value of the infant in initiating or eliciting caretaking.

2.5. Establishment of the Mother–Infant Bond

Clearly, behavioral and physical characteristics of both the mother and the infant contribute to the establishment of the mother–infant bond. The complementarity of these maternal and neonatal mechanisms is reflected in the fact that when one of these components is missing (e.g., the presence of a responsive infant), the presence of the other will often still result in the formation of an attachment bond. Reports of mothers carrying, grooming, and handling dead infants for several days demonstrate the effectiveness of the infant as a stimulus without its behavioral characteristics, as long as the mother's behavioral and hormonal systems are intact (Jay, 1963; Rahaman and Parthasarathy, 1969; Rumbaugh, 1965; Struthsaker, 1971; van Lawick-Goodall, 1968).

Experimental manipulations of the infant's normal perceptual or behavioral repertoire often result in compensatory reactions by the mother and establishment or maintenance of a bond. Berkson (1970) partially blinded two infant cynomolgus macaques and observed their behavior when placed back into their feral troop with their mothers. The infants were able to forage for themselves and could stay close to the troop, but when the troop was threatened and fled, the infants were unable to follow. The mothers of the blind infants compensated for the deficiencies in their infants by carrying them away when the group fled or by returning to retrieve the infant when it was left behind. Both of these compensating behaviors were very atypical of adult behavior in the studied band of monkeys. Berkson (1974) repeated this procedure in a feral group of rhesus macaques living in a more sheltered environment, and again found compensation in the mother's behavior toward the infants throughout the first year, and in two cases, beyond the birth of a new sibling. Group members other than the mother were also observed to compensate for the infants' deficiencies.

Using a species that shows less active maternal involvement in caretaking, Rumbaugh (1965) bound the arms of an infant squirrel monkey so that it could not cling to the mother's fur, and found that the mother squirrel monkey cradled the infant, and even carried it in both arms, walking bipedally as it did so. That the mother was compensating for a specific deficiency in the infant was indicated by the fact that when the infant's arms were released, the mother made no attempt to pick up or carry the infant.

When the mother is deficient in her caretaking behavior, the infant can compensate and will become responsible for establishing and maintaining proximity (Harlow, 1963; H. F. Harlow and M. K. Harlow, 1965). Infants of indifferent or punitive "motherless mothers" (isolate-reared females) were very persistent in their attempts to cling and nurse, even after a history of consistent rebuffs by the mother. Rosenblum and Harlow (1963) found that rhesus infants reared on a punitive cloth surrogate mother, one that intermittently delivered aversive blasts of compressed air during contact, spent more time on their surrogates than did a control group of infants reared on standard, nonpunitive cloth surrogates.

However, in the case of deficiencies in both mother and infant, establishment of an attachment bond may not occur. Meier (1965) found that isolate-reared rhesus monkeys would not accept caesarean-delivered infants, even when the infant was their second offspring, and they had reacted appropriately to their vaginally delivered firstborn infants. Feral-reared monkeys, on the other hand, would accept and react appropriately to either vaginally delivered or caesarean-delivered infants. Infants delivered by caesarean section are typically less active and much less vocal than normally delivered infants (Meier, 1964), suggesting that since neither the isolate-reared mother nor the caesarean-delivered infant could overcome the deficiencies in the other, the attachment bond could not be established.

In an experimental study of the compensatory contributions of group-living mother and infant bonnet macaques at various developmental stages, Rosenblum and Youngstein (1974) found that the maintenance of the bond was a function of mother and infant, and that the amount of contribution and compensation from both mother and infant was a function of the age of the infant. In their study, they anesthetized one partner of the mother–infant pair and observed the behavior of the other partner when both were returned to their group living cage. Their results showed that when the infant was young (4–6 months), both mother and infant made contact with the incapacitated partner, and both to the same degree. When the infants were older (16–19 months), they contacted their anesthetized mothers, but the mothers made little contact with their unconscious infants.

The reciprocal nature of the mother–infant bond was demonstrated by the finding that when both members were conscious following a stressful manipulation (i.e., the handling control manipulation which involved separation of mother and infant from their group, injection of a control saline solution into one of the partners, and subsequent reintroduction to the group), much more contact occurred between them than occurred following the anesthesization of one member (the experimental condition). Age differences of the same nature as above also appeared in the handling control condition: the mothers of younger infants showed more contact with their infants following handling than did the mothers of older infants.

2.6. Recognition of the Mother and Specificity of Partner Response

In order for a specific attachment, as we define it, to emerge, early in the mother–infant relationship mutual recognition must develop between the two members of the dyad (see also Ainsworth, 1972, for the significance of the specificity of response in attachment in human infants). For example, after brief exploratory forays, the infant's active role in reestablishing contact with its own mother in a group situation containing many similar-appearing adult females implies that the infant comes to recognize its own mother, that is, to respond differentially and preferentially to her over other adult females. Early experiments dealing with the question of individual recognition suggested that mother macaques recognize their own infants as early as 13 days of age, but that infants do not recognize their own mother even at 7 months. Jensen (1965) found that pigtail mothers could immediately distinguish their own infants from older infants, and that by the time they had been housed alone with their infants for 13 days postpartum, they could discriminate their own infants from same-aged peers. In contrast, Jensen and Tolman (1962) found that two pigtail macaque infants, one 5 months old and the other 7 months old, responded positively to unfamiliar adult females as well as to their own mothers, and only learned to avoid strangers after being punished by the strangers for approaching them. The mothers, however, responded positively only to their own infants.

It is important to note here that Jensen and Tolman's mother–infant pairs were reared in isolation from other animals. As Rosenblum and Alpert's (1974) results suggest, the social living conditions of mother and infant may be important in determining the development of recognition of mother. Rosenblum and Alpert used a two-choice procedure similar to Sackett's multiple-choice-selection circus, with the important addition that the subject animal could not be seen by the stimulus animals. Infants were placed in the center of a 2.4-m-long chamber that provided the infant with visual access to stimulus animals at both ends of the chamber. The infants were given choices between the mother and an unfamiliar conspecific female (stranger), the mother and an empty chamber, and the stranger and an empty chamber. Thus, through the infant's position in the center runway, assessment could be made of preference for and avoidance of particular social stimuli.

Group-reared bonnet infants did not show a preference for the mother until 3 months of age; however, beyond 12 weeks there was a clear preference for the mother over a stranger. Even following the emergence of this preference, however, the infant would often choose a stranger over an empty chamber, an indication perhaps of some general affiliative tendency (similar, perhaps, to the approach to conspecific adults reported for rhesus by Sackett, 1970). Superimposed on these results was a sex difference: at all ages, including an early period from 7 to 11 weeks, female infants showed a stronger preference for the mother than did male infants. In addition, although the infants, in general, preferred the stranger to the empty chamber, the time spent away from the stranger on these trials, compared with the time spent away from the mother in mother versus empty chamber trials, indicated some avoidance of the stranger. Females showed a great deal of avoidance of the stranger from 12 to 60 weeks of age, whereas males showed very little avoidance throughout the first year. When pigtail macaque infants were tested, a species difference emerged: group-reared pigtail macaque infants showed some preference for the mother by 12 weeks of age; however, the preference was not strong, nor did it increase over age as it did in the bonnet macaque.

Hansen (1976) has presented evidence for auditory recognition of the mother in rhesus monkeys. Group-reared rhesus juveniles separated from their mothers at 16–17 months of age responded differentially to the vocalizations of their own mothers over those of other females when given only auditory access to the female adults. These animals showed significantly higher rates of "coo" vocalizations and of locomotion in response to calls emitted by their own mothers in comparison to their rates of responding in the presence of calls from other adult females.

Olfactory cues may also play a role in the development of recognition of the mother in some nonhuman primate species. In a series of studies designed to investigate the relative contribution of visual and olfactory cues to mother recognition in the infant squirrel monkey, Kaplan (1977; Kaplan et al., 1977; Kaplan and Russell, 1974) has obtained evidence suggesting that naturally occurring olfactory cues may be fundamental to recognition in this species. There is also some suggestion from his data that olfactory recognition may develop more easily or earlier than visual recognition in squirrel monkey infants.

There is strong evidence, then, that the infant monkey does indeed come to recognize the mother and can differentiate her from other individuals. The next questions are: what cues does the infant use in its recognition of mother, how do these cues combine to provide recognition, what is the developmental progression of this recognition, and what kinds of social experience are necessary for its development? One way to begin to answer these questions is to present the infant with a limited set of cues (e.g., only visual cues), and to determine whether the infant can use the available cues to discriminate the mother.

In an experimental attempt to separate some of the perceptual cues used in recognition of the attachment figure, Mason et al. (1971, 1974) reared rhesus macaque infants on inanimate surrogate mothers in a distinctive environment. Once monthly, from 2 weeks through 3 months of age, the infants were presented with a series of "recognition" tests of surrogates and rearing environments. Mason et al. found that from the beginning of testing, the infants differentiated the familiar environment from an unfamiliar environment, a visually based discrimination, but that they did not differentiate familiar from unfamiliar surrogates when allowed only visual access to them. The two surrogates were differentiated at 2 months of age when both visual and tactual access were provided.

The data of Mason *et al.* would suggest that young macaques cannot discriminate the mother solely on the basis of visual cues; however, Sackett (1966) has shown that rhesus macaque infants can, by 2 or 3 months of age, make use of purely visual information in responding to potential social partners. Isolate-reared rhesus infants were presented with pictures of adult and infant rhesus monkeys engaged in several different classes of behavior. Sackett found that these infants showed play behaviors directed toward pictures of infants, and fear responses in reaction to pictures of threatening adults; all other pictures were ignored. The above differential behaviors showed a developmental progression consisting of few responses early in the infant's life followed by a peak of differential responding at 2–3 months of age and a later decline.

In order to determine whether infant macaques could use the visual information available in two-dimensional stimulus arrays to recognize their mothers, Swartz (1977) presented socially reared stumptail macaque infants with pictures of the mother, a familiar stumptail adult female, and an unfamiliar stumptail adult female. The results of the experiment indicated that by 4 months of age, the infants could discriminate pictures of individuals in a two-picture simultaneous-presentation-choice test, but that they did not respond differentially to successively presented single pictures of the individuals. In a choice test with live animal stimuli that allowed interaction between the infant subjects and the stimulus animals, the infants chose their own mothers over familiar and unfamiliar conspecific females. When the mothers of these infants were shown pictures of stumptail infants, they showed evidence of recognition of their own infants over familiar and unfamiliar same-aged conspecific infants in the single-picture-presentation procedure.

One aspect of the Swartz (1977) data of interest here is the finding that infant stumptails as old as 6 months of age did not discriminate pictures of individual adult females including the mother, when these pictures were presented successively. There are two possible explanations for this finding. First, it is possible that these infants did not learn to recognize their own mothers. Although this possibility cannot be completely ruled out, the results of the live animal choice tests and proximity data obtained in home cage observations suggest that the infants did act discriminatively toward the adult females, spending more time in proximity to their own mothers than to other adult females. These data imply that the infants recognized their own mothers. A second possibility is that these young infants were not cognitively capable of extracting information from a single two-dimensional array and then holding such information in memory long enough to compare that picture with the next picture or with the infant's own cognitive representation or memory of its mother's appearance. That this sort of developmental deficiency may have been operating in these infants is suggested by the finding that the mothers of these infants did show evidence of recognition of their own infants with successively presented pictures.

It should be noted that in recent work, using color-videotaped moving stimuli, presented approximately life-sized, eight 12-month-old pigtail infants showed significant differences in vocalization and approach patterns to mother, familiar female, and unfamiliar female stimuli (Rosenblum and Paully, 1980). Thus, further maturation and/or the inclusion of size and movement cues may permit consistent differentiation on the basis of visual information alone.

It may be the case that the relative reliance on particular sensory modalities in recognition of the mother by infant monkeys depends upon the type and number of cues available to the infant, as well as the species of the infant. In the case of the macaque infant, for example, visual cues may be central to partner recognition only if tactual cues are not available, (cf. Sackett's results above) whereas when both are available (cf. Mason et al.,1971, 1974) the presence of tactual cues may overshadow or retard the use of visual information. These differences, in turn, may appear only in the case of noninteractive rearing conditions in which the infant is reared in the presence of an artificial surrogate or no social partners. When an interacting live mother is present, single-modality cues may be sufficient for recognition to occur (cf. Rosenblum and Alpert, 1974). To take this argument one step further, there may also be cases in which the presence of a particular type of cue results in an alteration or a breakdown of recognition. Such a suggestion is supported by experimental data reported by Kaplan (Kaplan et al., 1979; Redican and Kaplan, 1978) in squirrel monkey infants. Although it appears that squirrel monkey infants rely on olfactory cues in recognition of live and surrogate mothers (Kaplan et al., 1977; Kaplan and Russell, 1974), olfactory recognition may break down and be replaced by visual recognition when the mothers are scented with artificial odors.

3. The Role of the Mother's Past and Present Social History

The first and most immediate social influence on the developing infant is that of the mother. Even in species in which the mother shows little active caretaking [e.g., galagos (Doyle et al., 1969); squirrel monkeys, (Baldwin, 1969; Rosenblum, 1968)], characteristics of the mother's current and past social history can influence the infant simply by virtue of the type of social situation presented by the mother's choice of companions. Such individual characteristics of the mother as her status in the group dominance hierarchy, her parity, and her own past rearing history will affect not only the behavior she directs toward her infant, but also the quality and quantity of social interaction provided the infant by the mother's immediate social companions. Similarly, species differences in maternal behavior patterns may affect the infant's access to social partners and, consequently, the course of the infant's socialization.

3.1. Maternal Rank

In macaque groups, the adult dominance status achieved by an animal is often a function of the relative dominance status of its mother (Cheney, 1977; Koford, 1963; Sade, 1967). Field studies of macaques have shown that female infants stay with their natal group throughout their life-span, maintaining a dominance rank just below that of their mothers. The resulting social structure over a long period of time becomes a set of matriarchies which differ in rank, such that the members of a dominant matriarchal unit will all be individually dominant to the individual members of a subordinate matriarchy (Kawai, 1958; Kawamura, 1958; Koyama, 1967). Male infants retain a rank just below that of their mothers until they reach maturity, at which time they either leave the natal group or become peripheral and variable in rank among the adult males of the group (Sade, 1967). Although there is more variability in the relationship between an adult male's rank in his natal group and that of his mother, Koford (1963) has observed cases in which the male offspring of the most dominant female of rhesus troop achieved, at a very young age, very high status in the group.

The relative dominance status of the mother does, then, affect the environment in which the infant is reared, From birth, infants of high-ranking females are treated differently from infants of low-ranking females. Gouzoules (1975) found that stumptail infants of high-ranking females received more grooming and general social contact during the first 6 months of life than did infants of low-ranking mothers. Once the infants were old enough to leave the mother consistently (over 2 months old), infants of high-ranking females received fewer physical and visual threats from other group members and showed less submissive behavior to others than did infants of low-ranking mothers. These data as well as other studies of the acquisition of social rank (Cheney, 1977; Marsden, 1968; Sade, 1967) suggest that the nonhuman primate infant acquires its rank at least in part through the differential treatment it receives from birth by other members of the social group. Not only do offspring of high-ranking females receive more aid from their mothers during aggressive encounters, but they also receive more aid from other members of the group than do offspring of low-ranking females (Cheney, 1977), and they receive more submissive behaviors from low-ranking group members (Gouzoules, 1975). Infants of high-ranking females also consistently win more fights with age peers during the developmental period than do lower-ranking infants (Sade, 1967.

Maternal rank is an important factor in the social development of the infant not only from the standpoint of immediate social interactions, but also in regard to viability of the infant and later reproductive success of female offspring. In a 10-year survey of the reproductive history of free-ranging rhesus monkeys in Puerto Rico, Drickamer (1974) found that (1) a higher percentage of high-ranking females gave birth each year than did lower-ranking females, (2) infants born to high-ranking females had a higher survival rate than infants of low-ranking females, and (3) daughters of high-ranking females gave birth to their first infants at a significantly earlier age than did daughters of low-ranking females.

3.2. Maternal Parity

The effects of maternal parity on the socialization of the infant has not typically been an active research question. In the wild, maternal parity has a dramatic effect on the viability of the infant. Drickamer (1974) reported that 40–50 % of first- and second-born rhesus infants did not survive the first year, whereas only 9% of fourth-born infants died during the first year of life. Laboratory studies of parity, however, have failed to demonstrate profound differences in maternal behavior or subsequent infant behavior across primiparous and multiparous rhesus mothers (Mitchell and Brandt, 1970; Mitchell and Stevens, 1969; Seay, 1966). In experimental test situations, primiparous mothers were more responsive to unfamiliar social stimuli, stroked their infants more, and were less likely to reject their infants than were multiparous mothers; however, these differences in maternal behavior did not substantively affect subsequent social behavior when the infants were observed later (Mitchell *et al.,* 1966). Although primiparous mothers have often been described as more "anxious" in their caretaking than multiparous mothers (Mitchell and Stevens, 1969), it is not clear that such laboratory-obtained behavioral differences are contributing to higher infant mortality in primiparious feral-living females.

3.3. Social Rearing History of the Mother

Adult females who themselves were not provided with adequate social experience during infancy do not typically demonstrate active maternal interest in or caretaking of their own infants. As Harlow and his associates have demonstread (Arling and Harlow, 1967; Seay *et al..,* 1964), rhesus females who were socially isolated during the first 6 months or year of life were at best indifferent mothers, and showed little protection or caretaking of their infants, to the extent that their infants often had to be removed and fed by the experimenters to avoid infant starvation. Other isolate-reared females physically abused their infants, necessitating experimenter intervention in some cases and occasionally resulting in permanent maiming or death of an infant. There was some interaction of social rearing history with maternal parity. Females who were inadequate mothers with their first-born infants often showed adequate maternal care of second-born infants. That this effect might be a function of the social influence of the first infant, rather than of hormonal changes associated with parity or age, is suggested by the finding that social-isolate females who had experience with infants prior to the birth of their own infants were adequate mothers (Seay *et al.,* 1964). The social rearing history of the mother clearly affects her maternal behavior. The question, then, becomes: to what extent do these deficiencies in maternal behavior affect the social behavior of the developing infant? In comparisons of social behavior in infants reared by the socially inadequate "motherless mothers" versus those reared by socially normal females, Seay *et al.*, (1964) found no differences in infant–infant play or social interaction. Thus, although these infants were provided with less than adequate maternal care, daily interactions with similarly reared age peers seemingly compensated for deficiencies in maternal socialization and allowed appropriate social behaviors to develop during the first 6 months of life.

3.4. The Absence of the Mother: Mother–Infant Separation

The socialization of the nonhuman primate infant is often dramatically affected by the absence of the mother, particularly in cases of short-term maternal deprivation brought about by mother–infant separation. Although studies of the removal or absence of the mother provide no direct information about parental behavior, the infant's behavioral response to the *absence* of maternal behavior may suggest some ideas about the role of maternal behavior in socialization. For example, the strength of the mother–infant bond is evident from the disruption of the infant's behavior that occurs when the two are separated from one another. This disruption suggests that the infant attributes a unique role to the mother and that it retains some representation of her role in her prolonged absence (see Hofer, this volume, for another possible interpretation).

The effects of mother–infant separation on the infant and, subsequently, on the mother–infant relationship, have primarily been studied experimentally using macaques as subjects. Typically, the separated macaque infant shows the first two stages of the three-stage separation reaction described in human infants by Bowlby (1973). That is, the separated macaque infant generally shows an initial agitation, or protest, stage followed by a stage of general inactivity and depression of play and social behavior (usually termed the depression or despair stage). Although some human infants also show an aloofness or withdrawal from the mother upon reunion, which Bowlby called the detachment stage, this stage has not formally been observed in nonhuman primates. Following the occurrence of the protest and depression stages, the macaque infant may, however, demonstrate a third set of behaviors, if the separation still continues, that could be termed a stage of recovery. During this period, activity levels and social play increase but do not return to preseparation levels. Finally, upon reunion with the mother, the infant shows more and closer contact with the mother, often reverting to patterns of contact typical of much younger infants.

Although the above pattern of responses is the most typical reaction to separation in the nonhuman primate infant, situational and subject variables do serve to modify the pattern. The variability in the separation response has been used as an argument against the value of the nonhuman primate model for separation to human infants (Lewis *et al.*, 1976); however, others (Suomi and Harlow, 1977) have suggested that the observed variability may be a function of several variables that have not been systematically studied or controlled. For example, such factors as the early rearing history of the infant and the nature of the mother–infant relationship prior to separation can affect the infant's response to the mother's absence. In particular, the species of the infant and concomitant species variations in social organization and social experience of the infant can dramatically alter the pattern of response. Such situational factors as the separation environment and accessibility to the mother or other social partners during the separation can also play a part in emphasizing or alleviating the infant's depressive response. In the following discussion, we will present an overview of the separation response in nonhuman primate infants, with an emphasis on the consistent patterns that emerge and the factors that may affect the pattern within and across species.

3.4.1. Experimental Studies of Mother–Infant Separation

Hinde and his associates (Hinde *et al.,* 1966; Spencer-Booth and Hinde, 1971*a,b*) have systematically studied the effects of separation on rhesus mother–infant pairs. Separations were effected by removing the mother and infant from their social group (consisting of a male, two to four females, and their young), separating the mother from the infant, and returning the infant to the group for the duration of the separation period. At the end of the separation, the mother was returned to the group. Observations were made of the mother and infant before and after separation, and of the infant during separation. Their results with 21-week-, 25- to 26-week-, and 30- to 32-week-old infants separated for 6 days were as follows. On the first day of separation, there was a significant increase in distress vocalizations and a decrease in activity reflected by an increase in time spent sitting still. As the separation period progressed the infants became even less active; the amount of locomotion, distance covered in each locomotive sequence, amount of rough and tumble play, noncontact social play, and nonsocial manipulative play all decreased relative to preseparation levels and remained at these low levels throughout the separation period (there was no evidence of a recovery stage in these infants). When the mothers were returned to the group, the infants spent more time on the mother, stayed closer to her when off her, and were more responsible for maintaining contact than they had been before separation. During the week following reunion the infants showed some recovery in activity and play, with most of the measures returning to preseparation level; two weeks after reunion there were no significant differences between pre-and postseparation measures of activity level, types of activity, proximity, or contact.

Infants separated for a longer period of time (13 days rather than 6 days) showed essentially the same qualitative effects (including the lack of a recovery stage during the separation period), with the exception that recovery took longer following the longer separation (Spencer-Booth and Hinde, 1971*b*). Immediately following reunion, the long-separation infants vocalized more than did short-separation infants, and failed to show the rather rapid recovery shown by short-separation infants. For at least 2 weeks following reunion, the long-separation infants were more responsible for maintaining proximity to the mother and were less active than they had been during the preseparation period.

Repeated separations did not appear either to aggravate or to alleviate separation effects in rhesus infants: two separations of 6 days each did not appear to affect the infants more or less than a single 6-day separation. Infants separated once at 21 or 25–26 weeks and again at 30–32 weeks did not react differently to their second separation than they had to the first, nor did they differ from infants separated for the first time at the later age (Spencer-Booth and Hinde, 1971a).

Seay et al. (1962) found somewhat the same separation reaction in non-group-reared rhesus infants separated for 3 weeks, although their results were less dramatic than the above. Their infants were housed individually with their mothers in a Harlow "playpen" living arrangement that allowed the infants access to a peer playmate and the playmate's mother as well as the infant's own mother. During the separation period the infants were housed together in the central playpen area while the mothers remained in the attached home cage. Physical access to the mother was prevented by a plexiglass partition, but visual, auditory, and olfactory contact were still available. These infants, who were 24–29 weeks old at separation, showed a sharp increase in vocalization (the highest point was the day of separation) coupled with a decrease in social play throughout the separation period. On the day of reunion, infant–mother clinging and mother–infant cradling were significantly higher than preseparation levels. In the 3 weeks following reunion, three of the four infants showed increases in amount of contact with the mother, but these scores were not significantly different from preseparation period scores. In a similar experiment Schlottman and Seay (1972) obtained the same separation reaction in 29- to 36-week-old cynomolgus macaques, and Seay and Harlow (1965) reported that removing the mother from the room during the separation period also resulted in essentially the same qualitative reaction on the part of the rhesus infants.

Rosenblum and his colleagues (Rosenblum and Kaufman, 1968; Kaufman and Rosenblum, 1967a,b, 1969) separated 17- to 19-week-old group-living pig-tailed macaque infants from their mothers for a period of 4 weeks (these separations were effected by removing the mother from the social group). These infants showed, very clearly, three stages of reaction to separation. For the first 24–36 hr, the infants were highly agitated; there were increased activity in the form of pacing, brief bouts of play, approach toward other group members, and distress cries. These infants also showed a dramatic rise in self-directed behavior such as digit-sucking and manipulating other body parts.

Following this initial stage, the infants significantly decreased their activity levels, level of social play, and level of nonsocial exercise play. These infants remained solitary and assumed the hunched-over posture, also described by Hinde, which is indicative of depression or "despair." Some recovery began after about 6 days of this extreme depression; the infants began to explore the environment and to engage in some social play, although they still showed evidence of depression. By the end of the separation period, play levels had almost reached their preseparation level.

Upon reunion, the mother–infant relationship was much closer than just prior to separation. The infant showed increased clinging and nipple contact, and maintained closer proximity to the mother when not on her. The mothers also showed an increase in protective enclosure and a decrease in punitive deterrence (a behavior which usually results in prevention of contact). Contrary to the rather quick recovery of separated rhesus infants, some of the above changes were still in evidence in the pigtail infants 3 months following reunion.

Rosenblum (1978) has demonstrated that the depression exhibited by pigtail macaque infants is a function of the mother's absence, and that it can be reinstated in the infant by presenting the mother to the infant but making her inaccessible. Group-living pigtail infants whose mothers had been removed were presented with the mother in a restraining cage on the floor of the home pen several times throughout the recovery stage of the separation period. When an infant's mother was presented, the infant did not approach her, but rather would exhibit the behavioral depressive responses typical of the second stage of the separation reaction. Activity and locomotion would decrease in the infant, and many infants assumed a huddled depressive posture. These responses would occur only when the infant's own mother was returned, and occurred only for the duration of the mother's presence. When the mother was removed, the infant's behavior would return to the level of recovery the infant had been demonstrating prior to the mother's introduction.

In striking contrast to rhesus, cynomolgus, and pigtail macaques, bonnet macaques separated for 4 weeks *do not* show a period of depression following the initial agitation response to separation (Kaufman and Rosenblum, 1969; Rosenblum and Kaufman, 1968; Rosenblum, 1971*b*). After their initial protest, separated bonnet infants (who were 2–10 months old) approached and contacted adult females remaining in the social group. In most cases, the adult females and even the males were accepting of the infants, thus providing substitute caretakers for the infant in the mother's absence. These "adoptions" of separated infants were in some cases quite complete, with the infants maintaining a degree of ventral contact with the adoptive mother equivalent to that maintained with the biological mother prior to separation. In a few cases, this adoptive relationship retarded or prevented reestablishment of contact with the biological mother upon her return to the group. The infant's readiness to approach group members other than the mother, and group members' active acceptance of the infants are possibly products of the relaxed, highly interactive social structure of the bonnet macaque (Rosenblum and Kaufman, 1968; Rosenblum *et al.,* 1964).

In a similar fashion, Preston *et al.* (1970) found that 26- to 34-week-old patas infants whose mothers were removed showed an initial agitation, including increased cooing and activity, but demonstrated no evidence of the depression stage typical of rhesus and pigtail macaques. Although these patas infants did not live in a social group, the separated infants were accessible to one another in pairs, for an hour a day, throughout the 3-week separation period. During the separation, these infants showed a decrease in play activity and an increase in nonspecific contact with one another. The lack of depression in the infants was attributed to the permissive nature of the mother–infant relationship prior to separation, as well as to the presence of the other infant during separation.

Cross-species comparisons of maternal behavior patterns and correlated qualitative differences in the infant's separation reaction suggest that the nature of the mother–infant interaction, as well as the availability of social partners during separation, can affect the infant's response to loss of the mother. Similarly, Hinde and Spencer-Booth (1970) have suggested that qualitative aspects of the mother–infant relationship within a species may affect an individual infant's reaction to separation. They found that rhesus infants who were rejected most by their mothers and who had taken more responsibility for maintaining contact with the mother prior to separation were more negatively affected during the separation period than were infants who had less active roles in the mother–infant relationship. Unfortunately, these individual differences in the nature of the mother–infant relationship and their effects on subsequent infant behavior have not been systematically studied in macaques.

However, McGinnis (1978, 1979; Hinde and McGinnis, 1977) has provided a first step in an analysis of individual differences in the infant's response to separation. In a series of studies with rhesus monkeys McGinnis systematically evaluated (during separation) the effects of removal of the mother, the infant, or both from their social group. Her results demonstrated that removing either the mother or infant from the familiar social group setting with transfer to a novel environment had a disruptive effect on that animal. The disruption occurred not only during the removal period but also during the reunion period when the subject was required to reintegrate itself into the group social structure. Upon reintroduction to the group, mothers who had been removed attended primarily to interactions with group companions, whereas the infants who had been removed attended primarily to reestablishing contact with the mother. Removing the infant from the social group during maternal separation had a greater disruptive effect on the infant itself, both during and after separation, than did removing the mother from the social group (Hinde and Davies, 1972). However, during reunion the mother's preoccupation with other group members was also disturbing to the infant, regardless of whether the infant was also removed from the group. Such results reiterate the notion that the mother–infant relationship is reciprocal. Such data may also provide insight into the contribution of individual differences in both members of the dyad to the nature of the attachment relationship.

3.5.   Interspecific Variations in Maternal Care: A Congeneric Example

Just as past ontogonetic influences may structure specific patterns of maternal behavior, the phylogeny of the species in question also plays a crucial role (see also Klopfer, this volume). Unfortunately, because of the diversity of behavioral measures, environmental, and social living conditions across experiments and across laboratory and field observations, species comparisons of mother–infant interactions are very difficult to make. Rosenblum (Rosenblum, 1971*b,c*; Rosenblum and Kaufman, 1968; Rosenblum *et al.*, 1964), however, has systematically observed bonnet and pigtail macaques in comparable (laboratory) environments over a period of many years, and has uncovered several striking species differences between these two species of macaques. An examination of the behavioral variation demonstrated by these two species will provide an example of the manner in which the species variable may play an important part in the infant's social development. First, laboratory-housed social groups of these two species show a difference in individual distance (or proximity) patterns: bonnet macaques spend a great deal of time sitting huddled together in "passive contact" (body contact made with a group member without additional social interaction) with other group members. Pigtail macaques, in contrast, are usually found distributed individually across the living pens, and tend to interact with other group members only in dynamic situations such as grooming. These seeming differences in social contact may not be the result so much of absolute differences in levels of contact, but rather in the social partners to whom they are directed. Pigtail macaques form strict kinship groups, within which occur high levels of social contact, grooming, and infant play. Bonnets do not restrict these behaviors to immediate family members, but, rather, distribute them across all members of the social group. Thus, both species tend to engage in comparable levels of social contact, but pigtails are more selective in their interactions. (It should be noted here that these differences in kinship patterns in the laboratory cannot as yet be generalized to the behavior of these species in the wild, inasmuch as appropriate field data are not yet available.)

In regard to the mother–infant interaction, bonnet mothers are permissive mothers, both in terms of the freedom they allow their infants and in terms of the freedom they allow others in interacting with their infants. Bonnet mothers often make contact with group members immediately after parturition, and they will allow others to touch and groom the neonate, although they do not allow carrying of the newborn. As the bonnet infant grows, it is permitted to interact freely with other infants and adults. The mother shows low levels of restraint and guarding of the infant, and other adults engage in active interaction with the infant (e.g., grooming infants, allowing infants to explore and climb on them, and mutual clinging).

In comparison, newly parturient pigtail mothers are highly protective of their infants, showing high levels of aggression toward other animals who approach. As the infant develops, the mother restricts its interactions with other group members. Pigtail mothers show much higher levels of departure restraint and retrieval during the first 4 months of the infant's life than do bonnet mothers.

As discussed previously, these species differences in maternal behavior patterns are also reflected in differences in the infant's response to maternal separation. Such profound species differences in two phylogenetically closely related species pose a problem for investigators interested in the evolution of dyadic patterns and the development of an animal model of a particular process such as the mother–infant attachment relationship. Empirical measurements of attachment within and across species have provided little consistency. Consider, for example, three measures of attachment typically used in studies of attachment in macaques: proximity measures, separation reaction, and recognition of (or preference for) the mother. Separation measures of attachment suggest that to the extent that the infant protests separation from its mother, it has developed an attachment to her. Indeed, most macaque infants show dramatic behavioral manifestations of distress when the mother is taken away; however, the failure of bonnet macaque infants to show extended behavioral disruption following separation from the mother presents a paradox in the measurement of attachment. On the one hand, the bonnet infant's relative lack of distress upon separation from the mother, and its willingness to approach and contact other adult group members during the mother's absence would suggest that the bonnet macaque infant forms a very weak attachment to its mother. On the other hand, other measures of attachment such as preference for the mother in a choice situation (Rosenblum and Alpert, 1974, 1977) suggest that the bonnet infant may be more strongly attached to its mother than is the pigtail macaque infant. Home-cage proximity measures of attachment provide little clarification of this problem. Although bonnet macaque infants spend less time in ventral contact with the mother and more time away from her during the first year of life than do pigtail macaque infants (Rosenblum, 1971a), this difference, as suggested, is probably more a function of the mother's behavior than of the infant's tendency to maintain contact with its attachment figure. The problem of combining various measures of attachment in order to provide an accurate account of the development of the attachment bond is a major problem in the area of attachment (see Gubernick, this volume, for further discussion). Although currently existing data do little to clarify the above inconsistencies, comparative studies that attempt to specify organismic and situational variables affecting individual empirical measures can provide insight into theoretical principles underlying the development and behavioral manifestation of a process like attachment.

4. The Role of Other Socializing Agents

Although the mother's influence on the nonhuman primate infant is generally the most immediate and pervasive, the infant's interactions with social-group
members other than the mother also contribute to its social development. In the
following section the contribution of various categories of social-group companions
will be discussed in the context of infant socialization.

4.1. "Aunting"

Interest in, or caretaking of, the infant by females other than the infant's
mother has often been referred to as "aunting" behavior. The term has been used
to encompass a wide range of behaviors, from merely approaching the infant to
active carrying and protection of the infant, and it has been applied to a wide age
range of females in the group, from juveniles through adult nonpregnant females.
In general, this active interest in the infant occurs across the primate order and
has been specifically noted in captive rhesus and stumptail macaque groups
(Rhine and Hendy-Neely, 1978; Rowell *et al.*, 1964; Spencer-Booth, 1968), in
wild vervet (Lancaster, 1971) and langur (Jay, 1963) groups, and in captive squirrel monkeys (Du Mond, 1968; Hunt *et al.*, 1978; Rosenblum, 1972). Even before
puberty female primates typically show more interest in infants than do males
(Chamove *et al.*, 1967; Rosenblum, 1972), and this interest appears to continue
as the females mature.

The type and extent of aunting behaviors shown appear to vary across species. Squirrel monkey aunts show a great deal of sustained interest and rather
active caretaking of the infant, although they do not protect the infant as much as
the mother does (Du Mond, 1968; Hunt *et al.*, 1978). Langurs actively share
infants with one another, and will often pass an infant around from one female
to another (Jay, 1963). Vervets, as well as langurs, will often permit a young aunt
to care for the infant while the mother forages or explores at a distance from her
infant and its temporary caretaker (Lancaster, 1971). In contrast to the above,
aunting behavior in other species such as macaques may be limited to occasional
carrying, grooming, and touching of the infant by the aunt, and such behaviors
are often restricted by the mother.

The age of the infant also affects the degree of contact with the infant that
the mother permits an interested female. Aunting has not been observed frequently
in laboratory studies prior to about 3 weeks of age (see, however, McKenna, this
volume). In both macaques and squirrel monkeys the mother is quite restrictive
during the first few weeks of the infant's life, but by the 2nd and 3rd months of
life the infant is permitted more freedom to explore the inanimate and the social
environment, and it is during this time that the greatest amount of contact with
females in the group occurs (Hunt *et al.*, 1978; Rowell *et al.*, 1964). By the second
half of the infant's 1st year, the aunt's interest has usually dropped off. Thus, it

appears that the type or amount of infant-directed behavior shown by females in the group is a joint function of the age-related attractiveness of the infant (younger infants are attended to much more than older infants) and the access to the infant permitted by the mother. Other variables such as the mother's relationship with the aunt prior to the infant's birth may also serve to affect the amount of access to the infant provided by the mother.

The interesting aspect of aunting is not that it occurs, but the function it serves in primate social structure. Several functions have been suggested, some from the standpoint of the adaptive nature of such behavior for the individuals showing the behavior and others in regard to the advantages of such behavior to the developing infant. Clearly, both aspects are important. One clear advantage provided by the presence of aunts is in the case of the mother's death while the infant is still dependent on an adult caretaker. Such an advantage has been demonstrated in captive rhesus monkeys (Rowell et al., 1964) and in free-ranging chimpanzees (van Lawick-Goodall, 1968). Lancaster (1971) has also suggested that the aunting behavior shown by juvenile female vervets may serve to provide young females with practice in caretaking so that they will provide adequate, efficient caretaking of their own infants later in life. Finally, the presence of aunts provides the infant with a set of familiar social partners who function as a cluster of close social relationships within the broader social structure of the group. Such relationships may serve the same function as kinship groups, particularly in captive groups that may have suffered the destruction of kinship lineages. Even in species in which kinship plays a major part in defining or restricting social relationships, the close social cluster that surrounds the infant may be based on similar mechanisms such as attraction to infants and the importance of ontogenetically based social familiarity in determining social interactions. Kinship, in fact, has been suggested as one factor in aunting (Lancaster, 1971; Spencer-Booth, 1968; for further discussion see McKenna, this volume).

Aunting, then, not only demonstrates the attractiveness of the infant and its effectiveness in eliciting caretaking, but also serves multiple functions in the ongoing social group and in defining the developing infant's social world.

4.2. Caretaking by Males

The range of interest in infants demonstrated by adult males is extremely wide across the primate order, varying from no contact and no interest (e.g., squirrel monkeys, Rosenblum, 1972) to active caretaking and interest even beyond the weaning period [e.g., marmosets (Stellar, 1960)]. Immature males of species that typically show little adult interest in infants have demonstrated exploratory interest in and care of young infants; however, at adulthood this interest breaks down (Brandt et al., 1970; Hunt et al., 1978; Rosenblum, 1972), and even when positive behavior is directed toward infants by immature males, it differs in quantity and quality from the infant-directed behaviors of similarly aged immature females (Chamove et al., 1967; Rosenblum, 1972).

Under certain circumstances it is possible to obtain caretaking behavior in adult males who typically would show little interest in infants. For example, Redican and Mitchell (1973) reported a case of some caretaking and positive affiliative behaviors directed toward an infant rhesus by an adult male rhesus when the two were housed alone together for 7 months. Similarly, Kaufman and Rosenblum (1969) observed caretaking by the dominant male of an infant bonnet macaque whose mother had been removed from the social group.

The most interesting question in regard to infant socialization, however, is: When (i.e., under what social circumstances) is caretaking naturally demonstrated by adult males? (see also Kleiman and Malcolm, this volume). In some species, primarily those demonstrating monogamous male–female pairs as the basic social structure, the male contributes actively to the care and socialization of the young (see Redican, 1976). In single-male, multiple-female social groups, the male takes a less active role, but still exhibits interest in infants (Redican, 1976; Tilford and Nadler, 1978). The amount of male interest in and male care of infants in multiple-male social groups varies, within and across species. In a comparison of male caretaking across four species of macaques, Brandt *et al.* (1970) found that cynomolgus and stumptail males showed more caretaking and more positive physical contact with infants than did bonnet or rhesus males. Other reports of stumptail males (Bertrand, 1969; Gouzoules, 1975; Hendy-Neely and Rhine, 1977) have variously found tolerance, wariness, and caretaking of infants by stumptail males. Rhesus males living in a nuclear family arrangement played with infants but showed no infant caretaking (Suomi, 1977). In a study of Barbary macaques, Burton (1972) suggested that the males were responsible for a great deal of the infant's socialization through the active caretaking role the males manifested, which included some selective social interactions that served to shape the infant's social behavior with the mother and other group members.

It appears, then, that there is a continuum of male caretaking that occurs across the genus *Macaca,* with rhesus at one end and Barbary macaques at the other end. Clearly, the amount and type of interaction provided to the young infant by adult males will affect the course of the infant's socialization, if not its outcome; however, the data are not clear in regard to when or how adult males of any macaque species demonstrate caretaking. Redican (1976) has suggested that there may be some correspondence between maternal restrictiveness and level of male interest in infants such that infants of a species with permissive maternal styles may be cared for more by males than infants of a species with restrictive mothering. Such an analysis would suggest that in some species the socialization of the infant is a group process, whereas in other species the mother has primary responsibility. This suggestion is provocative, particularly in regard to its implications for other aspects of the relationship of social structure to infant socialization across all primate species. Although data addressed to this question are currently sparse, this suggestion should be pursued conceptually and empirically. Such an analysis could supplement Trivers' (1972) notion that parental investment may be related to the degree to which kinship is assured, a notion that has

been applied to male caretaking behavior across primate species (Tilford and Nadler, 1978; for a discussion of the evolution of male parental investment, see Kleiman and Malcolm, this volume). It should be noted at this point that infant-
directed behaviors exhibited by adult males are not always positive. In particular,
Sugiyama (1967) and Hrdy (1974) have noted the deaths of infants at the time of
a new male takeover of single-male langur troops. Although the mechanism
underlying this phenomenon is as yet unresolved (see Angst and Thommen, 1977,
for an excellent analysis of infant deaths in captive groups), it is clear that at times
the infant may suffer from aggression exhibited by adult males.

4.3. Peer Groups

In the wild, the developing infant has social access not only to the mother
and other adults, but also to a set of same-aged peer playmates who serve to provide age-level-appropriate social interactions. Such mechanisms are suggested by
observations of play in infants who have access to an age peer in their social group
versus those who do not (Kuyk et al., 1976). The role of peer groups in the infant's
socialization has been experimentally explored in laboratory studies with macaque
subjects. Harlow and his associates (Chamove et al., 1973; Harlow, 1969) have
reared infant rhesus with other infants in pairs or in groups of four and have
observed various social behaviors across the infants' lives. In comparison with laboratory-reared infants who had access to the mother and to age peers, infants
reared only with peers showed significantly more ventral–ventral clinging and less
social play in their home cage as well as in novel environments. Infants reared in
groups of four showed less mutual clinging with age than did pairs of infants
reared together. The suggestion, at least for the first 6 months of life, is that without the influence of the mother the young infant will maintain immature clinging
patterns, even in a familiar environment, that prohibit the development of normal
social play patterns. Furthermore, such impoverished social contact may contribute to inadequate social behavior in adulthood. Chamove et al. (1973) found that
infants reared in pairs were sexually inadequate when tested at 6 years of age,
although infants reared in groups of four could not be distinguished from normally
reared animals.

In a further extension of the question of relative contributions of peer groups
on the infant's social development, Rosenblum and Plimpton (1979) addressed the
question of whether the presence or absence of adults would affect peer interaction. Their subjects were 1-year-old socially reared bonnet macaques who were
formed into three groups of four animals each. The infants in each group were
strangers to one another. In one group, the mother of each infant was also included
(mother–peer group). In a second group, four adult females who were strangers
to all the infants were included (peer–other-adult group) and in the third group,
no adults were included (peer-only group). The presence of unfamiliar adults constrained play and mobility in the infants of the peer–other-adult group relative to
the other two groups, with the mother–peer group intermediate, and the largest

amount of play and physical mobility demonstrated by the peer-only group. Thus, it appears that in regard to ongoing active behaviors, the presence of adults can have a restrictive effect, with unfamiliar adults providing the most constraint, on infant peer interactions.

Finally, Coe and Rosenblum (1974) have demonstrated that the presence of adults will structure the social organization of a group of subadult squirrel monkeys into the sexually segregated composition typical of squirrel monkeys. In the absence of adults, subadults showed no tendency to segregate into separate gender groups. Ontogenetic changes in social distance along gender lines in young squirrel monkeys were in fact attributed to social factors related to the presence of adults and their behavior toward the young animals, rather than to the influence of ontogenetic hormonal changes in the young animals themselves.

Thus, although the presence of peer playmates appears to be an important factor in the infant's socialization, the influence of adults serves to structure peer–peer interactions in a fashion that restricts and shapes the infant's social repertoire into a pattern of behaviors appropriate for the social group and species in question.

4.4. Kinship Groups

Kinship, as a variable affecting the socialization of the young infant, may have its effect by limiting interaction to a few genetically related individuals. In some species, such as pigtail macaques (Rosenblum, 1971c), the importance of kinship relations in determining close social interactions is paramount, whereas in other species, such as bonnet macaques, kinship may have little direct effect. Several investigators (e.g., Alexander and Bowers, 1969; Kurland, 1977; Sade, 1972; Yamada, 1963) have demonstrated that in group-living macaques each individual's patterns of grooming, proximity, and contact are limited to a few other individuals who are closely related (e.g., mother, siblings, biological aunts). Such limited patterns of contact early in an infant's life should lead to social restriction in the infant similar to that shown by its mother, and might functionally lead to the type of kinship-determined protection and altruism incorporated in the concept of "inclusive fitness" (Hamilton, 1964; Trivers, 1972). Indeed, Alexander and Bowers (1969) reported that in Japanese macaques defense of an infant against attack by a group member occurred primarily in animals who were related to, or who had formed some sort of attachment relationship with, the infant. Some of the defensive patterns reported by Alexander and Bowers involved adults who had formed strong affiliative relationships with orphaned infants and who were not necessarily genetically related to the infants. Thus it would appear that kinship-based affiliative and defensive behavior patterns are not necessarily based on genetic kinship, but, rather, are a function of limited social interaction among a small group of individuals during the infant's early life. In most cases, such limitation occurs along kinship lines; however, such patterns can appear in nonrelated individuals (see Bekoff, this volume).

Kinship may further serve to shape the social development of the young infant not only by restricting the social partners immediately available, but conversely, also by making some subset of the entire social group available from birth (e.g., members of the matriarchal lineage, such as biological aunts and their offspring). The availability of a selected set of familiar others, who may represent a range of age and gender combinations, can provide the infant with a wide variety of social interactions starting early in life that may significantly contribute to the infant's subsequent social adaptability as an adult. Questions such as those regarding the importance of individual familiarity in determining social interactions in the infant, and the ontogeny of recognition of specific familiar others or categories of familiar others in the developing infant, are of tremendous importance and yet have not been well studied. Sackett's (1970; Sackett and Ruppenthal, 1973, 1974) social choice data provide the best picture of phylogenetic and ontogenetic contributions to social categorization and preference; however, much more work in this area is needed (cf. Bekoff, this volume).

4.5. Isolation of the Mother–Infant Dyad

Even in the midst of a highly interactive social group, the mother is the primary agent of socialization. The question remaining, however, is: What is the effect on the infant's social development of isolation of the mother–infant dyad? From a learning theory perspective (e.g., Cairns, 1966) it would appear that the primary effect on the infant would be an increase in intensity of the mother–infant attachment relationship such that the mother, as the most salient social object in the infant's environment, should be highly discriminable to the infant at an early age. This theoretical suggestion, however, is not empirically supported; rather, the converse appears to be true. In a study of social choices in infant squirrel monkeys, Kaplan and Schusterman (1972) found that group-reared infants preferred their own mother to an unfamiliar adult female, whereas infants reared in isolation with their mothers did not show a consistent preference. Similarly, Rosenblum and Alpert (1977) have demonstrated that bonnet infants who were reared in isolation with only their mothers rarely showed differential response to their own mothers in a choice situation, at least throughout the first 47 weeks of life. Not only did they show equal preference for the mother and the stranger, they also failed to show fear or avoidance of the stranger and they were unable to differentiate the mother from the stranger when both were physically available in the same room. This finding suggests that the infant must have access to other adults in order to learn to identify or discriminate the attachment figure.

Another set of preference data obtained by Sackett et al. (1967), however, seemingly contradicts the above finding. Sackett et al. tested several groups of mother–infant pairs in Sackett's self-selection apparatus to determine preference for own mother or own infant. They found that 8-month-old rhesus infants rotated from one mother to another every 2 weeks showed no preference for their biological mothers, while infants who were separated for 2 hr and then reunited with their mothers every 2 weeks (control group) preferred their own mothers.

With one major exception, the data from the mothers paralleled the above results. Although the infants of motherless mothers showed the biggest preference of all the infants for their own mothers, the motherless mothers showed a preference for infants other than their own.

Thus, the rhesus infants in the Sackett et al. experiment that were reared alone with their mothers showed recognition of and preference for their own mothers over other adult females. Such a finding can be explained in the cases of the control mother–infant pairs by the facts that mothers do come to recognize their own infants (Jensen and Tolman, 1962) and that Sackett's apparatus allows interaction between subject and stimulus animals. Therefore, the control infants could each have been approaching the only female who responded positively in the apparatus. Such an explanation, however, cannot apply to infants of motherless mothers, for the motherless mothers showed a preference for infants other than their own.

Three major differences immediately separate the two studies in question (Rosenblum and Alpert, 1977; Sackett et al., 1967), and any one of them could account for the obtained contradiction that in one case infants reared alone with their mothers showed no recognition of the mother (Rosenblum and Alpert, 1977) during the first 47 weeks of life, whereas in the other case similarly reared infants did recognize their mothers at 8 months of age (Sackett et al., 1967). First, the species used were different (bonnet and rhesus macaques), and Rosenblum and Alpert (1977) have demonstrated the importance of the species variable in such experiments. It should be noted here that since bonnet infants typically encounter and actively interact with other group members more often than do rhesus infants in the typical group-living situation, these differences in experimental choice data may reflect differences between the group social structures of the two species. A second difference between the two studies was the type of apparatus used (two-choice apparatus with one-way vision glass versus multiple-choice apparatus with transparent partitions), as well as the number of choices provided. The importance of such specific task variables is as yet undetermined. Finally, the mother–infant interaction must, of necessity, have differed between the normal bonnet mothers and the punitive rhesus motherless mothers. Rosenblum and Harlow's (1963) results suggest that punishment enhances the infant's response to the mother, leaving open the possibility that other aspects of the mother–infant interaction, including recognition or preference, might similarly be affected. At any rate, the explanation for the differences in results across the two experiments still remains open. It is clear that under some circumstances isolation of the mother–infant dyad from other members of the social group will produce an infant who fails to respond discriminatively to familiar social partners (although it should be noted that this discriminative responding may only be developmentally delayed rather than failing to appear at all; since the bonnet infants were tested only until 47 weeks, it is unclear whether such recognition might have developed during the second year). The variables that lead to such behavior are yet to be specified; however, it does appear that at least in bonnet macaques some interaction with group members other than the mother may be necessary for the development of early discrimination of social partners.

6. Recapitulation: Infant Socialization and an Analysis of the "Adequate Social Rearing Environment"

To the present point, we have attempted to describe the contributions of a variety of social companions to the socialization of an infant who has access to all types of social companions. The unique contribution of members from each social category was only inferred through observations of their social interactions with the developing infant, and in no case did we specify how such interactions formed the infant's future social behavior. Although we have attempted to characterize the diversity of experience provided to the socially reared infant, we are still left with the question: What is the adequate social rearing environment for the nonhuman primate infant? That is, does the infant *require* interaction with each of the various social categories we have described, and if not, what social experiences, or combination of social experiences, are in fact necessary to produce a socially adequate adult?

A major problem in attempting an analysis of the adequate social rearing environment is that one must carefully define not only the independent variables but also the dependent variables of importance. Thus, for example, an early experience manipulation that seriously impairs sexual functioning in an adult primate (Mason, 1963) may have little or no effect on learning in adults of the same species (Griffin and Harlow, 1966). Similarly, a manipulation of the early social environment that produces severe effects in one species may produce mild, although somewhat similar, effects in another species (Sackett *et al.*, 1976). A second point crucial to the problem of adequately defining the dependent measure of interest in studies of social rearing effects on adult behavior is related to the complexity inherent in any primate behavioral system. Not only is one required to define carefully the behavioral system of interest (e.g., sexual behavior), but one must also consider the effects of early experience on several discrete aspects of that general class of behaviors (e.g., foot-clasp mounts; cf. Goldfoot, 1977; Wallen *et al.*, 1977).

Social isolation experiments, conducted primarily with rhesus macaque subjects, have been the primary source of data related to the above question. Thus, for example, it is clear that total removal of social companions during most of the infant's first year of life produces an adult who is deficient in several behavioral systems, including communication patterns, sexual behavior, maternal behavior, social choice behavior, and aggression (Arling and Harlow, 1967; Mason, 1960, 1961a,b; Mason and Green, 1962; Miller *et al.*, 1967; Mitchell, 1970; Sackett, 1965; Sackett and Ruppenthal, 1973). It is unclear, however, which aspects of the entire social configuration are important to the development of adequate adult behavior in each of the above domains. Studies that have attempted to provide some definable subset of social companions have demonstrated, for example, that the presence of peer playmates can alleviate some of the later behavioral disruption (cf. Harlow, 1969; Mitchell, 1970). It is not clear, however, how much peer contact is necessary, and further, whether the presence of peers alone is sufficient to produce a fully socialized adult. The role of the presence or absence of other social companions has also not been well specified in such studies. Clearly, the situation is complex, a notion well supported by the failure to find adequate theoretical explanations for the generalized disruption of social functioning brought about by early social deprivation (cf., e.g., Mason, 1968; Novak and Harlow, 1975; Sackett, 1965).

The situation is not as bleak as it may appear, however. The study of socialization in nonhuman primate infants is currently quite active, and we are slowly approaching the point at which we can construct theories of infant socialization. The present chapter represents a step toward the qualitative specification of social influences that act on the developing infant. Some attempts have been made to point out underlying principles as well as to suggest fruitful areas for future research. Although it may be too early for the development of precise theories of infant socialization in nonhuman primates, future research should be guided by theoretical considerations derived from existing literature. It may be that more multivariate approaches will be necessary in our experimental studies. The design of studies from the standpoint of an integrated conceptual framework should provide empirical data that will allow the construction of useful theories, and ultimately, an understanding of the socialization process in the developing nonhuman primate infant.

References

Ainsworth, M. D. S., 1972, Attachment and dependency: A comparison, in: *Attachment and Dependency* (J. Gewirtz, ed.), pp. 97–137, Wiley, New York.

Alexander, B. K., and Bowers, J. M., 1969, Social organization of a troop of Japanese macaques in a two-acre enclosure, *Folia Primatol.* **10**:230–242.

Angst, W., and Thommen, D., 1977, New data and a discussion of infant killing in Old World monkeys and apes, *Folia Primatol.* **27**:198–229.

Arling, G. L., and Harlow, H. F., 1967, Effects of social deprivation on maternal behavior of rhesus monkeys, *J. Comp. Physiol. Psychol.* **64**:371–378.

Baldwin, J. D., 1969, The ontogeny of social behavior of squirrel monkeys *(Saimiri sciureus)* in a seminatural environment, *Folia Primatol.* **11**:35–79.

Berkson, G., 1970, Defective infants in a feral monkey group, *Folia Primatol.* **12**:284–289.

Berkson, G., 1974, Social responses of animals to infants with defects, in: *The Effect of the Infant on Its Caregiver* (M. Lewis and L. A. Rosenblum, eds.), pp. 233–249, Wiley, New York.

Bertrand, M. 1969, The behavioral repertoire of the stumptail macaque, *Bibl. Primatol.* **11**:1–123.

Bo, W. J., 1971, Parturition, in: *Comparative Reproduction of Nonhuman Primates* (E. S. E. Hafez, ed.), pp. 61–95, Thomas, Springfield, Ill.

Bowden, D., Winter, P., and Ploog, D., 1967, Pregnancy and delivery behavior in the squirrel monkey *(Saimiri sciureus)* and other primates, *Folia Primatol.* **5**:1–42.

Bowlby, J., 1973, *Separation, Attachment and Loss,* Basic Books, New York.

Brandt, E. M., and Mitchell, G., 1971, Parturition in primates: Behavior related to birth, in: *Primate Behavior,* Vol. 2 (L. A. Rosenblum, ed.), pp. 177–223, Academic Press, New York.

Brandt, E. M., Irons, R., and Mitchell, G., 1970, Paternalistic behavior in four species of macaques, *Brain Behav. Evol.* **3**:415–420.

Burton, F. D., 1972, The integration of biology and behavior in the socialization of *Macaca sylvana* of Gibralter, in: *Primate Socialization* (F. E. Poirier, ed.), pp. 29–62, Random House, New York.

Cairns, R. B., 1966, Attachment behavior of mammals, *Psychol. Rev.* **73**:409–426.

Cairns, R. B., 1972, Attachment and dependency: A psychobiological and social learning synthesis, in: *Attachment and Dependency* (J. L. Gewirtz, ed.), pp. 29–80, Wiley, New York.

Chamove, A., Harlow, H. F., and Mitchell, G., 1967, Sex differences in the infant-directed behavior of preadolescent rhesus monkeys, *Child Dev.* **38**:329–335.

Chamove, A. S., Rosenblum, L. A., and Harlow, H. F., 1973, Monkeys *(Macaca mulatta)* raised only with peers: A pilot study, *Anim. Behav.* **21**:316–325.

Cheney, D. L., 1977, The acquisition of rank and the development of reciprocal alliances among free-ranging immature baboons, *Behav. Ecol. Sociobiol.* **2**:303–318.

Coe, C. L., and Rosenblum, L. A., 1974, Sexual segregation and its ontogeny in squirrel monkey social structure, *J. Hum. Evol.* **3**:551–561.

Cross, H. A., and Harlow, H. F., 1963, Observation of infant monkeys by female monkeys, *Percept. Mot. Skills* **16**:11–15.

DeVore, I., 1963, Mother-infant relations in free-ranging baboons, in: *Maternal Behavior in Mammals* (H. L. Rheingold, ed.), pp. 305–335, Wiley, New York.

Doyle, G. A., Andersson, A., and Bearder, S. K., 1969, Maternal behaviour in the lesser bushbaby *(Galago senegalensis moholi)* under semi-natural conditions, *Folia Primatol.* **11**:215–238.

Drickamer, L. A., 1974, A ten-year summary of reproductive data for free-ranging *Macaca mulatta, Folia Primatol.* **21**:61–80.

Du Mond, F. V., 1968, The squirrel monkey in a seminatural environment, in: *The Squirrel Monkey* (L. A. Rosenblum and R. W. Cooper, eds.), pp. 88–145, Academic Press, New York.

Goldfoot, D. A., 1977, Sociosexual behaviors of nonhuman primates during development and maturity: Social and hormonal relationships, in: *Behavioral Primatology* (A. M. Schrier, ed.), pp. 139–184, Erlbaum, Hillsdale, N.J.

Gouzoules, H., 1975, Maternal rank and early social interactions of infant stumptail macaques, *Macaca arctoides, Primates* **16**:405–418.

Griffin, G. A., and Harlow, H. F., 1966, Effects of three months of total social deprivation on social adjustment and learning in the rhesus monkey, *Child Dev.* **37**:533–547.

Hamilton, W. D., 1964, The genetical evolution of social behavior, *J. Theor. Biol.* **7**:1–51.

Hansen, E. W., 1966, The development of maternal and infant behavior in the rhesus monkey, *Behaviour* **27**:107–149.

Hansen, E. W., 1976, Selective responding by recently separated juvenile rhesus monkeys to the calls of their mothers, *Dev. Psychol.* **9**:83–88.

Harlow, H. F., 1963, The maternal affectional system, in: *Determinants of Infant Behaviour,* Vol. II (B. M. Foss, ed.), pp. 3–38, Methuen, London.

Harlow, H. F., 1969, Age-mate or peer affectional system, in: *Advances in the Study of Behavior,* Vol. 2 (D. S. Lehrman, R. A. Hinde, and E. Shaw, eds.), pp. 333–383, Academic Press, New York.

Harlow, H. F., and Harlow, M. K., 1965, Effects of various mother–infant relationships on rhesus monkey behaviors, in: *Determinants of Infant Behaviour,* Vol. IV (B. M. Foss, ed.), pp. 15–36, Methuen, London.

Hartman, C. G., 1928, The period of gestation in the monkey, *M. rhesus,* first description of parturition in monkeys, size and behavior of young, *J. Mammal.* **9**:181–194.

Hendy-Neely, H., and Rhine, R. J., 1977, Social development of stumptail macaques *(Macaca arctoides)*: Momentary touching and other interactions with adult males during the infants' first 60 days of life, *Primates* **18**:589–600.

Hinde, R. A., and Davies, L. M., 1972, Removing infant rhesus from mother for 13 days compared with removing mother from infant, *J. Child Psychol. Psychiatry* **13**:227–237.

Hinde, R. A., and McGinnis, L. M., 1977, Some factors influencing the effects of temporary mother-infant separation—Some experiments with rhesus monkeys, *Psychol. Med.* **7**:197–212.

Hinde, R. A., and Spencer-Booth, Y., 1970, Individual differences in the responses of rhesus monkeys to a period of separation from their mothers, *J. Child Psychol. Psychiatry* **11**:159–176.

Hinde, R. A. Rowell, T. E., and Spencer-Booth, Y., 1964, Behaviour of socially living rhesus monkeys in their first six months, *Proc. Zool. Soc. Lond.* **143**:609–649.

Hinde, R. A., Spencer-Booth, Y., and Bruce, M., 1966, Effects of 6-day maternal deprivation on rhesus monkey infants, *Nature (London)* **210**:1021–1033.

Hines, M., 1942, The development and regression of reflexes, postures, and progression in the young macaque, *Contrib. Embryol. Carneg. Inst.* **196**:155–209.

Hrdy, S. B., 1974, Male–male competition and infanticide among the langurs *(Presbytis entellus)* of Abu, Rajasthan, *Folia Primatol.* **22**:19–58.

Hunt, S. M., Gamache, K. M., and Lockard, J. S., 1978, Babysitting behavior by age/sex classification in squirrel monkeys *(Saimiri sciureus)*, *Primates* **19**:179–186.

Jay, P., 1963, Mother-infant relations in langurs, in: *Maternal Behavior in Mammals* (H. L. Rheingold, ed.), pp. 282–304, Wiley, New York.

Jensen, G. D., 1965, Mother–infant relationship in the monkey *Macaca nemestrina:* Development of specificity of internal response to own infant, *J. Comp. Physiol. Psychol.* **59**:305–308.

Jensen, G. D., and Tolman, C. W., 1962, Mother-infant relationship in the monkey *Macaca nemestrina:* The effect of brief separation and mother-infant specificity, *J. Comp. Physiol. Psychol.* **55**:131–133.

Kaplan, J., 1977, Perceptual properties of attachment in surrogate-reared and mother-reared squirrel monkeys, in: *Primate Bio-Social Development* (S. Chevalier-Skolnikoff and F. E. Poirier, eds.), pp. 225–234, Garland, New York.

Kaplan, J. and Russell, M., 1974, Olfactory recognition in the infant squirrel monkey, *Dev. Psychobiol.* **7**:15–19.

Kaplan, J., and Schusterman, R. J., 1972, Social preferences of mother and infant squirrel monkeys following different rearing experiences, *Dev. Psychobiol.* **5**:53–59.

Kaplan, J., Cubicciotti;, D. D., III, and Redican, W. K., 1977, Olfactory discrimination of squirrel monkey mothers by their infants, *Dev. Psychobiol.* **10**:447–453.

Kaplan, J., Cubicciotti, D. D., III, and Redican, W. K., 1979, Olfactory and visual differentiation of synthetically scented surrogates by infant squirrel monkeys, *Dev. Psychobiol.* **12**:1–10.

Kaufman, I. C., and Rosenblum, L. A., 1967*a*, Depression in infant monkeys separated from their mothers, *Science* **155**:1030–1031.

Kaufman, I. C., and Rosenblum, L. A., 1967*b*, The reaction to separation in infant monkeys: Anaclitic depression and conservation-withdrawal, *Psychosom. Med.* **29**:648–675.

Kaufman, I. C., and Rosenblum, L. A., 1969, Effects of separation from mother on the emotional behavior of infant monkeys, *Ann. N. Y. Acad. Sci.* **159**:681–695.

Kawai, M., 1958, On the rank system in a natural group of Japanese monkeys, I and II, *Primates* **1**:111–148.

Kawamura, S., 1958, The matriarchal social order in the Minoo-B group, *Primates* **1**:149–158.

Koford, C. B., 1963, Rank of mothers and sons in bands of rhesus monkeys, *Science* **141**:356–357.

Koyama, N., 1967, On dominance rank and kinship of a wild Japanese monkey troop in Arashujama, *Primates* **8**:189–216.

Kurland, J. A., 1977, Kin selection in the Japanese monkey, *Contrib. Primatol.* No. 12.

Kuyk, K., Dazey, J., and Erwin, J., 1976, Play patterns of pigtail monkey infants: Effects of age and peer presence, *J. Biol. Psychol.* **18**:20–23.

Lahiri, R. K., and Southwick, C. H., 1966, Parental care in *Macaca sylvana, Folia Primatol.* **4**:257–264.

Lancaster, J. B., 1971, Play-mothering: The relations between juvenile females and young infants among free-ranging vervet monkeys *(Cercopithecus aethiops)*, *Folia Primatol.* **15**:161–182.

Lewis, J. K., McKinney, W. T., Jr., Young, L. D., and Kraemer, G. W., 1976, Mother-infant separation in rhesus as a model of human depression, *Arch. Gen. Psychiatry* **33**:699–705.

Marsden, H. M., 1968, Agonistic behaviour of young rhesus monkeys after changes induced in social rank of their mothers, *Anim. Behav.* **16**:38–44.

Mason, W. A., 1960, The effects of social restriction on the behavior of rhesus monkeys: I. Free social behavior, *J. Comp. Physiol. Psychol.* **53**:583–589.

Mason, W. A., 1961*a*, The effects of social restriction on the behavior of rhesus monkeys: II. Tests of gregariousness, *J. Comp. Physiol. Psychol.* **54**:287–290.

Mason, W. A., 1961*b*, The effects of social restriction on the behavior of rhesus monkeys: III. Dominance tests, *J. Comp. Physiol. Psychol.* **54**:794–799.

Mason, W. A., 1963, The effects of environmental restriction on the social development of rhesus monkeys, in: *Primate Social Behavior* (C. H. Southwick, ed.), pp. 161–173, Van Nostrand, Princeton, N.J.

Mason, W. A., 1968, Early social deprivation in the nonhuman primates: Implications for human behavior, in: *Environmental Influences* (D. C. Glass, ed.), pp. 70–101, Rockefeller University Press, New York.

Mason, W. A., and Green, P. C., 1962, The effects of social restriction on the behavior of rhesus monkeys: IV. Responses to a novel environment and to an alien species, *J. Comp. Physiol. Psychol.* **55**:363–368.

Mason, W. A., Hill, S. D., and Thomsen, C. E., 1971, Perceptual factors in the development of filial attachment, in: *Proceedings of the Third International Congress of Primatology* (J. Biegert, ed.), pp. 85–89, Karger, Basel.

Mason, W. A., Hill, S. D., and Thomsen, C. E., 1974, Perceptual aspects of filial attachment in monkeys, in: *Ethology and Psychiatry* (N. F. White, ed.), pp. 84–93, University of Toronto Press, Toronto.

McGinnis, L. M., 1978, Maternal separation in rhesus monkeys within a social context, *J. Child Psychol. Psychiatry* **19**:313–327.

McGinnis, L. M., 1979, Maternal separation vs. removal from group companions in rhesus monkeys, *J. Child Psychol. Psychiatry* **20**:15–27.

Meier, G. W., 1964, Behavior of infant rhesus monkeys: Differences attributable to mode of birth, *Science* **143**:968–970.

Meier, G. W., 1964, Maternal behaviour of feral and laboratory-reared monkeys following the surgical delivery of their infants, *Nature (London)* **206**:492–493.

Miller, R. E., Caul, W. F., and Mirsky, I. A., 1967, Communication of affects between feral and socially isolated monkeys, *J. Pers. Soc. Psychol.* **7**:231–240.

Mitchell, G., 1970, Abnormal behavior in primates, in: *Primate Behavior,* Vol. I (L. A. Rosenblum, ed.), pp. 195–249, Academic Press, New York.

Mitchell, G., and Stevens, C. W., 1969, Primiparous and multiparous monkey mothers in a mildly stressful social situation: First three months, *Dev. Psychobiol.* **1**:280–286.

Mitchell, G., and Brandt, E. M., 1970, Behavioral differences related to experience of mother and sex of infant in the rhesus monkey, *Dev. Psychol.* **3**:149.

Mitchell, G., Ruppenthal, G. C., Raymond, E. J., and Harlow, H. F., 1966, Long-term effects of multiparous and primiparous monkey mother rearing, *Child Dev.* **37**:781–791.

Mowbray, J. B., and Cadell, T. E., 1962, Early behavior patterns in rhesus monkeys, *J. Comp. Physiol. Psychol.* **55**:350–357.

Novak, M. A., and Harlow, H. F., 1975, Social recovery of monkeys isolated for the first year of life: I. Rehabilitation and therapy, *Dev. Psychol.* **11**:453–465.

Poirier, F. E., 1968, The Nilgiri langur *(Presbytis johnii)* mother-infant dyad, *Primates* **9**:45–68.

Preston, D. G., Baker, R. P., and Seay, B. M., 1970, Mother-infant separation in Patas monkeys, *Dev. Psychol.* **3**:298–306.

Rahaman, H., and Parthasarathy, M. D., 1969, Studies on the social behavior of bonnet monkeys, *Primates* **10**:149–162.

Ransom, T. W., and Rowell, T. E., 1972, Early social development of feral baboons, in: *Primate Socialization* (F. E. Poirier, ed.), pp. 105–144, Random House, New York.

Redican, W. K., 1976, Adult male-infant interactions in nonhuman primates, in: *The Role of the Father in Child Development* (M. E. Lamb, ed.), pp. 345–385, Wiley, New York.

Redican, W. K., and Kaplan, J. N., 1978, Effects of synthetic odors on filial attachment in infant squirrel monkeys, *Physiol. Behav.* **20**:79–85.

Redican, W. K., and Mitchell, G., 1973, A longitudinal study of paternal behavior in adult male rhesus monkeys. I. Observations on the first dyad, *Dev. Psychol.* **8**:135–136.

Rhine, R. J., and Hendy-Neely, H., 1978, Social development of stumptail macaques *(Macaca arctoides):* Momentary touching, play, and other interactions with aunts and immatures during the infants' first 60 days of life, *Primates* **19**:115–123.

Rosenblatt, J. S., and Lehrman, D. S., 1963, Maternal behavior of the laboratory rat, in: *Maternal Behavior in Mammals* (H. L. Rheingold, ed.), pp. 8–57, Wiley, New York.

Rosenblum, L. A., 1968, Mother–infant relations and early behavioral development in the squirrel monkey, in: *The Squirrel Monkey* (L. A. Rosenblum and R. W. Cooper, eds.), pp. 207–233, Academic Press, New York.

Rosenblum, L. A., 1971*a*, The ontogeny of mother–infant relations in macaques, in: *The Ontogeny of Vertebrate Behavior* (H. Moltz, ed.), pp. 315–367, Academic Press, New York.

Rosenblum, L. A., 1971*b*, Infant attachment in monkeys, in: *The Origins of Human Social Relations* (H. R. Schaffer, ed.), pp. 85–113, Academic Press, New York.

Rosenblum, L. A., 1971*c*, Kinship interaction patterns in pigtail and bonnet macaques, in: *Proceedings of the Third International Congress of Primatology*, Vol. 3 (J. Biegert, ed.), pp. 79–84, Karger, Basel.

Rosenblum, L. A., 1972, Sex and age differences in response to infant squirrel monkeys, *Brain Behav. Evol.* **5**:30–40.

Rosenblum, L. A., 1978, Affective maturation and the mother–infant relationship, in: *The Development of Affect* (M. Lewis and L. A. Rosenblum, eds.), pp. 1–10, Plenum Press, New York.

Rosenblum, L. A., and Harlow, H. F., 1963, Approach-avoidance conflict in the mother-surrogate situation, *Psychol. Rep.* **12**:83–85.

Rosenblum, L. A., and Kaufman, I. C., 1968, Variations in infant development and response to maternal loss in monkeys, *Am. J. Orthopsychiatry* **38**:418–426.

Rosenblum, L. A., and Alpert, S., 1974, Fear of strangers and specificity of attachment in monkeys, in: *The Origins of Fear* (M. Lewis and L. A. Rosenblum, eds.), pp. 165–193, Wiley, New York.

Rosenblum, L. A., and Youngstein, K. P., 1974, Developmental changes in compensatory dyadic response in mother and infant monkeys, in: *The Effect of the Infant on Its Caregiver* (M. Lewis and L. A. Rosenblum, eds.), pp. 141–161, Wiley, New York.

Rosenblum, L. A., and Alpert, S., 1977, Response to mother and stranger: A first step in socialization, in: *Primate Bio-Social Development* (S. Chevalier-Skolnikoff and F. E. Poirier, eds.), pp. 463–447, Garland, New York.

Rosenblum, L. A., and Plimpton, E. H., 1979, The effect of adults on peer interactions, in: *The Child and Its Family* (M. Lewis and L. A. Rosenblum, eds.), pp. 1–8, Plenum Press, New York.

Rosenblum, L. A., and Paully, P., 1980, The social milieu of the developing monkey: Studies of the development of social perception, *Reprod. Nutr. Devel.* **20**:827–841.

Rosenblum, L. A., Kaufman, I. C., and Stynes, A. J., 1964, Individual distance in two species of macaque, *Anim. Behav.* **12**:338–342.

Rowell, T. E., and Hinde, R. A., 1962, Vocal communication by the rhesus monkey, *Proc. Zool. Soc. London* **138**:279–294.

Rowell, T. E., Hinde, R. A., and Spencer-Booth, Y., 1964, "Aunt"-infant interaction in captive rhesus monkeys, *Anim. Behav.* **12**:219–226.

Rumbaugh, D. M., 1965, Maternal care in relation to infant behavior in the squirrel monkey, *Psychol. Rep.* **16**:171–176.

Sackett, G. P., 1965, Effects of rearing conditions upon the behavior of rhesus monkeys *(Macaca mulatta)*, *Child Dev.* **36**:855–868.

Sackett, G. P., 1966, Monkeys reared in isolation with pictures as visual input: Evidence for an innate releasing mechanism, *Science,* **154**:1468–1473.

Sackett, G. P., 1970, Unlearned responses, differential rearing experiences, and the development of social attachments by rhesus monkeys, in: *Primate Behavior* (L. A. Rosenblum, ed.), pp. 111–140, Academic Press, New York.

Sackett, G. P., and Ruppenthal, G. C., 1973, Development of monkeys after varied experiences during infancy, in: *Ethology and Development* (S. A. Barnett, ed), pp. 52–87, Lippincott, Philadelphia.

Sackett, G. P., and Ruppenthal, G. C., 1974, Some factors influencing the attraction of adult female macaque monkeys to neonates, in: *The Effect of the Infant on Its Caregiver* (M. Lewis and L. A. Rosenblum, eds.), pp. 163–185, Wiley, New York.

Sacket, G. P., Griffin, G. A., Pratt, C., Joslyn, W. D., and Ruppenthal, G., 1967, Mother–infant and adult female choice behavior in rhesus monkeys after various rearing experiences, *J. Comp. Physiol. Psychol.* **63**:376–381.

Sackett, G. P., Holm, R. A., and Ruppenthal, G. C., 1976, Social isolation rearing: Species differences in behavior of macaque monkeys, *Dev. Psychol.* **12**:283–288.

Sade, D. S., 1967, Determinants of dominance in a group of free-ranging rhesus monkeys, in: *Social Communication among Primates* (S. A. Altmann, ed.), pp. 99–114, University of Chicago Press, Chicago.

Sade, D. S., 1972, Sociometrics of *Macaca mulatta.* I. Linkages and cliques in grooming matrices, *Folia Primatol.* **18**:537–542.

Schaller, G. B., 1963, *The Mountain Gorilla,* University of Chicago Press, Chicago.

Schlottman, R. S., and Seay, B., 1972, Mother-infant separation in the Java monkey *(Macaca irus)*, *J. Comp. Physiol. Psychol.* **79**:334–340.

Seay, B. M., 1966, Maternal behavior in primiparous and multiparous rhesus monkeys, *Folia Primatol.* **4**:146–168.

Seay, B., and Harlow, H. F., 1965, Maternal separation in the rhesus monkey, *J. Nerv. Ment. Dis.* **140**:434–441.

Seay, B., Hansen, E., and Harlow, H. F., 1962, Mother-infant separation in monkeys, *J. Child Psychol. Psychiatry* **3**:123–132.

Seay, B. M., Alexander, B. K., and Harlow, H. F., 1964, Maternal behavior of socially deprived rhesus monkeys, *J. Abnorm. Soc. Psychol.* **69**:345–354.

Spencer-Booth, Y., 1968, The behavior of group companions towards rhesus monkey infants, *Anim. Behav.* **16**:541–557.

Spencer-Booth, Y., and Hinde, R. A., 1971*a*, Effects of 6 days separation from mother on 18- to 32-week-old rhesus monkeys, *Anim. Behav.* **19**:174–191.

Spencer-Booth, Y., and Hinde, R. A., 1971, The effects of 13 days maternal separation on infant rhesus monkeys compared with those of shorter and repeated separations, *Anim. Behav.* **19**:595–605.

Stellar, E., 1960, The marmoset as a laboratory animal: Maintenance, general observations of behavior, and simple learning, *J. Comp. Physiol. Psychol.* **53**:1–10.

Struthsaker, T. T., 1971, Social behaviour of mother and infant vervet monkeys *(Cercopithecus aethiops)*, *Anim. Behav.* **19**:233–250.

Sugiyama, Y., 1967, Social Organization of Hanuman langurs, in: *Social Communication among Primates* (S. A. Altmann, ed.), pp. 221–236, University of Chicago Press, Chicago.

Suomi, S. J., 1977, Adult male-infant interactions among monkeys living in nuclear families, *Child Dev.* **48**:1255–1270.

Suomi, S. J., and Harlow, H. F., 1977, Production and alleviation of depressive behaviors in monkeys, in: *Psychopathology: Experimental Models* (J. D. Maser and M. E. P. Seligman, eds.), pp. 131–173, W. H. Freeman, San Francisco.

Swartz, K. B., 1977, Species and individual recognition in infant macaques. Ph.D. dissertation, Brown University.

Tilford, B. L., and Nadler, R. D., 1978, Male parental behavior in a captive group of lowland gorilla *(Gorilla gorilla gorilla)*, *Folia Primatol.* **29**:218–228.

Tinklepaugh, O. L., and Hartman, C. G., 1932, Behavior and Maternal care of the newborn monkey *(Macaca mulatta—M. rhesus)*, *J. Genet. Psychol.* **40**:257–286.

Trivers, R. L., 1972, Parental investment and sexual selection, in: *Sexual Selection and the Descent of Man* (B. Campbell, ed.), pp. 136–179, Aldine, Chicago.

van Lawick-Goodall, J., 1967, Mother-offspring relationships in free-ranging chimpanzees, in: *Primate Ethology* (D. Morris, ed.) pp. 287–346, Methuen, London.

van Lawick-Goodall, J., 1968, The behavior of free-living chimpanzees in the Gombe Stream Reserve, *Anim. Behav. Monogr.* **1**:161–311.

Wallen, K., Bielert, C., and Slimp, J., 1977, Foot clasp mounting in the prepubertal rhesus monkey: Social and hormonal influences, in: *Primate Bio-Social Development* (S. Chevalier-Skolnikoff and F. E. Poirier, eds.), pp. 439–461, Garland, New York.

Yamada, M., 1963, A study of blood-relationship in the natural society of the Japanese macaque, *Primates* **4**:43–66.

Yerkes, R. M., and Tomlin, M. I., 1935, Mother-infant relations in chimpanzee, *J. Comp. Psychol.* **20**:321–359.

Index